Lecture Notes in Physics

Editorial Board

R. Beig, Vienna, Austria
J. Ehlers, Potsdam, Germany
U. Frisch, Nice, France
K. Hepp, Zürich, Switzerland
R. L. Jaffe, Cambridge, MA, USA
R. Kippenhahn, Göttingen, Germany
I. Ojima, Kyoto, Japan
H. A. Weidenmüller, Heidelberg, Germany
J. Wess, München, Germany
J. Zittartz, Köln, Germany

Managing Editor

W. Beiglböck
Physics Editorial Department
Tiergartenstrasse 17, D-69121 Heidelberg, Germany

Springer
Berlin
Heidelberg
New York
Barcelona
Hong Kong
London
Milan
Paris
Singapore
Tokyo

The Editorial Policy for Proceedings

The series Lecture Notes in Physics reports new developments in physical research and teaching – quickly, informally, and at a high level. The proceedings to be considered for publication in this series should be limited to only a few areas of research, and these should be closely related to each other. The contributions should be of a high standard and should avoid lengthy redraftings of papers already published or about to be published elsewhere. As a whole, the proceedings should aim for a balanced presentation of the theme of the conference including a description of the techniques used and enough motivation for a broad readership. It should not be assumed that the published proceedings must reflect the conference in its entirety. (A listing or abstracts of papers presented at the meeting but not included in the proceedings could be added as an appendix.)
When applying for publication in the series Lecture Notes in Physics the volume's editor(s) should submit sufficient material to enable the series editors and their referees to make a fairly accurate evaluation (e.g. a complete list of speakers and titles of papers to be presented and abstracts). If, based on this information, the proceedings are (tentatively) accepted, the volume's editor(s), whose name(s) will appear on the title pages, should select the papers suitable for publication and have them refereed (as for a journal) when appropriate. As a rule discussions will not be accepted. The series editors and Springer-Verlag will normally not interfere with the detailed editing except in fairly obvious cases or on technical matters.
Final acceptance is expressed by the series editor in charge, in consultation with Springer-Verlag only after receiving the complete manuscript. It might help to send a copy of the authors' manuscripts in advance to the editor in charge to discuss possible revisions with him. As a general rule, the series editor will confirm his tentative acceptance if the final manuscript corresponds to the original concept discussed, if the quality of the contribution meets the requirements of the series, and if the final size of the manuscript does not greatly exceed the number of pages originally agreed upon. The manuscript should be forwarded to Springer-Verlag shortly after the meeting. In cases of extreme delay (more than six months after the conference) the series editors will check once more the timeliness of the papers. Therefore, the volume's editor(s) should establish strict deadlines, or collect the articles during the conference and have them revised on the spot. If a delay is unavoidable, one should encourage the authors to update their contributions if appropriate. The editors of proceedings are strongly advised to inform contributors about these points at an early stage.
The final manuscript should contain a table of contents and an informative introduction accessible also to readers not particularly familiar with the topic of the conference. The contributions should be in English. The volume's editor(s) should check the contributions for the correct use of language. At Springer-Verlag only the prefaces will be checked by a copy-editor for language and style. Grave linguistic or technical shortcomings may lead to the rejection of contributions by the series editors. A conference report should not exceed a total of 500 pages. Keeping the size within this bound should be achieved by a stricter selection of articles and not by imposing an upper limit to the length of the individual papers. Editors receive jointly 30 complimentary copies of their book. They are entitled to purchase further copies of their book at a reduced rate. As a rule no reprints of individual contributions can be supplied. No royalty is paid on Lecture Notes in Physics volumes. Commitment to publish is made by letter of interest rather than by signing a formal contract. Springer-Verlag secures the copyright for each volume.

The Production Process

The books are hardbound, and the publisher will select quality paper appropriate to the needs of the author(s). Publication time is about ten weeks. More than twenty years of experience guarantee authors the best possible service. To reach the goal of rapid publication at a low price the technique of photographic reproduction from a camera-ready manuscript was chosen. This process shifts the main responsibility for the technical quality considerably from the publisher to the authors. We therefore urge all authors and editors of proceedings to observe very carefully the essentials for the preparation of camera-ready manuscripts, which we will supply on request. This applies especially to the quality of figures and halftones submitted for publication. In addition, it might be useful to look at some of the volumes already published. As a special service, we offer free of charge LATEX and TEX macro packages to format the text according to Springer-Verlag's quality requirements. We strongly recommend that you make use of this offer, since the result will be a book of considerably improved technical quality. To avoid mistakes and time-consuming correspondence during the production period the conference editors should request special instructions from the publisher well before the beginning of the conference. Manuscripts not meeting the technical standard of the series will have to be returned for improvement.

For further information please contact Springer-Verlag, Physics Editorial Department II, Tiergartenstrasse 17, D-69121 Heidelberg, Germany

Henri Benisty Jean-Michel Gérard
Romuald Houdré John Rarity
Claude Weisbuch (Eds.)

Confined Photon Systems

Fundamentals and Applications

Lectures from the Summerschool
Held in Cargèse, Corsica, 3–15 August 1998

 Springer

Editors

Henri Benisty
Claude Weisbuch
École Polytechnique
Laboratoire de Physique de la Matière Condensée
F-91128 Palaiseau cedex, France

Jean-Michel Gérard
France Telecom, CNET
196 avenue Henri Ravera, F-92220 Bagneux, France

Romuald Houdré
EPFL
Institut de Micro et Opto-électronique
CH-1015 Lausanne, Switzerland

John Rarity
DERA Malvern, St. Andrews Road
WR14 3PS Malvern, UK

Library of Congress Cataloging-in-Publication Data.
Die Deutsche Bibliothek - CIP-Einheitsaufnahme

Confined photon systems : fundamentals and applications ; lectures from the summerschool, held in Cargèse, Corsica, 3 - 15 August 1998 / Henri Benisty ... (ed.). - Berlin ; Heidelberg ; New York ; Barcelona ; Hong Kong ; London ; Milan ; Paris ; Singapore ; Tokyo : Springer, 1999
(Lecture notes in physics ; Vol. 531)
ISBN 3-540-66435-1

ISSN 0075-8450
ISBN 3-540-66435-1 Springer-Verlag Berlin Heidelberg New York

This work is subject to copyright. All rights are reserved, whether the whole or part of the material is concerned, specifically the rights of translation, reprinting, reuse of illustrations, recitation, broadcasting, reproduction on microfilm or in any other way, and storage in data banks. Duplication of this publication or parts thereof is permitted only under the provisions of the German Copyright Law of September 9, 1965, in its current version, and permission for use must always be obtained from Springer-Verlag. Violations are liable for prosecution under the German Copyright Law.

© Springer-Verlag Berlin Heidelberg 1999
Printed in Germany

The use of general descriptive names, registered names, trademarks, etc. in this publication does not imply, even in the absence of a specific statement, that such names are exempt from the relevant protective laws and regulations and therefore free for general use.

Typesetting: Camera-ready by the authors/editors
Cover design: *design & production*, Heidelberg
SPIN: 10720466   55/3144/du - 5 4 3 2 1 0 - Printed on acid-free paper

Preface

This volume contains the lectures notes of the summerschool on "Confined Photon Systems, Fundamentals and Applications" held at the "Institut d'Etudes Scientifiques" at Cargèse, Corsica, August 3–15, 1998.

Confined photon systems such as microcavities and photonic crystals are being studied worldwide for many reasons :

1. They lead to low-dimensional photonic systems, with single electromagnetic mode behaviour exhibiting outstanding properties such as field enhancement and localization, ultra-high finesse electromagnetic modes, lossless resonators...
2. They open the way to modified light–matter interaction such as the strong-coupling between atoms and cavities, or between excitons and photons in all-solid-state semiconductor microcavities.
3. They are a main ground for the demonstration and use of many quantum optics phenomena such as single photon generation, squeezed light, low-power non-linear switching, quantum state entanglement, non-local quantum measurements, and conceivably quantum computation.
4. They are on the verge of yielding new, high-performance optical devices for large-scale industries such as telecommunications, lighting and displays... based on novel concepts : high-efficiency microcavity LEDs, photonic crystal integrated systems.

The need for a new school on these topics arose for various reasons : The many students entering the field need to be exposed to the main protagonists who constitute a very international and dispersed crowd due to the multidisciplinary aspects of the field. Also, in the past four years, we have witnessed many spectacular results which justify the hopes placed in the field, but also require wide, international dissemination.

The school aimed at giving students a working knowledge of a new, burgeoning field with widescale applications in both fundamental and applied sciences. The format and schedule of the school were drafted with that in mind. The originality of this school was its multi-disciplinarity, extending to the emphasis put on the mixing of fundamental and applied knowledge, as the field is indeed one of the few where deep, fundamental insight can quickly lead to widescale applications in devices of unsurpassed performance.

In the same manner as the school, these lecture notes are organized in a didactic way : We first have the four series of lectures on the basics of the field : properties of photon states by Fabre, electromagnetic properties of waveguides and structured media by Baets, properties of electronic states in matter by Koch, and introduction to photonic crystals by Joannopoulos.

The main knowledge is then developed in the lectures by Savona on optical properties of microcavities, Ho on the properties of low-dimensional optical systems, Hood on cavity QED effects in atomic physics, and Imamoglu on quantum optics of semiconductor optoelectronic systems.

This body of "tools" is put to use in the various specialized lectures : Gérard illustrates the properties of cavity modes by their effect on spontaneous recombination rate (the Purcell effect) ; Rarity discusses the needs and approaches to single photon emitters ; Benisty describes the use of planar microcavities to yield high-efficiency LEDs ; Labilloy shows how to measure quantitatively 2D photonic crystals and how the measured parameters agree well with theory.

The school, and this book, end with two papers that look towards the future, one on limitations to optical communications by Midwinter, the other, by DiVincenzo, on quantum computation.

The school was organized by a committee comprising Claude Weisbuch (Polytechnique, France) - Director, John Rarity (DRA-GB) - Co-Director, Jean-Michel Gérard (CNET-France) - Treasurer and Romuald Houdré (EPFL-Switzerland) - Scientific Secretary. The programme was set up with a scientific committee comprising Henri Benisty (Polytechnique-France), Elisabeth Giacobino (LKB-France), John Hegarty (Trinity-Ireland), John Joannopoulos (MIT-USA), Jürgen Mlynek (Konstanz-FRG).

We acknowledge generous financial support by the TMR programme of the European Union, and additional support by the PHANTOMS European network, CNRS (France), DGA (French ministry of defense), Ministère de l'Education Nationale, de la Recherche et de la Technologie (France), Office of Naval Research (USA), Ecole Polytechnique Fédérale de Lausanne (Switzerland), France Telecom/CNET, Alcatel (France), Cables Pirelli (Italy), Thomson CSF/LCR (France), Siemens (FRG).

Finally, we wish to thank the school secretary Karin Hoeldrich for extremely efficient help before, during and after the school, and the organizing staff at Cargèse (Elisabeth Dubois-Violette, Chantal Ariano, Brigitte Cassegrain, Vittoria and Claudine Conforto) who make every event at Cargèse an unforgetable one for all the participants.

Palaiseau, March 1999

Henri Benisty
Jean-Michel Gérard
Romuald Houdré
John Rarity
Claude Weisbuch

Contents

**Basics of Quantum Optics
and Cavity Quantum Electrodynamics**
C. Fabre 1

Basics of Dipole Emission from a Planar Cavity
R. Baets, P. Bienstman and R. Bockstaele 38

**Microscopic Theory of the Optical Semiconductor Response
Near the Fundamental Absorption Edge**
S. W. Koch 80

An Introduction to Photonic Crystals
J. D. Joannopoulos 150

**Linear Optical Properties of Semiconductor Microcavities
with Embedded Quantum Wells**
V. Savona 173

Spontaneous Emission Control and Microcavity Light Emitters
S.-T. Ho, L. Wang and S. Park 243

Cavity QED - Where's the Q?
C. J. Hood, T. W. Lynn, M. S. Chapman, H. Mabuchi, J. Ye
and H. J. Kimble 298

Quantum Optics in Semiconductors
A. Imamoglu 310

**Semiconductor Microcavities, Quantum Boxes,
and the Purcell Effect**
J.-M. Gérard and B. Gayral 331

Single Photon Sources and Applications
J. G. Rarity, S. C. Kitson and P. R. Tapster 352

Photonic Crystals for Nonlinear Optical Frequency Conversion
V. Berger 366

Physics of Light Extraction Efficiency in Planar Microcavity Light-Emitting Diodes
H. Benisty 393

Measuring the Optical Properties of Two-Dimensional Photonic Crystals in the Near-Infrared
D. Labilloy, H. Benisty, C. Weisbuch, T. F. Krauss,
C. J. M. Smith, R. M. De La Rue, D. Cassagne, C. Jouanin,
R. Houdré and U. Oesterle 406

Limitations to Optical Communications
J.E. Midwinter 426

Thoughts on Quantum Computing
D. P. DiVincenzo 482

Index 493

Basics of Quantum Optics and Cavity Quantum Electrodynamics

Claude FABRE

Laboratoire Kastler Brossel, Université Pierre et Marie Curie
Case 74, 75252 Paris cedex 05, France

The purpose of these lectures is to provide a brief survey of the domain of quantum optics, which is the study of light, and of the interaction between light and matter, at a microscopic level of understanding. As such, it requires the use of quantum mechanics, which can be used to describe the matter itself, and also to describe the electromagnetic field. In principle, in order to get a perfect understanding of the phenomena, one needs to use the complete quantum theory of both matter, light and their relative interactions. But, as often in physics, less sophisticated and exhaustive approaches of the system under study can give a better physical insight and simpler calculations when some conditions are fulfilled. In some regions of the parameter space, a totally classical, and apparently old-fashioned, approach can give an accurate description of reality, and a semiclassical (or semiquantum) approach in others (classical treatment of the field, quantum treatment of the matter). One of the aim of this overview is to precise the validity domains of these different possible approaches. More detailed descriptions of this domain of physics can be found in different textbooks [1, 2, 3, 4, 5, 6, 7, 8].

1 Light matter interaction : classical or quantum ?

To simplify this presentation, we will restrict ourselves to the case of a monochromatic plane wave of frequency ω, having an electric field at the location $\vec{r_{at}}$ of the atom given by $\vec{E} = E_0 \vec{\varepsilon_0} \cos \omega t$, interacting with an atom with only two levels, labeled 1 and 2, and a corresponding Bohr Frequency $\omega_0 = (E_2 - E_1)/\hbar$. We assume a quasi-resonant interaction ($\omega \approx \omega_0$) described by the following electric-dipole term in the hamiltonian

$$V_{int} = -\vec{E} \cdot \vec{D} \tag{1}$$

where $\vec{D}$ is the atomic dipole.

1.1 Lorentz model

The first phenomenological microscopic description of matter-light interaction was given by Lorentz one century ago [9]. It assumes that the atom is a classical dipole, in which the two opposite charges q and $-q$, of masses M

and m, are linked by an harmonic force, and oscillate freely with a frequency ω_0. When this classical atom is submitted to the electric field of the wave, the Newton equation for the electron motion implies that $\vec{D}$ obeys the following equation (when $M \gg m$)

$$\frac{d^2 \vec{D}}{dt^2} = -\omega_0^2 \vec{D} + \frac{q^2}{m} E_0 \vec{\varepsilon_0} \cos \omega t \tag{2}$$

Let us write the time dependent dipole as the product of a fast modulation $e^{-i\omega t}$ and of a slow envelope $\widetilde{D}(t)$

$$\vec{D}(t) = \operatorname{Re}\left[\vec{\varepsilon_0} \widetilde{D}(t) e^{-i\omega t}\right] \tag{3}$$

Then, using the quasi-resonant approximation to neglect the term $d^2 \widetilde{D}/dt^2$, one obtains the following first-order evolution equation of the dipole envelope

$$\frac{d\widetilde{D}}{dt} = i(\omega - \omega_0)\widetilde{D} + i\frac{q^2}{2m\omega_0} E_0 \tag{4}$$

This equation has a simple stationary solution (corresponding to the forced regime)

$$\widetilde{D} = \frac{q^2}{2m\omega_0(\omega - \omega_0)} E_0 \tag{5}$$

Reminding that the power P transferred from the atom to the field has the classical expression

$$P = -\overline{\vec{E} \cdot d\vec{D}/dt} = -\frac{\omega}{2} E_0 \operatorname{Im}\left[\widetilde{D}\right] \tag{6}$$

one sees that, since in this model $\widetilde{D}$ is real, there is no power exchanged between the atom and the field : only dispersive effects (index changes) are described by equation (4).

In order to account for power exchanges, one needs to introduce *dissipative processes*, such as collisions, or spontaneous emission. They will induce a *decay* of the atomic dipole, and a new term $-\gamma \widetilde{D}$ must be added in the dipole evolution equation (4). The new stationary value for the dipole is now

$$\widetilde{D} = \frac{q^2}{2m\omega_0[(\omega - \omega_0) - i\gamma]} E_0 \tag{7}$$

The imaginary part of $\widetilde{D}$ is now nonzero, but always positive : the Lorentz model gives rise only to energy transfers from the field to the atom (absorption process). Let us also notice that $\widetilde{D}$ is strictly proportional to the applied field : there are no saturation effects.

Despite its naivety (one knows that the electron is not harmonically bound in the atom), this model gives astonishingly accurate and quantitatively exact predictions in the atom-matter interaction in many situations, with only two adjustable parameters ω_0 and γ. We will see later the reason of this surprising success.

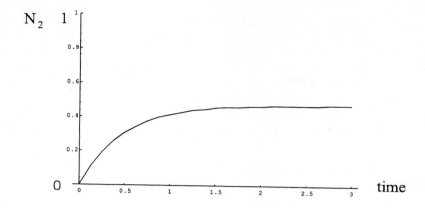

Fig. 1. Evolution of the upper state population in the Einstein model

1.2 Einstein model

Another kind of phenomenological approach of matter-wave interaction, based on Bohr's quantum theory, was introduced later by Einstein [10]. Let N_1 and N_2 be the "populations" of the ground state 1 and of the excited state 2 (i.e. the proportion of the atoms being in the corresponding state, with $N_1 + N_2 = 1$) and $u(\omega)$ the field energy density at frequency ω (equal to $\frac{1}{2}\varepsilon_0 E_0^2$). The evolution equation of the populations is in this model

$$\frac{dN_2}{dt} = -\frac{dN_1}{dt} = -A_{21} N_2 + B\left(N_1 - N_2\right) u\left(\omega_0\right) \tag{8}$$

A_{21} accounts for spontaneous emission and is the analog of the rate γ introduced in the previous section. The second term corresponds to the well-known stimulated absorption and emission processes. The steady state for the upper state population is

$$N_2 = \frac{Bu\left(\omega_0\right)}{2Bu\left(\omega_0\right) + A_{21}} \tag{9}$$

One sees on equation (9) that N_2 is proportional the light power when $Bu \ll A_{21}$. At high powers, N_2 saturates to the value $1/2$, corresponding to a balance between stimulated emission and absorption processes. Figure 1 gives the time dependence of N_2 when the system starts at time $t = 0$ from the ground state : the evolution towards the steady state is a monotonous exponential on a typical time A_{21}^{-1}.

As one photon is emitted for each transition $2 \to 1$, the power in the field exchanged in the process in given in this model by

$$P = -\hbar\omega_0 \frac{dN_2}{dt} \tag{10}$$

It can be either positive or negative, depending on the respective contributions of the stimulated emission and absorption processes.

The Einstein model successfully explains light amplification effects and also saturation effects. It is for this reason widely used in laser theory. It accounts furthermore for the properties of radiation in equilibrium with a thermal bath of atoms (Planck's law). But it says *nothing about the phase of the electromagnetic field* when it interacts with matter, and on the evolution of the system when the field is not exactly resonant with the atoms ($\omega \neq \omega_0$).

1.3 Semiclassical (or semiquantum) model : Bloch-Maxwell equations

We now turn to an ab-initio model (in contrast to the two previous ones, which were phenomenological approaches), in which the atom is treated by quantum mechanics, and the field by the Maxwell equations. The hamiltonian ruling the atomic evolution is

$$\hat{H} = \hat{H}_{atom} - E_0 \vec{\varepsilon_0} \cdot \widehat{\vec{D}} \cos \omega t \qquad (11)$$

where $\widehat{\vec{D}}$ is the atom dipole quantum operator. The equations of motion for the mean dipole and for the atomic populations can be derived by using the Ehrenfest theorem. They can be written in terms of the dipole slowly varying envelope, defined in a way analogous to (3)

$$\langle \psi(t) | \widehat{\vec{D}} | \psi(t) \rangle = \text{Re}\left[\vec{\varepsilon_0} \widetilde{D}(t) e^{-i\omega t} \right] \qquad (12)$$

In the limit $\omega \approx \omega_0$, one finds the following set of equations

$$\frac{d\widetilde{D}}{dt} = i(\omega - \omega_0)\widetilde{D} + if \frac{q^2}{m\omega_0} E_0 (N_1 - N_2) \qquad (13)$$

$$\frac{dN_2}{dt} = -\frac{dN_1}{dt} = \frac{1}{2\hbar} E_0 \, \text{Im}\left[\widetilde{D}\right] \qquad (14)$$

where $d = \langle 1 | \widehat{\vec{D}} \cdot \vec{\varepsilon_0} | 2 \rangle$ is the dipole matrix element and $f = 2m\omega_0 d^2/q^2\hbar$ is a dimensionless parameter called the *oscillator strength* of the transition. Equation (13) has a form very similar to (4) derived in the Lorentz model. It is strictly equivalent to (4) when the oscillator strength is 1, and more importantly when $N_2 \approx 0$, and therefore $N_1 \approx 1$. We thereby find the very important result that, within a numerical factor f, *the Lorentz model is equivalent to the semi-classical approach for a weakly excited atom*. This occurs for example when it is excited by a weak field generated by a discharge lamp or a thermal field.

The power exchanged P is given by $-\hbar\omega_0 dN_2/dt$, like in the Einstein model, and hence using (14)

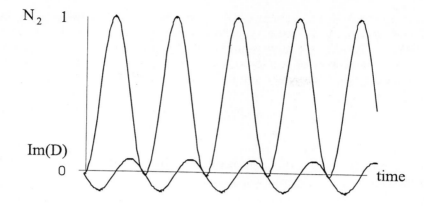

Fig. 2. Evolution of the upper state population (upper curve), and of the mean atomic dipole (lower curve) in the semiclassical model without relaxation

$$P = -\frac{\omega_0}{2} E_0 \operatorname{Im}\left[\widetilde{D}\right] \tag{15}$$

One retrieves an expression similar to the classical expression (6) of the Lorentz model. Figure 2 gives the time dependence of N_2 and $\operatorname{Im}\left[\widetilde{D}\right]$ for an atom initially in its ground state and for a field at exact resonance ($\omega = \omega_0$). One finds for the two quantities the well-known Rabi oscillations, with a period $T_{Rabi} = h/dE_0$.

An important feature of this model is the possibility of creating *coherent superposition of atomic states*. For example, if the resonant field is stopped abruptly at times $T = T_R/4$ ("$\frac{\pi}{2}$ pulse"), or $T = 3T_R/4$, the atom is respectively in the states

$$|\psi(T_R/4)\rangle = (|1\rangle - i|2\rangle)/\sqrt{2} \quad ; \quad |\psi(3T_R)/4\rangle = (|1\rangle + i|2\rangle)/\sqrt{2} \tag{16}$$

These states are very different from classical statistical superpositions of the upper and lower states with equal probabilities : they have a mean dipole D (depending on the relative phase of the complex coefficients of states $|1\rangle$ and $|2\rangle$) and are not stationary, whereas the classical superpositions have no dipole and are stationary. Coherent superposition of atomic states (which of course cannot be described in the Lorentz model) are quantum objects with many interesting applications. Let us mention the Ramsey fringes [11] and their use in metrology [12], the "black resonances" [13] and their application to laser cooling of atoms [14], and the existence of lasers without inversion [15].

Note that, in this model, the atomic system does not evolve to a stationary state. This comes from the fact that we have neglected dissipative phenomena, like in the first approach of the Lorentz model (Eq(4)). To take them into account, one needs to use a quantum description of the system in terms of density matrices, not of wavefunctions. Quantum theory of relaxation [6] implies the existence of two decay terms in the evolution equations of the mean values, which become

$$\frac{d}{dt}\widetilde{D} = [i(\omega - \omega_0) - \gamma_2]\widetilde{D} + if\frac{q^2}{2m\omega_0}E_0(N_1 - N_2) \qquad (17)$$

$$\frac{dN_2}{dt} = -\frac{dN_1}{dt} = -\gamma_1 N_2 + \frac{1}{2\hbar}E_0 \operatorname{Im}\left[\widetilde{D}\right] \qquad (18)$$

These equations are the Bloch equations, which together with the Maxwell equations for the field, rule the system evolution within the semiclassical approach. The decay rates γ_1 and γ_2 depend on the exact relaxation processes present in the system. If spontaneous emission is the only dissipative process, then the dipole decay rate γ_2 is $\frac{1}{2}$ of the population decay rate γ_1. If other relaxation processes take place, such as collisions, or lattice vibration, they randomize the phase and therefore destroy the dipole more quickly than they evacuate the energy out of the system, and often, in such non dilute systems, one has $\gamma_2 \gg \gamma_1$.

Figure 3 gives the time dependence of N_2 and $\operatorname{Im}\left[\widetilde{D}\right]$ for three relative values of the decay constants. One firstly notes than a steady state is now reached, after a time of the order of γ_2^{-1} and that the Rabi oscillation and the atomic dipole are quickly damped before reaching the stationary state. Fig 3c) gives an evolution of N_2 which is quite similar to the one obtained by the Einstein model. Indeed, if one *adiabatically eliminates* the quickly relaxing dipole (neglecting the $\frac{d}{dt}\widetilde{D}$ in equation (17)), one gets the following equations for the populations

$$\frac{dN_2}{dt} = -\frac{dN_1}{dt} = -\gamma_1 N_2 + \frac{\gamma_2}{\gamma_2^2 + (\omega_0 - \omega)^2}\frac{d^2 E_0^2}{2\hbar^2}(N_1 - N_2) \qquad (19)$$

One retrieves the Einstein equation (8) with explicit values of the Einstein coefficients A and B. This equation is also valid in the case where ω is not strictly equal to ω_0. We are then led to the conclusion that *the Einstein model correctly describes the matter-light interaction (within the semi-classical approach) when the dipole decays very quickly* ($\gamma_2 \gg \gamma_1$), which is the case when efficient non-radiative relaxation processes take place (or when the atom interacts with a light source with a broad spectrum, for which the dipole induced by the different frequency components of the source interfere destructively).

Note finally that when relaxation is introduced, the system evolves to a stationary state where the mean dipole is zero, and the populations are nonzero : this state precisely corresponds to a statistical mixture of atoms in states 1 and 2 with probabilities N_1 and N_2. The role of relaxation is thereby

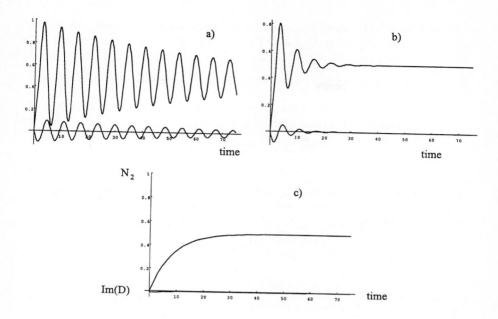

Fig. 3. Evolution of the upper state population and of the mean dipole for different values of the ratio between the population and the dipole relaxation constants. (a) :$\gamma_1 = 0.5\gamma_2$; (b) :$\gamma_1 = 30\gamma_2$; (c) :$\gamma_1 = 200\gamma_2$

to destroy the quantum coherences, i.e. the superpositions of atomic states like (16) created by the coherent interaction with the monochromatic field, and to transform the system into a classical statistical mixture.

1.4 Full quantum model

We now know that the Lorentz and Einstein approaches correspond to two special cases of the semi-classical model. The Maxwell-Bloch equations, derived in this model, are widely used in quantum optics and successfully account for an impressive amount of phenomena. But they are necessarily an approximation, because the electromagnetic field is indeed a quantum object. Before going to the full quantum model of the whole matter+field system, we need to know more about the quantum description of the free electromagnetic field itself. This is what is done in the next section.

2 Quantum description of light

This section is devoted to a brief introduction to the quantum description of light. Its aim is to underline the main physical new features which appear at this level of understanding, not to give a rigorous and comprehensive overview of the domain, which can be found in different textbooks [1].

2.1 Modal decomposition of the classical electromagnetic field

Inside a volume V (that we will take cubic of side L), any vector field $\vec{E}(\mathbf{r},t)$ can be expanded as a linear combination of *modes* $\vec{u_\ell}(\mathbf{r})$

$$\vec{E}(\vec{r},t) = \sum_\ell \alpha(t)\, \vec{u_\ell}(\mathbf{r}) \tag{20}$$

The summation extending only over discrete values of ℓ, because the volume V in which the field is described is finite. The modes $\{\vec{u_\ell}\}$ form an orthonormal basis, that is

$$\int_V \vec{u_\ell}^*(\mathbf{r}) \cdot \vec{u_{\ell'}}(\mathbf{r}) d^3r = \delta_{\ell\ell'} \qquad \sum_\ell \vec{u_\ell}^*(\mathbf{r}') \cdot \vec{u_\ell}(\mathbf{r}) = \delta(\mathbf{r}-\mathbf{r}') \tag{21}$$

Different bases $\{\vec{u_\ell}\}$ can be chosen, according to the configuration under study. The simplest is the basis of plane waves

$$\vec{u_\ell}(\vec{r}) = L^{-3/2} \vec{\varepsilon_\ell}\, e^{i\mathbf{k}_\ell \cdot \vec{r}} \qquad \ell = \{(n_x, n_y, n_z) \in Z^3, m = \pm 1\} \tag{22}$$

where $\vec{\varepsilon_\ell}$ is one of the two possible orthogonal polarization vectors ($m = \pm 1$) in the plane orthogonal to $\mathbf{k}_\ell$, and where $\mathbf{k}_\ell$ has components depending on relative integers n_x, n_y and n_z ($k_x = \frac{2\pi}{L}n_x, k_y = \frac{2\pi}{L}n_y, k_z = \frac{2\pi}{L}n_z$). If one deals with laser beams, the basis of Gaussian modes [16] is more convenient. A basis made of stationary waves can also be useful when one uses real boxes with perfectly conducting walls to confine the field, for example in cavity quantum electrodynamics.

2.2 Quantum field operators

In the case of the electron, its quantum description is obtained by replacing in the different relevant physical quantities of the classical theory c-numbers by q-numbers, i.e. numbers by operators with well defined commutation relations. The procedure is the same for the electromagnetic field. All the field operators are functions of basic non-Hermitian operators $\hat{a}_\ell$, named annihilation operators, defined for each mode ℓ, which fulfill the following commutator algebra

$$[\hat{a}_\ell, \hat{a}_{\ell'}] = [\hat{a}_\ell^+, \hat{a}_{\ell'}^+] = 0 \qquad ; \qquad [\hat{a}_\ell, \hat{a}_{\ell'}^+] = \delta_{\ell\ell'} \tag{23}$$

Actually, all the quantum properties of light arise from the non-commutativity of $\hat{a}_\ell$ and $\hat{a}_\ell^+$. The main observables of the quantum theory of light have then the following forms [1] in the Schrödinger representation

- Hamiltonian : $\hat{H} = \sum_\ell \hbar \omega_\ell\, (\hat{a}_\ell^+ \hat{a}_\ell + \frac{1}{2})$, where $\omega_\ell = c|\mathbf{k}_\ell|$

- Field momentum : $\widehat{\vec{P}} = \sum_\ell \hbar \mathbf{k}_\ell \, \widehat{a}_\ell^+ \widehat{a}_\ell$
- Photon number in mode ℓ : $\widehat{N}_\ell = \widehat{a}_\ell^+ \widehat{a}_\ell$
- Total photon number : $\widehat{N} = \sum_\ell \widehat{N}_\ell$
- Electric field : $\widehat{\vec{E}}(\mathbf{r}) = \widehat{\vec{E}}^{(+)}(\mathbf{r}) + \left[\widehat{\vec{E}}^{(+)}(\mathbf{r})\right]^+$
- Complex electric field: $\widehat{\vec{E}}^{(+)}(\mathbf{r}) = i \sum_\ell C_\ell \widehat{a}_\ell \, \vec{u_\ell}(\mathbf{r})$, with $C_\ell = (\hbar \omega_\ell / 2\varepsilon_o)^{1/2}$.

In the case of the plane wave basis, one has in particular

$$\widehat{\vec{E}}^{(+)}(\mathbf{r}) = i \sum_\ell \mathcal{E}_\ell \, \boldsymbol{\varepsilon}_\ell e^{i\mathbf{k}_\ell \cdot \mathbf{r}} \widehat{a}_\ell \tag{24}$$

where $\mathcal{E}_\ell = \sqrt{\hbar \omega_\ell / 2\varepsilon_o L^3}$ gives the order of magnitude of the electric field corresponding to one photon in the quantization box L^3.

It is often more convenient to work in *the Heisenberg representation* (time-dependent operators, time-independent field states). In the case of the free field (and only in this case), this amounts to replacing $\widehat{a}_\ell$ by $\widehat{a}_\ell e^{-i\omega_\ell t}$ in all the previous expressions. It is then possible to write the time-dependent electric field operator as

$$\widehat{\vec{E}}(\mathbf{r}, t) = i \sum_\ell \boldsymbol{\varepsilon}_\ell \left(\widehat{E}_{p\ell} \cos \psi_\ell + \widehat{E}_{q\ell} \sin \psi_\ell\right) \tag{25}$$

where ψ_ℓ is the usual propagation phase $\psi_\ell = \omega_\ell t - \mathbf{k}_\ell \cdot \mathbf{r}$ and

$$\widehat{E}_{p\ell} = i\mathcal{E}_\ell \xi_\ell (\widehat{a}_\ell - \widehat{a}_\ell^+) \qquad \widehat{E}_{q\ell} = \mathcal{E}_\ell (\widehat{a}_\ell + \widehat{a}_\ell^+) \tag{26}$$

are Hermitian operators called "quadrature operators", which do not commute

$$\left[\widehat{E}_{p\ell}, \widehat{E}_{q\ell'}\right] = 2i\mathcal{E}_\ell^2 \delta_{\ell\ell'} \tag{27}$$

This relation is somehow similar to the well-known relation $[q, p] = i\hbar$ valid for an electron of momentum p and position q.

2.3 Quantum field states

Starting from the commutation relations (23), one easily shows [17] that the spectrum of operator $\widehat{N}_\ell$ is the set of non-negative integers $n_\ell = 0, 1, 2, ...$. Let us name $|n_\ell\rangle$ the corresponding eigenvectors

$$\widehat{N}_\ell |n_\ell\rangle = n_\ell |n_\ell\rangle \tag{28}$$

They correspond to a field containing exactly n_ℓ photons in mode ℓ and in volume V, and are called number states, or Fock states. One shows that the well-known relations of the harmonic oscillator hold also here

$$\widehat{a}_\ell |n_\ell\rangle = \sqrt{n_\ell} |n_\ell - 1\rangle \quad \widehat{a}_\ell |0\rangle = 0 \quad \widehat{a}_\ell^+ |n_\ell\rangle = \sqrt{n_\ell + 1} |n_\ell + 1\rangle \qquad (29)$$

The eigenstates of $\widehat{H}$ and $\widehat{\mathbf{P}}$ are tensor products of states $|n_\ell\rangle$ for all the different modes : $|n_1\rangle \otimes |n_2\rangle \otimes ... \otimes |n_\ell\rangle$, that we will write $|n_1, n_2, ..., n_\ell\rangle$. They form a basis of the total Hilbert space of field states. Therefore, the most general field state can be written as

$$|\psi\rangle = \sum_{n_1=0}^{\infty} \sum_{n_2=0}^{\infty} ... \sum_{n_\ell=0}^{\infty} ... C_{n_1,n_2,...n_\ell...} |n_1, n_2, ..., n_\ell, ...\rangle \qquad (30)$$

with the constraint $\langle\psi|\psi\rangle = 1$. The Hilbert space spanned by vectors like (30) has a gigantic size, almost impossible to imagine : each number n_ℓ runs from 0 to ∞, and there is an infinity of modes ℓ. As a result, the variety of field states is extraordinary, whereas the classical states of the field depend on a single set of complex numbers α_ℓ, each one giving the complex amplitude in one of the modes. One may encounter in quantum optics much more different situations than in classical optics, as we will see. This immense world begins only to be explored.

Among all the states, the state $|n_1 = 0, n_2 = 0, ..., n_\ell = 0, ...\rangle = |vac\rangle$, called "vacuum", plays a particular role, as it constitutes the ground state of the system. It could have been also called "darkness", as it is a state from which no light energy can be extracted. We will see later that it has strange and interesting properties.

2.4 Single mode case

To simplify further, we will consider a single mode ℓ, with a given polarization. We can then omit the polarization vector, and work with a scalar field $E(\mathbf{r}, t)$.

Classically speaking, the general form of a single mode field is

$$E(\mathbf{r}, t) = E_o \cos(\omega_\ell t - \mathbf{k}_\ell \cdot \mathbf{r} + \varphi) = E_{p\ell} \cos \psi_\ell + E_{q\ell} \sin \psi_\ell \qquad (31)$$

It depends on two real quantities (E_o, φ) or $(E_{p\ell} = E_o \cos \varphi, E_{q\ell} = E_o \sin \varphi)$ and can be described by a point in a plane (the "'Fresnel plane") of Cartesian coordinates (E_{pl}, E_{ql}) or polar coordinates (E_o, φ).

In the quantum description, a single mode field is defined by a vector belonging to the subspace corresponding to this mode ℓ

$$|\psi\rangle = \sum_{n=0}^{\infty} C_n |n_1 = 0, n_2 = 0, ..., n_\ell = n, ...\rangle = \sum_{n=0}^{\infty} C_n |n\rangle \qquad (32)$$

Mean values It is easy to calculate the mean value of the electric field in such a state. One finds

$$\langle\psi|\widehat{E}(\mathbf{r},t)|\psi\rangle = E_o \cos(\omega_\ell t - \mathbf{k}_\ell \cdot \mathbf{r} + \varphi) \qquad (33)$$

with E_o and φ determined by the relation $E_o e^{i\varphi} = 2i\mathcal{E}_\ell \langle\psi|\widehat{a}_\ell|\psi\rangle$. We see that expressions (33) and (31) coincide : there is no difference between classical and quantum behaviors as far as *mean values* are concerned. As we will see later, differences do exist, but only for *higher order moments*, i.e. for the *fluctuations* around the classical mean values and for the *correlations* between different measurements.

Quantum field fluctuations The electric field operator restricted to mode ℓ is

$$\widehat{E}(\mathbf{r},t) = \widehat{E}_{p\ell} \cos\psi_\ell + \widehat{E}_{q\ell} \sin\psi_\ell \qquad (34)$$

with

$$[\widehat{E}_{p\ell}, \widehat{E}_{q\ell}] = 2i\,\mathcal{E}_\ell^2 \qquad (35)$$

$\widehat{E}(\mathbf{r},t)$ coincides with the quadrature operator $\widehat{E}_{p\ell}$ when $\psi_\ell = 0$ (2π) and with $\widehat{E}_{q\ell}$ when $\psi_\ell = \frac{\pi}{2}$ (2π). As these two operators do not commute, they have no common eigenstates, and therefore it is not possible to find a state $|\psi\rangle$ which would be an eigenstate of $\widehat{E}(\mathbf{r},t)$ for any value of $\psi_\ell = \omega_\ell t - \mathbf{k}_\ell \cdot \mathbf{r}$, i.e. at any point and at any time. In any state $|\psi\rangle$, the electric field will not be perfectly defined, and will have therefore *quantum fluctuations, or quantum noise*, from one measurement to another. In particular, the variances of these fluctuations are constrained by the Heisenberg inequality

$$\Delta E_{p\ell}\, \Delta E_{q\ell} \geq \mathcal{E}_\ell^2 \qquad (36)$$

deduced from (27). In the Fresnel plane, the point representing the mode is necessarily fuzzy. Relation (36) states that the area of uncertainty in the Fresnel plane has a lower bound. This is in particular true for the vacuum state, for which one easily shows that $\Delta E_{p\ell} = \Delta E_{q\ell} = \mathcal{E}_\ell$. The vacuum state does not therefore correspond to perfect darkness : it has a field with a zero mean value, but nonzero fluctuations, which are called "vacuum fluctuations".

Relation (36) can be written also in terms of phase and intensity noise. The rigorous definition of a phase operator is a difficult task in quantum optics [18]. However, for a state $|\psi\rangle$ corresponding to an "intense" field ($\langle\psi|\widehat{N}_\ell|\psi\rangle \gg 1$), of zero mean classical phase φ ($\langle\psi|\widehat{E}_{q\ell}|\psi\rangle = 0$), one can show that the phase fluctuations are proportional to the fluctuations of $\widehat{E}_{q\ell}$ around 0. Relation (36) then implies that in this case

$$\Delta\varphi\, \Delta H \geq \frac{\hbar\omega_\ell}{2} \quad \text{or} \quad \Delta\varphi\, \Delta N_\ell \geq \frac{1}{2} \qquad (37)$$

For intense fields, phase and energy, or phase and photon number, are conjugate quantities : the exact knowledge of one prevents the simultaneous exact knowledge of the other.

Coherent states They are defined as eigenstates of the annihilation operator $\widehat{a}_\ell$

$$\widehat{a}_\ell \left| \alpha_\ell \right\rangle = \alpha_\ell \left| \widehat{\alpha}_\ell \right\rangle \tag{38}$$

$\widehat{\alpha}_\ell$ being any complex number. These states, introduced by Schrödinger, have been extensively studied [19]. They are not eigenstates of the Hamiltonian, and have a Poissonian distribution over the different $|n_\ell\rangle$ states. One shows that

$$\begin{aligned}
\left\langle \widehat{N}_\ell \right\rangle &= |\alpha_\ell|^2 \quad & \Delta N_\ell &= |\alpha_\ell| = \sqrt{\langle N_\ell \rangle} \\
\left\langle \widehat{E}_{p\ell} \right\rangle &= \mathrm{Re}(2i\mathcal{E}_\ell \alpha_\ell) \quad & \left\langle \widehat{E}_{p\ell} \right\rangle &= \mathrm{Im}(2i\mathcal{E}_\ell \alpha_\ell) \\
\Delta E_{p\ell} &= \Delta E_{q\ell} = \Delta E(\mathbf{r},t) = \mathcal{E}_\ell & & \\
\Delta \varphi &= \frac{1}{2|\alpha_\ell|} \quad & \text{if } |\alpha_\ell| &\gg 1
\end{aligned} \tag{39}$$

The variance of field fluctuations in a coherent state, equal to $\mathcal{E}_\ell$, is independent of time and location of measurement and of the mean field amplitude and phase : it is equal to the vacuum fluctuations and corresponds to the minimum value in the Heisenberg inequality (36) : coherent states are minimum uncertainty states. In particular intense coherent states have quantum fluctuations which are much smaller than the mean value : they behave more and more classically as $|\alpha_\ell|$ grows.

From an experimental point of view, it turns out that single mode usual lasers well above threshold produce light beams which can be described by such coherent states, and that the other *classical light sources* (thermal sources, discharges,...) *have more fluctuations on any observable than coherent states*

$$\begin{aligned}
(\Delta E_{p\ell})_{\mathrm{clas}} &> \mathcal{E}_\ell \quad & (\Delta E_{q\ell})_{\mathrm{clas}} &> \mathcal{E}_\ell \\
(\Delta N)_{\mathrm{clas}} &> \sqrt{\langle N_\ell \rangle} \quad & (\Delta \varphi)_{\mathrm{clas}} &> \frac{1}{2\sqrt{\langle N_\ell \rangle}}
\end{aligned} \tag{40}$$

One usually calls *"standard quantum noise"* the fluctuations of coherent states. They constitute the *lower limit for light fluctuations of classical sources*.

There exist other light sources, producing "nonclassical states of light", which have fluctuations below the standard quantum noise given by (40). However, they cannot violate (36) and therefore must have $\Delta E_{p\ell} > \mathcal{E}_\ell$ if $\Delta E_{q\ell} < \mathcal{E}_\ell$, or $\Delta \varphi > 1/2\sqrt{\langle N_\ell \rangle}$ if $(\Delta N) < \sqrt{\langle N_\ell \rangle}$ (sub-Poissonian distribution). Generally speaking, they are more difficult to generate than the coherent states. Actually, most states described by (32) or even (35) have never been produced.

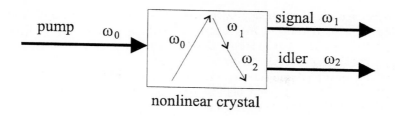

Fig. 4. Parametric splitting of light

Squeezed states They are formally defined by the relation

$$|\alpha_\ell, S\rangle = e^{S(\hat{a}_\ell^2 - \hat{a}_\ell^{+2})/2} |\alpha_\ell\rangle \qquad (41)$$

where S is real and $|\alpha_\ell\rangle$ a coherent state. One can show [20] that for these states

$$(\Delta E(\mathbf{r},t))^2 = \mathcal{E}_\ell^2 (\cos^2 \psi_\ell e^{2S} + \sin^2 \psi_\ell e^{-2S});$$
$$\Delta E_{p\ell} = \mathcal{E}_\ell e^S; \qquad \Delta E_{q\ell} = \mathcal{E}_\ell e^{-S} \qquad (42)$$

Squeezed states are still minimum uncertainty states with respect to (36), but, in contrast with coherent states, they have an unequal share of the fluctuations among the two quadratures. The field fluctuations are therefore smaller than the vacuum fluctuations at some points or at some times, if $S > 0$ when $\psi_\ell \equiv 0(\pi)$ for example. The remaining quantum fluctuations can be as small as desired, provided one takes S large enough. But in such a case, the field fluctuations will be very large at the same time t a quarter wavelength further, or at the same point $\mathbf{r}$, a quarter of a period later. A special interesting case is the "*squeezed vacuum state*" $|\alpha_\ell = 0, S\rangle$ in which the mean electric field is 0, like in vacuum, but with fluctuations below and above $\mathcal{E}_\ell$ as a function of ψ_ℓ. Another interesting squeezed state is the so-called "*sub-Poissonian state*", for which $\langle E_{p\ell} \rangle >> \mathcal{E}_\ell$, $\langle E_{q\ell} \rangle = 0$, $\Delta E_{p\ell} < \mathcal{E}_\ell$, and therefore, $\Delta E_{q\ell} > \mathcal{E}_\ell$. In this state, one easily shows that $\Delta N < \sqrt{\langle N_\ell \rangle}$, which corresponds to a sub-Poissonian distribution of photons.

These states have been experimentally produced for the first time in 1985 and have been the focus of a lot of theoretical and experimental attention [21, 22]. They require some kind of optical nonlinearity to be produced : parametric processes, frequency doubling, optical Kerr effect, phase conjugation have been successfully used to produce these states. Some lasers with reduced pumping noise have also been shown to generate sub-Poissonian states. To date, the lowest quantum noise variances experimentally observed are of the order of 20% of the vacuum level.

One of the simplest phenomena in nonlinear optics is the parametric splitting of light occuring in crystals with $\chi^{(2)}$ nonlinearity, sketched in figure 4

: a pump photon of frequency ω_0 is split into a signal photon of frequency ω_1 and an idler photon of frequency ω_2 (with $\omega_0 = \omega_1 + \omega_2$). A possible system likely to produce squeezed states is the "degenerate parametric amplifier", consisting of such a crystal, submitted to a beam of frequency ω_0 and producing signal and idler photons of frequency $\omega_0/2$. One can show [20] that if one injects a coherent field at frequency $\omega_0/2$ at the entrance of the crystal, this process leads to a squeezed state at the output. In particular, if one injects "nothing", i.e. the vacuum state, one gets a squeezed vacuum. In this configuration, the squeezing parameter S is significant only for very intense pumps (MW/cm^2), achievable only in the pulsed regime [22]. The squeezing effect can be enhanced if one inserts the crystal in a cavity resonant at frequency $\omega_0/2$, i.e. if one builds a set-up called "degenerate optical parametric oscillator". In theory, S goes to infinity as one approaches from below the oscillation threshold of the parametric oscillator. In experiments, squeezed vacuum with a noise reduction of 80% have been obtained in this configuration with only 100 mW of c.w. pump power [22].

Number states The energy eigenstates $|n_\ell\rangle$ are interesting because, as we will see later, they give rise to photodetection signals without any fluctuations. In such states, one easily shows that

$$\langle n_\ell | E(\mathbf{r},t) | n_\ell \rangle = 0 \qquad \Delta E(\mathbf{r},t) = \sqrt{2n_\ell + 1}\mathcal{E}_\ell \qquad (43)$$

The mean field is zero, like in vacuum or thermal fields ; however, the fluctuations around zero are very large when the number of photons n_ℓ is large. Since $\Delta N_\ell = 0$, then $\Delta\varphi = \infty$, according to (37) : number states correspond to a field with a perfectly defined amplitude E_0 and a totally random phase φ.

Except in the trivial case $n_\ell = 0$, *number states are highly nonclassical* and very difficult to produce. To generate the one photon state $|n_\ell = 1\rangle$, one technique is to take a single excited atom (or ion) in vacuum : If one waits a time long compared to the spontaneous emission lifetime, the atom ends up in the ground state and the field in a one photon state in some spherical wave mode [6]. Another technique is to make use of the parametric splitting effect, already encountered to generate squeezing (figure 4). If the $\chi^{(2)}$ crystal is submitted to a pump wave of frequency ω_0 and wavevector $\mathbf{k}_0$, spontaneous parametric splitting takes place, creating a field state $|\psi\rangle$ of the form

$$|\psi\rangle = C_0 |vac\rangle + \sum_{\mathbf{k}_1,\omega_1} C_{\mathbf{k}_1,\omega_1} |1:\mathbf{k}_1,\omega_1;1:\mathbf{k}_2,\omega_2\rangle \qquad (44)$$

containing *pairs of correlated photons* of frequencies ω_1 and ω_2 such that $\omega_1 + \omega_2 = \omega_0$, and wavevectors $\mathbf{k}_1$ and $\mathbf{k}_2$ such that $\mathbf{k}_1 + \mathbf{k}_2 = \mathbf{k}_0$. One then takes a photodetector with a small aperture and a narrow frequency filter which can detect only light of frequency ω_2 and wavevector $\mathbf{k}_2$. When this detector produces a nonzero signal, the state $|\psi\rangle$ projects onto the corresponding state

$|1:\mathbf{k}_1,\omega_1;1:\mathbf{k}_2,\omega_2\rangle$, and one is left with the one photon state $|1:\mathbf{k}_1,\omega_1\rangle$ which can be used in various experiments. This kind of technique is now currently used in many quantum optics experiments at the photon counting regime and has led to the discovery of very interesting phenomena (quantum cryptography, quantum teleportation...)[2].

Number states $|n_\ell\rangle$ with more than one photon, and in particular with $n_\ell \gg 1$, have never been produced so far. Their production would also lead to new kinds of quantum effects.

Other nonclassical states Let us briefly mention another class of nonclassical states, consisting in linear superposition of coherent states, such as

$$|\psi_1\rangle = |0\rangle + |\alpha\rangle \qquad |\psi_2\rangle = |\alpha\rangle + |-\alpha\rangle \qquad (45)$$

(within some normalization factor). When $|\alpha|^2 \gg 1$, they are called "Schrödinger cats", because they are quantum linear superpositions of two states of classical character, analogous to the superposition of a dead cat and a living cat considered by Schrödinger. These states have been recently produced (with $|\alpha|^2 \approx 10$) and allowed studies of the decoherence time in a quantum measurement process [23].

2.5 Two mode case

The general state is in this case

$$|\psi\rangle = \sum_{n_1 n_2} C_{n_1 n_2} |n_1, n_2\rangle \qquad (46)$$

It can give rise to nonclassical fluctuations (below the standard quantum limit) if one makes a measurement on mode 1 or on mode 2, like in the single mode case. But now, a new quantum feature arises when $|\psi\rangle$ cannot be written as a tensor product of two states in modes 1 and 2

$$|\psi\rangle \neq \left(\sum_{n_1} C_{n_1} |n_1\rangle\right) \otimes \left(\sum_{n_2} C_{n_2} |n_2\rangle\right) \qquad (47)$$

In this case, a measurement on mode 1 induces a state projection which changes also the result of later measurements on mode 2 : this shows that the two modes are correlated at the quantum level, in a way which in some cases cannot be accounted for by classical correlations, and gives rise for example to the violation of Bell inequalities [24].

A first technique to produce these states is to use spontaneous parametric down-conversion (eq.(44)), restricted to the case of two modes. One then produces the state

$$|\psi\rangle = C_0 |0,0\rangle + C_1 |1,1\rangle \qquad (48)$$

This is obviously a non factorizable state (named a "twin photon state"). In real experiments, the parametric process has a low efficiency, so that $|C_1|^2 \ll |C_0|^2$: the photon pairs are produced at a very low rate (typically 10^3 per second) and photon counting techniques are required. In a way similar to the generation of squeezed state in subsection (2.4), the process can be enhanced by inserting the crystal in a cavity resonant for modes 1 and 2 (nondegenerate optical parametric oscillator configuration). Above the oscillation threshold, the system generates the state

$$|\psi\rangle = \sum_n C_n |n_1 = n , n_2 = n\rangle \qquad (49)$$

with $C_n \neq 0$ even for large values of n. This state, called "twin beam" state, is non factorizable. It is an eigenstate of $\widehat{N}_1 - \widehat{N}_2$ with eigenvalue 0 : this means that the photon number fluctuations are strictly the same in the two modes, and exactly cancel in the difference [25].

2.6 More than two modes

Three-mode quantum correlations have been theoretically considered [26] and may have very interesting quantum properties . They have never been produced in experiments so far, nor quantum states with more than three states. This shows that the space of quantum light states is essentially unexplored, and that much more work remains to be done in quantum optics to investigate it.

3 Quantum measurements on light

In section 2, we have introduced different quantum operators for the electromagnetic field, such as the local instantaneous electric field $\widehat{E}(\overrightarrow{r}, t)$, or the quadrature operators $\widehat{E}_{p\ell}$ and $\widehat{E}_{q\ell}$. Even though they are called "observables", these quantities seem to be very difficult to determine experimentally, because so far there are no photodetectors able to follow the very fast oscillation, at the femtosecond scale, of the optical electric field. This section is devoted to the study of measurement processes with actual photodetectors, and to the quantum effects which can be observed with these detectors.

3.1 Direct photodetection

In the optical range, photodetectors are either photomultipliers or photodiodes. Both rely on a quantum process in which a bound electron is promoted to a continuum under the influence of the incoming light. The free electron gives rise to a photocurrent which is measured and analyzed by electronic means. Glauber [27] made the quantum analysis of this process and found that the mean photocurrent is given by

$$\langle i(t) \rangle = \alpha \iint\limits_{(S)} dx\, dy \left\langle \widehat{E}^{(+)}(\overrightarrow{r},t)^+ \widehat{E}^{(+)}(\overrightarrow{r},t) \right\rangle \tag{50}$$

where α is a constant depending on the detector, (x,y) the detector plane, (S) its area, and $\widehat{E}^{(+)}$ the complex electric field operator introduced in (24). For a photodetector of quantum efficiency 1, which totally absorbs the light and converts all of it into electrons (1 electron $\rightarrow$ 1 photon), one can define a photocurrent operator $\widehat{i}$

$$\widehat{i} = \frac{2q\varepsilon_0 c}{\hbar \omega_0} \iint\limits_{(S)} dx\, dy\, \widehat{E}^{(+)}(\overrightarrow{r},t)^+ \widehat{E}^{(+)}(\overrightarrow{r},t) \tag{51}$$

which will enable us to determine photocurrent mean values, but also the quantum fluctuations around the mean. For a single mode field $\widehat{E}^{(+)}$ is proportional to $\widehat{a}_\ell$, so that $\widehat{i}$ is proportional to $\widehat{N}_\ell = \widehat{a}_\ell^+ \widehat{a}_\ell$: the photodetector is then a *photon counter*, or an energy-meter. The information lying in the phase of the field is completely lost in direct photodetection.

Two qualitatively different regimes occur, according to the mean intensity of the measured light

- The photon counting regime, when $\langle N_\ell \rangle \approx 1$: the photocurrent appears as a series of isolated peaks, associated with the arrival of individual photons.

- The analog regime, when $\langle N_\ell \rangle \gg 1$: the different peaks overlap and give rise to a "macroscopic" photocurrent with a nonzero mean, around which exist quantum fluctuations, which are still the consequence of the randomness of arrival times of the incoming photons.

As we will see, quantum effects are not restricted to the photon counting regime. They are still present at the "macroscopic" level for light beams, where they can be observed on the photocurrent fluctuations.

If one assumes that the different photons are statistically independent, the distribution of photocurrent peaks is Poissonian, and one gets fluctuations of photocounts ΔN, or a variance of the photocurrent Δi, which is given by

$$\Delta N = \sqrt{\langle N \rangle} \qquad \Delta i = \sqrt{2q \langle i \rangle B} \tag{52}$$

where B is the frequency bandwidth of the detection system. This noise, called "shot noise", has long been considered to be the lower limit in photodetection noise.

Noise spectral density As seen in expression (52), the variance depends on B, i.e. on the detection properties. A more intrinsic and detailed characterization of the photocurrent fluctuations is its *noise spectral density* $S_i(\Omega)$, defined as the Fourier transform of the current autocorrelation function

$$S_i(\Omega) = \int_{-\infty}^{+\infty} d\tau\, e^{i\Omega\tau} \left[\langle i(0) i(\tau) \rangle - \langle i \rangle^2 \right] \tag{53}$$

$S_i(\Omega)$ is also related to the Fourier components $\delta i(\Omega)$ of the photocurrent fluctuations $\delta i(t) = i(t) - \langle i \rangle$ measured by a system of frequency bandwidth B around Ω by

$$\left\langle (\delta i(\Omega))^2 \right\rangle = 2B \, S_i(\Omega) \tag{54}$$

$S_i(\Omega)$ is easily measurable by using a spectrum analyzer.

Let us now assume that the field arriving at the detector is in a coherent state. Using expression (53) and the properties of the coherent state, one easily shows that

$$S_i(\Omega) = q \langle i \rangle \tag{55}$$

This rigorously shows that in a coherent state one gets a shot noise limited photocurrent, and enforces the naive vision of a coherent state as a beam composed of randomly distributed photons.

Generation of sub-Poissonian states As seen in subsection (2.2), there exist quantum states of light, called sub-Poissonian states, which have less fluctuations on N_ℓ than the coherent state (and more on the phase). When these states impinge on a photodetector, they produce a photocurrent with sub-shot noise fluctuations, allowing direct optical measurements with better signal to noise ratios and better sensitivity than with classical light.

In a laser, nothing determines the mean phase of the emitted field and sets it to a well defined value. As a result, the instantaneous phase evolves randomly under the influence of all the noise sources coupled to it, mainly spontaneous emission : this is the well-known *Schawlow-Townes phase diffusion phenomenon,* responsible for the ultimate laser linewidth Γ_{laser} [28]. The phenomenon has another consequence : it produces an *infinite phase noise* for measurement times long compared to Γ_{laser}^{-1}. According to eq. (37), nothing then prevents the intensity noise in a laser to be zero at the same time scale. Nothing thereby prevents a laser to generate a nonclassical state of light, in the form of a sub-Poissonian state. Actually, in real lasers [29], there are many noise sources which also act on the output intensity, and hide the intrinsic sub-Poissonian effect, mainly the almost inevitable noise present in the pumping process of the gain medium.

Yamamoto [30] has shown that in semiconductor lasers, the pump noise is nothing else than the injection current noise, equal to $2k_B T/R$, where R is the equivalent resistance of the injection circuit (power supply + diode effective resistance). By increasing R, one can reduce this noise at will. One is then left with the other sources of noise, coming either from spontaneous emission (negligible well above threshold) or from the existence of other laser modes close to the oscillation threshold. Experiments have been performed, which produce sub-Poissonian light with rather simple set-ups. The noise reduction below shot noise lies in the 80 % range for cooled samples [31] or 30 % range for room temperature devices [32].

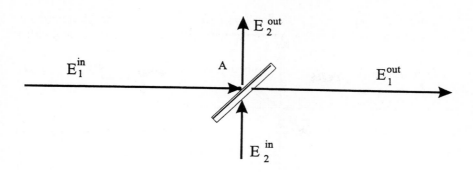

Fig. 5. Sketch of beam coupling on a beamsplitter

These sub-Poissonian sources have been used to enhance the sensitivity of optical experiments, for example to measure very weak spectroscopic signals, which would have been unobservable by using classical light sources of identical intensity [33].

3.2 Quantum theory of the linear optical coupler

To go further in this discussion, we need to analyze in quantum words a familiar object in optics : the linear coupler, such as a beamsplitter or a polarizing beamsplitter (see figure 5). It is a system which couples in a linear way light modes of different directions or different polarizations. Classical optics shows that the output complex amplitudes defined at point A are related to the input ones (defined at the same point) by the relations

$$\begin{cases} E_1^{out} = t\, E_1^{in} + r\, E_2^{in} \\ E_2^{out} = -r\, E_1^{in} + t\, E_2^{in} \end{cases} \quad (56)$$

where r and t are the real reflection and transmission coefficients of the device ($r^2 + t^2 = 1$). The minus sign in the second equation ensures the equality between the total input light power ($P_1^{in} + P_2^{in}$) and the total output power ($P_1^{out} + P_2^{out}$).

One can show that the same relation holds for the quantum operators $\widehat{E}^{(+)}$ introduced in (2.2). In the particular case of single mode fields, it reduces to the simple relation

$$\begin{cases} \widehat{a}_1^{out} = t\, a_1^{in} + r\, a_2^{in} \\ \widehat{a}_2^{out} = -r\, \widehat{a}_1^{in} + t\, \widehat{a}_2^{in} \end{cases} \quad (57)$$

between the annihilation operators of the two modes coupled by the device. It is easy to show that the quantum input-output relation (57) is canonical, i.e. preserves the commutation relations (23) and (27). We will often use relation (57) in the subsequent subsections.

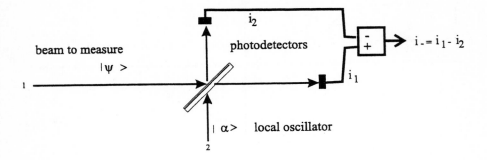

Fig. 6. Homodyne detection scheme

3.3 Homodyne detection

Let us consider the homodyne detection set-up sketched in figure 6 : a coherent state $|\alpha = |\alpha| e^{i\varphi}\rangle$ in mode 2 (named *local oscillator*) is mixed on a 50 % beamsplitter ($r = t = 1/\sqrt{2}$) with an unknown single mode field state $|\Psi\rangle$ in mode 1. Both modes have the same frequency and transverse variation. The two resulting beams are measured by two photodetectors with unity quantum efficiency. One then records the difference $i_- = i_1 - i_2$ between the two photocurrents, proportional to the difference $N_- = N_1 - N_2$ between the photon numbers in the two modes after the beamsplitter. Formulae (57) imply that the corresponding operator $\widehat{N}_-$ is given in terms of the operators before the beamsplitter by

$$\widehat{N}_- = \widehat{a}_1 \widehat{a}_2^+ + \widehat{a}_1^+ \widehat{a}_2 \tag{58}$$

The mean value of $\widehat{N}_-$ is then

$$\left\langle \widehat{N}_- \right\rangle = |\alpha| \langle \Psi | \widehat{a}_1 \, e^{-i\varphi} + \widehat{a}_1^+ \, e^{i\varphi} | \Psi \rangle \tag{59}$$

This quantity is proportional to $\left\langle \widehat{E}_{p1} \right\rangle$ when $\varphi = 0$ and to $\left\langle \widehat{E}_{q1} \right\rangle$ when $\varphi = \frac{\pi}{2}$. One can also show that the noise on i_- is proportional to the quantum fluctuations of the operators E_{p1} and E_{q1} in the case of a strong local oscillator (i.e. when $|\alpha| \gg \left| \langle \Psi | \widehat{E} | \Psi \rangle \right|$).

This shows that the quadrature components $\widehat{E}_{p\ell}$ and $\widehat{E}_{q\ell}$ of the electric field are really measurable quantities, even with detectors that are unable to measure instantaneous values of the optical field. The technique used, consisting in mixing the signal to measure with another signal of same frequency and variable phase, is similar to stroboscopy, or to holography.

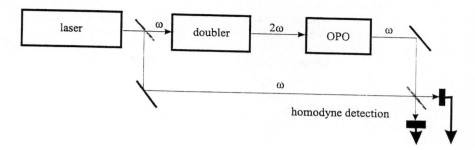

Fig. 7. Set-up for the generation and detection of squeezed vacuum

It is therefore possible to measure the noise spectral densities of the two quadrature operators $S_{E_p}(\Omega)$ and $S_{E_q}(\Omega)$. One shows [34] that the commutation relation (27) implies that

$$\forall \Omega \quad S_{E_p}(\Omega) S_{E_q}(\Omega) \geqslant (\hbar\omega_0/2\varepsilon_0 \, cA)^2 \qquad (60)$$

where A is the transverse area of the mode. This relation generalizes the Heisenberg inequality (36). In contrast to (36), containing the unphysical quantization volume V in $\mathcal{E}_\ell$, relation (60) contains only measurable quantities. It also concerns all the noise frequencies Ω.

If $|\Psi\rangle$ is a coherent state, one easily shows that

$$S_{E_p}(\Omega) = S_{E_q}(\Omega) = \hbar\omega_0/2\varepsilon_0 \, cA \qquad (61)$$

Like for the variances, a coherent state is a minimum state with respect to inequality (60), and has equal noises densities on the two quadratures. Eq (61) states also that this noise, also called "standard quantum noise", is the same for all noise Fourier frequencies Ω: it is a "white noise". In a way similar to subsection (2.4), we will then define a nonclassical state of light as a state for which, for example $S_{E_p}(\Omega)$ is smaller than the standard quantum limit within some interval $[\Omega_1, \Omega_2]$ of the Fourier spectrum.

Figure 7 shows the complete (but simplified) set-up of an experiment in which a squeezed vacuum state is produced and measured [35]. It consists in a laser source at frequency ω, which is frequency doubled in a first nonlinear device. The resulting light at frequency 2ω pumps a sub-threshold degenerate OPO, and produces squeezed vacuum at frequency ω (see section 2.4). To measure it, one uses the homodyne detection scheme of the previous subsection with a local oscillator coming from the initial laser. When one records the noise on i_- on a spectrum analyzer as a function of the local oscillator phase (which can be changed by varying the corresponding optical path), one observes a phase-dependent noise, alternatively above and below the standard quantum noise level (measured by switching off the beam coming from the OPO in the homodyne detector). In the experiment described here [35], the squeezed quadrature had a noise reduced by 70 % below the standard quantum noise level.

Such a squeezed vacuum beam can be used to improve the sensitivity in interferometric measurements, as theoretically shown by [36]. The technique consists in mixing the squeezed vacuum generated by the OPO with the input laser light (like in figure 7) on the beamsplitter used in the Michelson interferometer : the interference fringes recorded when the optical path is varied will have less noise around some values of this path (at mid-fringe), and in this configuration the signal to noise ratio is shown to be enhanced by the squeezing factor. This scheme has been successfully implemented experimentally [37]; it has important potential applications in the measurement of gravitational waves by interferometers with arm lengths of several kilometers, which are looking for ultra-weak interferometric signals very close to, or even below, the shot noise "limit".

3.4 Coincidence detection

Let us now turn to another kind of detection on light, which is almost always performed in the photon counting regime : it makes use of two photodetectors 1 and 2 of areas S_1 and S_2 around points $\vec{r}_1$ and $\vec{r}_2$ and a delay line introducing a delay τ on the photocurrent coming from detector 2. The coincidence signal $G_{12}(\tau)$ is the product of the two photocurrents : it is nonzero only when a photon arrives at some time t on detector 1, and another photon at the time $t + \tau$ on detector 2 (coincidence count). Glauber has shown that the count rate of such double counts is proportional to

$$\iint_{S_1} d^2 r_1 \iint_{S_2} d^2 r_2 \left\langle \widehat{E}^{(+)}(\vec{r}_1, 0)^+ \widehat{E}^{(+)}(\vec{r}_2, \tau)^+ \widehat{E}^{(+)}(\vec{r}_1, 0) \widehat{E}^{(+)}(\vec{r}_2, \tau) \right\rangle$$
(62)

If a single mode, labeled 1, is detected on S_1, and another single mode, labeled 2, on S_2, the complex field operators are simply proportional to the corresponding annihilation operators, and one has

$$G_{12}(\tau) \propto \left\langle a_1^+(0) a_2^+(\tau) a_1(0) a_2(\tau) \right\rangle$$
(63)

In order to get rid of the unknown proportionality constants it is useful to divide G_{12} by the single photodetection count rates measured on each detector N_1 and N_2 to obtain the normalized second order correlation function $g^{(2)}(\tau)$

$$g^{(2)}(\tau) = \frac{G_{12}(\tau)}{N_1 N_2}$$

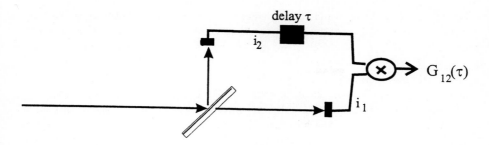

Fig. 8. Set-up for the measurement of the second-order correlation function

$$= \frac{G_{12}(\tau)}{\iint_{S_1} d^2r_1 \left\langle \widehat{E}^{(+)}(\vec{r}_1,0)^+ \widehat{E}^{(+)}(\vec{r}_1,0) \right\rangle \iint_{S_2} d^2r_2 \left\langle \widehat{E}^{(+)}(\vec{r}_2,0)^+ \widehat{E}^{(+)}(\vec{r}_2,0) \right\rangle} \tag{64}$$

This quantity has a classical analog, defined for classical stochastic fields, which is obtained simply by replacing quantum means by ensemble means in expressions (63,64). Using the Cauchy-Schwartz inequality, one shows that for classical correlation functions

$$G_{12}(\tau) \geqslant N_1 N_2 \quad \text{or} \quad g^{(2)}_{classical}(\tau) \geqslant 1 \quad \forall \tau \tag{65}$$

Using the quantum definition of $g^{(2)}(\tau)$, it is easy to show that, if $|\Psi\rangle$ is in a coherent state, and therefore is an eigenstate of $\widehat{E}^{(+)}(t)$, then $g^{(2)}(\tau) = 1$ for any value of τ. The value 1, frontier between the classical and pure quantum properties in this peculiar measurement device, is reached when one uses a coherent state. We will call it, as previously, the standard quantum limit in this kind of measurement.

Photon bunching Let us now consider the set-up of figure 8, which uses a beamsplitter with equal transmission and reflections ($r = t = 1/\sqrt{2}$ in equations (56) and (57)). This configuration was introduced in the late 50's by Hanbury-Brown and Twiss in radio-astronomy, to characterize various kinds of electromagnetic waves. Using expression (63) for G_{12} and relation (57) for the quantum transformation of operators on the beamsplitter, it is straightforward to calculate the result of coincidence measurement after the beamsplitter.

Let us firstly take the case of two coherent states $|\alpha_1\rangle$ and $|\alpha_2\rangle$ arriving on both sides of the beamsplitter (the configuration of a single nonzero coherent state on side (1) and nothing on side (2) is included in this case, as nothing is the vacuum state, which is a special case of coherent state $\alpha_2 = 0$). The operator mean values must then be taken in state $|\alpha_1\rangle \otimes |\alpha_2\rangle$. One finds in this case

$$G_{12}(\tau) = N_1 N_2 \quad ; \quad g^{(2)}(\tau) = 1 \qquad (66)$$

One still gets the standard quantum limit, which can be simply understood when one knows the following result, specific of coherent states : if one sends two uncorrelated coherent states $|\alpha_1\rangle \otimes |\alpha_2\rangle$ at the input of a beamsplitter, one gets at its output the quantum state $|\alpha_1^{out} = t\alpha_1 + r\alpha_2\rangle \otimes |\alpha_2^{out} = -r\alpha_1 + t\alpha_2\rangle$ made of two other coherent states, but still *uncorrelated*. If one sends a classical field on the beamsplitter (and nothing on the other input), one will find in general

$$g_{classical}^{(2)}(0) > 1 \qquad (67)$$

($g^{(2)}(0) = 2$ in particular for a thermal field). This phenomenon, called "photon-bunching", shows that photons arrive preferably by pairs in classical sources, and has been extensively studied in the 60's [1, 2].

Photon antibunching The quantum region ($0 \leq g^{(2)}(\tau) \leq 1$) corresponds to a phenomenon of "photon antibunching" (photons arrive more likely isolated), specific of nonclassical fields. If one sends for example on one side of the beamsplitter a single photon number state $|n_1 = 1\rangle$ (and the vacuum on the other side), the full quantum calculation gives straightforwardly the following result

$$N_1 = \frac{1}{2} \quad N_2 = \frac{1}{2} \quad G_{12}(2) = 0 \Rightarrow g^{(2)} = 0 \qquad (68)$$

This result of measurement cannot be obtained by classical fluctuating fields, but it can be simply explained in terms of photons behaving as classical particles : the photon is a particle which cannot be split by the beamsplitter, which then randomly transmits it or reflects it with equal probabilities. This implies that $N_1 = N_2 = 1/2$. But the single particle cannot be detected twice at two different locations or times, and therefore $G_{12} = 0$. One finds here a perfect example of wave-particle duality for the light : the result of a coincidence measurement for a single photon state cannot be explained by classical waves, even with stochastic fluctuations, but can be perfectly well understood in the language of classical particles.

As already explained in subsection (2.4), a single atom in an excited state is a perfect source of a single photon state. Actually the first experiment able to put in evidence a nonclassical effect on light [38] has been the measurement of a $g^{(2)}(0)$ value smaller than 1 in the resonance fluorescence of single atoms. The experimental value was $g^{(2)}(0) = 0.6$ in this celebrated "premiere" performed in 1977. More refined experiments performed later [39, 40], for example with single ions trapped in electromagnetic traps, yielded the value 0 for $g^{(2)}(0)$, and an impressive agreement between the experiment and the predictions of quantum theory.

Case of twin photons Let us now replace at the input of the beamsplitter tensor products of uncorrelated states, even nonclassical, by *correlated two-mode states*. We have shown in subsection (2.5) that it was possible to produce the following twin photon state by spontaneous parametric down conversion

$$|\Psi\rangle = c_1 |0,0\rangle + c_2 |1,1\rangle \qquad (69)$$

The quantum calculation of photodetection signals with such a state is still straightforward. One finds

$$N_1 = 1 \quad N_2 = 1 \quad \text{and} \quad G_{12} = 0 \qquad (70)$$

The value 1 for N_1 and N_2 comes from the fact that two photons are present in the set-up on the average, and that they are equally distributed on the two output ports of the beamsplitter. But the value 0 for G_{12} is difficult to understand, precisely because there are now 2 photons present in the system, which do not a priori rule out the possibility of a simultaneous double count.

Let us suppose for a while that photons are classical particles. Then there is a probability 1/4 of the two photons being measured both in output port 1, 1/4 of the two photons being both in output port 2 and 1/2 of finding one photon in port 1 and one in port 2. "Particle mechanics" gives therefore the value $0,5$ for $g^{(2)}(0)$ in this case, which lies outside the classical wave domain, but does not correspond to the exact quantum calculation.

One is therefore led to the following important statement : light, as it is described by the full quantum theory, sometimes looks like a classical wave (actually most of the time), sometimes looks like a classical particle, the photon, and sometimes (hopefully seldom) looks like nothing classical. There is in the quantum aspects of light more than the famous wave-particle duality (which should be called "classical wave-classical particle duality), and some phenomena in optics look really "strange" as compared to the classical world of waves and particles.

Actually a simple quantum "explanation" of the fact that $g_{12}(0) = 0$ in this experiment can be given. The double count situation comes from two different events : (photon (1) transmitted + photon (2) transmitted) and (photon (1) reflected + photon (2) reflected), having equal classical probabilities 1/4. As these two "paths" cannot be distinguished by any experiment, we must *add the probability amplitudes and square the result* to get the right quantum mechanical probability of the event. The probability amplitudes of reflection and transmission are equal r and t [1]. Because of the change of sign for r in the two possible reflections (eqs(56)), the probability amplitudes for the two paths are respectively $\frac{1}{\sqrt{2}} \times \frac{1}{\sqrt{2}}$ and $\left(\frac{1}{\sqrt{2}}\right)\left(-\frac{1}{\sqrt{2}}\right)$, the sum of which is 0. The zero value of G_{12} comes thereby from a *destructive interference between probability amplitudes*, a well known feature in quantum mechanics.

The first experiment check of this phenomenon has been performed by Mandel and coworkers in 87 [41] using parametric downconversion in a KDP crystal. In order to get the perfect probability amplitude cancelation giving rise to $G_{12} = 0$, the paths taken by the twin-photons from their "birth" in the cristal to the beamsplitter must be strictly equal. When this configuration is experimentally realized, one gets an almost perfect cancelation of the coincidence rate. This set-up has been refined and complicated in more recent experiments, in order to test other puzzling predictions of quantum optics [2, 4].

4 Interaction between matter and quantized light

We can now finally turn our attention to the problem of the full quantum description of light-matter interaction. In order to reduce the problem to its essence, and also to be able to compare the exact theory with the approximate treatments given in section 1, we will make the same simplifying assumptions, namely that the matter is made of a two level atom, of Bohr frequency ω_0, and that the field is in a single mode of frequency ω, which is almost resonant with the atom ($\omega \approx \omega_0$). This can be in particular obtained if the field is confined inside a real resonant cavity of frequency ω, so that one often labels "*cavity quantum electrodynamics*" the physics described in this section.

4.1 Jaynes-Cummings model

The Hilbert space in which the atom-field system is now described is the tensor product of the atomic Hilbert space, spanned by vectors that we will call here $|g\rangle$ (for the ground state), and $|e\rangle$ (for the excited state), and of the field Hilbert space, spanned in the single mode case by the number states $|n\rangle$. The general quantum state is then

$$|\Psi\rangle = \sum_{j=e,g} \sum_{n=0}^{\infty} c_{jn} |j\rangle \otimes |n\rangle = \sum_{j,n} c_{jn} |j, n\rangle \qquad (71)$$

Generally speaking, this state cannot be factorized into an atomic component and a field component : when matter and light interact, quantum correlations do appear between them and it is not possible to speak rigorously in terms of a field and of an atom with independent properties, even long after the interaction between them has ended. C. Cohen-Tannoudji [6] has introduced for this kind of state (71) the terminology of "dressed atom" (atom dressed by the surrounding photons). But it could as well have been named "dressed field" (field dressed by the existing atoms).

The hamiltonian $\widehat{H}$ describing the evolution of this "dressed system" is the sum of three terms

$$\widehat{H} = \widehat{H}_{at} + \widehat{H}_{field} + \widehat{H}_{int} \qquad (72)$$

At the dipole approximation, and treating the center of mass $\vec{r}_{at}$ of the atom as a classical variable, one has

$$\widehat{H}_{at} = \hbar\omega_0 |e\rangle\langle e|$$
$$\widehat{H}_{field} = \hbar\omega \widehat{a}^+\widehat{a} \qquad (73)$$
$$\widehat{H}_{int} = -\vec{E}\left(\vec{r_{at}}, t\right).\widehat{\vec{D}}$$

in which the origin of energies has been chosen to the energy of the atomic ground state $|g\rangle$ in presence of the vacuum field state $|n=0\rangle$, and $\widehat{\vec{D}}$ is the dipole atomic operator, defined like in subsection 1.3, and written here as

$$\widehat{\vec{D}} = \vec{d}\left(|g\rangle\langle e| + |e\rangle\langle g|\right) \qquad (74)$$

$\widehat{\vec{D}}$ has no diagonal elements, proportional to $|e\rangle\langle e|$ or $|g\rangle\langle g|$, because of the parity conservation symmetry of atoms. The interaction hamiltonian can be also written as

$$\widehat{H}_{int} = i\hbar\Omega_0 \left(|e\rangle\langle g| + |g\rangle\langle e|\right)\left(\widehat{a} - \widehat{a}^+\right) \qquad (75)$$

assuming the atom to be at position $\vec{r_{at}} = \vec{0}$, and setting

$$\Omega_0 = \mathcal{E}_\ell \vec{d} \cdot \vec{\epsilon_\ell}/\hbar \qquad (76)$$

The coupling constant Ω_0 is often written g, especially in solid state physics. Let us note first that $\widehat{H}$ is a time independent hamiltonian, which was not the case for the semiclassical hamiltonian (eq.(11)). This means that, if one considers the interaction term $\widehat{H}_{int}$ as a small perturbation compared to the first two terms, $\widehat{H}_{int}$ will induce transitions between levels $|j,n\rangle$ that have the same energy. For example, $\widehat{H}_{int}$ connects states $|g,n\rangle$ to states $|e, n\pm 1\rangle$ and describes the excitation of the atom with a change of the number of photons by 1 (and the reverse process). But the transition will be efficient only in the resonant case, i.e. between states $|g,n\rangle$ and $|e, n-1\rangle$ when $\omega = \omega_0$. Conversely, the states $|e,n\rangle$ and $|g, n+1\rangle$ will be also resonantly coupled. In the resonant or quasi-resonant configuration, we can therefore restrict ourselves to the approximate interaction term

$$\widehat{H}'_{int} = i\hbar\Omega_0\left(|e\rangle\langle g|\widehat{a} - |g\rangle\langle e|\widehat{a}^+\right) \qquad (77)$$

which describes only the resonant (or quasi-resonant) part of the processes. The new total hamiltonian

$$\widehat{H}' = \widehat{H}_{at} + \widehat{H}_{field} + \widehat{H}'_{int} = \hbar\left[\omega_0|e\rangle\langle e| + \omega\widehat{a}^+\widehat{a} + i\Omega_0\left(|e\rangle\langle g|\widehat{a} - |g\rangle\langle e|\widehat{a}^+\right)\right] \qquad (78)$$

has been first introduced by Jaynes and Cummings [43] and has been extensively studied as the prototype of light-matter interaction at the quantum level. It is a good approximation of the physical reality in two limiting configurations :

- a two-level atom confined in an electromagnetic cavity with perfectly reflecting boundaries having a resonant frequency close to the atomic Bohr frequency.

- a two-level atom submitted in free space to a strong single mode quasi resonant laser beam : the coupling between the atom and the mode "filled" by the laser light overwhelms the coupling to all the other modes, provided that all effects related to spontaneous emission of the atom are neglected.

In both cases, the finite quality factor of the cavity, or the existence of atomic spontaneous emission, will introduce a dissipative term, and therefore will lead to a physical behaviour of the system departing from the exact Jaynes-Cummings model. We will consider these modifications in subsection 4.4.

4.2 Energies and eigenvectors of the atom-field system

$\widehat{H}'$ has eigenstates and eigenvectors which can be easily determined because $\widehat{H}'$ is a tensor product of 2×2 matrices in subspaces $\{|g, n\rangle, |e, n-1\rangle\}$, plus a pure diagonal term in the 1-D subspace corresponding to vector $|g, 0\rangle$. One finds for the energies

$$E_0 = 0 \quad ; \quad E_{\pm n} = \hbar \left(n\omega - \frac{\delta}{2} \pm \frac{1}{2}\sqrt{n\Omega_0^2 + \delta^2} \right) \quad n > 0 \qquad (79)$$

and for the eigenvectors

$$\begin{aligned}
|\Psi_0\rangle &= |g, 0\rangle \\
|\Psi_{+n}\rangle &= i\cos\theta_n |g, n\rangle + \sin\theta_n |e, n-1\rangle \\
|\Psi_{-n}\rangle &= -i\sin\theta_n |g, n\rangle + \cos\theta_n |e, n-1\rangle
\end{aligned} \qquad (80)$$

in which $\tan 2\theta_n = -\Omega_0\sqrt{n}/(\omega - \omega_0)$.

Let us now focus our discussion on the exact resonance configuration ($\omega = \omega_0 : \theta_n = \frac{\pi}{4}$), for which the quantum effects are maximized. In absence of the coupling term (i.e. when the atoms are far from the electromagnetic field mode), the two states $|g, n\rangle$ and $|e, n-1\rangle$ have the same energy. The effect of the coupling is to remove this degeneracy and to give rise to new eigenstates : $\frac{1}{\sqrt{2}}(\pm i|g, n\rangle + |e, n-1\rangle)$ separated by an energy difference equal to $\hbar\Omega_0\sqrt{n}$. These states which are not factorizable ("entangled") are precisely the dressed atom states. They will allow us to determine the time evolution of the atom+field system for any initial configuration.

4.3 Time evolution (resonant case $\omega = \omega_0$)

Let us first assume that the atom has been prepared in the excited state, and that the field mode is empty. The initial state of the system $|\Psi(0)\rangle = |e, 0\rangle$ is not an eigenstate of $\widehat{H}$, and changes with time. One easily finds

$$|\Psi(t)\rangle = e^{-i\omega_0 t}\left[|e,0\rangle \cos\frac{\Omega_0 t}{2} + |g,1\rangle \sin\frac{\Omega_0 t}{2}\right] \tag{81}$$

The probability to find the atom in the excited state is $N_e = \cos^2\frac{\Omega_0 t}{2}$, and therefore has an oscillatory behaviour. This situation is very different to what is encountered for the same excited atom in vacuum $|e,0\rangle$, but in the free space, with an infinity of field modes with which the atom is coupled : one finds in this case the usual exponential decay $N_e = e^{-\Gamma t}$ characteristic of spontaneous emission : the presence of cavity walls thereby strongly affects the phenomenon of spontaneous emission.

Let us now take the system in the state $|\Psi(0)\rangle = |g,n_0\rangle$, i.e. the ground state atom in presence of a highly non classical field, described by a number state. One finds that in this case, the system state becomes later

$$|\Psi(t)\rangle = e^{-in_0\omega_0 t}\left[|g,n_0\rangle \cos\Omega_0\sqrt{n_0}\frac{t}{2} - |e,n_0-1\rangle \sin\Omega_0\sqrt{n_0}\frac{t}{2}\right] \tag{82}$$

Now N_g is equal to $\cos^2\Omega_0\sqrt{n_0}\frac{t}{2}$, and one finds again that the atomic population is submitted to a periodic oscillation : it is the "quantum Rabi oscillation", reminiscent of the Rabi oscillation found in the semi-classical model (subsection 1.3). But the system behaves in a much different way from what is predicted in the semi-classical theory. Let us take for example a "$\frac{\pi}{2}$ pulse" like in subsection (1.3). At the end of the pulse, the system is in the following quantum state (within a global phase factor)

$$|\Psi_{\frac{\pi}{2}}\rangle = \frac{1}{\sqrt{2}}(|g,n_0\rangle - |e,n_0-1\rangle) \tag{83}$$

which has nothing to do with the coherent superposition of atomic states found in eq (16) : one can show that the mean dipole and mean field is zero in this state. In addition, there are strong correlations between the atomic and field part : if one is able to measure the atomic state and find, say, $|g\rangle$, one knows that the field is precisely in the number state $|n_0\rangle$. This quantum correlation between the atomic and field parts is perfect even though the interaction has ended, because for example the atom has moved outside the field mode. In the semi-classical theory, the field is imposed from the outside and does not change when it interacts with the atom. This highly nonclassical situation is difficult to achieve experimentally for $n_0 \geqslant 2$.

Let us now consider a physical situation which is much simpler to realize experimentally : the atom is initially in state $|g\rangle$, and the field in a coherent state $|\alpha\rangle = \sum_n c_n |n\rangle$, with $c_n = \frac{\alpha^n}{\sqrt{n!}}e^{-|\alpha|^2/2}$. One finds in this case

$$|\Psi(t)\rangle = c_0|g,0\rangle + \sum_{n\geqslant 1} e^{-in\omega_0 t} c_n \left(|g,n\rangle \cos\frac{\Omega_0}{2}\sqrt{n}t - |e,n-1\rangle \sin\frac{\Omega_0}{2}\sqrt{n}t\right) \tag{84}$$

So that the atomic ground state population is now

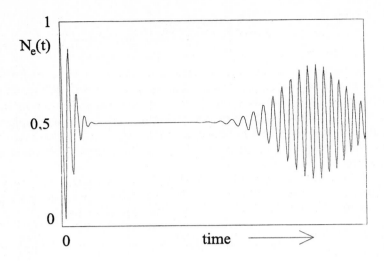

Fig. 9. Time evolution of the atomic upper state population for a two-level atom interacting with a coherent state ($\alpha = 4$)

$$N_g = \sum_{n=0}^{\infty} |c_n|^2 \cos^2 \frac{\Omega_0}{2} \sqrt{n} t \qquad (85)$$

$N_g(t)$ is a mixing of quantum Rabi oscillations with different frequencies and different weights. Figure 9 gives the time variation of the atomic population in the case $\alpha = 4$. One observes that the Rabi oscillations are quickly damped, and then reappear for a while after a delay. The damping is due to the non monochromatic character of the Rabi oscillations, the "revival" is a quantum effect due to the finite number of sinusoïdal oscillations contributing to N_g, which interfere again constructively when the different oscillations are agin in phase.

One therefore sees that when the atom interacts with a coherent state of weak intensity ($|\alpha| \approx 1$), specific quantum features on the atomic variables appear which are not accounted for by the semiclassical model. The same is true for the field, which has also a nonclassical character, especially during the time when $N_g \approx 0.5$, when the field is found to be somehow similar to a Schrödinger cat state (as defined in subsection 2.4). This kind of evolution has been experimentally observed for Rydberg atoms in interaction with cavity modes in the microwave domain [47].

The situation is drastically simplified in the case of an intense coherent field $||\alpha| e^{i\varphi}\rangle$ with $|\alpha| \gg 1$. One easily shows then that, neglecting terms of the order $\frac{1}{|\alpha|}$, the system state factorizes into a field part and an atomic part

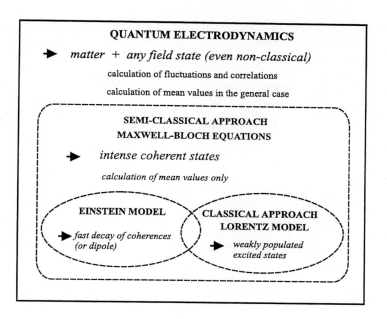

Fig. 10. Validity domains of the different approaches used in quantum optics

$$|\Psi(t)\rangle \approx |\alpha e^{-i\omega_0 t}\rangle \otimes |\Psi_{at}(t)\rangle$$
$$|\Psi_{at}(t)\rangle = |g\rangle \cos\frac{|\alpha|\Omega_0}{2}t - e^{-i\omega_0 t + i\varphi}|e\rangle \sin\frac{|\alpha|\Omega_0}{2}t \qquad (86)$$

which has three physical implications :

(i) there is a "disentanglement" between the two parts, and therefore there are no longer correlations between the atomic and field measurements ;

(ii) the field part evolves like a free coherent state, unaffected by the interaction with the atom ;

(iii) the atomic part *evolves in a semi-classical way* : one exactly finds at this limit an atomic evolution ruled by the semiclassical equations (13) and (14).

We have thus been able to bridge the gap between the exact quantum evolution and the semi-classical one, which is found to be only valid when the system interacts with strong coherent beams. This is fortunately the case in a vast majority of experimental situations. Figure 10 gives in a schematic way the validity domains of the different approximations that we have discussed in these lectures.

4.4 Influence of relaxation

In realistic experimental situations, cavity walls are not perfectly reflecting, and atomic excited levels are not stable. The previously calculated time evolution of the system is then only valid for times short compared to the atomic lifetime and to the cavity decay time. Dissipation has actually not only deleterious effects on the system : it couples it to the outside world, and allows us to have informations about the intracavity system, by the way of atomic fluorescence or field transmission through the cavity limits. To take relaxation into account, one must describe the system by a density matrix for the total atom-field system. Its evolution is ruled by the master equation

$$\frac{d}{dt}\rho = \frac{1}{i\hbar}\left[\widehat{H}',\rho\right]+\gamma\left(2\widehat{\sigma}_-\rho\widehat{\sigma}_+ - \widehat{\sigma}_+\widehat{\sigma}_-\rho - \rho\widehat{\sigma}_+\widehat{\sigma}_-\right)+\kappa\left(2\widehat{a}\rho\widehat{a}^+ - \widehat{a}^+\widehat{a}\rho - \rho\widehat{a}^+\widehat{a}\right) \quad (87)$$

where the first term describes the energy conserving evolution under the influence of the Jaynes-Cummings hamiltonian (eq (77)), the second the atomic decay (γ : dipole decay constant, $\widehat{\sigma}_+ = |g\rangle\langle e|$, $\widehat{\sigma}_- = |e\rangle\langle g|$), and the third the field decay (κ : intracavity electric field decay rate). This complicated equation has no simple solution, except in some limiting cases, that we will now examine.

4.5 Low excitation limit : weak and strong coupling cases

In this case, one can assume that the system evolves only in the "low energy subspace", spanned by vectors $\{|g,0\rangle, |e,0\rangle, |g,1\rangle\}$. Then, from (87), one can deduce the following simple equations for the mean values of $\widehat{a}$ (proportional to the complex electric field) and $\widehat{\sigma}_-$ (proportional to the atomic dipole) [44]

$$\begin{cases} \frac{d}{dt}\langle\widehat{a}\rangle = (i\omega_0 - \kappa)\langle\widehat{a}\rangle + \Omega_0\langle\widehat{\sigma}_-\rangle \\ \frac{d}{dt}\langle\widehat{\sigma}_-\rangle = (i\omega_0 - \gamma)\langle\widehat{\sigma}_-\rangle - \Omega_0\langle\widehat{a}\rangle \end{cases} \quad (88)$$

These equations are identical to the evolution equations of two damped classical oscillators with the same oscillation frequency ω_0, which are linearly coupled to each other. As is well-known, they lead to the onset of new *uncoupled normal modes* of the system, evolving like $e^{i\lambda t}$. Two different regimes can be found :

(i) *the weak-coupling regime*, when $\Omega_0 \leqslant \Omega_{critical} = |\kappa - \gamma|/2$, for which the system still oscillates at frequency ω_0, but with *different decay rates*

$$\mathrm{Im}\,\lambda = \frac{\kappa + \gamma}{2} \pm \sqrt{\frac{(\kappa-\gamma)^2}{4} - \Omega_0^2} \quad (89)$$

(ii) *the strong-coupling regime*, when $\Omega_0 \geqslant \Omega_{critical}$, for which the system is damped with a fixed average rate $(\kappa + \gamma)/2$, but now with *two different oscillation frequencies*

$$\mathrm{Re}\,\lambda = \omega_0 \pm \Omega_R \text{ with } \Omega_R = \Omega_0\sqrt{1 - \frac{(\kappa-\gamma)^2}{4\Omega_0^2}} \quad (90)$$

This removal of degeneracy appears for example in the form of a *periodic exchange of oscillatory energy* between the dipole and the electric field at frequency Ω_R i.e. of a Rabi oscillation.

Note here that these conclusions are changed if more than one atom interact with the cavity mode. In this case, one must replace $\hat{\sigma}_\pm$ by a collective dipole operator $\widehat{\Sigma} = \sum_{i:atoms} (\hat{\sigma}_+)_i$. One can show that this leads to a change of the Rabi oscillation frequency Ω_0, which must be replaced by $\Omega_0 \sqrt{\overline{N_{at}}}$, where $\overline{N_{at}}$ is a mean number of interacting atoms, defined by

$$\overline{N_{at}} = \sum_i |u_\ell(\vec{r}_i)|^2 \tag{91}$$

where $u_\ell(\vec{r})$ is the modal amplitude of the intracavity field defined in eq(20). The bifurcation point between the weak and strong coupling regimes is then equal to

$$\Omega_{critical} = \frac{|\kappa - \gamma|}{2\sqrt{\overline{N_{at}}}} \tag{92}$$

This qualitative change in the system behaviour when one crosses $\Omega_0 = \Omega_{critical}$ can be seen either on the spectrum of the light transmitted through the atom-cavity system, or on the spectrum of the fluorescence light emitted by the atom when it is in its excited state. In both configurations, one observes a single peak with a variable width when $\Omega_0 < \Omega_{critical}$ (this width tending to κ for the cavity spectrum, to γ for the fluorescence spectrum when $\Omega_0 \to 0$), and two peaks with the same width and a splitting varying with Ω_0 and $\overline{N_{at}}$ (but not with the probe field intensity) when $\Omega_0 > \Omega_{critical}$. This features have been experimentally observed by Kimble and coworkers for Cesium atoms traversing a high finesse optical cavity tuned on the resonance line of the Cesium atom [45]. They observed in particular the variation of the normal mode splitting with the mean number of interacting atoms on the transmission spectrum of the Fabry Perot cavity.

4.6 Bad cavity limit

Let us now consider the case when the cavity decay time is much shorter than the atomic decay time and than the Rabi period Ω_0^{-1}. The cavity mode can no longer be considered as a single mode field, but rather a continuum of modes with a width κ to which the two level atom is coupled. In this case, the Fermi's golden rule apply in order to calculate the transition probability, but with a density of states $\rho(E) = 1/\hbar\kappa$ (one mode in a energy band width of $\hbar\kappa$). One therefore gets a *modification of the atomic spontaneous emission rate*, which becomes

$$\gamma_{cav} = \gamma(1 + C_0) \text{ where } C_0 = \frac{2\Omega_0^2}{\kappa\gamma} \tag{93}$$

C_0 is usually called the "single atom cooperativity parameter", and can be much larger than 1 when $\Omega_0 \gg \gamma$. In this case, using expression (76) of Ω_0 and the detailed expressions of κ and γ, one finds for C_0 an expression independent of the dipole coupling d (present in Ω_0^2 and γ), which represents a purely geometry enhancement factor of the spontaneous emission rate, known as the Purcell factor

$$\gamma_{cav} = \gamma \frac{3}{4\pi^2} Q \frac{\lambda^3}{V} \qquad (94)$$

This factor is significant when the cavity quality factor Q is large, or when the cavity volume V is comparable to λ^3.

This cavity-induced enhancement of the spontaneous emission rate has been observed in various experiments :

(i) using microwave fields (for which λ is large) firstly, interacting with highly excited atomic states (Rydberg states), which have very large dipole moments because of their large dimensions. In [47] Na Rydberg states $|n = 23, S\rangle$ were inserted in a resonant cavity ($\nu_{res} \sim 100 GHZ$) of $Q = 10^5$. An enhancement of spontaneous emission by a factor 500 was observed.

(ii) using optical fields and high finesse Fabry-Perot cavities (tuned to the resonance line of the atom). In [48], Rb atoms ($\lambda = 0.56\ \mu m$) crossed a Fabry-Perot of finesse $F = 500$. A decrease by 20 % of the atomic lifetime was observed.

For the sake of briefness, we have restricted the present analysis to the quasi-resonant case. It can be shown also that far from resonance, a high Q cavity changes also the spontaneous emission : it enhances the lifetime by a large factor, because in some way, the accessible density of states $\rho(E)$ for the spontaneous emission light is now much smaller in detuned cavities than in free space. In this configuration, one must be careful to "forbid" to spontaneous photons all light modes of the free space and to use a closed cavity, or at least a large aperture cavity. One will find in [49, 50] a review of this subject and a description of the different experiments which have successfully confirmed this effect.

4.7 Generation of nonclassical states of light in cavity quantum electrodynamics

We have seen so far that the atom-field coupling leads to strong changes of the atomic dynamics. As we will see now, it also strongly affects the field, and permits the degeneration of nonclassical states in the cavity mode.

Antibunching In the low excitation limit, the intracavity field projects essentially on states $|n = 0\rangle$ and $|n = 1\rangle$. As such, it exhibits an *antibunching character*, which is observable by the method explained in subsection 2.4 on the weak probe field transmitted through the system. Furthermore, in the strong coupling regime, the normal mode splitting manifests itself as oscillations in the correlation function $g^{(2)}$ around 1. This feature has been observed by [46] in the case of small numbers of atoms ($\overline{N_{at}} \approx 20$) interacting with the cavity mode, where a non-classical value of $g^{(2)}(0)$ of 0.8 was measured in the experiment.

Squeezing When one increases the input probe field, it is possible to observe on the transmitted light intensity as a function of input light intensity a hysteresis loop : one reaches a bistability region, well known for nonlinear systems inserted in the feedback loop consisting of the cavity. This occurs when the cooperativity parameter $C = \overline{N_{at}} C_0$ is large enough, which can be reached either by increasing the cavity finesse, or the number interacting atoms. The last solution is generally easier : even with cavities of moderate finesse ($F \approx 50$), the bistability region is reached with less than a mW of input light if the number of atoms is of the order of 10^6.

These conditions have been achieved in [51] using a magneto-optical trap to confine Cs atoms inside the mode of a Fabry-Perot cavity. It is then possible to find detuning parameters between the cavity mode, the probe field and the atoms, for which squeezing can be observed on the probe beam reflected by the cavity. In [51] a quadrature squeezing of 40 % was observed. Other squeezing experiments using atomic beams instead of trapped atoms, travelling in the cavity mode, have generated squeezed states by using the atom + cavity mode system [52, 53].

5 Conclusion

This very rapid overview of quantum electrodynamics in the optical range has allowed us to give a first insight concerning some basic physical phenomena of the domain, and to precise the validity conditions of the different approximations. It has also missed many others, which could not fit within such a short series of lectures. Let us mention briefly some of them :

- the non resonant atom-cavity interaction, leading to electromagnetic field phaseshifts dependent on the atomic state, or to atomic coherence phase factors dependent on the field state.

- the tailoring of atom-field quantum correlations, leading to quantum non demolition measurement of field observables and to the very active domain of "quantum computing".

- the radiative forces due to the atom-field coupling, which have also a nonclassical character (Casimir force between mirrors, trapping force due to the vacuum field, ...).

- the micromaser, in which excited single atoms are continously sent in the cavity mode.

All these fascinating subjects (and some others) are presented in the references [4, 6, 49, 50, 44].

References

1. R. Loudon, "The quantum theory of light" Clarendon Press (Oxford, 1973)
2. L. Mandel, E. Wolf, "Optical coherence and quantum optics", Cambridge University Press (1995)
3. P. Meystre, M. Sargent, "Elements of quantum optics" Springer-Verlag (Berlin 1990)
4. M. Scully, Zubairy, "Quantum optics" Cambridge University Press (New-York 1997)
5. C. Cohen-Tannoudji, J. Dupont-Roc, G. Grynberg, "Photon and atoms" Wiley (New York 1989)
6. C. Cohen-Tannoudji, J. Dupont-Roc, G. Grynberg, "Atom-photon interactions" Wiley (New York 1992)
7. G. Grynberg, A. Aspect, C. Fabre, "Introduction aux lasers et à l'optique quantique" (in french), Ellipses, (Paris 1998), english translation to be published (Cambridge University Press)
8. Physics Reports **219**, special issue on the Solvay conference on Quantum Optics (October 1992)
9. H. Lorentz, "Theory of electrons", Teubner (1908), and Dover (1952)
10. A. Einstein, Z. Phys. **18**, 121 (1917)
11. L. Allen, J. Eberly, "Optical resonance and two-level atoms", Wiley (New York 1975)
12. A. Clairon, C. Salomon, S. Guellati, W. Phillips, Europhys. Lett. **16**, 165 (1991)
13. G. Alzetta, A. Gozzini, L. Moi, G. Orriols, Nuovo Cimento **B52**, 209 (1979)
14. A. Aspect, E. Arimondo, R. Kaiser, N. Vansteenkiste, C. Cohen-Tannoudji, Phys. Rev. Lett. **61**, 826 (1988)
15. O. Kocharovskaya, Physics Reports **219** 175 (1992)
16. A. Siegman, "Lasers" University Science Books (New-York 1986)
17. C. Cohen-Tannoudji, B. Diu, F. Laloe, "Quantum mechanics" Wiley Interscience (New York 1977)
18. S. Barnett, D. Pegg, J. Mod. Optics **36** 7 (1989)
19. W. Louisell, "Quantum statistical properties of radiation" Wiley (New York 1973)
20. D. Walls, G. Milburn, "Quantum optics", Springer (Berlin 1994)
21. H. Kimble, p. 227, and C. Fabre, p. 215, Physics Reports **219** (1992)
22. see for example H. Bachor, "Guide to experiments in quantum optics" VCH Verlag (1997)
23. M. Brune, E. Hagley, J. Dreyer, X. Maître, A. Maali, C. Wunderlich, J.M. Raimond, S. Haroche, Phys. Rev. Lett., **77**, 4887 (1996);
24. A. Aspect, P. Grangier, G. Roger, Phys. Rev. Lett. **47**, 460 (1981)
25. J. Mertz, T. Debuisschert, A. Heidmann, C. Fabre, E. Giacobino, Optics Lett. **16**, 1234 (1991)
26. D. Greenberger, M. Horne, A. Shimony, A. Zeilinger, Am. J. Phys **58**, 1131 (1990)
27. R. Glauber, in "Quantum optics and electronics", C. De Witt, A. Blandin, C. Cohen-Tannoudji eds, Gordon and Breach (New-York 1965)
28. A.Schawlow, C. Townes, , Phys. Rev. **112** 1940 (1958)
29. M. Kolobov, L. Davidovich, E. Giacobino, C. Fabre, Phys. Rev. **A46**, 1630 (1993)

30. Y. Yamamoto, S. Machida, Phys. Rev. **A35,** 5114 (1987)
31. D. Kilper, D. Steel, R. Craig, D. Scifres, Optics Lett. **21** 1283 (1996)
32. C. Becher, K. Boller, Optics Commun. **147** 366 (1998)
33. F. Marin, A. Bramati, V. Jost, E. Giacobino, Optics Commun. **140** 146 (1997)
34. C. Fabre, "Quantum fluctuations in light beams", p. 181, Les Houches Session 63, North-Holland (Amsterdam 1997)
35. L-A Wu, H. Kimble, J. Hall, H. Wu Phys. Rev. Lett. **57** 2720 (1986)
36. C. Caves, , Phys. Rev. **D26** 1817 (1982)
37. M. Xiao, L-A Wu, H. Kimble Phys. Rev. Lett. **59** 278 (1987)
38. H. Kimble, M. Dagenais, L. Mandel, Phys. Rev. Lett. **39** 691 (1977)
39. F. Diedrich, H. Walther, Phys. Rev. Lett. **58** 203 (1987)
40. M. Teich, E. Saleh, Progress in Optics, **26** (1988)
41. C. Hong, K. Ou, L. Mandel, , Phys. Rev. Lett. **59** 2044 (1987)
42. M. Brune, F. Schmidt-Kaler, A. Maali, J. Dreyer, E. Hagley, J-M Raimond, S. Haroche, Phys. Rev. Lett. **76**, 1800 (1996)
43. E. Jaynes, F. Cummings, Proc. IEEE **51**, 89 (1963)
44. H. Kimble, "Cavity Quantum Electrodynamics", p. 203, P. Berman ed., Academic Press (Boston 1994)
45. R. Thompson, G. Rempe, H. Kimble, Phys. Rev. Lett. **68**, 1133 (1992)
46. G. Rempe, R. Thompson, R. Brecha, W. Lee, H. Kimble, Phys. Rev. Lett. **67**, 1727 (1991)
47. P. Goy, J-M Raimond, M. Gross, S. Haroche, Phys. Rev. Lett. **50**, 1903 (1983)
48. D. Heinzen, J. Childs, J. Thomas, M. Feld, Phys. Rev. Lett. **58**, 1320 (1987)
49. S. Haroche, "Cavity Quantum Electrodynamics" p. 767, Les Houches session 53, North-Holland (Amsterdam 1992)
50. J-M Raimond, S. Haroche, "Cavity Quantum Electrodynamics" p. 309, Les Houches session 63, North-Holland (Amsterdam 1995)
51. A. Lambrecht, T. Coudreau, A. Steinberg, E. Giacobino, Europhys. Lett. **36**, 93 (1996)
52. M. Raizen, L. Orozco, Min Xiao, R. Brecha, H. Kimble, Phys. Rev. Lett. **59**, 198 (1987)
53. D. Hope, D. Mc Clelland, H. Bachor, A. Stevenson, Applied Phys. **B55**, 210 (1992)

Basics of Dipole Emission from a Planar Cavity

R. Baets, P. Bienstman and R. Bockstaele

University of Gent - IMEC, Department of Information Technology (INTEC)
Sint-Pietersnieuwstraat 41, B-9000 Gent, Belgium

Abstract

This chapter discusses, in a tutorial way, the basic theory of electromagnetic emission by a dipole in a planar cavity, which is representative for the spontaneous emission in micro-cavity LED's. We start from the expansion of a point source field into plane waves. Then the enhancement and inhibition effects of a cavity upon plane wave components are introduced. Next, the vectorial aspects of dealing with a dipole field are discussed, as well as the effects caused by the use of realistic mirrors. Finally we describe the effect of the cavity upon the carrier lifetime and give a discussion of guided modes. More in particular the plane wave decomposition and normal mode decomposition are confronted with each other.

1. Introduction

Light emitting diodes (LED's) are among the most widely used semiconductor optoelectronic components and the worldwide production volume of these devices is huge. Their applications are wide ranging, from simple indicator lamps to optical communication sources, from lamps for lighting to elements for large colour displays. When compared to incandescent lamps, they are very compact and reliable and offer a relatively high efficiency, brightness and modulation bandwidth. On most aspects they are outperformed by semiconductor laser diodes, but they are simpler to make and hence cost less. In many volume applications the incoherence of LED's, both temporally and spatially, is more of an advantage than a disadvantage. This is certainly true in applications where the human eye is exposed to the light beam, in terms of safety as well as in terms of the inconvenience of speckle typical for coherent light.

In view of their commercial importance, LED's have obviously been optimised thoroughly. Nevertheless, improvements of the various performance aspects of LED's would still be very welcome. One such aspect is efficiency. Although the internal power conversion efficiency of LED's - the ratio of internally produced light power to electrical drive power - can approach 100% in high quality semiconductor diodes, the external power efficiency - being the internal efficiency multiplied with the light extraction efficiency from the semiconductor into air - is rather poor. The high refractive index of the light generating semiconductor layer is responsible for this. This can be understood from fig. 1, where a simple situation with an isotropic point source near a plane semiconductor-air interface is considered.

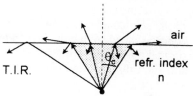

Fig. 1: Refraction of rays from a point source at a semiconductor-air interface, showing the extraction cone.

The point source emits rays - or plane waves - with uniform power density in all directions. Of all rays only those with an angle θ (with respect to the normal to the interface) smaller than the critical angle for total internal reflection θ_c can escape from the semiconductor into air, whereby θ_c is given by

$$sin\,\theta_c = \frac{1}{n} \qquad (1)$$

with n the refractive index of the semiconductor.

The solid angle Ω of the cone corresponding to those escaping rays is given by:

$$\Omega = 2\pi(1 - cos\,\theta_c)$$
$$\approx \pi\,sin^2\,\theta_c \quad \text{for } \theta_c \text{ small} \qquad (2)$$
$$= \frac{\pi}{n^2}$$

Even if the rays in this cone could fully escape (which is not the case due to partial reflection at the interface), the extraction efficiency would still be limited to

$$\eta_{extraction} \leq \frac{1}{4n^2} \qquad (3)$$

For a typical semiconductor like GaAs, with $n \approx 3.6$, this means that $\theta_c \approx 18°$ and $\eta_{extraction} \leq 2\%$. If the simple semiconductor-air structure is replaced by a planar multilayer structure, with a semiconductor half-space holding the point source at one side, the situation remains more or less the same. An anti-reflection coating for example can help to boost the transmission for normal incidence to 100%, but cannot have any impact on the extraction cone.

An obvious way to circumvent the problem of this low extraction efficiency is to put the point source at the centre of a semiconductor ball with a spherical semiconductor-air interface. In that case the entire spherical wave from the point source hits the interface in normal direction. A simple anti-reflective coating then allows to have an extraction efficiency of 100%. Unfortunately, the fabrication of such a semiconductor ball would be very difficult. Several less ideal but more practical approaches have been adopted to improve the extraction efficiency. The most obvious approach is to embed the semiconductor chip in a transparent plastic with a dome shape. Since the refractive index of the plastic is higher than that of air, the extraction cone in the semiconductor is larger and hence is the extraction efficiency. The plastic is dome-shaped so that the rays in the plastic hit the plastic-air interface at close-to-normal angles, allowing for easy extraction. Actually, the dome shape is often made such that the rays are more or less collimated upon refraction at the plastic-air interface.

A second approach is to make use of LED-chips with a semiconductor substrate that is transparent to the generated light. Light is extracted from all six sides of the chip. The extracted light is then collected and to a certain degree collimated by curved mirrors, possibly in combination with a lens. Extraction is further improved by roughening the semiconductor surfaces. This provides a randomisation of the rays such that every ray has - after a multiplicity of reflections within the chip - a finite probability of hitting a surface under escape conditions. The transparent substrate approach has led to devices with external quantum efficiencies in the range of 20 to 40%. While these efficiency levels are quite respectable, the brightness suffers from this approach. Brightness - or radiance - (of a more or less directional light beam) is defined as the ratio of the (wavelength integrated) beam power to the product of its effective emission area and the solid angle of its radiation pattern. Since in the transparent substrate approach, light leaves the chip from all sides with a more or less Lambertian radiation profile the brightness cannot be very high. This basically means that it is difficult to focus this light beam to a small spot or to couple it into an optical system with limited aperture or numerical aperture. The transparent substrate approach is also unsuitable for array devices.

To obtain a high brightness/high efficiency LED there are basically two approaches, both of which are based on local enhancement of the extraction efficiency, so that they are suitable for LED-arrays. In the first approach the LED active layer is embedded between a specular mirror and a roughened mirror, separated by no more than a few micrometers. This leads again to a randomisation mechanism as in the transparent substrate device, but now in a very localised way. This approach has led to very high efficiencies [Schnitzer, 1993], [Windisch, 1998], but it requires substrate removal and a non-trivial roughening technology. The second - and technologically more mature - approach is the micro-cavity (MC-LED) or resonant cavity (RC-LED) approach. In these devices the active layer is embedded in a small cavity with dimensions of the order of the wavelength of the emitted light. Under those circumstances the spontaneous emission process itself is modified, such that the emission is no longer isotropic. By a proper design the emission pattern can be tailored to enlarge the extraction efficiency, which leads to an improved overall efficiency. This can basically be understood as an interference effect. Consider a point source between two mirrors, as shown in fig. 2. The point source emits waves in all directions. For certain directions, as is the case in fig. 2a, the contributions to the total output wave from consecutive reflections in the cavity happen to be in phase. This means that the total radiated power is strong in that direction, which also means that the point source emits a lot of power in that direction. For certain other directions the opposite happens (fig. 2b): the various contributions to the total beam interfere destructively leading to a small beam power. This implies that the source emits little power in that direction.

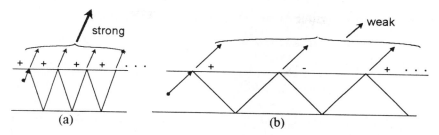

Fig.2: Figure showing how the cavity can lead to enhancement (a) or inhibition of emission (b).

In other words, the cavity induces a non-isotropic radiation pattern even if the point source is intrinsically isotropic. The basic design principle of an MC-LED is to enhance emission in directions or - more generally speaking - into electromagnetic modes that can be extracted from the cavity and to inhibit those directions or modes that cannot be extracted. The micro-cavity effect can not only be used to boost the efficiency of LED's but also to narrow the emission pattern and the emission spectrum, but a combination of all of these is difficult to achieve. It is worth noting that interference effects do not only influence the radiation pattern in a two-mirror cavity but also in the case of a single mirror configuration. The enhancement and inhibition effects are however less pronounced.

The effect of mirrors and cavities on spontaneous emission has been known since many years [Purcell, 1946]

The theory of dipole emission in layered structures has been rigorously described by Lukosz et al [Lukosz, 1977a, 1977b, 1979, 1980]. The experimental interest in micro-cavity LED's has risen considerably since the early nineties and results have been reported on MC-LED's in a variety of material systems and for a variety of wavelengths (for a review, see [De Neve, 1997] and [Blondelle, 1997]). The design issues in these devices have also been elaborated in detail [Benisty, 1998, a,b and c].

The vast majority of these results is based on planar (or one-dimensional) MC-LED's, in which the device structure is basically a planar layered structure (apart from a provision to confine current injection to a finite area). It is generally believed that three-dimensional cavities (with wavelength scale dimensions in all directions) would perform much better in a number of ways, but such structures are difficult to make (and also difficult to model) and reports are still scarce.

The problem of spontaneous emission in a cavity can in the so-called weak-coupling regime be described accurately as an electromagnetic problem with a dipole of given dipole strength radiating in the cavity. In te next section a strongly related but simpler problem will be analysed first, i.e. the problem of an isotropic point source radiating in a cavity whereby a scalar wave equation is assumed. Following that the consequences of working with dipole sources and vectorial wave equations will be derived. Next the various kinds of mirrors that can be used in a planar microcavity LED will be described, followed by a discussion of dipole emission in a planar cavity. The last paragraph discusses some aspects of guided modes.

2. Expansion of a point source into plane waves

Consider a wave function $\psi(x, y, z, t)$ that satisfies a scalar wave equation in uniform space with refractive index n:

$$\nabla^2 \psi(x,y,z,t) - \frac{n^2}{c^2}\frac{\partial^2 \psi(x,y,z,t)}{\partial t^2} = sources \qquad (4)$$

with c being the speed of light. This equation can be seen as an approximation of Maxwell's equations whereby the field quantity ψ represents any of the components of either electric or magnetic field. In a uniform medium this equation is actually correct for any of the 6 field components. Assuming operation at a single (circular) frequency ω the field $\psi(x, y, z, t)$ can be described through its complex phasor $\psi(x, y, z)$

$$\psi(x,y,z,t) = Re[\psi(x,y,z)exp(+j\omega t)] \qquad (5)$$

whereby for simplicity the same notation ψ is used for both the field and its phasor representation. The phasor then satisfies the time independent wave equation:

$$\nabla^2 \psi(x,y,z) - k_0^2 n^2 \psi(x,y,z) = sources \qquad (6)$$

whereby $k_0 = \frac{\omega}{c} = \frac{2\pi}{\lambda}$ and λ is the (vacuum) wavelength.

In general two types of sources need to be considered in the right hand side of (6). First there are sources that are independent of the field ψ at the location of the source. An example is spontaneous emission in a semiconductor. Secondly there are sources that are induced by the local field, often in a linear way. Examples are absorption of light (an "annihilation source" providing attenuation) and stimulated emission of light (a "generation source" providing amplification). In as far that the induced sources are proportional to the field, they can be combined with the term $k_0^2 n^2 \psi$ in the left hand of equation (6), whereby the refractive index n needs to be replaced by a new complex refractive index $n = n_r + jn_i$ with n_i representing loss or gain. With the convention of equation (5), loss and gain are represented by a negative and positive value of n_i respectively.

We now consider the generic case of a (field-independent) point source:

$$sources = \delta(x,y,z) \qquad (7)$$

The field solution for this point source problem is given by:

$$\psi_\delta(x,y,z) = -\frac{1}{4\pi}\frac{exp(-jk_0 nr)}{r} \text{ with } r = \sqrt{x^2 + y^2 + z^2} \qquad (8)$$

The question now considered is whether the point source field can be decomposed into plane waves ψ_k of the form:

$$\psi_k = A\exp(-j\mathbf{k}.\mathbf{r}) \qquad (9)$$

whereby $|\mathbf{k}| = nk_0$ so as to satisfy the sourceless wave equation. The **k**-vector is normal to the phasefront and its direction is that of the energy flow. Plane waves form a set of orthogonal eigenfunctions of a uniform sourceless medium.

In order to derive this expansion the two-dimensional Fourier transform of $\psi_\delta(x, y, z)$ is taken with respect to x and y:

$$\psi_\delta(k_x, k_y, z) = \int_{-\infty}^{+\infty}\int_{-\infty}^{+\infty} \psi_\delta(x, y, z)\exp(+j(k_x x + k_y y))dx\,dy \qquad (10)$$

$\psi_\delta(k_x, k_y, z)$ denotes the Fourier transform of $\psi_\delta(x, y, z)$, as is clear from the arguments of the function. The spatial frequencies k_x and k_y are real and have a range from $-\infty$ to $+\infty$. The inverse Fourier transform is:

$$\psi_\delta(x, y, z) = \frac{1}{(2\pi)^2}\int_{-\infty}^{+\infty}\int_{-\infty}^{+\infty}\psi_\delta(k_x, k_y, z)\exp(-j(k_x x + k_y y))dk_x\,dk_y \qquad (11)$$

This equation can be written as:

$$\psi_\delta(x, y, z) = \frac{1}{(2\pi)^2}\int_{-\infty}^{+\infty}\int_{-\infty}^{+\infty}\psi_\delta(k_x, k_y, z)\exp(+jk_z z)\exp(-j(k_x x + k_y y + k_z z))dk_x\,dk_y$$

with $\quad k_z = \pm\sqrt{n^2 k_0^2 - k_x^2 - k_y^2} = \pm\sqrt{n^2 k_0^2 - k_\parallel^2}$

whereby $\quad k_\parallel^2 = k_x^2 + k_y^2$

$$(12)$$

This expression clearly has the meaning of a plane wave expansion. The term $\exp(-j(k_x x + k_y y + k_z z))$ is a plane wave with **k**-vector (k_x, k_y, k_z)[1]. It is important to realise however that the expansion contains two kinds of plane waves. For

$$k_\parallel^2 \leq n^2 k_0^2 \qquad (13)$$

the resulting k_z is real and we deal with a normal propagative plane wave. For

$$k_\parallel^2 > n^2 k_0^2 \qquad (14)$$

the resulting k_z is imaginary (at least for lossless media with real refractive index) and the corresponding plane waves are called evanescent plane waves. The term "wave" is actually not very appropriate for these contributions, since they do not

---

[1] The conventional definition of Fourier transform uses a minus-sign in front of the argument $j(k_x x + k_y y)$ for the forward transform and a plus-sign for the inverse transform. Here we use the opposite convention so that a spatial frequency component with positive k_x and k_y corresponds to a plane wave travelling in the positive x- and y-direction. This choice is related to the (arbitrary) sign choice of the argument $+j\omega t$ in equation 5.

propagate. They are necessary however for a complete description of the field near the point source where the field amplitude varies very rapidly. The evanescent waves are high spatial frequency functions (in the x- and y-direction) capable of describing these rapid variations. Away from the point source the evanescent waves decay and they do not contribute to the far field at a large distance from the source, at least not in a uniform medium. In a layered medium it is possible that an evanescent wave generated by the dipole is transformed into a propagative wave upon incidence on a layer with high refractive index. In this case the dipole does radiate power, tunneled via the evanescent mode, into the propagative mode.

The expression $k_z = \pm\sqrt{n^2 k_0^2 - k_x^2 - k_y^2}$ suggests an uncertainty about the sign of k_z in the plane wave interpretation of the 2D Fourier transform. This sign can be determined from a priori knowledge about the field. In the point source case it is obvious that the waves propagate upward for $z>0$ and downward for $z<0$. When a priori knowledge is lacking information about $\frac{d\psi(x,y,z)}{dz}$ is needed to determine the sign of k_z (and more generally the relative amplitudes of both contributions).

The two types of plane waves can be drawn pictorially in two ways, as shown in fig. 3.

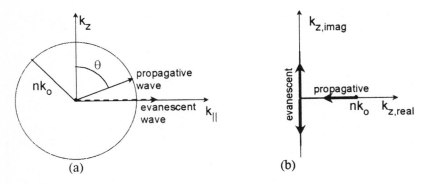

Fig. 3: Pictorial representations of plane waves in k-space
(a: $(k_{||}, k_z)$-space, b: complex k_z-plane).

Fig. 3a shows a $(k_{||}, k_z)$ representation, in which the propagative waves have k-vectors on a circle and the evanescent waves have a real $k_{||}$-value outside the circle and a (not shown) imaginary k_z-value. In fig. 3b the complex k_z-plane is shown with the propagative waves along the real axis and the evanescent waves along the imaginary axis.

It is worth emphasising here that the 2D Fourier decomposition has a relevant interpretation as plane wave decomposition only if the medium is uniform along the (x, y)-plane (for a given z). In media with a non-uniformity in the (x, y)-plane plane waves are no longer eigenfunctions of the wave equation and therefore the 2D Fourier transform is no more a privileged tool in a wave context as it would require to additionally solve the coupling between in-plane waves. In a layered structure with planar interfaces ⊥z however, the field can in each layer be expanded in plane waves. This is extremely convenient since it is trivial to describe the propagation of a plane wave through such a layer and the reflection and transmission of a plane wave at a planar interface is also very simple. Therefore the problem of a point source in a layered structure, e.g. a planar cavity, becomes very simple through a plane wave expansion.

An analytical expression for $\psi_\delta(k_x, k_y, z)$ can be found either by calculating the Fourier transform of $\psi_\delta(x, y, z)$ directly or - more conveniently - by taking the 2D Fourier transform of the wave equation (6):

$$\frac{d^2}{dz^2}\psi_\delta(k_x,k_y,z) + (k_0^2 n^2 - k_x^2 - k_y^2)\psi_\delta(k_x,k_y,z) = \delta(z) \qquad (15)$$

the solution of which is:

$$\psi_\delta(k_x,k_y,z) = \frac{j}{2\sqrt{n^2 k_0^2 - k_x^2 - k_y^2}} \exp\left(-j\sqrt{n^2 k_0^2 - k_x^2 - k_y^2}|z|\right) \qquad (16)$$

in which for the square root the solution with zero or negative imaginary part is chosen.

Both $\psi_\delta(x, y, z)$ and $\psi_\delta(k_x, k_y, z)$ are shown in fig. 4, for two values of z.

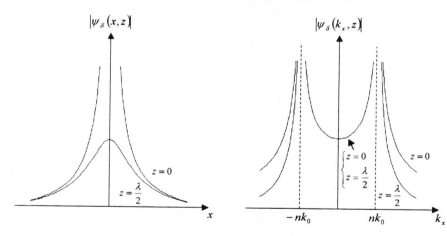

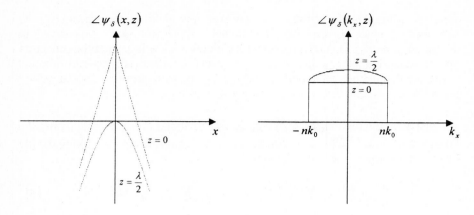

Fig. 4: ψ_δ in real space and in k-space (for y=0 and k_y=0).

The figure of $\psi_\delta(k_x, k_y, z)$ clearly shows that the propagative waves have a constant amplitude but varying phase as a function of z, while the evanescent waves decay as |z| increases. It may be surprising that the amplitude of the propagative waves depends on (k_x, k_y), while the source is isotropic. For $k_x^2 + k_y^2 = n^2 k_o^2$ the amplitude is even singular. All of this is a consequence of the fact that the solid angle corresponding to $dk_x dk_y$ is a non-linear function of k_x and k_y. The plane wave amplitude under consideration is a density per unit of spatial frequency $dk_x dk_y$ and not per unit of solid angle. We can clarify this further by the following manipulations. Inserting the specific form of (16) into (11) leads to:

$$\psi_\delta(x,y,z) = \frac{1}{(2\pi)^2} \int_{-\infty}^{+\infty}\int_{-\infty}^{+\infty} \frac{j}{2\sqrt{n^2 k_0^2 - k_x^2 - k_y^2}} \exp\left(-j\left(k_x x + k_y y + \sqrt{n^2 k_0^2 - k_x^2 - k_y^2}|z|\right)\right) dk_x dk_y \quad (17)$$

This equation (17) can be rewritten in terms of the elevation θ and the azimuth φ, defined as:

$$\begin{aligned} k_x &= nk_0 \sin\theta \cos\varphi \\ k_y &= nk_0 \sin\theta \sin\varphi \\ k_z &= nk_0 \cos\theta \end{aligned} \quad (18)$$

or alternatively in terms of $k_\parallel$ and φ_0.
Using the Wronskian the conversion from one coordinate system to another can be done:

$$dk_x dk_y = \begin{vmatrix} \frac{\partial k_x}{\partial \theta} & \frac{\partial k_x}{\partial \varphi} \\ \frac{\partial k_y}{\partial \theta} & \frac{\partial k_y}{\partial \varphi} \end{vmatrix} d\theta d\varphi = n^2 k_0^2 \cos\theta \sin\theta d\theta d\varphi = nk_0 k_z \sin\theta d\theta d\varphi$$

$$= \begin{vmatrix} \frac{\partial k_x}{\partial k_\|} & \frac{\partial k_x}{\partial \varphi} \\ \frac{\partial k_y}{\partial k_\|} & \frac{\partial k_y}{\partial \varphi} \end{vmatrix} dk_\| d\varphi = k_\| dk_\| d\varphi$$

(19)

The solid angle $d\Omega$ associated with (dk_x, dk_y) or with the equivalent $(d\theta, d\varphi)$ or $(dk_\|, d\varphi)$ is given by:

$$d\Omega = \sin\theta d\theta d\varphi = \frac{1}{nk_0 k_z} dk_x dk_y = \frac{1}{nk_0} \frac{k_\|}{k_z} dk_\| d\varphi \tag{20}$$

The boundaries of the integration are given in the following table.

	k_x, k_y	θ, φ	$k_\|, \varphi$
propagative	$\|k_x\| \le nk_0$ and $\|k_y\| \le \sqrt{n^2 k_0^2 - k_x^2}$	$0 \le \theta \le \pi/2$ $0 \le \varphi \le 2\pi$	$0 \le k_\| \le nk_0$ $0 \le \varphi \le 2\pi$
evanescent	$\|k_x\| > nk_0$ or $\|k_y\| > \sqrt{n^2 k_0^2 - k_x^2}$	$\pi/2 + j0 \le \theta \le \pi/2 + j\infty$ $0 \le \varphi \le 2\pi$	$k_\| > nk_0$ $0 \le \varphi \le 2\pi$

This leads to the following alternative formulations of equation (17):

$$\psi_\delta(x,y,z)$$
$$= \int_{-\infty}^{+\infty}\int_{-\infty}^{+\infty} \left[\frac{jnk_0}{2(2\pi)^2}\right] \left[\exp(j(k_x x + k_y y + k_z |z|))\right] \left[\frac{1}{nk_0 k_z} dk_x dk_y\right]$$
$$= \left(\int_0^{\pi/2} + \int_{\pi/2+j0}^{\pi/2+j\infty}\right)\int_0^{2\pi} \left[\frac{jnk_0}{2(2\pi)^2}\right] \left[\exp(-j(k_x x + k_y y + k_z |z|))\right] [\sin\theta d\theta d\varphi] \tag{21}$$
$$= \int_0^{+\infty}\int_0^{2\pi} \left[\frac{jnk_0}{2(2\pi)^2}\right] \left[\exp(-j(k_x x + k_y y + k_z |z|))\right] \left[\frac{1}{nk_0}\frac{k_\|}{k_z} dk_\| d\varphi\right]$$

In each of these expressions the first term between brackets has the meaning of plane wave amplitude per unit of solid angle or mode density or density of states, the second term is a plane wave and the third term is the solid angle $d\Omega$. Expressed in this way, the equations show that the mode density is independent of *k*-vector direction, as expected for an isotropic point source. Furthermore the mode density is proportional to refractive index and to frequency. The formulation in terms of θ and φ is elegant for the propagative part of the field, but less elegant for the evanescent part because θ becomes a complex number.

If the medium surrounding the point source has a complex refractive index, whereby the imaginary part represents loss or gain, the same Fourier transform approach for plane wave decomposition can still be followed. The plane waves in this expansion are somewhat uncommon however since their phase fronts do not coincide with their iso-amplitude planes. The latter are parallel to the (x,y)-plane (by choosing k_x and k_y to be real in the Fourier transform), while the phase fronts are perpendicular to the vector $(k_x, k_y, Re(k_z))$. Nevertheless these plane waves satisfy the wave equation, since $k_x^2 + k_y^2 + k_z^2 = n^2 k_o^2$. Furthermore the clear distinction between propagative and evanescent waves vanishes in the complex case.

From a physical point of view it would be more logical to expand the field into plane waves of the form $exp(-j(nk_xx+nk_yy+nk_zz))$ with k_x, k_y and k_z real and $k_x^2 + k_y^2 + k_z^2 = k_o^2$. While these plane waves have a constant amplitude along their phase front, they do not have a constant amplitude in the (x,y)-plane and therefore cannot be used as basis for the Fourier transform along x and y, as needed for the treatment of layered structures.

3. Point source near a single mirror

Consider the simple case of a mirror with field reflectivity r and a distance d away from an isotropic point source (Fig. 5).

The point source field can be expanded into plane waves, as described in the previous paragraph, and we consider the components with angle θ and $\pi-\theta$ respectively (and arbitrary value for φ). These components have the same (k_x, k_y) and opposite k_z. Upon reflection the downward propagating plane wave will get the same k-vector as the upward propagating wave and hence the total upward field will be the sum of both.

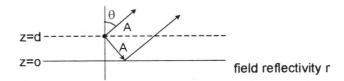

Fig. 5: Point source near a single mirror.

The total field for these components can thus be written as:

$$0 \leq z \leq d : A exp(-j(k_x x + k_y y))[exp(+jk_z(z-d)) + r exp(-jk_z(z+d))]$$
$$z \geq d : \quad A exp(-j(k_x x + k_y y))[exp(-jk_z(z-d)) + r exp(-jk_z(z+d))] \quad (22)$$
$$\text{with } k_z = k_0 n \cos\theta$$

In the region between the point source and the mirror each plane wave component gives rise to a standing wave pattern. In the other region there are only outward plane waves. The phase difference between the direct and the reflected contribution is given by $2\phi + phase(r)$, with

$$\phi = k_0 nd \cos\theta \qquad (23)$$

In Fig. 6 a few situations are shown for the case where $r = -1$. For combinations of λ, n, d and θ for which $\phi = \pi/2$ $(+m\pi)$, the standing wave pattern has an antinode position at the point source location and both outward propagating contributions interfere constructively.

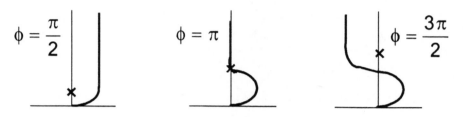

Fig. 6: Total field for $r = -1$ and different ϕ-values. X denotes the position of the point source.

In the opposite case for which $\phi = \pi + m\pi$ the point source is located at a node of the standing wave pattern and the two outward contributions interfere destructively. This means that in this particular direction the point source does not radiate any power. In other words the presence of the mirror has a dramatic impact on the radiation profile of the point source, that ceases to behave as an isotropic radiator. From this discussion one can conclude that emission by the point source in a particular direction will be strong if, and only if, the total field caused by the point source emission in that direction (and its corresponding direction) is strong at the position of the point source itself.

To elaborate this point we calculate the normalized radiation profile of the point source in the upward direction:

$$I(\theta) = \frac{\text{Power flux }(\theta)\text{ with mirror}}{\text{Power flux }(\theta)\text{ without mirror}} = |1 + r\exp(-j2\phi)|^2 \qquad (24)$$
$$= 1 + r^2 + 2r\cos 2\phi \text{ (for r real)}$$

The maximum, minimum and average of $I(\theta)$ over all θ-values are given by:

$$\text{maximum} = (1 + |r|)^2$$
$$\text{minimum} = (1 - |r|)^2 \qquad (25)$$
$$\text{average} = 1 + |r|^2$$

This shows that for a perfect mirror ($|r| = 1$), the power radiated by the point source in a particular (upward) direction can be anything between zero and four times the power radiated without the presence of the mirror. The average power is twice that without mirror, as expected.

4. Point source within a simple cavity

As a next step we consider the case of a point source in between two parallel mirrors, as shown in fig. 7.

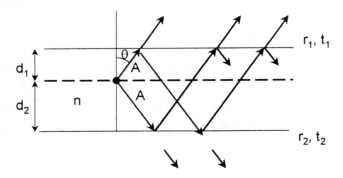

Fig. 7: *Point source within a simple cavity.*

As in the simple mirror case the point source field is decomposed in plane waves. The total field amplitude of the plane wave component in the direction θ and in the plane of the point source, caused by point source emission in the directions θ and $\pi-\theta$, is given by

$$A\left[1+r_1r_2\,exp(-j2\phi)+r_1^2r_2^2(-4j\phi)...\right] \\ +Ar_2\,exp(-j2\phi_2)\left[1+r_1r_2\,exp(-j2\phi)+r_1^2r_2^2(-4j\phi)...\right] \\ =\frac{A[1+r_2\,exp(-2j\phi_2)]}{1-r_1r_2\,exp(-j2\phi)}\,for|r_1r_2|<1 \qquad (26)$$

with

$$\phi_1 = k_0 n d_1 \cos\theta$$
$$\phi_2 = k_0 n d_2 \cos\theta$$
$$\phi = \phi_1 + \phi_2$$

For $r_1r_2=1$ the infinite series in equation (26) has no meaning as a function since it does not converge. In a distributional sense one can write however:

$$A[1+r_2\,exp(-2j\phi_2)][1+exp(-j2\phi)+exp(-j4\phi)+...] \\ = A[1+r_2\,exp(-2j\phi_2)][1+cos(2\phi)+cos(4\phi)+...-j(sin(2\phi)+sin(4\phi)+...)] \\ = A[1+r_2\,exp(-2j\phi_2)]\left[\frac{1}{2}+\frac{\pi}{2}\sum_{n=-\infty}^{+\infty}\delta(\phi-n\pi)-\frac{j}{2}P.V.(cot(\phi))\right]$$

(27)

whereby P.V. stands for the principal value of the cot-function around each singularity. One should note that the expression for $r_1r_2=1$ is not simply the limit for r_1r_2 towards 1 of the expression for $|r_1r_2|<1$. A series of dirac-functions needs to be added at the singular points. This is relevant in the context of guided modes, as will be discussed in paragraph 8.

For $|r_1r_2|<1$ the radiation pattern (within the cavity) for upward emission in the direction θ is then given by:

$$I(\theta) = \frac{\text{Power flux }(\theta) \text{ with cavity}}{\text{Power flux }(\theta) \text{ without cavity}} = \frac{\left(1+r_2^2+2r_2\cos 2\phi_2\right)}{\left|1+r_1r_2\exp(-j2\phi)\right|^2} \text{ (for } r_2 \text{ real)} \quad (28)$$

This equation indicates that two factors have an impact on the point source radiation profile. The denominator

$$\frac{1}{\left|1-r_1r_2\exp(-j2\phi)\right|^2} \quad (29)$$

is called the cavity enhancement factor or Airy factor. It depends on r_1, r_2 and ϕ but *not* on the point source position. The numerator

$$1+r_2^2+2r_2\cos 2\phi_2 \quad (30)$$

is very similar to the radiation pattern of a single mirror and can be called the standing wave factor. It obviously depends strongly on the position of the point source and leads to a similar conclusion as in the single mirror case: the radiated emission in a particular direction is high if the standing wave field strength at the point source is high.

At this point the concept of resonance can be introduced. A resonance - or resonant mode - can be described as an electromagnetic field of which the cavity enhancement factor goes through a maximum. In a given planar cavity this will happen for particular combinations of wavelength and radiation direction. For a simple situation in which r_1 and r_2 are real, of equal sign and independent of λ and θ, resonance occurs for

$$2\phi = 2m\pi \quad (31)$$

with m a positive or negative integer.
This condition can be rewritten as

$$n_r d \cos\theta = m\frac{\lambda}{2} \quad (32)$$

where n_r is the real part of n, or as

$$k_z = m\frac{\pi}{d} \quad (33)$$

where k_z is the z-component of a plane wave **k**-vector in the direction θ. This latter form of the resonance condition calls for a simple graphical representation in k-space as shown in fig. 8.

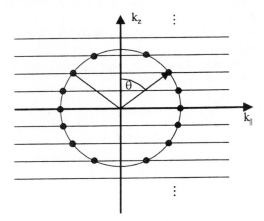

*Fig. 8: **k**-space representation of emission in cavity.
The dots represent cavity resonances (a dot in the upper space and its symmetric dot in the lower space belong to the same resonance).*

Resonance occurs along the planes $k_z = m\dfrac{\pi}{d}$. However for a given wavelength of emission the **k**-vector satisfies the dispersion equation

$$k_x^2 + k_y^2 + k_z^2 = \left(n\frac{\omega}{c}\right)^2 \qquad (34)$$

which is represented by a sphere. Therefore resonance can only occur at the crossing of the resonance planes and the sphere, as indicated by the dots in the figure. The resonant plane waves therefore form cones of **k**-vectors, one of which is indicated in the figure. It is clear that at a given wavelength the resonant modes form a discrete set as a function of θ and a continuous set as a function of φ. From the figure and from equation (33) it is clear that the number of resonant modes (or rather of resonant cones) will increase as the cavity thickness is increased. Vice versa, by decreasing the thickness the number of cones can be decreased to 1 (or even 0 in special cases).

An alternative way to write equation (31)-(33) is through the quantity m_c, being the cavity thickness expressed in number of half wavelengths:

$$m_c = \frac{d}{\lambda/2n} \qquad (35)$$

The resonance condition is then rewritten as

$$m_c \cos\theta = m \qquad (36)$$

From this equation it is clear that the number of discrete resonances is limited to $[m_c]$ (= integer(m_c)). The integer quantity $[m_c]$ is called the cavity order. If the resonance for $m=0$ ($\theta = 90°$) is also counted the number of resonances actually equals $[m_c]+1$. Such a mode can only exist when the field is independent of z in the cavity. This is only possible in the theoretical case where $r_1=r_2=+1$. So in practice the $m=0$ mode is not very relevant.

The resonator can be either "perfect" or "imperfect". A "perfect" resonance mode is a mode that will continue to exist without amplitude decay after removal of the source that excited the resonance. In a planar cavity a perfect resonance arises when

$$\left| r_1 r_2 \exp(-j2\phi) \right| = 1 \qquad (37)$$

This means that

$$\left| r_1 r_2 \exp\left(\frac{4\pi}{\lambda} n_i d \cos\theta\right) \right| = 1 \qquad (38)$$

with n_i the imaginary part of n.

Such a "perfect" resonance occurs either when the cavity is lossless and has perfect mirrors or when the medium exhibits enough gain to compensate for the losses. This latter situation is what happens in a laser and the resonant mode is then called a laser mode.

The function $I(\theta)$ becomes singular in the case of a perfect resonance and contains an odd cotangent-part together with an even delta-part. This may seem unphysical but it is not, since the singularity only occurs for discrete angles each carrying an infinitesimal fraction of the source power.

In the case of an "imperfect" resonance, the resonant mode will decay with time in the absence of a source to sustain it. The reason for the loss of power is either absorption or partial transmission through the mirrors. The quality of the resonace is commonly described by the quality factor Q or by the cavity finesse F, both of which are defined in the next figure.

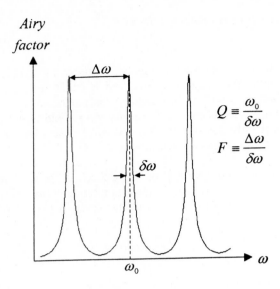

Fig.9: Definition of Q-factor (Q) and finesse (F). $\delta\omega$ is the spectral width at half maximum of the resonance peak.

In this figure, ω_0 is the central resonance angular frequency, $\Delta\omega$ is the separation between two adjacent resonances and $\delta\omega$ is the full-width half-maximum value of the resonance peak. The quality factor Q is defined as the ratio of ω_0 to $\delta\omega$, while the finesse F is the ratio of $\Delta\omega$ to $\delta\omega$. In other words a finesse of, say, 100 means that 1% of the angular space (for a given wavelength) or 1% of the spectral space (for a given direction) is enhanced by the cavity.

For imperfect resonances the *k*-space picture of fig.8 should be slightly modified. Due to the finite finesse the resonance planes should be broadened to slabs. If on top of that the point source is not monochromatic but has a non-zero spectral width (as in the case of spontaneous emission from a semiconductor), the dispersion sphere should be broadened to a shell. This leads to fig. 10 in which the volumetric overlap between the slabs and the shell identifies the resonant regions.

The *k*-vectors in this figure are *k*-vectors in the medium of the cavity with refractive index n. Transmission to the outside world (generally with refractive index of 1) can only occur for *k*-vectors within the extraction cone, also indicated in fig. 9 and described by the equation

$$k_z = \frac{k_{||}}{\sqrt{n^2 - 1}} \qquad (39)$$

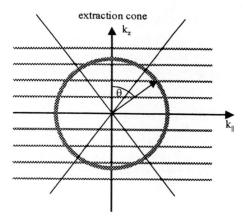

*Fig. 10: **k**-space representation of emission in an imperfect cavity, with thermally broadened emission.*

This means that resonant modes can be divided in two types: those that can escape and those that cannot. The first give rise to a standing wave pattern within the cavity, and propagative waves outside the cavity. They are often called Fabry-Perot modes or also longitudinal cavity modes. The second also give rise to a standing wave pattern within the cavity but to evanescent waves outside the cavity. In other words no power leaks away to the outside of the cavity and hence these resonances are called guided modes. Fig. 11 shows a typical field amplitude for both a Fabry-Perot mode and a guided mode.

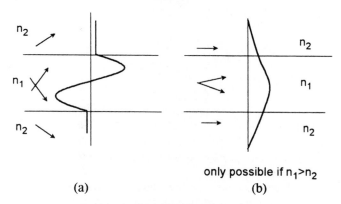

*Fig. 11: The two types of resonant modes:
Fabry-Perot modes (a) and waveguide modes (b).*

It is worth emphasising once more that, what we call a resonant mode here, is a plane wave field distribution that is subject to cavity enhancement. This enhancement takes place over a finite range of plane wave directions for each mode. This point is elaborated further in paragraph 8.

5. Transition to a vectorial electromagnetic problem

So far we have considered a scalar wave approximation of emission from a planar cavity. This implied not only that the fields were considered to be scalar but also that the point source was represented by a scalar source term. In reality we do have to consider a vectorial electromagnetic problem with vectorial fields and vectorial sources.

In a uniform layer, Maxwell's equation can be rewritten in phasor notation as:

$$\nabla^2 E + k_0^2 n^2 E = \textbf{\textit{sources}}$$
$$H = \frac{j}{\omega \mu_0} \nabla \times E \tag{40}$$

While in the wave equation for E the three field components of E are decoupled, coupling does arise from the continuity equations at abrupt interfaces between layers, where the tangential components of E and H need to be continuous. In a layered structure (with layers perpendicular to the z-axis), this means that the x- and y-components need to be continuous. As in the scalar case, any field in a uniform layer can be decomposed into plane waves. In the vectorial case these plane waves have a polarisation, i.e. a particular time evolution of the E-field orientation. For monochromatic waves the most general type of polarisation is an elliptic polarisation. This means that the end point of the electric field vector will follow an elliptic trajectory in time (with one revolution per period) in the plane normal to the k-vector (as shown in fig. 12). An elliptically polarised field can however be seen as the sum of two linearly polarised fields with orthogonal polarisation.

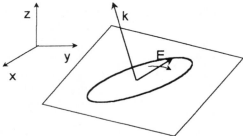

Fig. 12: General plane wave with elliptic polarisation.

In principle an arbitrary choice can be made for the direction of the linear polarisations in this decomposition. In layered structures it is very elegant to make a particular choice, known as TE-TM decomposition The TE (Transverse Electric) plane wave (also known as E-wave or s-wave) has its E-field in the (x-y)-plane and orthogonal to k. This means that the E-field is transverse (meaning orthogonal) to the plane of incidence with respect to the layer structure, being the plane formed by the k-vector and the z-axis. The TM-wave (or H-wave or p-wave) has its H-field transverse to the plane of incidence. The names TE and TM are widely used but nevertheless misleading since both TE- and TM-waves are TEM-plane waves in uniform space (transverse electric and magnetic waves with both E and H-field orthogonal to k). In fig. 13 the field orientations of both TE and TM-waves are shown.

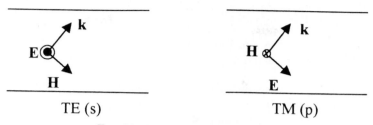

Fig. 13: Definition of TE and TM.

The reason for the particular choice of TE and TM-waves to decompose an arbitrary polarised plane wave is that these linear polarisations are the only ones that maintain their polarisation after reflection or refraction at a planar interface. In other words: a TE-field incident upon a planar interface generates a TE-reflection and a TE-transmission (and vice versa for TM). This means that the overall analysis of electromagnetic wave propagation in isotropic layered structures can be decomposed into two entirely uncoupled systems: a TE-system and a TM-system.

The second problem to deal with relates to the vectorial point source. In electromagnetic theory two types of vectorial point sources are generally considered: the electric dipole and the magnetic dipole. The first can be seen as a point-source-like current density whereas the second as a point-source-like loop current. It is well established that spontaneous emission in a semiconductor can be very well represented as electric dipole emission (at least in the weak-coupling limit). Any electric dipole source can be decomposed in a parallel electric dipole (with current density or dipole moment in the (x-y)-plane) and a perpendicular electric dipole (with current density or dipole moment along z). In bulk semiconductor material the dipole can have any orientation. This means that one third of the power generated by the dipoles is generated by perpendicular dipoles (z-direction) and two thirds by parallel dipoles (x and y-direction). In quantum wells there is a preference for emission through parallel dipoles [Yamanishi, 1984]. It is well known [Van Bladel, 1964] that a plane wave component of the field resulting from an electric dipole has an electric field in the plane of the dipole moment and of the **k**-vector. Furthermore the **E**-field is maximum for emission in the plane normal to the dipole moment and drops sinusoidally to 0 for emission in the direction of the dipole moment itself. All of this leads to the conclusion depicted in fig. 14.

The complete vectorial problem can be decomposed in three simple scalar problems ([De Neve, 1997], [Benisty, 1998a]):
- the TE-fields generated by a parallel dipole;
- the TM-fields generated by a parallel dipole;
- the TM-fields generated by a perpendicular dipole.

The perpendicular dipole does not generate any TE-fields.

In many practical micro-cavity LED's the perpendicular dipoles have virtually no contribution to the power coupled to air. The power emitted by the perpendicular dipoles into the small perpendicular extraction cone is obviously small (unless the cavity has a very strong resonant enhancement in this direction). Furthermore, the relative occurrence of perpendicular dipoles is small in quantum-well active layers. So, in many cases the full vectorial treatment reduces to two relevant scalar problems: one TE and one TM problem, both for a horizontal dipole.

In fig. 14 the amplitude of the plane wave contributions to the dipole fields is given. Only the propagating plane waves are shown. To these should be added the evanescent contributions. This can be done by reformulating the plane wave amplitude in terms of k. Expressed as a density per unit of solid angle, this leads to:

parallel dipole	perpendicular dipole
TE-field ampl.: $A \sin\varphi$	TE-field ampl.: 0
TM-field ampl.: $A(k_z/nk_0)\cos\varphi$	TM-field ampl.: $Ak_\parallel/nk_0$

whereby in these expressions φ ranges from 0 to 2π and $k_\parallel$ ranges from 0 to nk_0 for the propagative contributions and from nk_0 to infinity for the evanescent contributions.

In the parallel dipole case the dipoles are generally randomly oriented. This means that one can work with an average field amplitude over φ (in root mean square sense, since one considers an ensemble of non-correlated, random oriented dipoles implying that the total power density is the sum of the contributions from the individual dipoles), leading to the final result for the plane wave amplitude (expressed as a density per unit solid angle):

parallel dipole	perpendicular dipole
TE-field ampl.: $A/\sqrt{2}$	TE-field ampl.: 0
TM-field ampl.: $(A/\sqrt{2})k_z/nk_0$	TM-field ampl.: $Ak_\parallel/nk_0$

Again φ ranges from 0 to 2π and $k_\parallel$ ranges from 0 to nk_0 for the propagative contributions and from nk_0 to infinity for the evanescent contributions.

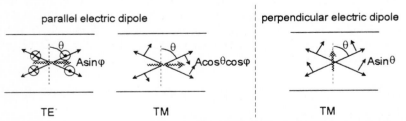

Fig. 14: Decomposition of a general vectorial dipole problem into a set of three scalar problems. The amplitude of the plane wave decomposition is given in each case. The plane shown is the plane where $\varphi=0$ or π.

6. Mirrors for planar cavities

In planar cavities one can basically make use of two types of mirrors: metallic mirrors or multilayer dielectric mirrors. Mirrors made of metals such as gold or silver provide high reflectivity over a wide angular and spectral range and therefore they are rather ideal mirrors. However they can only be used as rear reflector in a micro-cavity LED and not as a front reflector to couple light to the outside world. To use a metallic mirror as a front reflector the metal layer would need to be thin enough to keep the absorption of the transmissive wave low enough but thick enough to keep the reflectivity high enough. These two requirements are generally incompatible. When the metallic mirror is also used as an electrical contact to the semiconductor, a technological problem arises. To form a reliable electrical contact the metal-semiconductor interface is normally alloyed at high temperature. This however destroys the high reflectivity of the metal. Therefore a non-alloyed contact has to be used. Although good non-alloyed contacts can be made by using very highly doped semiconductor layers, the long term reliability of such contacts is unclear.

Multi-layer dielectric mirrors - in particular Distributed Bragg Reflectors (DBR-mirrors) - can provide very high reflectivity over a limited angular and spectral range. They are often rather thick which means that the effective thickness of a cavity with one or two DBR-mirrors can not be very small.

To analyse the properties of both metallic and dielectric mirrors, we first review the properties of single dielectric interfaces. In fig. 15 the reflection and refraction of a plane wave incident upon an interface from a medium with (complex) refractive index n_1 to a complex refractive index n_2 is considered

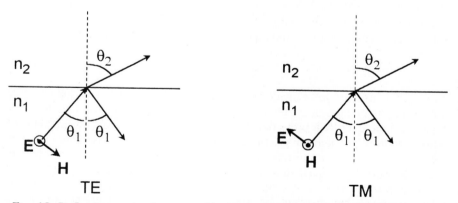

Fig. 15: Reflection and refraction of a plane wave at a planar interface (for both TE- and TM-case).

Continuity of the tangential field components (and hence equality of the $k_\parallel$-values in both media) leads to the following equations:

TE-case

$$n_1 \sin\theta_1 = n_2 \sin\theta_2$$
$$r_E = \frac{k_{1z} - k_{2z}}{k_{1z} + k_{2z}} \qquad t_E = \frac{2k_{1z}}{k_{1z} + k_{2z}}$$
$$R_E = |r_E|^2 \qquad T_E = \frac{Re(k_{2z})}{Re(k_{1z})}|t_E|^2 \qquad (41)$$
$$T_E = 1 - R_E$$

TM-case

$$n_1 \sin\theta_1 = n_2 \sin\theta_2$$
$$r_H = \frac{k_{1z}/n_1^2 - k_{2z}/n_2^2}{k_{1z}/n_1^2 + k_{2z}/n_2^2} \qquad t_H = \frac{2k_{1z}/n_1^2}{k_{1z}/n_1^2 + k_{2z}/n_2^2}$$
$$R_H = |r_H|^2 \qquad T_H = \frac{Re\left(\dfrac{k_{2z}}{n_2^2}\right)}{Re\left(\dfrac{k_{1z}}{n_1^2}\right)}|t_H|^2 \qquad (42)$$
$$T_H = 1 - R_H$$

r_E and r_H are the field reflections for E-field and H-field respectively and R and T denote the power reflectivity and transmittivity. k_{1z} and k_{2z} are given by:

$$k_{iz} = \sqrt{n_i^2 k_0^2 - k_\parallel^2} \qquad (43)$$

Two comments need to be made about the expression $T=1-R$. The quantity T is not the amplitude ratio of the transmitted Poynting vector to the incident Poynting vector, but rather of their projection onto the z-axis. In other words it refers to the ratio of power per unit of interface area. Secondly, the expression $T=1-R$ is actually incorrect when the refractive index of the medium of incidence has an imaginary part. In that case the total Poynting vector in the medium of incidence can no longer be written as the sum of a Poynting vector of incident and reflected field respectively but also contains cross-terms of both fields. Correct expressions for this case are:

$$TE: R+T+2\frac{Im(k_{1z})}{Re(k_{1z})}Im(r_E^*)=1$$

$$TM: R+T+2\frac{Im\left(\frac{k_{1z}}{n_1^2}\right)}{Re\left(\frac{k_{1z}}{n_1^2}\right)}Im(r_H^*)=1 \qquad (44)$$

For normal incidence the distinction between TE and TM vanishes and we get:

$$r_E = -r_H = \frac{n_1-n_2}{n_1+n_2}$$

$$t_E = \frac{2n_1}{n_1+n_2} \text{ and } t_H = \frac{2n_2}{n_1+n_2}$$

$$R_E = R_H = \left|\frac{n_1-n_2}{n_1+n_2}\right|^2 \qquad (45)$$

$$T_E = T_H = \frac{Re(n_2)}{Re(n_1)}\left|\frac{2n_1}{n_1+n_2}\right|^2$$

In figure 16 and 17 the amplitude and phase of r_E (TE-case) and r_H (TH-case) are shown as a function of angle of incidence for two widely used interfaces in semiconductor structures. The structure considered in fig. 16 is a simple semiconductor to air structure. The figure clearly shows the phenomena of total internal reflection (TIR) for θ larger than about 17°, the changing phase of r in the TIR regime and the Brewster effect for the TM-wave (for which $\theta_1+\theta_2=\pi/2$). The structure of fig. 17 is a typical semiconductor heterojunction, as used for example in a DBR-stack. The same phenomena can be seen but they are spread over a larger angular range. The very distinct behaviour of TE- and TM-waves do, among others, imply that TM-light can more easily escape from a semiconductor to air. On the other hand a TE-wave will generally exhibit a stronger cavity resonance.

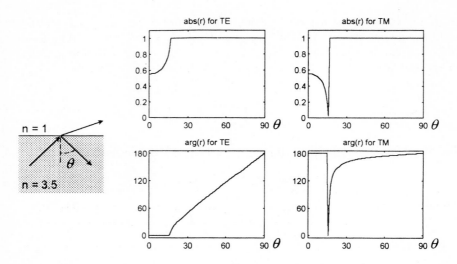

Fig. 16: Amplitude and phase of the field reflection coefficient for both TE and TM waves for a semiconductor (n = 3.5) to air interface.

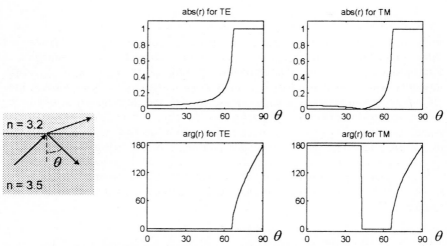

Fig. 17: Amplitude and phase of the field reflection coefficient for a semiconductor (n = 3.5) to semiconductor (n = 3.2) interface.

In fig. 18 we consider the case of a typical semiconductor to metal interface. The metal considered is gold with $n=0.2 - j6.5$. One can see that the field reflection amplitude is close to 1 for all angles of incidence and both for TE- and TM-waves. For normal incidence the phase of the reflection coefficient for the TE-field is about 120° (while it would be 180° for a perfect metal). This means that in a cavity with a gold mirror, the metal interface needs to be closer to the cavity center by an amount equal to about $\lambda/12n$ as compared to the equivalent cavity with a perfect metallic mirror. As a function of angle the phase of the reflection coefficient does behave distinctly different for the TE and TM-case. Because of this a cavity that is resonant for normal incidence (both for TE and TM) will move away from resonance more rapidly in the TE-case than in the TM-case.

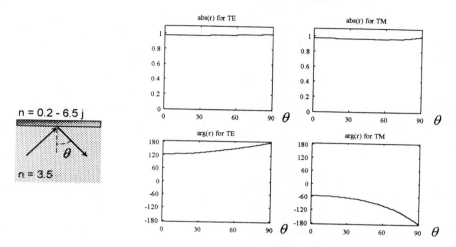

Fig. 18: Amplitude and phase of the field reflection coefficient for a semiconductor to metal interface.

We now turn our attention to multi-layer dielectric mirrors. The overall reflection and transmission of multi-layer structures upon incidence of a plane wave can be analysed in an elegant way with the transfer matrix method [Brekhovskikh, 1960], [Born, 1990]. For a structure as shown in fig. 19 this method leads to a linear 2x2 matrix relation between the complex amplitude (of E-field for TE and of H-field for TM) of an incident plane wave F_1 with angle of incidence θ_1 and the resulting reflected wave B_1 (also with angle θ_1) and the resulting transmitted wave F_N with angle θ_N:

$$\begin{pmatrix} F_1 \\ B_1 \end{pmatrix} = \begin{pmatrix} t_{FF} & t_{FB} \\ t_{BF} & t_{BB} \end{pmatrix} \begin{pmatrix} F_N \\ B_N \end{pmatrix} \qquad (46)$$

whereby B_N represents the amplitude of a backward incident wave in medium N (which is 0 if no wave is incident from this end). One should realize that in a layered structure with parallel interfaces and one incident plane wave all forward and backward plane waves in any layer have the same k_x an k_y values due to the continuity relations at the interfaces. Hence we can write:

$$n_1 \sin\theta_1 = n_2 \sin\theta_2 = n_i \sin\theta_i = n_j \sin\theta_j = n_N \sin\theta_N \qquad (47)$$

where n_i represents the refractive index (more precisely its real part) of layer i. The coefficients t_{FF}, t_{FB}, t_{BF} and t_{BB} depend on the layer structure (thicknesses, refractive indices), on the wavelength and angle of incidence of the incident plane wave and on its polarisation (TE or TM).

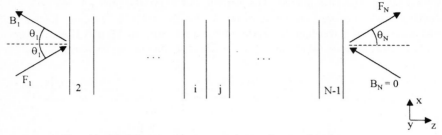

Fig. 19: Multilayer structure with incidence of a plane wave.

From this matrix equation the overall reflection and transmission coefficients can be readily derived:

$$r = \frac{B_1}{F_1} = \frac{t_{BF}}{t_{FF}}$$
$$t = \frac{F_N}{F_1} = \frac{1}{t_{FF}} \qquad (48)$$

A Distributed Bragg Reflector (DBR) or quarter-wavelength stack is a structure of alternating layers of high and low refractive index, each a quarter wavelength thick, as shown in fig. 20.

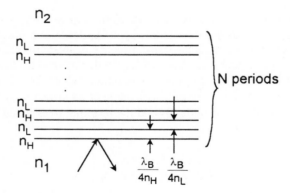

Fig. 20: Distributed Bragg reflector.

Upon incidence of a plane wave this structure gives rise to strong reflection when the Bragg condition for diffraction in periodic structures is fulfilled:

$$k_{diffracted} = k_{incident} + mK_{grating} \text{ (m integer)} \qquad (49)$$

together with the dispersion relation:

$$|k_{diffracted}| = |k_{incident}| = n_1 k_0 \qquad (50)$$

The grating vector $K_{grating}$ is oriented along the direction of periodicity and has a length:

$$\left| K_{grating} \right| = \frac{2\pi}{\text{period}} \quad (51)$$

This leads to the conclusion that the multilayer structure has a strong reflection for normal incidence when the period equals half the wavelength (or a multiple of that). The highest reflectivity is actually obtained when both the high and the low refractive index layer are each a quarter wavelength thick, i.e. $\lambda/4n_H$ and $\lambda/4n_L$ respectively. The wavelength for which this occurs is called the Bragg wavelength λ_B. For plane waves with oblique incidence the wavelength of maximum reflectivity λ_{max} shifts to shorter wavelengths according to (approximately):

$$\lambda_{max} = \lambda_B \cos\theta \quad (52)$$

Fig. 21 shows the field reflectivity (amplitude and phase) for a typical GaAs-AlAs DBR (n_H = 3.5, n_L = 2.9) for normal incidence as a function of wavelength.

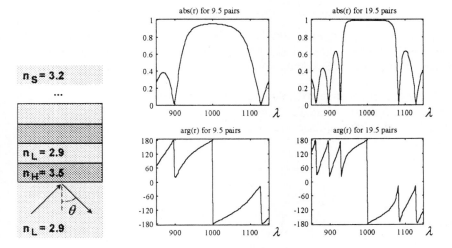

Fig. 21: Amplitude and phase of the field reflection coefficient for the DBR-mirror shown as a function of wavelength, with 9.5 and 19.5 layer pairs respectively.

The peak power reflectivity is given by an analytical formula [Macleod, 1986 and Yeh, 1988]:

$$R_L(\lambda = \lambda_B) = \left(\frac{1 - \frac{n_2}{n_1}\left(\frac{n_H}{n_L}\right)^{2N}}{1 + \frac{n_2}{n_1}\left(\frac{n_H}{n_L}\right)^{2N}} \right)^2 \quad (53)$$

if the first layer of the quarter-wave stack has refractive index n_H (in the opposite case n_H and n_L need to be reversed).

The spectral width $\Delta\lambda$ around λ_B for which the reflectivity is high increases with the contrast n_H-n_L and is given by [Yeh, 1988]:

$$\frac{\Delta\lambda}{\lambda_B} = \frac{4}{\pi}arcsin\left(\frac{n_H - n_L}{n_H + n_L}\right) \approx \frac{2}{\pi}\frac{\Delta n}{n} \tag{54}$$

at least if the number of pairs N is sufficiently large.

The phase of the reflection coefficient is not constant within this wavelength range but changes approximately in a linear way as a function of wavelength. In view of this, one can define an effective penetration depth of the DBR-mirror, defined as the depth (measured from the first DBR-interface) where a single mirror interface (with wavelength independent phase of reflection) should be positioned to give rise to the same phase slope around $\lambda = \lambda_B$ as the DBR itself. This penetration depth is approximately given by [Ram, 1995]:

$$L_{eff,\lambda} \approx \frac{\lambda_B}{4n_{ref}n_{ref}}\frac{n_H n_L}{n_H - n_L} \tag{55}$$

In this equation a reference refractive index n_{ref} is introduced, which is the (arbitrary) refractive index of the fictitious medium in front of the equivalent mirror, often chosen to be equal to the refractive index of the cavity core n_1.

In fig. 22 the reflection coefficient of a DBR-mirror with $n_H = 3.5$ and $n_L = 3.2$ is shown as a function of angle of incidence (for $\lambda = \lambda_B$). As expected from the Bragg-condition one can see that the reflectivity is high for a limited angular range only.

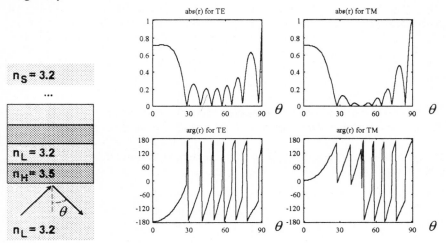

Fig. 22: Amplitude and phase of the field reflection coefficient for the DBR-mirror shown, as a function of angle of incidence, for both TE and TM polarisation.

Outside this range the DBR-mirror is not very reflective at all, but highly transmissive or "leaky". This is particularly so for the TM waves where the single interface reflectivity decreases with increasing angle of incidence, down to zero reflectivity at the Brewster point (which is the same for the n_H to n_L interface and for the n_L to n_H interface). The behaviour as a function of angle has a lot of resemblance (for small enough θ) with the behaviour as a function of wavelength (around λ_B). This can be understood from the fact that the optical thicknesses of the layers are proportional to $\cos\theta/\lambda$. Therefore a change of angle $\Delta(\cos\theta)$ (around $\theta=0$) is equivalent to a relative change of wavelength $\Delta\lambda/\lambda_B$ (around λ_B), if $\Delta(\cos\theta)=\Delta\lambda/\lambda_B$. So one can expect the angular width of high reflectivity to be given by:

$$2\left(1-\cos\frac{\Delta\theta}{2}\right) = \frac{\Delta\lambda}{\lambda_B} \approx \frac{n_H - n_L}{n_H + n_L} \qquad (56)$$

in which $\Delta\theta$ is the full width of the high reflection region.

The above argument of equivalence between the spectral and the angular domain is approximate for two reasons. First the reflectivities of the individual interfaces have a different spectral and angular behaviour. Secondly the change of $\cos\theta$ is slightly different for each layer due to refraction. Around $\theta=0$ one can derive from Snell's law that a change of $\cos\theta_i$ in layer i relates to a change of $\cos\theta_j$ in layer j as:

$$n_i^2 \Delta(\cos\theta_i) \approx n_j^2 \Delta(\cos\theta_j) \qquad (57)$$

From this discussion one can understand that the phase of the reflection coefficient varies with the angle of incidence in a cosinusoidal way, as can be seen in fig. 22. As in the spectral case one can define an effective penetration depth, now defined as the depth of a single interface (with angle independent phase of reflection) such that the phase slope (with angle) is the same as for the DBR itself. For sufficiently small refractive index difference n_H-n_L this "angular" penetration depth equals the "spectral" penetration depth

$$L_{eff,\theta} = L_{eff,\lambda} \qquad (58)$$

When n_H-n_L increases the error due to equation (57) increases. If again we assume that the imaginary material in front of the effective mirror has a refractive index n_{ref}, we can write:

$$n_H^2 \Delta(\cos\theta_H) = n_L^2 \Delta(\cos\theta_L) = n_{ref}^2 \Delta(\cos\theta_{eff}) \qquad (59)$$

Hence the average $\Delta(\cos\theta)$ in the actual DBR is given by:

$$\frac{\Delta(\cos\theta_H) + \Delta(\cos\theta_L)}{2} = \frac{1}{2}\left(\frac{1}{n_H^2} + \frac{1}{n_L^2}\right) n_{ref}^2 \Delta(\cos\theta_{eff}) \qquad (60)$$
$$> \Delta(\cos\theta_{eff})$$

From this we can derive a more accurate expression for $L_{eff,\theta}$ [Ram, 1995]:

$$L_{eff,\theta} = \frac{1}{2}\left(\frac{1}{n_H^2} + \frac{1}{n_L^2}\right) n_{ref}^2 L_{eff,\lambda} > L_{eff,\lambda} \tag{61}$$

For an extreme case where for example $n_H = n_{ref} = 3.5$ and $n_L = 1.5$, the correction factor is 3.2.

The requirements that DBR-mirrors have to satisfy strongly depend on the application. In VCSEL's the key requirement is very high reflectivity (well above 99%) for normal incidence. The main problem generally is to avoid losses due to scattering at interfaces with some residual roughness or due to absorption in the layers. for example due to doping of the semiconductor layers. The larger the refractive index contrast, the easier it is to achieve a very high reflectivity. In Micro-Cavity LED's the key problem is very different. In this device the mirror should be highly reflective for all angles of incidence and therefore the leaky angular region of the DBR should be as small as possible. Furthermore the cavity length should be small to reduce the number of resonances and therefore the penetration depth should be small. Both factors call for a DBR with high refractive index contrast. So both devices take advantage of a large refractive index contrast, but for very different reasons.

7. A dipole in a planar cavity

In paragraph 2.3 the case of a point source in a simple cavity was discussed. We now consider the more realistic case of a dipole in a cavity with planar mirrors. The total field can be derived by expanding the dipole source into plane waves as discussed earlier and by calculating, for each plane wave component, the total plane wave amplitude as a result of the cavity interference.

Consider fig. 23 with a dipole in an arbitrary planar cavity. The dipole is either a horizontal or a vertical dipole.

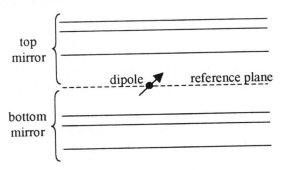

Fig. 23: A dipole in an arbitrary cavity.

A reference plane is defined through the dipole and parallel to the mirrors. The plane wave field amplitude of the dipole source is denoted as:

$$\psi_{dipole}^{up}(k_x, k_y) \tag{62}$$

for the field propagating towards the top mirror and

$$\psi_{dipole}^{down}(k_x, k_y) \tag{63}$$

for the downward field.
These represent the E-field amplitude in the TE-case and the H-field amplitude in the TM-case. Depending on the case (vertical or horizontal dipole ; TE or TM) and on the sign convention we can write:

$$\psi_{dipole}^{up}(k_x, k_y) = \pm \psi_{dipole}^{down}(k_x, k_y) \tag{64}$$

The total plane wave amplitude in the reference plane is denoted as

$$\psi_{total}^{up}(k_x, k_y) \text{ and } \psi_{total}^{down}(k_x, k_y) \tag{65}$$

respectively. The plane wave field reflectivity of the top and bottom mirror are denoted as:

$$r_{top}(k_x, k_y) \text{ and } r_{bottom}(k_x, k_y) \tag{66}$$

respectively. Continuity (or self-consistency) of the fields at the reference plane is then expressed as:

$$\left[\psi_{total}^{down}(k_x, k_y) r_{bottom}(k_x, k_y) + \psi_{dipole}^{up}(k_x, k_y) \right] r_{top}(k_x, k_y) + \psi_{dipole}^{down}(k_x, k_y) = \\ \psi_{total}^{down}(k_x, k_y) \tag{67}$$

From this equation the total downward field can be derived

$$\psi_{total}^{down} = \frac{\psi_{dipole}^{down} + \psi_{dipole}^{up} r_{top}}{1 - r_{top} r_{bottom}} \tag{68}$$

Likewise

$$\psi_{total}^{up} = \frac{\psi_{dipole}^{up} + \psi_{dipole}^{down} r_{bottom}}{1 - r_{top} r_{bottom}} \tag{69}$$

This is a generalisation of equation (26) and the term $1/(1 - r_{top} r_{bottom})$ in these expressions leads to a more general Airy factor.

Fig. 24 shows a generic example of a dipole in the middle of a semiconductor layer, one wavelength thick, surrounded by air. This is a cavity with rather modest mirror reflectivity (about 30%) for normal incidence and perfect reflectivity (100%) for oblique incidence in the total internal reflection regime.

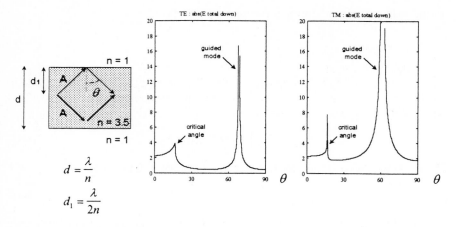

Fig. 24: Total plane wave field amplitude in a λ-cavity as a function of angle for both TE and TM wave.

For normal incidence the standing wave pattern has maxima at both interfaces and at the position of the dipole. Therefore a resonance for normal incidence can be expected. In fig. 24 the calculated total downward field is shown as a function of θ for both the TE- and TM-case.

In this figure the dipole field amplitude dependence on θ and ϕ (as shown in fig. 14) is not taken into account but can be included in a trivial way. One can see two resonances: the first is a broad resonance from $\theta = 0$ up to $\theta = \theta_c$. The second is a sharp guided wave resonance at about $\theta = 68°$ for TE and $\theta = 60°$ for TM. This resonance leads to a singularity in the total field (and hence an infinite quality factor), as expected for a lossless cavity with total internal reflection at both mirrors. The broad resonance shows a peculiar feature. While phase resonance obviously occurs for $\theta = 0$, the total field peaks for $\theta = \theta_c$. This can be understood easily. For $\theta = 0°$ the phase resonance is optimal but the reflectivities are low. For $\theta = \theta_c$ the phase resonance is not perfectly fulfilled but the reflectivity reaches 100% and hence the Airy factor is stronger than at $\theta = 0°$. This example demonstrates that the definition of resonance is somewhat ambiguous. The strongest resonant enhancement does not necessarily coincide with the conventional phase resonance operation point but can also occur at the operation point of smallest cavity loss (or somewhere in between). In special cases a single resonance can present itself by two peaks: one peak at phase resonance and one peak at the cavity loss minimum.

In fig. 25 the cavity length is changed from 1 to 2 wavelengths (while the dipole is kept at a distance half a wavelength away from one mirror). Both for the "lambda-cavity" and the "2 lambda"-cavity the total internal field amplitude is shown for the central wavelength as a function of angle and for normal incidence as a function of wavelength.

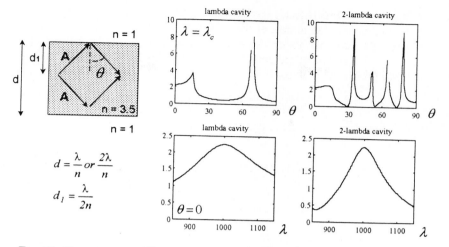

Fig. 25: Comparison of "lambda"- and "2 lambda"-cavity in terms of internal angular field distribution and normal emission spectrum. Note that the source is not centered in the cavity in the 2-lambda case.

These figures show that the number of resonances increases when the cavity length is increased. Very roughly speaking the radiated power of the dipole is equally distributed over all resonances. Since in this case only one resonance falls within the extraction cone, the extraction efficiency will drop as the cavity length increases. This explains why a micro-cavity LED needs to have a small cavity length. The figure also shows that the spectrum of normal emission ($\theta = 0$) narrows as the cavity length is increased. This may be advantageous in an application requiring a narrow spectrum but when the filter action of the cavity is narrower than the natural linewidth of the dipole emission, the overall extraction efficiency deteriorates.

An aspect not discussed so far is the impact of the cavity on the total power emitted by the dipole (for a given dipole strength). For a given density of dipoles per unit volume, a change in the total emitted power can, in case a dipole represents radiative electron-hole recombination, only be associated with a change in recombination rate and hence in lifetime. One can write

$$\frac{\frac{1}{\tau}}{\frac{1}{\tau_0}} = \frac{\text{power emitted by dipole in cavity}}{\text{power emitted by dipole in uniform space}} \qquad (70)$$

where τ and τ_o are the lifetimes with and without cavity. The change of carrier lifetime due to the presence of a cavity is known as the Purcell-factor.

The impact of a planar cavity on the change of spontaneous decay rate has been studied in detail ([Brorson, 1990], [Yokohama, 1992], [Björk, 1994] and [Abram, 1998]). In the case of a cavity with perfect (100% reflecting) mirrors an analytical expression for the Purcell-factor has been derived in [Abram, 1998] for the case of a horizontal dipole in the middle of the cavity. The results depend on the phase of the mirror reflection coefficient. For $r=+1$ there are $[m_c]+1$ modes but only $[m_c/2 +1]$ are excited. The others are not excited because the dipole is located at a zero of the mode profile. One finds:

$$\frac{1/\tau}{1/\tau_0} = \frac{6\left[\frac{m_c}{2}\right]+3}{4m_c} + \frac{4\left[\frac{m_c}{2}\right]^3 + 6\left[\frac{m_c}{2}\right]^2 + 2\left[\frac{m_c}{2}\right]}{2m_c^3} \tag{71}$$

For $r=-1$ there are $[m_c]$ modes, $[(m_c+1)/2]$ of which are excited and one finds:

$$\frac{1/\tau}{1/\tau_0} = \frac{6\left[\frac{m_c+1}{2}\right]}{4m_c} + \frac{4\left[\frac{m_c+1}{2}\right]^3 - \left[\frac{m_c+1}{2}\right]}{2m_c^3} \tag{72}$$

The terms in $1/m_c$ in these expressions correspond to TE-waves, whereas the terms in $1/m_c^3$ correspond to TM-waves.

The two expressions are shown in fig. 26 as a function of m_c. One can see that apart from the singular $1/m_c$ behaviour for small m_c in the $r=+1$ case, the maximum Purcell-factor is 3 and is obtained in a half-wavelength thick cavity with $r=-1$ (perfect metallic mirrors). For thick cavities the Purcell-factor converges to 1. In other words, thick cavities with many modes have a similar impact on the dipole as uniform space with a continuum of modes.

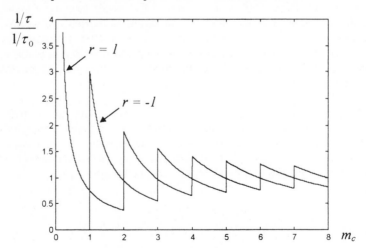

Fig.26: Modification of spontaneous emission.

Although one would expect intuitively that the cavities with 100% reflecting mirrors present the strongest decay rate enhancement, it has been shown in [Abram 98] that this is not the case. A cavity with two Au-mirrors can give rise to a Purcell-factor of 4.4 for $m_c = 0.7$. This behaviour is explained by the phase shift at the Au-interface which leads to a situation in between the $r=+1$ case (infinite

Purcell-factor for $m_c=0$) and the $r=-1$ case (Purcell-factor = 3 for $m_c=1$). A special feature of cavities with realistic metallic mirrors is also that part of the evanescent fields generated by the dipole can be coupled to and dissipated by the metal, thereby giving a contribution to the decay rate. The extent of this mechanism obviously depends on the distance between the dipole and the metal.

In a micro-cavity bound by DBR-mirrors the Purcell factor is generally close to 1 (between 0.8 and 1.2 typically) except if the m_c-value is close to 0. The reason for this is dual. First of all, the DBR is leaky over a wide angular range which means that in this range the situation is little different from free space. Secondly, the penetration depth of the DBR leads to an enlarged effective cavity thickness and hence an enlarged effective cavity order defined as:

$$[m_{c,eff}] = \left[\frac{d + L_{1,eff,\theta} + L_{2,eff,\theta}}{\lambda/2n} \right] \tag{73}$$

whereby $L_{1,eff,\theta}$ and $L_{2,eff,\theta}$ are the angular penetration depths of both mirrors (with n_{ref} chosen to be equal to n), while d and n are the thickness and the refractive index of the layer in between both DBR-mirrors.

In 3-dimensional cavities the Purcell-factor can be substantially larger than in planar cavities. In [Gérard 98] Purcell-factors up to 5 have been demonstrated in optically pumped micropillars.

8. Guided modes

In the discussion so far, the field produced by the dipole was decomposed into a continuum of plane waves, with k_x and k_y as independent variables. The resonances with evanescent fields outside the cavity were called "guided modes". The examples shown in the figures demonstrate that, as a function of $k_\parallel$ or θ, a discrete number of resonant peaks -"guided modes"- appeared. For each "guided mode" the cavity enhancement appears over a finite range of $k_\parallel$ or θ and becomes singular in the lossless case. The finite θ-range of enhancement may be surprising for anybody familiar with waveguide theory. It is well established indeed that the guided field can be represented as a sum of guided modes, each of which has a sharply defined propagation constant (corresponding to $k_\parallel$), rather than a finite range of propagation constants. The origin of this apparent paradox lies in the fact that the concept of a "guided mode" derived from a resonant plane wave is very much related to, but not rigorously identical to the classical concept of a guided mode.

The total field propagating in a waveguide structure - for example the field due to a radiating dipole - can be decomposed into base functions of the (sourceless) structure in two distinct ways. This is illustrated in fig. 27 where the same waveguide structure is shown with two different reference planes for decomposition. The (x,y,z) coordinate system has deliberately been chosen in a different way, such that the (x,y)- (or (x',y')-plane) is always the plane of decomposition. In the plane wave approach of fig. 27a the base functions are given by $exp(-j(k_x x + k_y y))$ and are not influenced by the waveguide structure. In each layer they represent a plane wave with a different direction. Both **E** and **H** fields can be expressed in terms of these base functions.

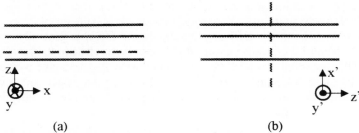

Fig. 27: Two methods to decompose a propagating field in a planar structure. The dashed line is the decomposition plane.

In the normal (waveguide) mode approach of fig. 27b the base functions are the waveguide modes. Such a waveguide mode is a solution of Maxwell's equations for this structure (without sources) of the form:

$$E(x',y',z') = exp(\pm jk_{z'}z')e_{k_{z'}}(x',y')$$
$$H(x',y',z') = exp(\pm jk_{z'}z')h_{k_{z'}}(x',y')$$
(74)

In other words a waveguide mode is a solution such that its (x',y')-dependence does not change upon propagation in the z'-direction. The propagation constant $k_{z'}$ is unique for each mode (apart from degeneracy effects). In planar structures the modes subdivide into two categories: the TE-waveguide modes with field components $E_{y'}$, $H_{x'}$ and $H_{z'}$ and the TM-modes with field components H_y, $E_{x'}$ and $E_{z'}$. These names are consistent with the TE/TM decomposition in the plane wave approach. It has been shown (see for example [Vassallo, 1991]) that the normal modes can be subdivided in a continuous set of radiation modes (either propagating or evanescent in the z'-direction) and a truly discrete set of guided modes (whereby a guided mode is defined as a normal mode which decays to zero for large $|x'|$ or $|y'|$).

The combination of the radiation modes and the guided modes forms a complete set. Furthermore the normal modes satisfy the general orthogonality relation:

$$\iint \left(e_{k_{z',1}}(x',y') \times h_{k_{z',2}}(x',y')\right) u_{z'} dx' dy' = 0$$
$$for\ k_{z',1} \neq k_{z',2}$$
(75)

For degenerate modes an orthogonalisation procedure has to be applied before one can write this relation. This orthogonality relation is always valid, irrespective of whether the structure is lossless or not. In lossless structures one can also write:

$$\iint \left(e_{k_{z',1}}(x',y') \times h^*_{k_{z',2}}(x',y')\right) u_{z'} dx' dy' = 0$$
$$for\ k_{z',1} \neq k_{z',2}$$
(76)

This latter relation is interesting because it can be seen as a power orthogonality relation: the total propagating power is the sum of the powers carried by the (non-evanescent) modes. Unfortunately such a statement can only be made in lossless structures (or in uniform lossy structures).

We return to the example shown in fig. 25 where a 1 wavelength thick waveguide with a core index of 3.5 and a cladding index of 1 was considered. The lowest order TE-mode for this waveguide is shown in fig. 28. The effective refractive index (k_z/k_0) for this mode is 3.248. Within the core of the waveguide this effective index corresponds to plane waves with elevation $\theta=\arcsin(n_{eff}/n_{core}) = \pm 68.1°$. As can be seen this angle corresponds well to the angle of the "guided mode" - resonance in fig. 25.

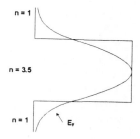

Fig. 28: E_y-component of TE-field for the waveguide shown in fig. 25.

At this point the basic difference between the plane wave decomposition in the (x,y)-plane and the normal mode decomposition in the (x',y')-plane can be understood. A dipole in a waveguide will excite guided and radiating normal modes propagating away from the dipole. This situation is depicted in fig. 29 (where only a single guided mode excitation is shown).

In the plane wave approach the dipole excites plane waves that span the entire (x,y)-plane, as shown in fig. 29(b) (where 4 plane waves with the same elevation as those in fig. 29(a) are shown).

Clearly the two situations are different and the only way to make them consistent with each other is to take a directionally broadened distribution of the plane waves of fig. 29(b) in the (k_x,k_y) domain, with complex coefficients such that the linear combination reproduces the different directions in the right and left half spaces around the dipole as in fig. 29(a). This is illustrated in fig. 29(c) where (only) two such plane waves of this distribution are drawn.

To understand the equivalence we write down the entire field solution in the normal mode decomposition

$$\psi_\delta(x',y',z') = \sum_i \psi_{\beta_i}(x') e^{-j\beta_i|z'|} + \int \psi_{k_{z'}}(x') e^{-jk_{z'}|z'|} dk_{z'}. \tag{77}$$

For ease of notation the solution was written down in a scalar way and incorporates only the wave components with *k*-vector in the (x',z') plane ($\varphi=0$). To these should be added all the components with other azimuthal values so as to arrive at a cylindrical solution.

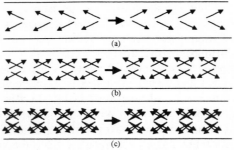

Fig. 29: (a) Guided normal mode decomposition of a dipole(1 guided mode shown) (b) The four plane waves that one would naively map onto the above field. Note however that each of the plane waves has the same direction throughout the horizontal plane (c) Solution of the "discrepancy" between figure (a) and (b): the plane waves of figure (b) are directionally broadened with coefficients as described in equation (81)

The first term in equation (77) represents the discrete guided modes, while the second represents the radiation modes, both propagating and evanescent (in the z'-direction). Equation (77) can be rewritten as

$$\psi_\delta(x',y',z') = \sum \psi_{\beta_i}(x')\left(e^{-j\beta_i z'} H(z') + e^{+j\beta_i z'} H(-z')\right) \\ + \int \psi_{k_{z'}}(x')\left(e^{-jk_{z'} z'} H(z') + e^{+jk_{z'} z'} H(-z')\right) \quad (78)$$

whereby $H(z)$ is the Heaviside or step function ($H(z')=0$ for $z'<0$ and $H(z')=1$ for $z'>0$).

The main point is that the distribution of plane waves that allows to reconstruct this field is just the Fourier Transform of equation (78) with respect to z' (and y'). Doing this for one discrete guided mode (and assuming a lossless structure with real β_i) leads to:

$$\psi_\delta(x',k_{y'},k_{z'}) = \psi_{\beta_i}(x')\left[\delta(k_{z'}-\beta_i) * H^F(k_{z'}) + \delta(k_{z'}+\beta_i) * \left(-H^F(-k_{z'})\right)\right] \quad (79)$$

in which $H^F(k_{z'})$ is the Fourier Transform of the Heaviside function, given by

$$H^F(k_{z'}) = \pi\delta(k_{z'}) - \frac{j}{k_{z'}} \quad (80)$$

This formula represents the useful link between the diverging aspect of the discrete mode approach of fig.29(a) and the uniform plane wave approach of fig. 29(b). Combination of equations (79) and (80) finally leads to:

$$\psi_\delta(x',k_{y'},k_{z'}) = \psi_{\beta_i}(x')\left[\pi\delta(k_{z'}-\beta_i) - \frac{j}{k_{z'}-\beta_i} - \pi\delta(k_{z'}+\beta_i) - \frac{j}{k_{z'}+\beta_i}\right] \quad (81)$$

This equation clarifies that a single guided mode excitation by a dipole is represented by a broadened set of plane waves. The broadening function $H^F(k_{z'})$

should then be related to the Airy function around the resonance peak (or rather to its special form for $r_1r_2=1$, described in equation 26). Both functions are not identical because equation (81) contains only a single guided mode while the Airy function contains all contributions. All other normal modes, and in particular the evanescent normal modes, do also contribute to the plane wave amplitude for all k_z-values. Nevertheless near a resonance peak the contribution of the guided mode should be dominant. The Airy function does (in the special lossless case) indeed have a $1/j(k_z-\beta_i)$ dependence near the resonance peak together with a delta-function at $k_z=\beta_i$, as is the case with the $H^F(k_z)$ -function.

In summary: the Airy function - which represents the plane wave decomposition of the field generated by a point - like source - has non-discrete resonance peaks so as to represent the excitation of discrete guided modes that propagate away from the point source (whereas the plane waves do only partly propagate away from the point source).

In a case where the structure is lossy, the guided modes remain discrete, but they have a complex propagation constant. Hence the Heaviside function needs to be multiplied with an exponentially decaying function. This leads to a further broadening of the peaks in the plane wave decomposition (and disappearance of the singularities), consistent with the behaviour of the Airy function in this case.

Acknowledgements

The authors thank their colleagues and former colleagues, who work or have worked in the field of micro-cavity LED's, in particular Hans De Neve, Johan Blondelle, Ingrid Moerman and Peter Van Daele, for very stimulating teamwork in this field. Furthermore they thank their colleagues of the electromagnetic theory group at University of Gent - IMEC (Daniel De Zutter, Frank Olyslager and Eric Laermans) as well as Roger Van Keer (of the Department of Mathematics at University of Gent) for contributions to some theoretical aspects.

Part of this work has been supported by the European Community through the Esprit SMILED project and the authors gratefully acknowledge the partners of this project. In particular they thank Claude Weisbuch, Henri Bénisty (Ecole Polytechnique Paris) and Ross Stanley (Ecole Polytechnique Fédérale Lausanne) for discussions on the subject. The Belgian DWTC-project "IUAP-13" is also acknowledged for support.

Peter Bienstman and Ronny Bockstaele acknowledge support from FWO-Vlaanderen and IWT respectively.

References

[Abram, 1998] I. Abram, I. Robert and R. Kuszelewicz, "Spontaneous Emmission Con trol in Semiconductor Microcavities with Metallic or Bragg Mirrors", IEEE-JQE Vol. 34-1 (1998), pp. 71-76.

[Benisty, 1998a] H. Benisty, R. Stanley and M. Mayer, "Method of source terms for dipole emission modification in modes of arbitrary planar strucures", J. Opt. Soc. Am. A Vol. 15-5 (1998), pp. 1192-1201.

[Benisty, 1998b] H. Benisty, H. De Neve and C. Weisbuch, "Impact of Planar Microcavity Effects on Light Extraction - Part I : Basic Concepts and Analytical Trends", IEEE-JQE Vol. 34-9 (1998), pp. 1612-1631.

[Benisty, 1998c] H. Benisty, H. De Neve and C. Weisbuch, "Impact of Planar Microcavity Effects on Light Extraction - Part II : Selected Exact Simulations and Role of Photon Recycling", IEEE-JQE Vol. 34-9 (1998), pp. 1632-1643.

[Bjork, 1994] G. Bjork, "On the Spontaneous Lifetime Change in an Ideal Planar Microcavity - Transition from a Mode Continuum to Quantized Modes", IEEE-JQE Vol. 30-10 (1994), pp. 2314-2318.

[Blondelle, 1997] J. Blondelle, "Realisation of high efficient substrate emitting InGaAs/(Al)GaAs microcavity LED's using MOCVD growth", June 1997.

[Born, 1990] M. Born and E. Wolf, "Principles of optics", Oxford, Pargamon 1980.

[Brekhovskikh, 1960] L.M. Brekhovskikh, "Waves in layered media", Academic Press, New York, 1960.

[Brorson, 1990] S.D. Brorson, H. Yokoyama, E.P. Ippen, "Spontaneous emission rate alteration in optical waveguide structures", IEEE J. Quantum Electronics, vol. 26-9 (1990), pp. 1492-1499.

[De Neve, 1997] H. De Neve, "Design and realisation of light emitting diodes based on the microcavity effect", May 1997.

[Gérard, 1998] J. M. Gérard, B. Sermage, B. Gayral, B. Legrand, E. Costard, V. Thierry-Mieg, "Enhanced Spontaneous Emission by Quantum Boxes in a Monolithic Optical Microcavity", Phys. Rev. Lett., vol. 81, no. 5, pp. 1110-1113, 1998.

[Lukosz, 1977a] W. Lukosz and R. E. Kunz, "Light Emission by Magnetic and Electric Dipoles Close to a Plane Interface. I Total radiated power", J. Opt. Soc. Am Vol. 67-12 (1977), pp. 1607-1615.

[Lukosz, 1977b] W. Lukosz and R. E. Kunz, "Light Emission by Magnetic and Electric Dipoles Close to a Plane Interface. I Radiation Patterns of Perpendicular Oriented Dipoles", J. Opt. Soc. Am Vol. 67-12 (1977), pp. 1615-1619.

[Lukosz, 1979] W. Lukosz and R. E. Kunz, "Light Emission by Magnetic and Electric Dipoles Close to a Plane Interface. I RadiationPatterns of Dipoles with Arbitrary Orientation", J. Opt. Soc. Am Vol. 69-11 (1979), pp. 1495-1503.

[Lukosz, 1980] W. Lukosz, "Theory of optical-environment-dependent spontaneous-emission rates for emitters in thin layers", Phys. Rev. B Vol. 22-6 (1980), pp. 3030-3037.

[Macleod, 1986] H.A. Macleod, "Thin film optical filters", Adam Hilger, Bristol (1969)

[Purcell, 1946] E.M. Purcell, "Spontaneous emission probabilities at radio frequencies", Phys. Rev., vol. 69, p. 681, 1946.

[Ram, 1995] R. Ram, D. Babic, R. York, J. Bowers, "Spontaneous emission in microcavities with distributed mirrors", IEEE J. Quantum Electronics, Vol. 31, pp. 399-410, 1995.

[Schnitzer, 1993] I. Schnitzer, E. Yablonovitch, C. Caneau, T. J. Gmitter and A. Scherer, "30 % external quantum efficiency from surface textured, thin-film light-emitting diodes", Appl. Phys. Lett. Vol. 63-16 (1993), pp. 2174-2176.

[Van Bladel, 1964] J. Van Bladel, "Electromagnetic fields", McGraw-Hill (1964)

[Vassallo, 1991], C. Vassallo, "Optical waveguide concepts", Elsevier, 1991.

[Windisch, 1998] R. Windisch, P. Heremans, B. Dutta, M. Kuijk, S. Schoberth, P. Kiesel, G. H. Dohler and G. Borghs, "High-efficiency non-resonant cavity light-emitting diodes", Electr. Lett. Vol. 34-11 (1998), pp. 1153-1154.

[Yamanishi, 1984] M. Yamanishi, I. Suemune, "Comment on polarization dependent momentum matrix elements in quantum well lasers, Jap. J. Appl. Phys., 23 (1984), pp. L35-L36

[Yeh, 1988] P. Yeh, "Optical waves in layered media', John Wiley and Sons (1988)

[Yokoyama, 1992] H. Yokoyama, "Physics and Device Applications of optical Microcavities", Science Vol. 256 (1992), pp. 66-70.

Microscopic Theory of the Optical Semiconductor Response near the Fundamental Absorption Edge

S. W. Koch

Department of Physics, University of Marburg, Renthof 5, 35032 Marburg/Germany

Abstract. This paper reviews the theory needed to describe the near bandgap optical properties of semiconductor quantum-well structures. After brief introductions of the basic concepts of electronic banstructures, quantum confinement effects, and optical susceptibility, a systematic derivation of the semiconductor Bloch equations is presented. In the regime of linear response the excitonic and Coulombic enhancement effects in the absorption spectra of direct-gap semiconductor structures are discussed. Nonlinear phenomena, such as absorption saturation, gap renormalization and excitation induced dephasing are analyzed using a systematic scheme of approximations. The resulting nonlinear semiconductor theory is self-consistently combined with solutions of Maxwell's equations to compute the optical properties of excitonic semiconductor microcavity systems.

1. Introduction
2. Basic Physics of Semiconductor Electrons
    2.1 Elementary Bandstructure Aspects
    2.2 Tight-Binding Approximation
    2.3 General Bandstructure Aspects
    2.4 Quantum Confinement Effects
    2.5 Second Quantization
3. Optical Response of Semiconductors
    3.1 Optical Response
    3.2 Free-Carrier Equations of Motion
    3.3 Semiconductor Bloch Equations
    3.4 Excitons
4. Optical Nonlinearities in Semiconductors
    4.1 Coulomb Correlation Effects
    4.2 Carrier Quantum Boltzmann Equation
    4.3 Dephasing
    4.4 Screened Hartree-Fock Approximation
5. Excitons in Semiconductor Microcavities
    5.1 Semiclassical Theory
    5.2 Nonlinear Exciton Saturation
    5.3 Microcavity Luminescence

1 Introduction

The optical response of semiconductors has been the subject of intense experimental and theoretical investigations for several decades. Consequently, many basic features are well understood by now, and some properties are even exploited in comercial devices such as light emitting diodes, semiconductor laser, optical switches, etc. To some degree it is the huge success of these devices, as well as the need of even faster operational speed and higher levels of on- and off-chip integration that drives the development of smaller and smaller semiconductor structures. A recent example with enormous application potential are the so-called VCSEL (vertical cavity surface emitting laser) structures, where semiconductor quantum wells between Bragg mirrors operate as highly efficient lasers with excellent performance characteristics and truely miniature geometry in the range of only a few μm (10^{-6}m).

Interestingly enough, the same heterostructures, i.e. quantum wells between high reflectivity mirrors, show very exciting light-matter coupling effects at low excitation conditions if the exciton resonance is in the spectral vicinity of the cavity resonance. The interaction leads to so-called "normal mode coupling" (NMC) which manifests itself in these microcavity systems, e.g. as anticrossing in the hybride exciton and light dispersion resulting in a double peak structure in transmission, reflection and luminescence. The basic physics behind these and related effects is one of the main topics of this summer school on "QED Phonomena and Applications of Microcavities and Photonic Crystals".

Recent research has shown that even the linear optical properties of real microcavity structures are not at all trivial, partly due to the unavoidable presence of structural disorder, which is always present at least on an atomic scale at the interfaces between quantum wells and the surrounding barrier material. Furthermore, unexpected optical nonlinearities have been observed, which gave rise to speculations about quantum-statistical condensation effects in such heterostruture systems. These and related experimental observations clearly demonstrate the need for a basic understanding of the physical mechanisms, governing the optical microcavity response.

These lecture notes attempt to summarize many of the important aspects needed to understand the elementary microcavity physics. After an introduction into some of the fundamental concepts, such as electronic band strutures, electron-hole excitations, and excitons, a microscopic theory is outlined that allows us to analyze the linear and nonlinear optical response of semiconductors under quasi-stationary and dynamic excitation conditions. These results serve as input for a self-consistent solution of Maxwell's wave equation for the spatially inhomogeneous microcavity heterostructure.

Most of the material summarized in these lecture notes can be found in textbooks where it is, however, often distributed over many chapters and/or combined with other, more general results. In order to provide a brief but fairly comprehensive introduction, the basic material is collected here in a condensed form. For this purpose I have used material from the textbooks by *W. Chow and S.W. Koch, Semiconductor Laser Fundamentals (Springer Verlag 1998)* and *H. Haug and S.W. Koch, Quantum Theory of the Optical and Electronic Properties of Semiconductors, (World Scientific Publ., 3rd ed. 1994)*, where many more details and results can be found. The application of the semiconductor many body theory to microcavity systems is based on the recent papers by Jahnke, Kira, Koch, Z. Physik **B 104**, 559 (1997) and Kira et al. (submitted).

It is my pleasure to thank F. Jahnke and M. Kira for ongoing theory collaborations and H. Gibbs, G. Khitrova and coworkers in Tucson/Arizona for theory/experiment collaborations. This work is supported by the Deutsche Forschungsgemeinschaft, partially through the Leibniz prize.

2 Basic Physics of Semiconductor Electrons

This chapter summarizes some important concepts needed to analyze the optical properties of semiconductor systems. The first section discusses the problem of a semiconductor electron in the periodic lattice potential of the ions. The resulting band structure, i.e. the occurrence of regions of allowed energy states and energetically forbidden regions is illustrated in Sec. 2.2 for the example of a simple tight-binding model. Sec. 2.3 summarizes general bandstructure results and Sec. 2.4 dicusses the basic quantum confinement effects occurring for electrons in an effectively two-dimensional quantum-well structure. Sec. 2.5 summarizes the fundamental rules of the particle number representation or second quantization approach which is extremely useful when one is dealing with systems where the occupation probabilities of the different quantum states are a dynamic variable.

2.1 Elementary Bandstructure Aspects

In a simple picture of a semiconductor, an electronic state is identified by its momentum, $\mathbf{k}$, and z-component of spin, s_z (or, more generally by its total angular momentum quantum number). As our model solid we take a perfect crystal, where the ions are arranged in a periodic lattice. We assume that the influence of this periodic arrangement of ions on a given crystal electron can be expressed in the form of an effective periodic *lattice potential* $V_0(\mathbf{r})$ which contains the mean field of the nuclei and all the other electrons. We are interested in the wavefunctions and allowed energy states of a single crystal electron, all many-electron effects will be discussed later.

Microscopically the lattice potential $V_0(\mathbf{r})$ results from the superposition of the Coulomb potentials of the nuclei and the inner electrons of the ions. However, we never need the explicit form of $V_0(\mathbf{r})$, we only utilize some general features, such as the symmetry and periodicity properties of the potential which reflect the structure of the crystal lattice.

The periodicity of the effective lattice potential is expressed by the translational symmetry:

$$V_0(\mathbf{r}) = V_0(\mathbf{r} + \mathbf{R}_n) \ , \tag{1}$$

where $\mathbf{R}_n$ is a lattice vector, i.e., a vector which connects two identical sites in an infinite lattice, which are n lattice cells apart. To make use of the lattice periodicity, we introduce an operator T_n which when acting on a function $f(\mathbf{r})$ adds a lattice vector $\mathbf{R}_n$ to the argument of the function. Applying T_n to the wavefunction Ψ of an electron in the periodic potential $V_0(\mathbf{r})$ yields

$$T_n \Psi(\mathbf{r}) = \Psi(\mathbf{r} + \mathbf{R}_n) = t_n \Psi(\mathbf{r}) \ , \tag{2}$$

where t_n is a phase factor, because the electron probability distributions $|\Psi(\mathbf{r})|^2$ and $|\Psi(\mathbf{r} + \mathbf{R}_n)|^2$ have to be identical. Since the Hamilton operator

$$H = \frac{p^2}{2m_0} + V_0(\mathbf{r}) \tag{3}$$

has the full lattice symmetry, the commutator of H and T_n vanishes:

$$[H, T_n] = HT_n - T_n H = 0 \ . \tag{4}$$

Under this condition a complete set of functions exists which are eigenfunctions to H and T_n:

$$H\Psi_\lambda(\mathbf{k}, \mathbf{r}) = E_\lambda \Psi_\lambda(\mathbf{k}, \mathbf{r}) \tag{5}$$

and Eq (2) have to be satisfied simultaneously. Here we identified $\mathbf{k}$ as the quantum number associated with the translation operator and

$$t_n = e^{i(\mathbf{k} \cdot \mathbf{R}_n + 2\pi N)} \ , \tag{6}$$

where 2π is an allowed additional factor because

$$e^{i2\pi N} = 1, \text{ for } N = \text{ integer } . \tag{7}$$

Considerations along the lines of Eqs. (1) - (7) led F. Bloch to formulate the following theorem, that is now known as the *Bloch theorem*:

$$e^{i\mathbf{k} \cdot \mathbf{R}_n} \Psi_\lambda(\mathbf{k}, \mathbf{r}) = \Psi_\lambda(\mathbf{k}, \mathbf{r} + \mathbf{R}_n) \ . \tag{8}$$

To satisfy relation (8), we make the ansatz

$$\Psi_\lambda(\mathbf{k}, \mathbf{r}) = \frac{e^{i\mathbf{k} \cdot \mathbf{r}}}{L^{3/2}} u_\lambda(\mathbf{k}, \mathbf{r}) \ , \tag{9}$$

where L^3 is the volume of the crystal. Eq. (9) defines the *Bloch wavefunction*. We see that Eq. (9) fulfills the Bloch theorem (8) only, if the Bloch function u_λ is periodic in real space

$$u_\lambda(\mathbf{k}, \mathbf{r}) = u_\lambda(\mathbf{k}, \mathbf{r} + \mathbf{R}_n) \ . \tag{10}$$

The function multiplying the lattice periodic function $u_\lambda(\mathbf{k}, \mathbf{r})$ in Eq. (9) is often denoted as *envelope function*. In the present case of a three-dimensional bulk material this envelope function is simply a plane wave, but it is significantly modified in semiconductor structures with a lower effective dimensionality, such as quantum wells, wires or dots (see Sec. 2-4 for more details).

2.2 Tight-Binding Approximation

To get some basic understanding for the occurrence of bandstructures, we discuss in this section a simple approximation for calculating the allowed energies and eigenfunctions of an electron in the crystal lattice. For this purpose we treat the so-called *tight-binding approximation* where we start from the electron wave-functions of the isolated atoms which form the crystal. We assume that the electrons stay close to the atomic sites and that the electronic wavefunctions centered around neighboring sites have little overlap. Consequently there is almost no overlap between wavefunctions for electrons that are separated by two or more atoms (next-nearest neighbors, next-next nearest neighbors, etc). The relevant overlap integrals decrease rapidly with increasing distance between the atoms at site m and l, so that only a few terms have to be taken into account.

The Schrödinger equation for a single atom located at the lattice point l is

$$H_0 \phi_\lambda(\mathbf{r} - \mathbf{R}_l) = \varepsilon_\lambda \phi_\lambda(\mathbf{r} - \mathbf{R}_l) \tag{11}$$

with the Hamiltonian

$$H_0 = -\frac{\hbar^2 \nabla^2}{2m_0} + V_0(\mathbf{r} - \mathbf{R}_l) \, , \tag{12}$$

where $V_0(\mathbf{r} - \mathbf{R}_l)$ is the potential of the l-th ion. The full problem of the periodic solid contains the sum of all individual ionic potentials,

$$\left(-\frac{\hbar^2 \nabla^2}{2m_0} + \sum_l V_0(\mathbf{r} - \mathbf{R}_l) - E_\lambda(\mathbf{k}) \right) \psi_\lambda(\mathbf{k}, \mathbf{r}) = 0 \, . \tag{13}$$

To solve Eq. (13) we make the ansatz

$$\psi_\lambda(\mathbf{k}, \mathbf{r}) = \sum_n \frac{e^{i \mathbf{k} \cdot \mathbf{R}_n}}{L^{3/2}} \phi_\lambda(\mathbf{r} - \mathbf{R}_n) \, . \tag{14}$$

This *tight-binding wavefunction* obviously satisfies the Bloch theorem, Eq. (8).

In order to compute the energy we have to evaluate

$$E_\lambda(\mathbf{k}) = \frac{\int d^3 r \, \psi_\lambda^\star(\mathbf{k}, \mathbf{r}) H \psi_\lambda(\mathbf{k}, \mathbf{r})}{\int d^3 r \, \psi_\lambda^\star(\mathbf{k}, \mathbf{r}) \psi_\lambda(\mathbf{k}, \mathbf{r})} = \frac{N}{D} \, , \tag{15}$$

where the numerator can be written as

$$N = \frac{1}{L^3} \sum_{n,m} e^{i \mathbf{k} \cdot (\mathbf{R}_n - \mathbf{R}_m)} \int d^3 r \, \phi_\lambda^\star(\mathbf{r} - \mathbf{R}_m) H \phi_\lambda(\mathbf{r} - \mathbf{R}_n) \tag{16}$$

and the denominator is

$$D = \frac{1}{L^3} \sum_{n,m} e^{i\mathbf{k}\cdot(\mathbf{R}_n - \mathbf{R}_m)} \int d^3r \phi_\lambda^\star(\mathbf{r} - \mathbf{R}_m) \phi_\lambda(\mathbf{r} - \mathbf{R}_n) \ . \tag{17}$$

Since we assume strongly localized electrons, the integrals decrease rapidly with increasing distance between sites n and m. The leading contribution is $n = m$, then $n = m \pm 1$, etc. In our final result we only want to keep the leading order of the complete expression (15). Therefore it is sufficient to approximate in the denominator:

$$\int d^3r \phi_\lambda^\star(\mathbf{r} - \mathbf{R}_m) \phi_\lambda(\mathbf{r} - \mathbf{R}_n) \simeq \delta_{n,m} \ , \tag{18}$$

so that

$$D = \sum_{n,m} \frac{\delta_{n,m}}{L^3} = \sum_n \frac{1}{L^3} = \frac{N}{L^3} \ . \tag{19}$$

We denote the integral in the numerator as

$$I = \int d^3r \phi_\lambda^\star(\mathbf{r} - \mathbf{R}_m) \left(-\frac{\hbar^2 \nabla^2}{2m_0} + \sum_l V_0(\mathbf{r} - \mathbf{R}_l) \right) \phi_\lambda(\mathbf{r} - \mathbf{R}_n) \ . \tag{20}$$

It can be approximated as follows

$$I = \delta_{n,m} \left(\sum_l \delta_{l,n} \varepsilon_\lambda + \sum_{l \neq n} \int d^3r \phi_\lambda^\star(\mathbf{r} - \mathbf{R}_n) V_0(\mathbf{r} - \mathbf{R}_l) \phi_\lambda(\mathbf{r} - \mathbf{R}_n) \right)$$
$$+ \delta_{n\pm1,m} \sum_l \int d^3r \phi_\lambda^\star(\mathbf{r} + \mathbf{R}_{n\pm1}) V_0(\mathbf{r} - \mathbf{R}_l) \phi_\lambda(\mathbf{r} + \mathbf{R}_n) + ...$$
$$\equiv \delta_{n,m} \varepsilon_\lambda' + \delta_{n\pm1,m} B_\lambda + ... \ , \tag{21}$$

where ε_λ' is the renormalized (shifted) atomic energy level and B_λ is the overlap integral. Energy shift and overlap integral for the states λ usually have to be determined numerically by evaluating the integral expressions in Eq. (21). The resulting numerical values and signs depend on details of the functions ϕ_λ and the potential V_0.

If we neglect the contributions of the next nearest neighbors in Eq. (21), we can write the total numerator as

$$N \simeq \frac{1}{L^3} \sum_{n,m} e^{i\mathbf{k}\cdot(\mathbf{R}_n - \mathbf{R}_m)} (\delta_{n,m} \varepsilon_\lambda' + \delta_{n\pm1,m} B_\lambda) \ . \tag{22}$$

Inserting Eqs. (19) and (22) into Eq. (15) we obtain

$$E_\lambda(\mathbf{k}) = \varepsilon'_\lambda + \frac{B_\lambda}{N} \sum_{n,m} \delta_{m,n\pm1} e^{i\mathbf{k}\cdot(\mathbf{R}_n - \mathbf{R}_m)} \ . \tag{23}$$

To analyze this result and to gain some insight into the formation of energy bands, we now restrict the discussion to the case of an ideal cubic lattice with lattice vector **a**, so that

$$\mathbf{R}_{n\pm1} = \mathbf{R}_n \pm \mathbf{a} \ . \tag{24}$$

Using Eq. (24) to evaluate the m summation in Eq. (23) we see that the exponentials combine as

$$e^{i\mathbf{k}\cdot\mathbf{a}} + e^{-i\mathbf{k}\cdot\mathbf{a}} = 2\cos(\mathbf{k}\cdot\mathbf{a}) \tag{25}$$

so that the n-summation simply yields a factor N and the final result for the energy is

$$E_\lambda(\mathbf{k}) = \varepsilon'_\lambda + 2B_\lambda \cos(\mathbf{k}\cdot\mathbf{a}) \ . \tag{26}$$

Eq. (26) describes the tight-binding cosine bands. Schematically two such bands are shown in Fig. 2.1, one for $B_\lambda > 0$ (lower band) and one for $B_\lambda < 0$ (upper band).

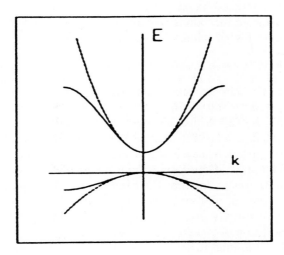

Fig. 2.1. Schematic drawing of the energy dispersion resulting from Eq. (26) for the cases of B_λ.) (lower band) and $B_\lambda < 0$ (upper band). The effective mass approximations, Eq. (27), are inserted as dashed lines. (From Haug and Koch 1994).

2.3 General Bandstructure Aspects

We can summarize some general bandstructure aspects in the following qualitative statements:

(i) The discrete atomic energy levels become quasi-continuous energy regions, called energy bands, with a certain band width.
(ii) There may be energy gaps between different bands.
(iii) Depending on the corresponding atomic functions, the bands $E_\lambda(\mathbf{k})$ may have positive or negative curvature around the band extrema.
(iv) In the vicinity of the band extrema one can often make a parabolic approximation

$$\varepsilon_{\lambda \mathbf{k}} \simeq \varepsilon_{\lambda,0} + \frac{\hbar^2 k^2}{2m_{\lambda,eff}}, \quad m_{\lambda,eff} = \frac{\hbar^2}{\frac{\partial^2 \varepsilon_{\lambda,\mathbf{k}}}{\partial k^2}|_{k=0}} . \tag{27}$$

In the regimes where the parabolic approximation is valid, the electrons can be considered quasi-free electrons but with an effective mass m_{eff}, which may be positive or negative, as indicated in Fig. 2.1. A large value of the overlap integral B_λ results in a wide band and correspondingly small effective mass $m_{\lambda,eff}$.

(v) Without considering correlation effects one can often assume that the states in the bands are filled according to the Pauli principle, beginning with the lowest states. The last completely filled band is called *valence band*. The next higher band is the *conduction band*.

There are three basic cases realized in nature:

(i) The conduction band is empty and separated by a large bandgap from the valence band. This defines an *insulator*. The electrons cannot be accelerated in an electric field since no empty states with slightly different $E(k)$ are available. Therefore we have no electrical conductivity.
(ii) An insulator with a relatively small bandgap is called a *semiconductor*. The definition of *small bandgap* is somewhat arbitrary, but a good operational definition is to say that the bandgap should be on the order of or less than an optical photon energy. In semiconductors electrons can be moved relatively easily from the valence band into the conduction band, e.g., by absorption of visible or infrared light.
(iii) If the conduction band is partly filled, we have a finite electrical conductivity and hence a *metal*.

Realistic bandstructure calculations are based on sophisticated schemes, often using self-consistent approximations for the ionic potentials and/or local density approximations for the electrons. As an example for the result of such a calculation we show in Fig. 2.2 the electronic bandstructure for GaAs. This

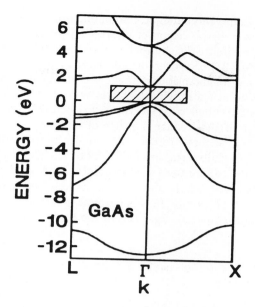

Fig. 2.2. GaAs bandstructure. The x axis shows the **k** values in different directions of the first Brillouin zone, where Γ, X and L are high symmetry points. The hatched region is the region of interest for most optical transitions. (From Chow and Koch 1998)

material is a *direct bandgap* semiconductor, for which the conduction-band energy minimum and the valence-band energy maximum both occur at $k = 0$. If the band extrema are at different momentum values, the semiconductor has an indirect bandgap. Most III-V and II-VI compounds (the numerals refer to columns in the Periodic Table) are direct bandgap materials, whereas Si and Ge are examples of indirect bandgap materials.

Fortunately, for optical transitions with frequencies in the visible or near infrared, it is often not necessary to deal with the complete semiconductor bandstructure. First of all, optical transitions are direct transitions, i.e., the momenta of the initial and final electronic states are essentially equal because of the smallness of the photon momentum

$$\hbar K = \frac{\hbar \omega n_b}{c} \ . \tag{28}$$

Here $\hbar\omega$ is the photon energy, K is the magnitude of the photon wavevector, n_b is the background refractive index of the semiconductor, and c is the speed of light. For GaAs, $\hbar\omega \approx 1.4 eV$ and $n_b \approx 3.6$, so that $K \approx 2.54 \cdot 10^7/m$, which is negligible on the scale ($\frac{2\pi}{a} \approx 10^{10}/m$) of the electronic band structure. Therefore, we only need to consider a small region of the band structure around the bandgap minimum, where optical transitions are most likely to occur. If the region of interest is sufficiently small, it is often reasonable to approximate the energy bands in that region using the parabolic approximation, Eq. (27). However, deviations from this simple approximation are often important for quantitative theory/experiment comparisons.

Another important simplifying factor is that all energetically low bands which are completely filled with electrons do not contribute directly to the optical transitions in the frequency range of interest. Hence, the electronic band structure that we have to consider usually involves only a very small portion of the entire band structure indicated by the hatched area in Fig. 2.2.

For many model calculations we can restrict our band model to one (or at most a few) valence bands and the conduction band, including only k-values around $k = 0$ (Γ point). In this parameter region we can often use the effective-mass approximation, Eq. (27), in the form

$$\varepsilon_{c\mathbf{k}} = \frac{\hbar^2 k^2}{2m_c} + \varepsilon_{g0} \;, \tag{29}$$

$$\varepsilon_{v\mathbf{k}} = \frac{\hbar^2 k^2}{2m_v} \;. \tag{30}$$

Here m_c and m_v are the effective masses of the electrons in the conduction and valence bands, respectively, and ε_{g0} is the bandgap energy in the absence of excited electrons.

Light absorption in a semiconductor promotes an electron out of the full valence band into the empty conduction band, leaving behind a missing electron in the valence band. For simplicity, we refer to the conduction electrons simply as electrons and the missing valence-band electrons as holes. The conduction band electron has the charge $(-e)$, and the hole, i.e. the missing electron can be characterized by the charge $-(-e) = +e$, i.e. opposite that of the electron.

If we ignore the photon momentum in an optical transition, the transition energy at the carrier momentum $\mathbf{k}$ is given by

$$\hbar \omega_\mathbf{k} = \varepsilon_{e\mathbf{k}} + \varepsilon_{h\mathbf{k}} + \varepsilon_{g0} \;, \tag{31}$$

where the electron and hole energies are

$$\varepsilon_{e\mathbf{k}} = \frac{\hbar^2 k^2}{2m_e} \;,$$

$$\varepsilon_{h\mathbf{k}} = \frac{\hbar^2 k^2}{2m_h} \;, \tag{32}$$

and m_e and m_h are the effective masses of the electron and hole, respectively. The electron mass m_e equals m_c. In this electron-hole description of a semiconductor, the energy of the hole is the energy of the completely filled valence band minus the energy of the valence band with a vacant electronic state. Hence, an increase in the hole momentum leads to an increase in the hole energy. Therefore, whereas the effective electron mass in the valence band is negative, the effective hole mass is positive.

The resonance energies for the optical transitions can be changed by the Coulomb interaction, which for low densities leads to the creation of *excitons* (see Sec. 3.4 for details). Here the Coulomb attraction can bind an excited electron and hole pair into an exciton, which is a hydrogen-like "atom" with a finite lifetime. The exciton lives are terminated through electron-hole recombination, which transfers the exciton energy to light (*radiative recombination*), or to the lattice, impurities, etc. (*nonradiative recombination*). By replacing the proton mass by the reduced electron-hole mass, we can use the Bohr hydrogen model to describe an exciton. The radius of the lowest exciton state is given by the exciton Bohr radius

$$a_0 = \frac{\hbar^2 \epsilon_b}{e^2 m_r}, \tag{33}$$

and the energy of the lowest state is given by the exciton Rydberg energy

$$\varepsilon_R = \frac{\hbar^2}{2 m_r a_0^2}, \tag{34}$$

where ϵ_b is the background dielectric constant and m_r is the reduced electron-hole mass defined by

$$\frac{1}{m_r} = \frac{1}{m_e} + \frac{1}{m_h}.$$

In GaAs, $a_0 = 124\text{Å}$ compared to 0.5Å in the H atom, and $\varepsilon_R = 4.2meV$, which is tiny compared with $13.6eV$ for the H atom and small compared to room temperature thermal energy $k_B T \approx 25meV$. Whether excitons are important in the description of semiconductor behavior depends on a_0 compared to the mean distance between electron-hole pairs and the screening length, and ε_R compared to $k_B T$. The screening length is a measure of the effectiveness of the screening of the Coulomb interaction between two carriers by other carriers. As the carrier density increases (due to an injection current or optical absorption), the Coulomb potential becomes increasingly screened, and for sufficiently high densities the excitons are completely ionized. Similarly, for increasing density the mean particle separation decreases, leading to increasing overlap of the electrons and holes in the excitons. Since electrons and holes are Fermions, each quantum state cannot be occupied by more than one particle (*Pauli exclusion principle*). Hence, different electrons (holes) compete for the available phase space. *Phase-space filling* effectively reduces the electron-hole attraction, quite similar to screening.

2.4 Quantum Confinement Effects

GaAs does not occur in nature and as such can be considered a "designer material". Modern crystal growth techniques make it possible to determine not only the composition of semiconductors with remarkable precision, but one can also determine their shape virtually on an atomic scale. In particular, it is possible to fabricate microstructures so small that their electronic and optical properties deviate substantially from those of bulk materials. The onset of pronounced quantum confinement effects occurs when one or more dimensions of a structure become comparable to the characteristic length scale of the elementary excitations. Quantum confinement may be in one spatial dimension, as in *quantum wells*, in two spatial dimensions as in *quantum wires*, or in all three spatial dimensions as in *quantum dots*. The confinement modifies the allowed energy states of the crystal electrons and changes the density of states. In this section, we introduce the basic properties of quantum-well structures. For the finer but still important modifications to the quantum-well bandstructure, we refer to the literature (Bastard 1988; Chow and Koch 1998).

A basic understanding of quantum-well confinement effects is obtained most easily by considering *ideal quantum confinement* conditions, for which the elementary excitations are completely confined inside the microstructure and the electronic wavefunctions vanish beyond the surfaces. For this idealized situation, we can write the confinement potential as

$$V_{con}(z) = \left\{ \begin{array}{l} 0 \mid z \mid < L_c/2 \\ \infty \mid z \mid > L_c/2 \end{array} \right\} . \tag{35}$$

In the xy plane there is no quantum confinement and the carriers can move freely. The electron eigenfunction (actually the envelope of the crystal electron eigenfunction) can be separated as

$$\psi_{n,\mathbf{k}_\perp}(\mathbf{r}) = \phi_{\mathbf{k}_\perp}(\mathbf{r}_\perp)\zeta_n(z) , \tag{36}$$

where the z and transverse components $\mathbf{r}_\perp$ (x,y) obey the Schrödinger equations

$$[-\frac{\hbar^2}{2m_z}\frac{d^2}{dz^2} + V_{con}(z)]\zeta_n(z) = E_n \zeta_n(z) , \tag{37}$$

and

$$-\frac{\hbar^2}{2m_\perp}\nabla_\perp^2 \phi_{\mathbf{k}_\perp}(\mathbf{r}_\perp) = E_{\mathbf{k}_\perp}\phi_{\mathbf{k}_\perp}(\mathbf{r}_\perp) , \tag{38}$$

respectively. For simplicity, we assume that the bulk-material bandstructure can be described by parabolic energy bands that are characterized by the effective masses m_z and $m_\perp$, which are equal for the conduction bands. However, for GaAs like materials they differ for the valence bands, which leads to the interesting property of mass reversal (see e.g. Bastard 1988). Equation (38) describes a two dimensional free particle (i.e., no external potential and not interacting with other particles) with eigenfunctions

$$\phi_{\mathbf{k}_\perp}(\mathbf{r}_\perp) = \frac{1}{\sqrt{L_c}} e^{\pm i \mathbf{k} \cdot \mathbf{r}_\perp} \tag{39}$$

and eigenvalue

$$E_{\mathbf{k}_\perp} = \frac{\hbar^2 k_\perp^2}{2 m_\perp} . \tag{40}$$

Because of the infinite confinement potential, we have the boundary conditions

$$\zeta_n \left(\frac{L_c}{2} \right) = \zeta_n \left(-\frac{L_c}{2} \right) = 0 , \tag{41}$$

which lead to the even and odd solutions of Eq. (37)

$$\zeta_n(z) = \sqrt{\frac{2}{L_c}} \cos(k_n z) \ , n \ even \ , \tag{42}$$

$$\zeta_n(z) = \sqrt{\frac{2}{L_c}} \sin(k_n z) \ , n \ odd \ , \tag{43}$$

where the wave numbers k_n are given by

$$k_n = \frac{n\pi}{L_c} , \tag{44}$$

and the bound state energies E_n are

$$E_n = \frac{\hbar^2 k_n^2}{2 m_z} = \frac{\pi^2 \hbar^2 n^2}{2 m_z L_c^2} . \tag{45}$$

Adding the energies of the motion in the xy plane and in the z-direction, we find the total energy of the electron subjected to one-dimensional quantum confinement to be

$$E = \frac{\pi^2 \hbar^2 n^2}{2 m_z L_c^2} + \frac{\hbar^2 k_\perp^2}{2 m_\perp} , \tag{46}$$

where $n = 1, 2, 3, ...$, indicating a succession of energy subbands, i.e., energy parabola $\hbar^2 k_\perp^2 / 2m_\perp$ separated by $\pi^2 \hbar^2 / 2 m_z L_c^2$. The different subbands are labeled by the quantum numbers n. Figure 2.3 depicts the energy eigenstates.

Realistically, we can only fabricate finite confinement potentials, so that

$$V_{con}(z) = \begin{Bmatrix} 0 \ | \ z \ |< L_c/2 \\ V_c \ | \ z \ |> L_c/2 \end{Bmatrix} . \tag{47}$$

The analysis for this case follows closely the treatment of the infinite potential, with the Schrödinger equation for the $x - y$ motion being unchanged. However, the solutions in the z-direction can no longer be determined analytically.

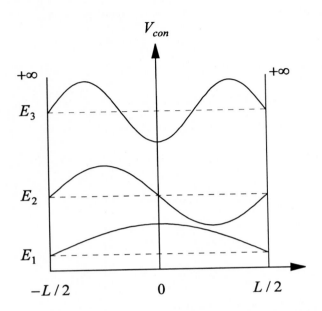

Fig. 2.3. The first three energy eigenstates of the one dimensional confinement potential, Eq. (35). (From Chow and Koch 1998)

2.5 Second Quantization

This section briefly summarizes the main rules of the second quantized (or Fock) representation that is very useful to compute the electronic and optical properties of the semiconductor medium. This method is well adapted to treat systems of indistinguishable particles with varying particle number.

We introduce the creation and annihilation operators for electrons in a state which we specify by the band index λ, the momentum **k**, and the z-component of the spin, s_z. The annihilation operator is then $a_{\lambda \mathbf{k} s_z}(t)$. Its Hermitean adjoint $a^{\dagger}_{\lambda \mathbf{k} s_z}$ *creates* an electron in the same state.

For the sake of clarity, we write the operators with all their indices appearing explicitly. In the later chapters we usually incorporate the spin variable into **k** for typographical simplicity. In that case, the subscript **k** represents the three-dimensional momentum vector **k**, and two possible spin directions $s_z = \pm \frac{1}{2}$. The summation over **k** then involves summations over k_x, k_y, k_z, and s_z.

Since the crystal electrons are Fermions, the creation and annihilation operators obey *anticommutation* relations. These relations are a consequence of the Pauli exclusion principle, which states that at most one Fermion can occupy any given state. The anticommutation relations for the electron creation and annihilation operators are

$$[a_{\lambda k s_z}, a_{\lambda' k' s'_z}]_+ = [a^\dagger_{\lambda k s_z}, a^\dagger_{\lambda' k' s'_z}]_+ = 0 \qquad (48)$$

$$[a_{\lambda k s_z}, a^\dagger_{\lambda' k' s'_z}]_+ = \delta_{\lambda,\lambda'} \delta_{k,k'} \delta_{s_z s'_z} , \qquad (49)$$

where for two operators, A and B, the anticommutator is defined by

$$[A, B]_+ = AB + BA .$$

The combination $a^\dagger_{\lambda k s_z} a_{\lambda k s_z}$ is the number operator for an electron in band λ with momentum $\mathbf{k}$ and s_z. The eigenstates for $a^\dagger_{\lambda k s_z} a_{\lambda k s_z}$ are $|0_{\lambda k s_z}\rangle$ and $|1_{\lambda k s_z}\rangle$, which are the states containing no electron and one electron, respectively. These eigenstates when operated on by the creation, annihilation and number operators give

$$a_{\lambda k s_z} |0_{\lambda k s_z}\rangle = a^\dagger_{\lambda k s_z} |1_{\lambda k s_z}\rangle = 0$$
$$a_{\lambda k s_z} |1_{\lambda k s_z}\rangle = |0_{\lambda k s_z}\rangle$$
$$a^\dagger_{\lambda k s_z} |0_{\lambda k s_z}\rangle = |1_{\lambda k s_z}\rangle$$
$$a^\dagger_{\lambda k s_z} a_{\lambda k s_z} |0_{\lambda k s_z}\rangle = 0$$
$$a^\dagger_{\lambda k s_z} a_{\lambda k s_z} |1_{\lambda k s_z}\rangle = |1_{\lambda k s_z}\rangle , \qquad (50)$$

where the first equation expresses the fact that it is impossible to create an electron in an already filled state.

In the electron-hole representation for a two-band model, we define the hole creation operator

$$b^\dagger_{-\mathbf{k},-s_z} = a_{v k s_z} . \qquad (51)$$

This equation indicates that the annihilation of a valence-band electron with a given momentum and z-component of spin corresponds to the creation of a hole with the opposite momentum and z-component of spin. Note that for clarity in Eq. (51) we use a comma between the $-\mathbf{k}$ and $-s_z$ subscripts, although it is probably clear without the comma since it does not make sense to subtract a spin quantum number from a wavevector $\mathbf{k}$. Similarly the hole annihilation operator is given by

$$b_{-\mathbf{k},-s_z} = a^\dagger_{v k s_z} . \qquad (52)$$

The hole operators also obey anticommutation relationships, so that the probability of finding a particular valence-band electron becomes

$$\langle a^\dagger_{v\mathbf{k}s_z} a_{v\mathbf{k}s_z} \rangle = 1 - \langle b^\dagger_{-\mathbf{k},-s_z} b_{-\mathbf{k},-s_z} \rangle \;, \tag{53}$$

where the brackets $\langle ... \rangle$ are used to indicate an expectation value. The probability of finding a valence electron is one *minus* the probability of finding a hole. The electron annihilation operator is

$$a_{\mathbf{k}s_z} = a_{c\mathbf{k}s_z} \tag{54}$$

and the electron creation operator is

$$a^\dagger_{\mathbf{k}s_z} = a^\dagger_{c\mathbf{k}s_z} \;. \tag{55}$$

In the second quantized representation, the Hamiltonian for N non-interacting electrons is

$$H_{kin} = \sum_\mathbf{k} \sum_{s_z} \left[\varepsilon_{c\mathbf{k}} a^\dagger_{c\mathbf{k}s_z} a_{c\mathbf{k}s_z} + \varepsilon_{v\mathbf{k}} a^\dagger_{v\mathbf{k}s_z} a_{v\mathbf{k}s_z} \right] \;. \tag{56}$$

where we used the two-band approximation i.e. $\lambda = c, v$. The index kin in Eq. (56) indicates that this is the kinetic energy part of the full Hamiltonian of the interacting system. In the electron-hole representation, Eq. (56) becomes

$$H_{kin} = \sum_\mathbf{k} \sum_{s_z} \left[\varepsilon_{c\mathbf{k}} a^\dagger_{\mathbf{k}s_z} a_{\mathbf{k}s_z} + \varepsilon_{v\mathbf{k}} \left(1 - b^\dagger_{-\mathbf{k},-s_z} b_{-\mathbf{k},-s_z} \right) \right] \;. \tag{57}$$

Since the origin of energy is arbitrary, the constant term, $\sum_{\mathbf{k}s_z} \varepsilon_{v\mathbf{k}}$ is usually left out. Then

$$H_{kin} = \sum_\mathbf{k} \sum_{s_z} (\varepsilon_{g0} + \varepsilon_{e\mathbf{k}}) a^\dagger_{\mathbf{k}s_z} a_{\mathbf{k}s_z} + \sum_\mathbf{k} \sum_{s_z} \varepsilon_{h\mathbf{k}} b^\dagger_{\mathbf{k}s_z} b_{\mathbf{k}s_z} \;, \tag{58}$$

where $\varepsilon_{e\mathbf{k}}$ and $\varepsilon_{h\mathbf{k}}$ are given by Eq. (32). In going from Eq. (56) to Eq. (58) we set $m_h = -m_v$, where m_v is the valence electron effective mass and m_h is the hole effective mass.

An example of a physical quantity that is represented by an operator is the particle number operator

$$N = \sum_{\mathbf{k},s_z} n_{\mathbf{k},s_z} = \sum_{\mathbf{k},s_z} a^\dagger_{\mathbf{k}s_z} a_{\mathbf{k}s_z} \;. \tag{59}$$

If we want to study the dynamics of the carrier distribution we have to solve the equation of motion for $n_{\mathbf{k},s_z}$. For this purpose we use the Heisenberg picture, where the operator O obeys the equation of motion (Heisenberg equation)

$$i\hbar \frac{dO}{dt} = [O, H] = OH - HO \;. \tag{60}$$

References

The material in this and the following two chapters is adapted from:

Chow, W. W. and S. W. Koch (1998), *Semiconductor Laser Fundamentals*, Springer Verlag, Berlin.

Haug, H., and S. W. Koch (1994), *Quantum Theory of the Optical and Electronic Properties of Semiconductors*, 3rd Edition, World Scientific Publ., Singapore.

Bandstructure theory, envelope function approximation and much more is discussed e.g. in:

Callaway, J. (1976), *Quantum Theory of the Solid State*, Academic Press, New York.

Bastard, G. (1988), *Wave Mechanics Applied to Semiconductor Heterostructures*, Les Editions de Physique, France.

3 Optical Response of Semiconductors

In our analysis of the optical semiconductor properties we often want to compute the optical absorption, refractive index etc. For this purpose we need the connection between the microscopic theory for the material and the solution of Maxwell's wave equation. For this purpose we introduce quantities such as the dielectric function and/or the complex susceptibility. More generally, and in particular for dynamical calculations we need the optical polarization $\mathbf{P}(\mathbf{r},t)$. The computation of this quantity is the main focus of the microscopic theory of the optical semiconductor medium properties. In this chapter we start with the elementary oscillator model to introduce the relevant optical quantities (Sec. 3.1). In Sec. 3.2 we compute the optical response for a system of noninteracting (free) carriers before we proceed to derive the semiconductor Bloch equations for the interacting electron-hole system (Sec. 3.3). In Sec. 3.4 we evaluate these equations in the linear regime to study the excitonic signatures of the semiconductor bandgap absorption.

3.1 Optical Response

In order to get some first insights we use the model of a single oscillator which allows us to introduce some basic properties of light-matter interaction. The oscillator is assumed to be a model for an electron with a charge e which can be displaced from its equilibrium position. For the case of an electric field polarized in x direction we introduce a polarization, defined as dipole moment per unit volume, as

$$P = -n_0 e x \; , \tag{1}$$

where ex is the electric dipole moment, and n_0 is the mean electron density per unit volume. Describing the electron under the influence of the electric field $E(t)$ (parallel to x) as a damped driven oscillator, we can write Newton's equation as

$$m\frac{d^2 x}{dt^2} = -2m\gamma \frac{dx}{dt} - m\omega_0^2 x - eE(t) \; , \tag{2}$$

where γ is the damping constant, and m and ω_0 are the mass and resonance frequency of the oscillator, respectively. Assuming a monochromatic field

$$E(t) = E(\omega)e^{-i\omega t} \; , \tag{3}$$

and using

$$x(t) = x(\omega)e^{-i\omega t} \tag{4}$$

we get from Eq. (2)

$$m(\omega^2 + i2\gamma\omega - \omega_0^2)x(\omega) = eE(\omega) \tag{5}$$

and from Eq. (1)

$$P(\omega) = -\frac{n_0 e^2}{m} \frac{1}{\omega^2 + i2\gamma\omega - \omega_0^2} E(\omega). \tag{6}$$

The coefficient between $P(\omega)$ and $E(\omega)$ is defined as the complex optical susceptibility $\chi(\omega)$. We obtain

$$\chi(\omega) = -\frac{n_0 e^2}{2m\omega_0} \left[\frac{1}{\omega + i\gamma - \omega_0'} - \frac{1}{\omega + i\gamma + \omega_0'} \right] \tag{7}$$

where

$$\omega_0' = \sqrt{\omega_0^2 - \gamma^2} \tag{8}$$

is the renormalized (shifted) resonance frequency of the damped harmonic oscillator.

In general, the optical susceptibility is a tensor relating different vector components of the polarization P_i and the electric field E_i. An important feature of $\chi(\omega)$ is that it becomes singular at

$$\omega = -i\gamma \pm \omega_0'. \tag{9}$$

This relation can only be satisfied if we formally consider complex frequencies $\omega = \omega' + i\omega''$. We see from Eq. (7) that $\chi(\omega)$ has poles in the lower half of the complex frequency plane, i.e. for $\omega'' < 0$, but it is an analytic function on the real frequency axis and in the whole upper half plane. This property of the susceptibility can be related to causality, i.e., to the fact that the polarization $P(t)$ at time t can only be influenced by fields $E(t-\tau)$ acting at earlier times, i.e., $\tau \geq 0$. For more details see e.g. Haug and Koch (1994).

The macroscopic Maxwell's equations can be written as

$$\text{curl } \mathbf{B}(\mathbf{r},t) = \frac{1}{c}\frac{\partial}{\partial t}\mathbf{D}(\mathbf{r},t) \tag{10}$$

$$\text{curl } \mathbf{E}(\mathbf{r},t) = \frac{1}{c}\frac{\partial}{\partial t}\mathbf{H}(\mathbf{r},t) \tag{11}$$

The Fourier transformed displacement field $D(\omega)$ can be expressed in terms of the polarization and electric field (in cgs units)

$$D(\omega) = E(\omega) + 4\pi P(\omega) = [1 + 4\pi\chi(\omega)] E(\omega) = \epsilon(\omega)E(\omega) \tag{12}$$

where the optical (or transverse) *dielectric function* $\epsilon(\omega)$ is obtained from the optical susceptibility (7) as

$$\epsilon(\omega) = 1 + 4\pi\chi(\omega) = 1 - \frac{\omega_{pl}^2}{2\omega_0}\left(\frac{1}{\omega + i\gamma - \omega_0'} - \frac{1}{\omega + i\gamma + \omega_0'}\right). \tag{13}$$

Here ω_{pl} denotes the plasma frequency of an electron plasma with mean density n_0,

$$\omega_{pl} = \sqrt{\frac{4\pi n_0 e^2}{m}} \ . \tag{14}$$

The plasma frequency is the eigenfrequency of the electron plasma density oscillations around the position of the ions.

The frequency dependent dielectric function $\epsilon(\omega)$ has poles at $\omega = \pm\omega_0' - i\gamma$, describing the resonant and the nonresonant part, respectively. If we are interested in the optical response in the spectral region around ω_0 and if ω_0 is sufficiently large, the nonresonant part gives only a small contribution and it is often a good approximation to neglect it completely. In order to simplify the resulting expressions we now consider only the resonant part of the dielectric function and assume $\omega_0 \gg \gamma$, so that $\omega_0 \cong \omega_0'$ and

$$\epsilon(\omega) = 1 - \frac{\omega_{pl}^2}{2\omega_0} \frac{1}{\omega + i\gamma - \omega_0} \ . \tag{15}$$

For the real part of the dielectric function we thus get the relation

$$\epsilon'(\omega) - 1 = -\frac{\omega_{pl}^2}{2\omega_0} \frac{\omega - \omega_0}{(\omega - \omega_0)^2 + \gamma^2} \ , \tag{16}$$

while the imaginary part has the following resonance structure

$$\epsilon''(\omega) = \frac{\omega_{pl}^2}{4\omega_0} \frac{2\gamma}{(\omega - \omega_0)^2 + \gamma^2} \ . \tag{17}$$

Examples of the spectral variations described by Eqs. (16) and (17) are shown in Fig. 3.1. The spectral shape of the imaginary part is determined by the Lorentzian lineshape function $2\gamma/((\omega - \omega_0)^2 + \gamma^2)$. It decreases asymptotically like $1/(\omega - \omega_0)^2$, while the real part of $\epsilon(\omega)$ decreases like $1/(\omega - \omega_0)$ far away from the resonance.

In order to understand the physical content of the formulae for $\epsilon'(\omega)$ and $\epsilon''(\omega)$, we consider how a light beam propagates in the dielectric medium. From Maxwell's equations (10) and (11) with $\mathbf{B}(\mathbf{r}, t) = \mathbf{H}(\mathbf{r}, t)$, which holds at optical frequencies, we obtain

$$\text{curl curl } \mathbf{E}(\mathbf{r}, t) = -\frac{\partial}{\partial t} \text{curl } \mathbf{H}(\mathbf{r}, t) = -\frac{1}{c^2} \frac{\partial^2}{\partial t^2} \mathbf{D}(\mathbf{r}, t) \ . \tag{18}$$

Using curl curl = grad div $-\Delta$, we get for a transverse electric field with div$\mathbf{E}(\mathbf{r}, t) = 0$, the wave equation

$$\Delta \mathbf{E}(\mathbf{r}, t) - \frac{1}{c^2} \frac{\partial^2}{\partial t^2} \mathbf{D}(\mathbf{r}, t) = 0 \ . \tag{19}$$

Here $\Delta \equiv \nabla^2$ is the Laplace operator. A Fourier transformation of Eq. (19) with respect to time yields

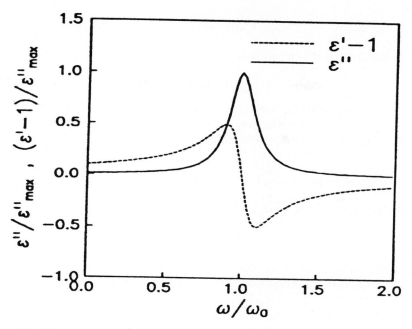

Fig. 3.1. Dispersion of the real and imaginary part of the dielectric function. (From Haug and Koch 1994)

$$\Delta \mathbf{E}(\mathbf{r},\omega) + \frac{\omega^2}{c^2}\epsilon'(\omega)\mathbf{E}(\mathbf{r},\omega) + i\frac{\omega^2}{c^2}\epsilon''(\omega)\mathbf{E}(\mathbf{r},\omega) = 0 \ . \qquad (20)$$

For a plane wave propagating with wavenumber $k(\omega)$ and extinction coefficient $\kappa(\omega)$ in the z direction,

$$\mathbf{E}(\mathbf{r},\omega) = \mathbf{E}_0(\omega)e^{i[k(\omega)+i\kappa(\omega)]z} \ , \qquad (21)$$

we get from Eq. (20)

$$[k(\omega) + i\kappa(\omega)]^2 = \frac{\omega^2}{c^2}[\epsilon'(\omega) + i\epsilon''(\omega)] \ . \qquad (22)$$

Separating real and imaginary part of this equation yields

$$k^2(\omega) - \kappa^2(\omega) = \frac{\omega^2}{c^2}\epsilon'(\omega) \ , \qquad (23)$$

$$2\kappa(\omega)k(\omega) = \frac{\omega^2}{c^2}\epsilon''(\omega) \ . \qquad (24)$$

Next, we introduce the index of refraction $n(\omega)$ as the ratio between the wavenumber $k(\omega)$ in the medium and the vacuum wavenumber $k_0 = \omega/c$

$$k(\omega) = n(\omega)\frac{\omega}{c} \tag{25}$$

and the absorption coefficient $\alpha(\omega)$ as

$$\alpha(\omega) = 2\kappa(\omega) \ . \tag{26}$$

The absorption coefficient determines the decay of the intensity $I \propto |E|^2$ in real space. $1/\alpha$ is the length, over which the intensity decreases by a factor $1/e$. From Eqs. (23) - (26) we obtain the index of refraction

$$n(\omega) = \sqrt{\frac{1}{2}\left[\epsilon'(\omega) + \sqrt{\epsilon'^2(\omega) + \epsilon''^2(\omega)}\right]} \tag{27}$$

and the absorption coefficient

$$\alpha(\omega) = \frac{\omega}{n(\omega)c}\epsilon''(\omega) \ . \tag{28}$$

Hence, Eqs. (17) and (28) yield a Lorentzian absorption line, and Eqs. (16) and (27) describe the corresponding frequency-dependent index of refraction. Note that for $\epsilon''(\omega) \ll \epsilon'(\omega)$, which is usually true in semiconductors, Eq. (27) simplifies to

$$n(\omega) \cong \sqrt{\epsilon'(\omega)} \ . \tag{29}$$

Furthermore, if the refractive index $n(\omega)$ is only weakly frequency dependent for the ω-values of interest, one may approximate Eq. (28) as

$$\alpha(\omega) \cong \frac{\omega}{n_b c}\epsilon''(\omega) = \frac{4\pi\omega}{n_b c}\chi''(\omega) \ , \tag{30}$$

where n_b is again the background refractive index.

3.2 Free-Carrier Equations of Motion

As a first example for a many electron system we compute the optical response of a system of noninteracting carriers. Clearly, since this free-carrier approximation ignores the important effect of the Coulomb interaction among the carriers it is never justified for real semiconductors, however, many of its results are useful for our full calculation in the following sections. The free-carrier Hamiltonian

$$\begin{aligned}H &= H_{kin} + H_{c-f} \\ &= \sum_{\mathbf{k}}\left[\varepsilon_{c\mathbf{k}}a^\dagger_{c\mathbf{k}}a_{c\mathbf{k}} + \varepsilon_{v\mathbf{k}}a^\dagger_{v\mathbf{k}}a_{v\mathbf{k}}\right] \\ &\quad - \sum_{\mathbf{k}}\left[\mu_{\mathbf{k}}a^\dagger_{c\mathbf{k}}a_{v\mathbf{k}} + \mu^*_{\mathbf{k}}a^\dagger_{v\mathbf{k}}a_{c\mathbf{k}}\right]E(z,t) \ ,\end{aligned} \tag{31}$$

where we used Eq. (2-56) for the kinetic part of the Hamiltonian. The quantity

$$\mu_{\mathbf{k}} = \mu_{cv}(\mathbf{k}) \tag{32}$$

in H_{c-f} is the dipole matrix element between the valence and conduction band. We assume a dipole interaction between the optical field and the carriers

$$H_{c-f} = -\mathbf{P} \cdot \mathbf{E} , \tag{33}$$

where the active medium polarization is given by the operator

$$\mathbf{P} = \sum_{\mathbf{k}} \left[\mu_{\mathbf{k}} a_{\mathbf{k}}^\dagger a_{v\mathbf{k}} + \mu_{\mathbf{k}}^* a_{v\mathbf{k}}^\dagger a_{\mathbf{k}} \right] . \tag{34}$$

In Eqs. (31)-(34), we used the two-band approximation, and absorbed the spin index into $\mathbf{k}$, so that $\sum_{\mathbf{k}}$ is actually $\sum_{\mathbf{k}s_z}$, and $a_{\alpha\mathbf{k}}$ and $a_{\alpha\mathbf{k}}^\dagger$ are actually $a_{\alpha\mathbf{k}s_z}$ and $a_{\alpha\mathbf{k}s_z}^\dagger$, respectively. In the electron-hole representation, using Eqs. (2-51) and (2-52) we can write

$$P = \sum_{\mathbf{k}} \left[\mu_{\mathbf{k}} a_{\mathbf{k}}^\dagger b_{-\mathbf{k}}^\dagger + \mu_{\mathbf{k}}^* b_{-\mathbf{k}} a_{\mathbf{k}} \right] , \tag{35}$$

and

$$H = \sum_{\mathbf{k}} \left[\left(\varepsilon_{g0} + \frac{\hbar^2 k^2}{2m_e} \right) a_{\mathbf{k}}^\dagger a_{\mathbf{k}} + \frac{\hbar^2 k^2}{2m_h} b_{-\mathbf{k}}^\dagger b_{-\mathbf{k}} \right]$$
$$- \sum_{\mathbf{k}} \left[\mu_{\mathbf{k}} a_{\mathbf{k}}^\dagger b_{-\mathbf{k}}^\dagger + \mu_{\mathbf{k}}^* b_{-\mathbf{k}} a_{\mathbf{k}} \right] E(z,t) , \tag{36}$$

where we assumed simple parabolic bands. We denote the unrenormalized bandgap energy ε_{g0} and reserve ε_g for the renormalized value obtained from many-body theory. With $\mu_{\mathbf{k}}$ without vector symbol we denote the projection of the dipole matrix element in field direction.

As discussed in the previous section of this chapter, we need the polarization, i.e. the expectation value of the polarization operator,

$$P(z,t) = \langle P \rangle \tag{37}$$

as link between the classical field and the quantum mechanical semiconductor medium. In the slowly varying amplitude and phase approximation we can write

$$P(z) = 2e^{-i[Kz - \nu t - \phi(z)]} \sum_{\mathbf{k}} \mu_{\mathbf{k}}^* p_{\mathbf{k}s_z} , \tag{38}$$

where

$$p_{\mathbf{k}} = \langle b_{-\mathbf{k}} a_{\mathbf{k}} \rangle . \tag{39}$$

The relevant quantities in the microscopic semiconductor medium theory are $p_\mathbf{k}$ together with the electron and hole occupation numbers

$$n_{e\mathbf{k}} = \langle a_\mathbf{k}^\dagger a_\mathbf{k} \rangle , \tag{40}$$

$$n_{h\mathbf{k}} = \langle b_{-\mathbf{k}}^\dagger b_{-\mathbf{k}} \rangle . \tag{41}$$

In the Heisenberg picture, the derivation of the equations of motion for the bilinear operators in Eqs. (39)-(41) involves the evaluation of the commutators as appearing in the Heisenberg equation of motion, Eq. (2-60). Performing the elementary calculations with the Hamiltonian (36) and the operators in Eqs. (40) and (41) we obtain

$$\frac{d}{dt} b_{-\mathbf{k}} a_\mathbf{k} = -i\omega_\mathbf{k} b_{-\mathbf{k}} a_\mathbf{k} - \frac{i}{\hbar} \mu_\mathbf{k} E(z,t) \left(a_\mathbf{k}^\dagger a_\mathbf{k} + b_{-\mathbf{k}}^\dagger b_{-\mathbf{k}} - 1 \right) \tag{42}$$

$$\frac{d}{dt} a_\mathbf{k}^\dagger a_\mathbf{k} = \frac{i}{\hbar} \left(\mu_\mathbf{k} a_\mathbf{k}^\dagger b_{-\mathbf{k}}^\dagger - \mu_\mathbf{k}^* b_{-\mathbf{k}} a_\mathbf{k} \right) E(z,t) \tag{43}$$

$$= \frac{d}{dt} b_{-\mathbf{k}}^\dagger b_{-\mathbf{k}} , \tag{44}$$

where the transition energy is

$$\hbar\omega_\mathbf{k} = \varepsilon_{g0} + \varepsilon_{e\mathbf{k}} + \varepsilon_{h\mathbf{k}} = \varepsilon_{g0} + \frac{\hbar^2 k^2}{2m_r} , \tag{45}$$

and m_r is the reduced mass:

$$\frac{1}{m_r} = \frac{1}{m_e} + \frac{1}{m_h} . \tag{46}$$

Taking the expectation values of Eqs. (42) through (44), yields

$$\frac{dp_\mathbf{k}}{dt} = -i\omega_\mathbf{k} p_\mathbf{k} - \frac{i}{\hbar} \mu_\mathbf{k} E(z,t) (n_{e\mathbf{k}} + n_{h\mathbf{k}} - 1) , \tag{47}$$

$$\frac{dn_{e\mathbf{k}}}{dt} = \frac{i}{\hbar} E(z,t) \left(\mu_\mathbf{k} p_\mathbf{k}^* - \mu_\mathbf{k}^* p_\mathbf{k} \right) , \tag{48}$$

$$= \frac{dn_{h\mathbf{k}}}{dt} . \tag{49}$$

In order to compute the free-carrier absorption spectrum we now assume that the carrier distribution is the quasi-equilibrium Fermi-Dirac distribution

$$n_{\alpha\mathbf{k}} = f_{\alpha\mathbf{k}} = \frac{1}{e^{\beta(\varepsilon_{\alpha\mathbf{k}} - \mu_\alpha)} + 1} , \tag{50}$$

where $\beta = 1/(k_B T)$, k_B is the Boltzmann constant, and μ_α is the quasi-chemical potential for electrons ($\alpha = e$) or holes ($\alpha = h$). The chemical potential is determined from the condition

$$N_\alpha = \sum_{\mathbf{k}} f_{\alpha\mathbf{k}} \Rightarrow \mu_\alpha(N_\alpha, T) \; , \qquad (51)$$

i.e. that the sum over the distribution function yields the total number of carriers. Usually, Eq. (51) has to be evaluated numerically, but analytical approximations exist for special situations (see e.g. Haug and Koch 1994). The carrier scattering processes leading to the quasi-equilibrium distribution (50) are discussed in Sec. 4.2. Here we simply take the carrier distribution as given and use the carrier density as an input parameter, which may be slowly time dependent if we are interested in such a situation. Hence, Eq. (47) in the quasi-equilibrium limit becomes

$$\frac{dp_\mathbf{k}}{dt} = -(i\omega_\mathbf{k} + \gamma)p_\mathbf{k} - \frac{i}{\hbar}\mu_\mathbf{k} E(z,t)(f_{e\mathbf{k}} + f_{h\mathbf{k}} - 1) \; . \qquad (52)$$

Here we phenomenologically added a finite damping rate γ which is the inverse dephasing time

$$T_2 = \frac{1}{\gamma} \; . \qquad (53)$$

Formally integrating Eq. (52) from $-\infty$ to t yields

$$p_\mathbf{k}(t) = -i \int_{-\infty}^t dt' e^{(i\omega_\mathbf{k}+\gamma)(t-t')} \frac{\mu_\mathbf{k} E(z,t')}{\hbar} [f_{e\mathbf{k}}(t') + f_{h\mathbf{k}}(t') - 1] \; . \qquad (54)$$

At this point we now make the *rate equation approximation* where we assume that the carrier distributions $f_{\alpha\mathbf{k}}(t)$ and the field envelope in

$$E(z,t) = \frac{E(z)}{2} e^{i[Kz - \nu t - \phi(z)]} \qquad (55)$$

vary little in the time T_2. Pulling these terms out of the time integral in Eq. (54) allows us to easily evaluate the remaining integration to get

$$p_\mathbf{k}(t) = -i\frac{\mu_\mathbf{k}}{2\hbar}[f_{e\mathbf{k}}(t) + f_{h\mathbf{k}}(t) - 1] E(z) \frac{e^{i[Kz-\nu t-\phi(z)]}}{i(\omega_\mathbf{k} - \nu) + \gamma} \; . \qquad (56)$$

Substituting this into Eq. (54) allows us to compute the optical polarization. Using the results of Sec. 3.1 we obtain the free-carrier absorption coefficient

$$\alpha_{free}(\nu) = \alpha_f \sum_\mathbf{k} |\mu_\mathbf{k}|^2 (1 - f_{e\mathbf{k}} - f_{h\mathbf{k}}) \delta(\omega_\mathbf{k} - \nu) \; , \qquad (57)$$

where α_f contains all the prefactors.

To see how this result depends on the dimensionality d of the system under consideration, we convert the sum over $\mathbf{k}$ to an integral over the energy $\varepsilon = \hbar^2 k^2/(2m)$. For $d = 3, 2, 1$ we get

$$\sum_{\mathbf{k}} = \sum_{\mathbf{k}} \frac{(\Delta k)^d}{(\Delta k)^d} \to 2 \left(\frac{L}{2\pi}\right)^d \int d^d k \ , \tag{58}$$

where the factor 2 results from the summation of s_z. In polar coordinates Eq. (58) can be written as

$$\sum_{\mathbf{k}} \to 2 \left(\frac{L}{2\pi}\right)^d \int d\Omega_d \int_0^\infty dk \, k^{d-1} \ , \tag{59}$$

where Ω_d is the space angle. For an isotropic integrand

$$\int d\Omega_d = \Omega_d = \begin{matrix} 4\pi \text{ for } d = 3 \\ 2\pi \text{ for } d = 2 \\ 2 \text{ for } d = 1 \end{matrix} \ . \tag{60}$$

Using

$$d\varepsilon = \frac{\hbar^2}{m} k \, dk \tag{61}$$

we finally get

$$\sum_{\mathbf{k}} F(k^2) \to \left(\frac{L}{2\pi}\right)^d \Omega_d \left(\frac{2m}{\hbar^2}\right)^{d/2} \int_0^\infty d\varepsilon \, \varepsilon^{(d-2)/2} F(\varepsilon) \ . \tag{62}$$

Using this result in Eq. (58) we obtain

$$\alpha_{free}(\nu) = \alpha'_f \Delta^{(d-2)/2} \Theta(\Delta) \left[1 - f_e(\nu) - f_h(\nu)\right] \ , \tag{63}$$

where

$$\Delta = \frac{\hbar\nu - \varepsilon_{g0} - E_0^{(d)}}{\varepsilon_R} \tag{64}$$

and α'_f again contains all irrelevant prefactors. The unit step function $\Theta(\Delta)$ in Eq. (63) makes sure that the absorption starts at

$$\hbar\nu = \varepsilon_{g0} + E_0^{(d)} \ , \tag{65}$$

i.e. at the bandgap energy plus the confinement energy in the d-dimensional system. Clearly, $E_0^{(d)} = 0$ for $d = 3$, and it is given by Eq. (2.46) with $n = 1$ and $k_\perp = 0$ for the two-dimensional quantum-well system. For a quadratic one-dimensional quantum wire we have twice the quantum-well confinement energy. An example of the resulting absorption coefficient is plotted in Fig. 3.2.

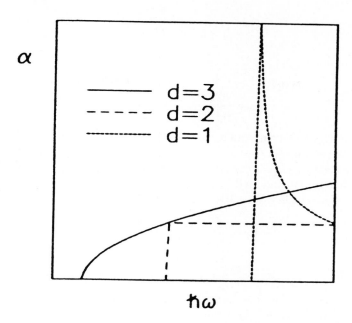

Fig. 3.2. Free carrier absorption spectra for semiconductors in which unrestricted carrier motion in three, two or one space dimension is possible. (From Haug and Koch 1990)

3.3 Semiconductor Bloch Equations

For an interacting electron system in a dielectric medium we have to include also the Coulomb interaction. Then the system Hamiltonian is

$$H = H_{kin} + H_{c-f} + H_C , \qquad (66)$$

where we write the Coulomb interaction energy among the electrons in two-band approximation as

$$H_C = \frac{1}{2} \sum_{\mathbf{k},\mathbf{k}'} \sum_{\mathbf{q} \neq 0} V_q \left[a^\dagger_{c,\mathbf{k}+\mathbf{q}} a^\dagger_{c,\mathbf{k}'-\mathbf{q}} a_{c\mathbf{k}'} a_{\mathbf{k}'} + a^\dagger_{v,\mathbf{k}+\mathbf{q}} a^\dagger_{v,\mathbf{k}'-\mathbf{q}} a_{v\mathbf{k}'} a_{v\mathbf{k}} \right.$$
$$\left. + 2 a^\dagger_{c,\mathbf{k}+\mathbf{q}} a^\dagger_{v,\mathbf{k}'-\mathbf{q}} a_{v\mathbf{k}'} a_{c\mathbf{k}} \right] . \qquad (67)$$

Here the first two terms describe the repulsive intraband carrier interaction, and the last term describes the repulsive interband Coulomb interaction between the electrons in the valence band and the electrons in the conduction band. The Coulomb interaction Hamiltonian (67) contains the Fourier transform of the Coulomb potential energy V_q, which for a bulk material is

$$V_q = \frac{1}{V} \int d^3r \, e^{-i\mathbf{q}\cdot\mathbf{r}} V(r)$$
$$= \frac{1}{V} \int d^3r \, e^{-i\mathbf{q}\cdot\mathbf{r}} \frac{e^2}{\epsilon_b r}$$
$$= \frac{4\pi e^2}{\epsilon_b V q^2} \,. \tag{68}$$

As discussed in Sec. 2.4 the confinement potential in quantum wells ideally restricts the carrier motion to the x-y plane. However, the Coulomb interaction between the electrons is not confined, i.e. it remains three dimensional. Hence, for an idealized quantum well system we have to take the two-dimensional Fourier transformation of the 3D Coulomb potential,

$$V_q = \frac{1}{V} \int d^2r \, e^{-i\mathbf{q}\cdot\mathbf{r}} \frac{e^2}{\epsilon_b r}$$
$$= \frac{2\pi e^2}{\epsilon_b q A} \,. \tag{69}$$

In Eq. (69), $A = V/w$ is the quantum well area, and w is its width.

In the derivation of Eq. (67), we use the fact that the Coulomb scattering does not alter the spin orientation of an electron and that the $\mathbf{q}=0$ contribution, which diverges, is cancelled by the $\mathbf{q}=0$ terms from the electron-ion and ion-ion Coulomb potentials. Furthermore, we omit Coulomb terms that fail to conserve the number of electrons in each band, since such terms involve interband transitions, which are very unfavorable energetically.

We transform to the electron-hole representation and restore normal ordering of all creation and annihilation operators. This gives the two-band Hamiltonian for interacting electrons and holes

$$H = \sum_\mathbf{k} \left[(\varepsilon_{e\mathbf{k}} + \varepsilon_{g0}) a_\mathbf{k}^\dagger a_\mathbf{k} + \varepsilon_{h\mathbf{k}} b_{-\mathbf{k}}^\dagger b_{-\mathbf{k}} \right]$$
$$+ \frac{1}{2} \sum_{\mathbf{k},\mathbf{k}'} \sum_{\mathbf{q} \neq 0} V_q \left[a_{\mathbf{k+q}}^\dagger a_{\mathbf{k'-q}}^\dagger a_{\mathbf{k'}} a_\mathbf{k} + b_{\mathbf{k+q}}^\dagger b_{\mathbf{k'-q}}^\dagger b_{\mathbf{k'}} b_\mathbf{k} - 2 a_{\mathbf{k+q}}^\dagger b_{\mathbf{k'-q}}^\dagger b_{\mathbf{k'}} a_\mathbf{k} \right]$$
$$- \sum_\mathbf{k} \left[\mu_\mathbf{k} a_{\mathbf{k} s_z}^\dagger b_{-\mathbf{k},-s_z}^\dagger + \mu_\mathbf{k}^* b_{-\mathbf{k},-s_z} a_{\mathbf{k} s_z} \right] E(z,t) \,, \tag{70}$$

where constant terms have been dropped because they only lead to an irrelevant shift of the reference energy. The kinetic energies in Eq. (70) are

$$\varepsilon_{e\mathbf{k}} = \frac{\hbar^2 k^2}{2m_e} \,, \tag{71}$$

$$\varepsilon_{h\mathbf{k}} = -\varepsilon_{v\mathbf{k}} + \sum_{\mathbf{q} \neq 0} V_q = \frac{\hbar^2 k^2}{2m_h} \,, \tag{72}$$

where the term containing V_q in $\varepsilon_{h\mathbf{k}}$ originates from the replacement of valence-band electron operators by hole operators in the interaction term

$$\frac{1}{2}\sum_{\mathbf{k},\mathbf{k'}}\sum_{\mathbf{q}\neq 0} V_q \, a^\dagger_{v,\mathbf{k+q}} a^\dagger_{v,\mathbf{k'-q}} a_{v\mathbf{k'}} a_{v\mathbf{k}}$$

of Eq. (67). Equation (72) differs from the free-carrier result in that the kinetic energy and therefore the hole energy includes the Coulomb energy of the full valence band.

Proceeding as in the free carrier model calculations, we derive coupled equations of motion for the electron and hole populations $n_{e\mathbf{k}}$ and $n_{h\mathbf{k}}$ and for the interband polarization $p_{\mathbf{k}}$. The derivation requires simple but lengthy operator commutations to reduce the commutators in the Heisenberg equations to

$$\frac{dp_{\mathbf{k}}}{dt} = -i\omega'_{\mathbf{k}} p_{\mathbf{k}} - i\hbar^{-1}\mu_{\mathbf{k}} E(z,t) [n_{e\mathbf{k}} + n_{h\mathbf{k}} - 1] + \frac{i}{\hbar}\sum_{\mathbf{k'},\mathbf{q}\neq 0} V_q$$
$$\times \Big[\langle a^\dagger_{\mathbf{k'+q}} b_{-\mathbf{k}} a_{\mathbf{k'}} a_{\mathbf{k+q}}\rangle + \langle b_{\mathbf{k'-q}} b_{\mathbf{k'}} a_{\mathbf{k}} b_{-\mathbf{k-q}}\rangle - \langle a^\dagger_{\mathbf{k'+q}} b_{-\mathbf{k+q}} a_{\mathbf{k'}} a_{\mathbf{k}}\rangle$$
$$- \langle b^\dagger_{\mathbf{k'-q}} b_{-\mathbf{k}} b_{\mathbf{k'}} a_{\mathbf{k-q}}\rangle + \langle b_{-\mathbf{k+q}} a_{\mathbf{k-q}}\rangle \delta_{\mathbf{k},\mathbf{k'}} \Big] \,, \tag{73}$$

$$\frac{dn_{e\mathbf{k}}}{dt} = \frac{i}{\hbar} E(z,t)[\mu_{\mathbf{k}} p^*_{\mathbf{k}} - \mu^*_{\mathbf{k}} p_{\mathbf{k}}]$$
$$+ \frac{i}{\hbar}\sum_{\mathbf{k'},\mathbf{q}\neq 0} V_q \Big[\langle a^\dagger_{\mathbf{k}} a^\dagger_{\mathbf{k'-q}} a_{\mathbf{k-q}} a_{\mathbf{k'}}\rangle - \langle a^\dagger_{\mathbf{k+q}} a^\dagger_{\mathbf{k'-q}} a_{\mathbf{k}} a_{\mathbf{k'}}\rangle$$
$$+ \langle a^\dagger_{\mathbf{k}} a_{\mathbf{k-q}} b^\dagger_{\mathbf{k'-q}} b_{\mathbf{k'}}\rangle - \langle a^\dagger_{\mathbf{k+q}} a_{\mathbf{k}} b_{\mathbf{k'-q}} b_{\mathbf{k'}}\rangle \Big] \,, \tag{74}$$

and

$$\frac{dn_{h\mathbf{k}}}{dt} = \frac{i}{\hbar} E(z,t)[\mu_{\mathbf{k}} p^*_{\mathbf{k}} - \mu^*_{\mathbf{k}} p_{\mathbf{k}}]$$
$$+ \frac{i}{\hbar}\sum_{\mathbf{k'},\mathbf{q}\neq 0} V_q \Big[\langle b^\dagger_{-\mathbf{k}} b^\dagger_{\mathbf{k'-q}} b_{-\mathbf{k-q}} b_{\mathbf{k'}}\rangle - \langle b^\dagger_{-\mathbf{k+q}} b^\dagger_{\mathbf{k'-q}} b_{-\mathbf{k}} b_{\mathbf{k'}}\rangle$$
$$+ \langle a^\dagger_{\mathbf{k'+q}} a_{\mathbf{k'}} b^\dagger_{-\mathbf{k}} b_{-\mathbf{k+q}}\rangle - \langle a^\dagger_{\mathbf{k'+q}} a_{\mathbf{k'}} b^\dagger_{-\mathbf{k-q}} b_{\mathbf{k}}\rangle \Big] \,, \tag{75}$$

where the transition energy is here denoted by

$$\hbar\omega'_{\mathbf{k}} = \varepsilon_{e\mathbf{k}} + \varepsilon_{h\mathbf{k}} + \varepsilon_{g0} \,. \tag{76}$$

We introduce the prime to distinguish between the unrenormalized energy $\hbar\omega'_{\mathbf{k}}$ from the renormalized one $\hbar\omega_{\mathbf{k}}$, which appears later. Equations (73) -

(75) show that the Coulomb interaction couples the two operator dynamics to four-operator terms. One way to proceed is to factorize these terms into products of two-operator terms, yielding the Hartree-Fock limit of the equations. To obtain a systematic hierarchy of equations, we separate out the Hartree-Fock contributions. For example, we write for a two-operator combination AB,

$$\frac{d}{dt}\langle AB\rangle = \frac{d}{dt}\langle AB\rangle_{HF} + \left[\frac{d}{dt}\langle AB\rangle - \frac{d}{dt}\langle AB\rangle_{HF}\right]$$
$$\equiv \frac{d}{dt}\langle AB\rangle_{HF} + \frac{d}{dt}\langle AB\rangle_{col} . \qquad (77)$$

Here HF indicates the Hartree-Fock contribution. The quantity inside the square bracket then contains both two and four-operator products, which we represent in general by $\langle ABCD\rangle$. These contributions beyond the Hartree-Fock approximation are often called collision (subscript col) or correlation contributions. They will be discussed in detail in the following chapter. Here we only mention that with the many-body Hamiltonian (70), the Heisenberg equation of motion gives for the equation of motion for $\langle ABCD\rangle$:

$$\frac{d}{dt}\langle ABCD\rangle = \frac{d}{dt}\langle ABCD\rangle_F + \left[\frac{d}{dt}\langle ABCD\rangle - \frac{d}{dt}\langle ABCD\rangle_F\right] , \qquad (78)$$

where $d\langle ABCD\rangle/dt$ contains expectation values of products of up to six operators, and we use the label F to indicate the result from a Hartree-Fock factorization of the four and six operator expectation values. We can continue by deriving the equation of motion for

$$\langle ABCDEF\rangle \equiv \left[\frac{d}{dt}\langle ABCD\rangle - \frac{d}{dt}\langle ABCD\rangle_F\right]$$

and so on. The result is a hierarchy of equations, where each succeeding equation describes a correlation among operators that is higher than the one before. In practice, we truncate the hierarchy at some point.

Returning to Eqs. (73) - (75), we first evaluate the Hartree-Fock contributions. To do so, we factorize all the expectation values of four operator products into *all possible* operator combinations leading to products of densities and/or polarizations. For example, for $\langle a_\mathbf{k}^\dagger a_{\mathbf{k}'}^\dagger a_1 a_{1'}\rangle$, we can have the two-operator combinations $\langle a_\mathbf{k}^\dagger a_{\mathbf{k}'}^\dagger\rangle\langle a_1 a_{1'}\rangle$, $\langle a_\mathbf{k}^\dagger a_1\rangle\langle a_{\mathbf{k}'}^\dagger a_{1'}\rangle$, and $\langle a_\mathbf{k}^\dagger a_{1'}\rangle\langle a_{\mathbf{k}'}^\dagger a_1\rangle$. Taking the anticommutation relations into account to get the proper signs between these combinations, we find

$$\langle a_\mathbf{k}^\dagger a_{\mathbf{k}'}^\dagger a_1 a_{1'}\rangle \simeq \langle a_\mathbf{k}^\dagger a_{\mathbf{k}'}^\dagger\rangle\langle a_1 a_{1'}\rangle - \langle a_\mathbf{k}^\dagger a_1\rangle\langle a_{\mathbf{k}'}^\dagger a_{1'}\rangle + \langle a_\mathbf{k}^\dagger a_{1'}\rangle\langle a_{\mathbf{k}'}^\dagger a_1\rangle$$
$$= 0 + [-\delta_{\mathbf{k},1}\delta_{\mathbf{k}',1'} + \delta_{\mathbf{k},1'}\delta_{\mathbf{k}',1}]n_{e\mathbf{k}}n_{e\mathbf{k}'} . \qquad (79)$$

Another example is

$$\langle a_{\mathbf{k'+q}}^\dagger b_{\mathbf{q-k}} a_{\mathbf{k'}} a_{\mathbf{k}} \rangle \simeq \delta_{\mathbf{k'},\mathbf{k-q}} n_{e\mathbf{k}} p_{\mathbf{k'}} \ . \tag{80}$$

Factorizing all the other four-operator products in this way we find the *semiconductor Bloch equations*

$$\frac{dp_{\mathbf{k}}}{dt} = -i\omega_{\mathbf{k}} p_{\mathbf{k}} - i\Omega_{\mathbf{k}}(z,t)\left[n_{e\mathbf{k}} + n_{h\mathbf{k}} - 1\right] + \left.\frac{\partial p_{\mathbf{k}}}{\partial t}\right|_{col} \tag{81}$$

$$\frac{dn_{e\mathbf{k}}}{dt} = \frac{i}{\hbar}\left[\Omega_{\mathbf{k}}(z,t)p_{\mathbf{k}}^* - \Omega_{\mathbf{k}}^*(z,t)p_{\mathbf{k}}\right] + \left.\frac{\partial n_{e\mathbf{k}}}{\partial t}\right|_{col} \tag{82}$$

$$\frac{dn_{h\mathbf{k}}}{dt} = \frac{i}{\hbar}\left[\Omega_{\mathbf{k}}(z,t)p_{\mathbf{k}}^* - \Omega_{\mathbf{k}}^*(z,t)p_{\mathbf{k}}\right] + \left.\frac{\partial n_{h\mathbf{k}}}{\partial t}\right|_{col} \ . \tag{83}$$

We have written the terms containing the Hartree-Fock contributions explicitly, while the higher order correlations are denoted formally by the partial derivatives $\partial/\partial t|_{col}$. The Hartree-Fock contributions in Eqs. (81)-(83) contain two important many-body effects, namely a density dependent contribution to the transition energy, and a renormalization of the electric-dipole interaction energy. Specifically, $\hbar\omega'_{\mathbf{k}}$ of Eq. (76) is replaced by the renormalized transition energy

$$\hbar\omega_{\mathbf{k}} = \hbar\omega'_{\mathbf{k}} - \sum_{\mathbf{k'}\neq\mathbf{k}} V_{|\mathbf{k}-\mathbf{k'}|}\left(n_{e\mathbf{k'}} + n_{h\mathbf{k'}}\right) \ , \tag{84}$$

and the Rabi frequency $\mu_{\mathbf{k}} E(z,t)/\hbar$ is renormalized as

$$\Omega_{\mathbf{k}}(z,t) = \frac{\mu_{\mathbf{k}} E(z,t)}{\hbar} + \frac{1}{\hbar}\sum_{\mathbf{k'}\neq\mathbf{k}} V_{|\mathbf{k}-\mathbf{k'}|} p_{\mathbf{k'}} \ , \tag{85}$$

where the Coulomb terms ($\propto V_{|\mathbf{k}-\mathbf{k'}|}$) in Eqs. (84) and (85) are called the *exchange shift*, and the *field renormalization*, respectively. The fact, that the Rabi frequency is multiplied by the population factor $[n_{e\mathbf{k}} + n_{h\mathbf{k}} - 1]$ in Eq. (81) leads to nonlinearities in the resulting optical response often denoted as *phase space filling* effects.

The Coulomb terms in Eq. (81) show a large degree of symmetry. To see this more clearly, we write the Hartree-Fock part of Eq. (81), i.e. without $\partial p_{\mathbf{k}}/\partial t|_{col}$ as

$$\left.\frac{dp_{\mathbf{k}}}{dt}\right|_{HF} = \sum_{\mathbf{k'}} \Theta_{\mathbf{k}\mathbf{k'}} p_{\mathbf{k'}} - \frac{i}{\hbar}\mu_{\mathbf{k}} E\left(n_{e\mathbf{k}} + n_{h\mathbf{k}} - 1\right) \ , \tag{86}$$

where for $\mathbf{k} = \mathbf{k'}$

$$\Theta_{\mathbf{k}\mathbf{k}} = -i\omega'_{\mathbf{k}} + \frac{i}{\hbar}\sum_{\mathbf{k''}\neq\mathbf{k}} V_{|\mathbf{k}-\mathbf{k''}|}\left(n_{e\mathbf{k''}} + n_{h\mathbf{k''}}\right) \ , \tag{87}$$

and for $\mathbf{k} \neq \mathbf{k}'$

$$\Theta_{\mathbf{k}\mathbf{k}'} = -\frac{i}{\hbar} V_{|\mathbf{k}-\mathbf{k}'|} \left(n_{e\mathbf{k}} + n_{h\mathbf{k}} - 1 \right) \ . \tag{88}$$

We see that the Coulomb terms appear with opposite signs on the diagonal and nondiagonal parts of the matrix Θ. This leads to compensation effects in the influence of these many-body terms on aspects of the optical spectra. For example, the excitation independence of the excitonic resonance frequency, seen in Figs. 5.5 and 5.6, results to a significant degree from cancellation effects between the density dependent bandgap renormalization (diagonal part of Θ) and the weakening of the exciton binding energy (nondiagonal part).

3.4 Excitons

In this section, we examine the low density limit of the semiconductor Bloch equations. In this limit, $n_{e\mathbf{k}} = n_{h\mathbf{k}} \simeq 0$, and the collision terms vanish because no scattering partners are available. Equation (81) reduces to

$$\frac{dp_{\mathbf{k}}}{dt} = -i\omega_{\mathbf{k}} p_{\mathbf{k}} + i\Omega_{\mathbf{k}} \ , \tag{89}$$

which efficiently isolates the influence of the renormalized electric-dipole interaction frequency $\Omega_{\mathbf{k}}$. Choosing a plane-wave optical field and making the rotating-wave approximation, we obtain

$$[\omega_{\mathbf{k}} - \nu + i\gamma] p_{\mathbf{k}} = \Omega_{\mathbf{k}} \ , \tag{90}$$

where γ is a small phenomenological damping coefficient. Fourier transforming Eq. (90) to coordinate space, we find

$$\left[-\frac{\hbar^2 \nabla_{\mathbf{r}}^2}{2m_r} - \frac{e^2}{\epsilon_b r} + \varepsilon_g - \hbar \left(\nu - i\gamma \right) \right] p(\mathbf{r}) = \mu E \delta^3(\mathbf{r}) V \ , \tag{91}$$

where we ignore the $\mathbf{k}$-dependence of the interband dipole matrix element, which is often a reasonable approximation as long as we are only interested in small $\mathbf{k}$-values and frequencies close to the fundamental absorption edge.

Equation (91) is an inhomogeneous differential equation, which may be solved by expanding $p(\mathbf{r})$ as a linear superposition of the solutions of the corresponding homogeneous equation

$$\left[-\frac{\hbar^2 \nabla_{\mathbf{r}}^2}{2m_r} - \frac{e^2}{\epsilon_b r} \right] \psi_n(\mathbf{r}) = \varepsilon_n \psi_n(\mathbf{r}) \ . \tag{92}$$

Equation (92) is the Schrödinger equation for the relative motion of an electron and a hole interacting via the attractive Coulomb potential. In semiconductor physics Eq. (92) is known as the *Wannier equation*. As already mentioned near the end of Sec. 2.3 there is a one-to-one correspondence between the electron-hole problem and the hydrogen atom if one replaces the proton by the valence-band hole. The solutions of the Wannier equation are therefore completely analogous to those of the hydrogen problem, which are discussed in most quantum mechanics textbooks. There are bound states called excitons, or more specifically *Wannier excitons*, and there are continuum states.

The bound and continuum eigenfunctions of the Wannier equation form a complete and orthonormal basis set, so that we can write

$$p(\mathbf{r}) = \sum_n p_n \psi_n(\mathbf{r}) \ . \tag{93}$$

Substituting Eq. (93) into Eq. (91), multiplying by $\psi_m^*(\mathbf{r})$ and integrating over $\mathbf{r}$ yields

$$p_m = -\frac{\mu V \psi_m^*(\mathbf{r}=0)}{\hbar(\nu - i\gamma) - \varepsilon_g - \varepsilon_m} E \ , \tag{94}$$

where we used the orthormality condition

$$\int d^3r \ \psi_m^*(\mathbf{r})\psi_n(\mathbf{r}) = \delta_{m,n} \ . \tag{95}$$

Inserting Eq. (94) into Eq. (93) gives

$$p(\mathbf{r}) = -\sum_n E \frac{\mu V \psi_n^*(\mathbf{r}=0)}{\hbar(\nu - i\gamma) - \varepsilon_g - \varepsilon_n} \psi_n(\mathbf{r}) \ , \tag{96}$$

which has the Fourier transform

$$p_{\mathbf{k}} = -\sum_n E \frac{\mu \psi_n^*(\mathbf{r}=0)}{\hbar(\nu - i\gamma) - \varepsilon_g - \varepsilon_n} \int d^3r \psi_n(\mathbf{r}) e^{i\mathbf{k}\cdot\mathbf{r}} \ . \tag{97}$$

For the space dependent optical polarization we thus obtain

$$P(\nu) = -2|\mu|^2 E \sum_n \frac{|\psi_n(\mathbf{r}=0)|^2}{\hbar(\nu - i\gamma) - \varepsilon_g - \varepsilon_n} \ , \tag{98}$$

where $|\psi_n(\mathbf{r}=0)|^2$ is the probability of finding the electron and the hole within the same atomic unit cell. The optical susceptibility

$$\chi(\nu) = \frac{P(\nu)}{E(\nu)} \tag{99}$$

is then given by

$$\chi(\nu) = -\frac{2|\mu|^2}{\epsilon_b} \sum_n \frac{|\psi_n(\mathbf{r}=0)|^2}{\hbar(\nu - i\gamma) - \varepsilon_g - \varepsilon_n} \ , \tag{100}$$

and the corresponding absorption coefficient is

$$\alpha(\nu) = \frac{4\pi\nu}{n_b c} \text{Im}\left[\chi(\nu)\right]$$

$$= \alpha_0 \left[\sum_{n=1}^{\infty} \frac{4\pi}{n^3} \delta\left(\Delta + \frac{1}{n^2}\right) + \Theta(\Delta) \pi e \frac{\pi/\sqrt{\Delta}}{\sinh\left(\pi/\sqrt{\Delta}\right)} \right], \quad (101)$$

where

$$\Delta = \frac{\hbar\nu - \varepsilon_g}{\varepsilon_R},$$

$$\alpha_0 = \frac{2|\mu|^2}{\hbar n_b c a_0^3},$$

and we have used the explicit form of the electron-hole pair eigenfunctions (see Haug and Koch 1994). Equation (101) is known as the *Elliott formula* and describes the bandgap absorption spectrum in an unexcited bulk semiconductor.

Equation (101) predicts that the absorption spectrum consists of a series of δ-functions at discrete energies. These resonances are the exciton peaks. The prefactor in front of the δ-functions in Eq. (101) shows that the exciton resonances have a rapidly decreasing oscillator strength $\propto n^{-3}$. The appearance of the exciton resonances in the absorption spectrum is a unique consequence of the electron-hole Coulomb attraction. The second term in Eq. (101), α_{cont}, describes the continuum absorption due to the ionized states. It can be written in terms of the free-carrier absorption, Eq. (63) for $d = 3$ and $f_e = f_h = 0$ as

$$\alpha_{cont}(\nu) = \alpha_{free}(\nu) \frac{\pi}{\sqrt{\Delta}} \frac{e^{\pi/\sqrt{\Delta}}}{\sinh\left(\pi/\sqrt{\Delta}\right)}, \quad (102)$$

where the correction to α_{free} is called the *Sommerfeld* or *Coulomb enhancement* factor. It is a simple exercise to verify that this factor approaches the value $2\pi/\sqrt{\Delta}$ for $\Delta \to 0$, which cancels the $\sqrt{\Delta}$ factor in the free-carrier absorption for a three-dimensional system and yields a constant value at the bandgap. This is strikingly different from the square-root law of the free-carrier absorption. Similar calculations for $d = 2$ and $d = 1$ show that the free-carrier absorption at the bandgap is enhanced by a factor 2 for the two-dimensional system, and the $1/\sqrt{\Delta}$ singularity of the one-dimensional system is replaced by a constant. (For more details see e.g. the discussion in Haug and Koch 1994.) Hence, we note that the free-carrier absorption results are significantly modified in real semiconductors, where the electron-hole Coulomb attraction is always present.

If one takes into account the broadening of the exciton resonances caused by, for example, the scattering of electron-hole pairs with phonons, then only a few bound states can be spectrally resolved. An example of an absorption spectrum predicted by the Elliott formula is depicted in Fig. 3.3. We see that the dominant feature is the 1s-exciton absorption peak. The 2s-exciton can also be resolved, but its height is only 1/8-th that of the 1s-resonance. The other exciton states in GaAs materials usually appear only as a collection of unresolvable peaks just below the bandgap.

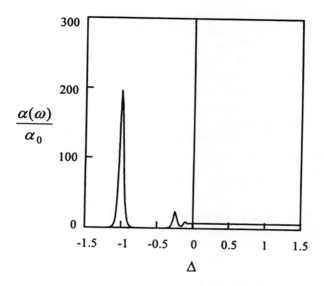

Fig. 3.3. Near bandgap absorption spectrum for a three dimensional semiconductor with parameters typical for good quality GaAs at low temperatures. (From Haug and Koch 1994)

References

For more details on the semiconductor Bloch equations and for further references see:

Binder, R. and Koch, S. W. Progress in Quantum Electronics, **19**, 307 (1995).
Koch, S.W., N. Peyghambarian, and M. Lindberg (1988), J. Phys. **C21**, 5229.
Haug, H. and S.W. Koch (1989), Phys. Rev. **A39**, 1887.
Haug, H. (1988), Ed., *Optical Nonlinearities and Instabilities in Semiconductors*, Academic, New York (1988).
Stahl, A. and I. Balslev (1987), *Electrodynamics of the Semiconductor Band Edge*, Springer Tracts in Modern Physics **110**, Springer Verlag, Berlin.
Haug, H. and S.W. Koch (1994), *Quantum Theory of the Optical and Electronic Properties of Semiconductors*, 3rd ed., World Scientific, Singapore.
Lindberg, M. and S.W. Koch (1988), Phys. Rev. **B38**, 3342.

4 Optical Nonlinearities in Semiconductors

In this chapter we summarize the basic ingredients of a microscopic theory of those semiconductor nonlinearities that are caused by resonantly excited electron-hole pairs. Since the properties of the electron-hole system are sensitively dependent on the excitation level (light intensity) and on the detailed excitation conditions, the resulting optical polarization is a nonlinear function of the light field. Hence, the various carrier interaction processes are observed experimentally as a nonlinear optical response. Besides the resonant optical nonlinearities discussed here, there are other nonlinear effects, such as two- or more photon absorption processes, effects due to band non parabolicities etc. which are not the subject of these lectures.

In Sec. 4.1 we outline the derivation of approximations for the correlation contributions entering the semiconductor Bloch equations. Sec. 4.2 discusses the Boltzmann equation as an approximation to the intraband (carrier) relaxation dynamics, and Sec. 4.3 presents the corresponding analysis of the interband polarization decay. In Sec. 4.4 we outline the screened Hartree-Fock approximation which for special situations leads to a simplified numerical analysis of the nonlinear semiconductor response.

4.1 Coulomb Correlation Effects

Our basic equations are the semiconductor Bloch equations, Eqs. (3-81) - (3-83) derived in the previous chapter. Even though these equations already at the Hartree-Fock level constitute a set of nonlinear integro-differential equations, the analysis of most resonant optical nonlinearities in semiconductors requires the inclusion of additional effects such as Coulomb screening, damping, and polarization dephasing. For a systematic investigation of these phenomena we have to evaluate the correlation contributions at some level of approximation.

To study these correlation contributions in the semiconductor Bloch equations, we start by deriving equations of motion for quantities describing the deviations of the full correlation terms from their corresponding Hartree-Fock factorized parts. For example,

$$\delta \left\langle a_{\mathbf{k}}^{\dagger} a_{\mathbf{k'}-\mathbf{q}}^{\dagger} a_{\mathbf{k}-\mathbf{q}} a_{\mathbf{k'}} \right\rangle = \left\langle a_{\mathbf{k}}^{\dagger} a_{\mathbf{k'}-\mathbf{q}}^{\dagger} a_{\mathbf{k}-\mathbf{q}} a_{\mathbf{k'}} \right\rangle - \left\langle a_{\mathbf{k}}^{\dagger} a_{\mathbf{k}} \right\rangle \left\langle a_{\mathbf{k'}-\mathbf{q}}^{\dagger} a_{\mathbf{k}-\mathbf{q}} \right\rangle \delta_{\mathbf{k},\mathbf{k'}} , \tag{1}$$

whose time derivative is

$$\frac{d}{dt} \delta \left\langle a_{\mathbf{k}}^{\dagger} a_{\mathbf{k'}-\mathbf{q}}^{\dagger} a_{\mathbf{k}-\mathbf{q}} a_{\mathbf{k'}} \right\rangle = \left\langle \frac{d}{dt} \left(a_{\mathbf{k}}^{\dagger} a_{\mathbf{k'}-\mathbf{q}}^{\dagger} a_{\mathbf{k}-\mathbf{q}} a_{\mathbf{k'}} \right) \right\rangle$$

$$-\left[\left\langle\frac{d}{dt}\left(a_{\mathbf{k}}^\dagger a_{\mathbf{k}}\right)\right\rangle\left\langle a_{\mathbf{k-q}}^\dagger a_{\mathbf{k-q}}\right\rangle + \left\langle a_{\mathbf{k}}^\dagger a_{\mathbf{k}}\right\rangle\left\langle\frac{d}{dt}\left(a_{\mathbf{k-q}}^\dagger a_{\mathbf{k-q}}\right)\right\rangle\right]\delta_{\mathbf{k},\mathbf{k}'} . \quad (2)$$

From the Heisenberg equation of motion, we find

$$\frac{d}{dt}\delta\left\langle a_{\mathbf{k}}^\dagger a_{\mathbf{k'-q}}^\dagger a_{\mathbf{k-q}} a_{\mathbf{k'}}\right\rangle$$
$$= \frac{i}{\hbar}\delta\left\langle a_{\mathbf{k}}^\dagger a_{\mathbf{k'-q}}^\dagger a_{\mathbf{k-q}} a_{\mathbf{k'}}\right\rangle\Delta\varepsilon_{e\mathbf{kk'q}} + \frac{d}{dt}\delta\left\langle a_{\mathbf{k}}^\dagger a_{\mathbf{k'-q}}^\dagger a_{\mathbf{k-q}} a_{\mathbf{k'}}\right\rangle\bigg|_{Coul} , \quad (3)$$

where we use the full electron-hole Hamiltonian except for the carrier-field interaction, since it does not play a role in the collisions. Furthermore, we introduced the abbreviations

$$\Delta\varepsilon_{e\mathbf{kk'q}} = \varepsilon_{e\mathbf{k}} + \varepsilon_{e,\mathbf{k'-q}} - \varepsilon_{e,\mathbf{k-q}} - \varepsilon_{e\mathbf{k'}} \quad (4)$$

and

$$i\hbar\frac{d}{dt}\delta\left\langle a_{\mathbf{k}}^\dagger a_{\mathbf{k'-q}}^\dagger a_{\mathbf{k-q}} a_{\mathbf{k'}}\right\rangle\bigg|_{Coul} = \left\langle\left[H_C, a_{\mathbf{k}}^\dagger a_{\mathbf{k'-q}}^\dagger a_{\mathbf{k-q}} a_{\mathbf{k'}}\right]\right\rangle . \quad (5)$$

Evaluation of the commutator in Eq. (5) leads to expressions containing products of up to six operators that are too lengthy to show here.

Formally integrating Eq. (3), we get

$$\delta\left\langle a_{\mathbf{k}}^\dagger a_{\mathbf{k'-q}}^\dagger a_{\mathbf{k-q}} a_{\mathbf{k'}}\right\rangle(t)$$
$$= \int_{-\infty}^t dt' \exp\left[\left(\frac{i}{\hbar}\Delta\varepsilon_{e\mathbf{kk'q}} - \gamma\right)(t-t')\right]\frac{d}{dt'}\delta\left\langle a_{\mathbf{k}}^\dagger a_{\mathbf{k'-q}}^\dagger a_{\mathbf{k-q}} a_{\mathbf{k'}}\right\rangle\bigg|_{Coul} , \quad (6)$$

where γ is a phenomenological decay constant added so that the integral vanishes at the lower boundary. In general, the correlation at time t depends on the evolution of the system from $-\infty$ to t. This is the Coulombic *memory effect*. Fortunately, it is often a reasonable approximation to assume that the memory time is not very long and to neglect memory effects altogether, making the *Markov approximation*. Technically, this amounts to assuming that the Coulomb contribution is slowly varying compared to the exponential, so that we can move it outside the integral in Eq. (6). The resulting integral can be readily evaluated to give

$$\delta\left\langle a_{\mathbf{k}}^\dagger a_{\mathbf{k'-q}}^\dagger a_{\mathbf{k-q}} a_{\mathbf{k'}}\right\rangle \simeq \frac{d}{dt}\delta\left\langle a_{\mathbf{k}}^\dagger a_{\mathbf{k'-q}}^\dagger a_{\mathbf{k-q}} a_{\mathbf{k'}}\right\rangle\bigg|_{Coul}\frac{1}{i\Delta\varepsilon_{e\mathbf{kk'q}}/\hbar - \gamma} . \quad (7)$$

A similar result is obtained for the other four-operator terms which appear in the semiconductor Bloch equations, (3-81)-(3-83).

At this stage we still do not have a closed set of equations because of the six-operator expectation values occuring in $\delta \left\langle a_{\mathbf{k}}^{\dagger} a_{\mathbf{k'-q}}^{\dagger} a_{\mathbf{k-q}} a_{\mathbf{k'}} \right\rangle / dt \Big|_{Coul}$. The equation of motion for these six-operator terms introduces eight-operator terms, i.e., the typical many-body hierarchy problem. In order to close the equations we again make a factorization approximation. We factorize all the six- and four-operator terms which occur in the Coulomb parts to obtain the simplest possible expression for the scattering terms. The detailed calculation yields the following results for the electron population equation:

$$\frac{\partial n_{e\mathbf{k}}}{\partial t}\bigg|_{col} = -n_{e\mathbf{k}} \Sigma_{e\mathbf{k}}^{out}\{n\} + (1 - n_{e\mathbf{k}}) \Sigma_{e\mathbf{k}}^{in}\{n\} , \qquad (8)$$

where, for simplicity, we ignored terms containing scattering contributions involving interband polarizations. These additional terms are discussed e.g., by Jahnke et al., (1997). The rates $\Sigma_{e\mathbf{k}}^{out}\{n\}$ and $\Sigma_{e\mathbf{k}}^{in}\{n\}$, describing the effective scattering out of and into the state $\mathbf{k}$, are given by

$$\Sigma_{e\mathbf{k}}^{out}\{n\} = \frac{\pi}{\hbar} \sum_{b=e,h} \sum_{\mathbf{q} \neq 0} \sum_{\mathbf{k'}} \left(2V_q^2 - \delta_{e,b} V_q V_{|\mathbf{k}-\mathbf{k'}+\mathbf{q}|}\right)$$
$$\times \delta(\varepsilon_{e,\mathbf{k}} + \varepsilon_{b,\mathbf{k'}} - \varepsilon_{e,\mathbf{k}+\mathbf{q}} - \varepsilon_{b,\mathbf{k'}-\mathbf{q}})$$
$$\times (1 - n_{e,\mathbf{k}+\mathbf{q}}) n_{b,\mathbf{k'}} (1 - n_{b,\mathbf{k'}-\mathbf{q}}) , \qquad (9)$$

and

$$\Sigma_{e\mathbf{k}}^{in}\{n\} = \frac{\pi}{\hbar} \sum_{b=e,h} \sum_{\mathbf{q} \neq 0} \sum_{\mathbf{k'}} \left(2V_q^2 - \delta_{e,b} V_q V_{|\mathbf{k}-\mathbf{k'}+\mathbf{q}|}\right)$$
$$\times \delta(\varepsilon_{e,\mathbf{k}} + \varepsilon_{b,\mathbf{k'}} - \varepsilon_{e,\mathbf{k}+\mathbf{q}} - \varepsilon_{b,\mathbf{k'}-\mathbf{q}})$$
$$\times n_{e,\mathbf{k}+\mathbf{q}} (1 - n_{b,\mathbf{k'}}) n_{b,\mathbf{k'}-\mathbf{q}} . \qquad (10)$$

The notation $\Sigma\{n\}$ symbolizes the functional dependence of these rates on the electron and hole distribution functions. The corresponding equations for the hole population $n_{h\mathbf{k}}$ are obtained by the interchange $e \rightleftharpoons h$ in Eqs. (8)-(10).

It is sometimes convenient to write the total relaxation rate of $n_{\alpha\mathbf{k}}$ as a single decay rate

$$\gamma_{\alpha\mathbf{k}}\{n\} \equiv \Sigma_{\alpha\mathbf{k}}^{out}\{n\} + \Sigma_{\alpha\mathbf{k}}^{in}\{n\} , \qquad (11)$$

in terms of which the scattering integral becomes

$$\frac{\partial n_{e\mathbf{k}}}{\partial t}\bigg|_{col} = -\gamma_{\alpha\mathbf{k}}\{n\} n_{\alpha\mathbf{k}} + \Sigma_{\alpha\mathbf{k}}^{in}\{n\} . \qquad (12)$$

Here we see explicitly that scattering both into and out of the state $\mathbf{k}$ are relaxation processes, although one increases the probability $n_{\alpha\mathbf{k}}$, while the other one decreases it.

For the collision terms in the equation for the interband polarization, we obtain

$$\left.\frac{\partial p_{\mathbf{k}}}{\partial t}\right|_{col} = -\sum_{\mathbf{k}'} \Lambda_{\mathbf{k}\mathbf{k}'} p_{\mathbf{k}'} , \qquad (13)$$

where, for notational simplicity, we show here only the terms that are linear in the polarization. For $\mathbf{k} = \mathbf{k}'$,

$$\Lambda_{\mathbf{k}\mathbf{k}} = \frac{1}{\hbar} \sum_{a,b=e,h} \sum_{\mathbf{k}''} \sum_{\mathbf{q} \neq 0} \left(2V_q^2 - \delta_{a,b} V_q V_{|\mathbf{k}-\mathbf{k}''+\mathbf{q}|}\right) g(\delta\varepsilon)$$
$$\times \left[n_{e,\mathbf{k}+\mathbf{q}} \left(1 - n_{b\mathbf{k}''}\right) n_{b,\mathbf{k}''-\mathbf{q}} + \left(1 - n_{e,\mathbf{k}+\mathbf{q}}\right) n_{b\mathbf{k}''} \left(1 - n_{b,\mathbf{k}''-\mathbf{q}}\right)\right] . \quad (14)$$

Here, we used the abbreviation

$$\delta\varepsilon = \varepsilon_{a\mathbf{k}} + \varepsilon_{b\mathbf{k}''} - \varepsilon_{a,\mathbf{k}+\mathbf{q}} - \varepsilon_{b,\mathbf{k}''-\mathbf{q}} , \qquad (15)$$

and the generalized δ-function (Heitler Zeta function) is

$$g(x) = \lim_{\gamma \to 0} \frac{i}{x + i\gamma} = \pi\delta(x) + iP\left(\frac{1}{x}\right) . \qquad (16)$$

For $\mathbf{k} \neq \mathbf{k}'$,

$$\Lambda_{\mathbf{k}\mathbf{k}'} = \frac{1}{\hbar} \sum_{a,b=e,h} \sum_{\mathbf{k}''} \left(2V_q^2 - \delta_{a,b} V_q V_{|\mathbf{k}-\mathbf{k}''+\mathbf{q}|}\right) g(-\delta\varepsilon)$$
$$\times \left[\left(1 - n_{a\mathbf{k}}\right)\left(1 - n_{b\mathbf{k}''}\right) n_{b,\mathbf{k}''-\mathbf{q}} + n_{a\mathbf{k}} n_{b\mathbf{k}''} \left(1 - n_{b,\mathbf{k}''-\mathbf{q}}\right)\right] , \quad (17)$$

where $\mathbf{q} = \mathbf{k}' - \mathbf{k}$.

Equation (13) has been written already in a form showing that the scattering matrix $\underline{\Lambda}$ adds to the Hartree-Fock matrix $\underline{\Theta}$ of Eq. (3-86). In fact, the diagonal and nondiagonal terms in $\underline{\Lambda}$ are the second order (in the Coulomb potential) contributions to the energy and field renormalization, respectively. In general, the matrix $\underline{\Lambda}$ contains also terms where one or more of the population factor are replaced by polarizations (Jahnke et al. 1997). Those terms are especially important under coherent nonlinear excitation conditions, where a large induced interband polarization is present.

4.2 Carrier Quantum Boltzmann Equation

The collison contributions in the carrier equation (8) are actually the simplest version of the famous *quantum Boltzmann scattering integral* for carrier-carrier collisions. One remarkable feature of Eq. (8) is that for $n_{\alpha\mathbf{k}} = f_{\alpha\mathbf{k}} = \frac{1}{e^{\beta(\varepsilon_{\alpha\mathbf{k}} - \mu_\alpha)} + 1}$ i.e. if the carriers are in Fermi-Dirac distributions, the Boltzmann scattering integral is identically zero. The fact that $\partial f / \partial t|_{col} = 0$ implies that

$$f_{\alpha\mathbf{k}} \Sigma_{\alpha\mathbf{k}}^{out}\{f\} = (1 - f_{\alpha\mathbf{k}}) \Sigma_{\alpha\mathbf{k}}^{in}\{f\} , \qquad (18)$$

for nonvanishing in- and out-scattering rates. Equation (18) describes the *detailed balance* condition, for which the scattering into each state is exactly balanced by the scattering out of that state. This is the *quasi-equilibrium situation* where the carrier plasma is characterized by a plasma temperature T and a quasi-chemical potential. Generally, the plasma temperature differs from the lattice temperature. It is important to realize that even though the distribution functions are time independent, i.e. Fermi-Dirac, this does not imply the absence of scattering events. The individual terms in Eq. (18) are non-zero and rather large. However, they exactly balance each other.

Generally, there are a number of physical quantities that are conserved in the carrier-carrier scattering processes. These conservation rules can be written as

$$\left.\frac{\partial}{\partial t}\right|_{col} \left[\sum_{\mathbf{k}} F_i(\mathbf{k}) n_{\alpha \mathbf{k}}\right] = 0, \ i = 1, 2, ...5; \ \alpha = e, h \qquad (19)$$

with

$$F_1 = 1 \qquad (20)$$
$$F_2 = k_x, \ F_3 = k_y, \ F_4 = k_z \qquad (21)$$
$$F_5 = k^2 \ . \qquad (22)$$

Equations (19) and (20) correspond to total particle number conservation, Eqs. (19) and (21) to total momentum conservation, and Eqs. (19) and (22) to total kinetic energy conservation, respectively.

Since one typically does not encounter a drifting plasma in a semiconductor medium, the total momentum is originally zero, and because of Eqs. (19) and (21) it will remain zero. To see the implications of the other conservation rules, let us consider the example of nonequilibrium carrier relaxation experiments performed using femtosecond ($10^{-15}s$) pulse excitation of semiconductor interband transitions. In these investigations, the initially prepared nonequilibrium carrier distribution is rapidly modified by carrier-carrier collisions so that it approaches the Fermi-Dirac distribution. Under most conditions, the carrier-carrier equilibration processes occur very rapidly, at a sub-picosecond timescale. This situation is depicted in Fig. 4.1, where the electron distributions are plotted at different times after the excitation. The curves are numerical solutions of the full carrier-carrier Boltzmann equation (8), including electron-electron, electron-hole, and hole-hole scattering. The carrier distributions immediately after the excitation pulse is plotted as the solid curve in Fig. 4.1. The other curves show the evolution of the carrier populations towards a Fermi-Dirac distribution. However, since the kinetic energy is conserved, the plasma temperature of the relaxed distribution is determined by the kinetic energy of the original nonequilibrium distribution. As a result, depending on the excess energy of the excitiation pulse (i.e., how high above the band minimum are the carriers generated), the effective plasma temperature is well above the lattice temperature. Relaxation of the electron and hole kinetic energies (plasma cooling) happens only by collisons with other quasi-particles, most importantly with phonons.

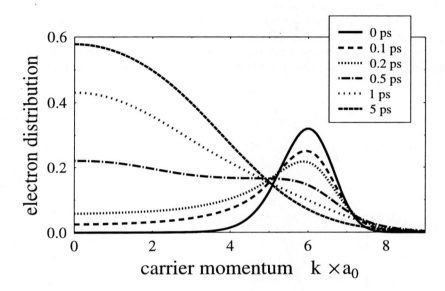

Fig. 4.1. Temporal relaxation of an electron distribution function. At $t = 0$ the carriers are generated by interband excitation with a short pulse. Snapshots of the distribuion function at different times after the generation are plotted until final equilibrium is reached ($t = 5ps$). For these results only carrier-carrier scattering was included. (From Jahnke and Koch 1995)

The carrier-phonon Boltzmann equation can be derived in a similar way as Eq. (8), where however, the carrier-phonon part has to be added in the total Hamiltonian. For the example of LO (longitudinal optical) phonons the resulting equation is

$$\left.\frac{dn_{e\mathbf{k}}}{dt}\right|_{e-p} = -2\pi \sum_{\mathbf{q},\pm} G_\mathbf{q}^2 \delta\left(\Delta_{\mathbf{k},\mathbf{q}}^\pm\right) \left[n_{e\mathbf{k}}\left(1 - n_{e,\mathbf{k}-\mathbf{q}}\right)\left(n_{ph,\pm\mathbf{q}} + \frac{1}{2} \pm \frac{1}{2}\right)\right.$$
$$\left. - (1 - n_{e\mathbf{k}})\, n_{e,\mathbf{k}-\mathbf{q}}\left(n_{ph,\pm\mathbf{q}} + \frac{1}{2} \mp \frac{1}{2}\right)\right], \tag{23}$$

which is the Boltzmann collision integral for electron-LO phonon scattering, with $g_{\pm\mathbf{q}}$ being the phonon population. Here we denote the frequency differences $\Delta_{\mathbf{k},\mathbf{q}}^\pm$ as

$$\hbar\Delta^{\pm}_{\mathbf{k},\mathbf{q}} = \varepsilon_{\mathbf{k}} - \varepsilon_{\mathbf{k}-\mathbf{q}} \pm \hbar\omega_{LO} \ , \tag{24}$$

and G_q^2 is the scattering matrix element [for its definition and further details see e.g. Chap. 21 in Haug and Koch (1994)]. The different terms in Eq. (23) describe the transition rates in and out of the state $\mathbf{k}$ under absorption or emission of LO-phonons. The first two terms are the transition $\mathbf{k} \to \mathbf{k} - \mathbf{q}$ under emission (upper sign) or absorption (lower sign) of a phonon. The second two terms describe the transition $\mathbf{k} - \mathbf{q} \to \mathbf{k}$ under absorption or emission of a phonon. The phonon population function $n_{ph,\mathbf{q}}$ in general has to be computed self-consistently. However, it is often possible to simplify the problem by assuming that the phonons are in thermal equilibrium, acting as a reservoir, so that

$$n_{ph,\mathbf{q}} = \frac{1}{e^{\beta\hbar\omega_{LO}} - 1} \ , \tag{25}$$

i.e., the phonon distribution is described by a thermal Bose function.

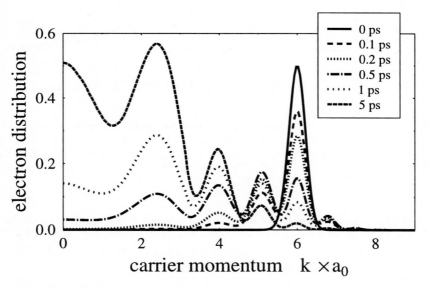

Fig. 4.2. Same as Fig. 4.1 but for the case where only carrier-LO phonon scattering is included. (From Jahnke and Koch 1995)

To illustrate the different scattering effects we plot the results obtained by solving the carrier-phonon Boltzmann equation alone (Fig. 4.2) and together with the carrier-carrier Boltzmann equation (8) (Fig. 4.3). As the LO phonon energy $\hbar\omega_{LO}$ has a discrete value, the carrier relaxation of an initial nonequilibrium distribution (solid line in Fig. 4.2), which occurs via successive emission of LO phonons, leads to the occurence sidebands in the carrier distribution at the energies $\varepsilon - n\hbar\omega_{LO}, n = 1, 2, \ldots$. If one simultaneously includes carrier-carrier scattering, as it occurs under realistic conditions, the discrete phonon sidebands are often suppressed, as shown in Fig. 4.3. However, now the carrier distributions relax to a Fermi-Dirac distribution at the lattice (LO-phonon) temperature.

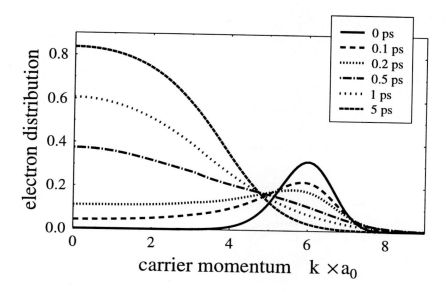

Fig. 4.3. Same as Fig. 4.1 but for the case where additionally also carrier-LO phonon scattering is included. (From Jahnke and Koch 1995)

In Fig. 4.4 we study a situation where the initial carrier distribution is basically a Fermi-Dirac distribution, which, however, is locally disturbed in k-space. Such a situation is relevant for lasers if we consider e.g. the situation of a single mode burning a kinetic hole into the distribution function. Fig. 4.4a shows the rapid relaxation of this disturbed distribution function back to quasi-equilibrium once the perturbation is switched off. Figure 4.4b shows the corresponding carrier-carrier scattering rates $\gamma_{\alpha k}$ defined in Eq. (11). We see that typical scattering times are of the order of 50-100fs.

To get an approximate expression for the carrier relaxation rate, we consider a nonequilibrium carrier distribution that is sufficiently close to the quasiequilibrium Fermi-Dirac distributions as in Fig. 4.4. In such a case we can set

$$\Sigma_{\alpha\mathbf{k}}^{in}\{n\} \simeq \Sigma_{\alpha\mathbf{k}}^{in}\{f\},$$

$$\Sigma^{out}_{\alpha\mathbf{k}}\{n\} \simeq \Sigma^{out}_{\alpha\mathbf{k}}\{f\} \; , \tag{26}$$

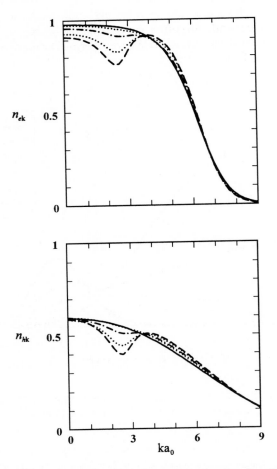

Fig. 4.4 (a). Relaxation of disturbed Fermi distribution functions for electrons (top) and holes (bottom) at a density $N = 3 \times 10^{18} cm^{-3}$ and temperature $T \simeq 300\,K$ obtained by numerically solving the Boltzmann equation using the dynamically screened Coulomb potential in RPA approximation. The times are: $t = 0$ (long dashed), 21 *fs* (dotted), 75 *fs* (dash-dotted), 147 *fs* (dotted), 796 *fs* (solid). (From Binder et al. 1992)

in Eq. (8). Furthermore, substituting

$$\Sigma^{in}_{\alpha\mathbf{k}}\{f\} = \gamma_{\alpha\mathbf{k}}\{f\}f_{\alpha\mathbf{k}} \; , \tag{27}$$

which is a simple rearrangement of Eq. (12) under detailed balance conditions, we find

$$\left.\frac{\partial n_{\alpha \mathbf{k}}}{\partial t}\right|_{col} \simeq -\gamma_{\alpha \mathbf{k}}\{f\}[n_{\alpha \mathbf{k}} - f_{\alpha \mathbf{k}}] \ . \tag{28}$$

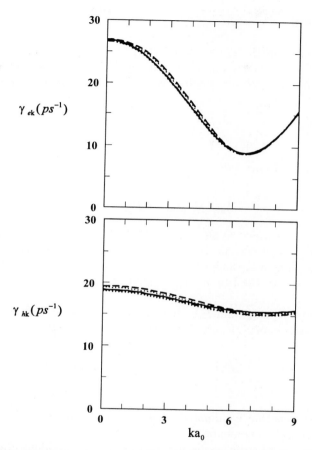

Fig. 4.4 (b). Carrier-carrier scattering rates extracted from Fig. 4.4a. (top) electrons, (bottom) holes. (From Binder et al. 1992)

This approximation fails (barely) to preserve the total carrier density N. To remedy this defect, we study interactions in the neighborhood of $\mathbf{k}_0$, and choose $\gamma_{\alpha \mathbf{k}_0}\{f\}$ instead of $\gamma_{\alpha \mathbf{k}}\{f\}$, that is

$$\left.\frac{\partial n_{\alpha \mathbf{k}}}{\partial t}\right|_{col} \simeq -\gamma_{\alpha \mathbf{k}_0}\{f\}[n_{\alpha \mathbf{k}} - f_{\alpha \mathbf{k}}] \ . \tag{29}$$

which is the carrier relaxation rate approximation. This expression conserves the total carrier density since $\sum_{\mathbf{k}} n_{\alpha \mathbf{k}} = \sum_{\mathbf{k}} f_{\alpha \mathbf{k}} = VN$ (N is the total number of carriers), and is often used to simplify collision terms in the carrier distribution equations of motion.

4.3 Dephasing

The collision terms in the dynamic equation for the interband polarization include such effects as screening of the Hartree-Fock terms and decay of the total polarization, i.e., optical dephasing. Using Eq. (16), we can write the diagonal part of $\underline{\Lambda}$ as

$$\Lambda_{\mathbf{kk}} = -i\Delta_{\mathbf{k}} + \Gamma_{\mathbf{k}} , \qquad (30)$$

and the nondiagonal part as

$$\Lambda_{\mathbf{kk'}} = i\Delta_{\mathbf{kk'}} + \Gamma_{\mathbf{kk'}} . \qquad (31)$$

Since $\Lambda_{\mathbf{kk}}$ has to be added to $\Theta_{\mathbf{kk}}$, Eq. (3-88), in the full dynamic Eq. (3-81) for the interband polarization, we see that formally, $\Gamma_{\mathbf{k}}$ describes a momentum dependent diagonal dephasing rate and $\Delta_{\mathbf{k}}$ yields the corresponding corrections ($\propto V_q^2$) to the Hartree-Fock renormalizations of the free-particle energies.

In the same way, $\Delta_{\mathbf{kk'}}$ and $\Gamma_{\mathbf{kk'}}$ yield momentum dependent nondiagonal damping and shift contributions. It is interesting to note that

$$\sum_{\mathbf{k}} \left.\frac{\partial p_{\mathbf{k}}}{\partial t}\right|_{col} = -\sum_{\mathbf{kk'}} \Lambda_{\mathbf{kk'}} p_{\mathbf{k'}} = 0 , \qquad (32)$$

which demonstrates that Coulomb dephasing of the interband polarization is a pure interference phenomenon.

In the following Chap. 5 we discuss the excitation dependent properties of excitons in semiconductor quantum wells and in microcavity systems. There we will see directly (e.g. Fig. 5.4) how the nondiagonal dephasing contributions have the effect of partially compensating the influence of the $\Gamma_{\mathbf{k}}$. Because this compensation is significant, it is crucial to treat both diagonal and nondiagonal terms symmetrically. The pure dephasing approximation cannot be justified at the microscopic level, so that an analysis of experimental results based on a purely diagonal description of dephasing can only be regarded as a *fit* of the microscopic results, with the dephasing rate treated as a phenomenological input parameter. As such it is important that generally one *should not* use Eq. (30) to compute a dephasing rate. Rather, the dephasing rate should be chosen to account for the effects of both the $\Gamma_{\mathbf{k}}$ and the $\Gamma_{\mathbf{kk'}}$ contributions.

4.4 Screened Hartree-Fock Approximation

If one wants to treat the nonlinear semiconductor response in a simpler approximation without solving the full equations of the previous sections, one often relies on an approach whose advantage is that it is considerably easier to implement numerically. This approach involves using a phenomenological relaxation time approximation to describe the carrier and polarization relaxation. Furthermore, the effects of plasma screening are included phenomenologically by replacing the bare Coulomb potential V_q by the screened Coulomb potential $V_{s,q}$. This treatment of screening effects leads to the *semiconductor Bloch equations* in the *screened Hartree-Fock approximation*. Note that the Coulomb interaction Hamiltonian with the bare Coulomb potential, Eq. (3-67), already contains the mechanism for plasma screening. Therefore, one should be concerned that an *ad hoc* replacement of V_q with $V_{s,q}$ might count some screening effects twice. However, such problems can be avoided within a systematic many-body approach. (Binder and Koch 1995)

The resulting semiconductor Bloch equations in screened Hartree-Fock approximation are

$$\frac{dp_\mathbf{k}}{dt} = -i\omega_{s,\mathbf{k}}p_\mathbf{k} - i\Omega_{s,\mathbf{k}}(z,t)\left[n_{e\mathbf{k}} + n_{h\mathbf{k}} - 1\right] - \Gamma_{eff,\mathbf{k}}p_\mathbf{k} , \qquad (33)$$

$$\frac{dn_{e\mathbf{k}}}{dt} = \frac{i}{\hbar}\left[\Omega_{s,\mathbf{k}}(z,t)p_\mathbf{k}^* - \Omega_{s,\mathbf{k}}^*(z,t)p_\mathbf{k}\right] - (n_{e\mathbf{k}} - f_{e\mathbf{k}})\gamma_{e\mathbf{k}_0} , \qquad (34)$$

and

$$\frac{dn_{h\mathbf{k}}}{dt} = \frac{i}{\hbar}\left[\Omega_{s,\mathbf{k}}(z,t)p_\mathbf{k}^* - \Omega_{s,\mathbf{k}}^*(z,t)p_\mathbf{k}\right] - (n_{h\mathbf{k}} - f_{h\mathbf{k}})\gamma_{h\mathbf{k}_0} , \qquad (35)$$

where

$$\hbar\omega_{s,\mathbf{k}} = \hbar\omega_\mathbf{k} - \sum_{\mathbf{k'}\neq\mathbf{k}} V_{s,|\mathbf{k}-\mathbf{k'}|}(n_{e\mathbf{k}} + n_{h\mathbf{k'}}) + \sum_{\mathbf{q}\neq 0}(V_{s,q} - V_q) \qquad (36)$$

and

$$\Omega_{s,\mathbf{k}}(z,t) = \frac{\mu_\mathbf{k}E(z,t)}{\hbar} + \frac{1}{\hbar}\sum_{\mathbf{k'}\neq\mathbf{k}} V_{s,|\mathbf{k}-\mathbf{k'}|}p_{\mathbf{k'}} . \qquad (37)$$

The screened Coulomb potential is

$$V_{s,q} = \frac{V_q}{\varepsilon(q)} , \qquad (38)$$

where $\varepsilon(q)$ is the longitudinal dielectric function. At the level of a static Hartree-Fock approximation $\varepsilon(q)$ is given by

$$\varepsilon(q) = 1 - V_q \sum_\mathbf{k} \sum_{\alpha=e,h} \frac{n_{\mathbf{k}-\mathbf{q}} - n_\mathbf{k}}{\varepsilon_{\mathbf{k}-\mathbf{q}} - \varepsilon_\mathbf{k}} . \qquad (39)$$

This equation is the static *Lindhard formula*, which is discussed in detail in the literature listed in the references.

References

The classical theory of plasma screening is discussed e.g. in:
Ashcroft, N.W. and N.D. Mermin (1976), *Solid State Theory*, Saunders College, Philadelphia.
Harrison, W. A. (1980), *Solid State Theory*, Dover Publ. New York

The screened Hartree-Fock approximation is discussed in:
Binder, R., and S.W. Koch, (1995), Progr. Quant. Electron. **19**, 307.

The carrier relaxation dynamics is studied e.g. in
Jahnke, F., and S.W. Koch, (1995), Appl. Phys. Lett. **67**, 2278 and Phys. Rev. **A 52**, 1712.
Binder, R., D. Scott, A. Paul, M. Lindberg, K. Henneberger, and S.W. Koch, (1992), Phys. Rev. **B 45**, 1107.

The correlation contributions of the Semiconductor Bloch Equations are evaluated in:
Jahnke, F., M. Kira and S.W. Koch, (1997), Z. Phys. **B 104**, 559.
See also Jahnke, F., et al., (1996), Phys Rev. Lett. **77**, 5257.

5 Excitons in Semiconductor Microcavities

In this chapter we apply the results of the previous sections to discuss the optical properties of semiconductors inside high finesse microcavities, i.e. Fabry-Perot type cavities with a length comparable to the light wavelength. We assume a structure completely made of semiconductor material where the high quality mirrors consist of pairs of quarter wavelength thick alternating layers of different background refractive index. Such a structure is shown schematically in Fig. 5.1. Since many details of such microcavity systems are discussed in the other articles of this lecture note volume, we concentrate here on the microscopic theory of excitonic features. In Sec. 5.1 we start at the level of a semiclassical theory where we describe the light field classically and use for the excitonic response the microscopic theory developed in the previous chapters. At this level we can analyze linear and nonlinear absorption, transmission and reflection properties, (Sec. 5.2). To study also the microcavity luminescence (Sec. 5.3), we then introduce also a quantization of the light field, yielding the semiconductor luminescence equations.

5.1 Semiclassical Theory

Our starting point is Maxwell's wave equation, Eq. (3-19). Using

$$\mathbf{D} = \mathbf{E} + 4\pi \mathbf{P} \tag{1}$$

we have

$$\Delta \mathbf{E}(\mathbf{r},t) - \frac{1}{c^2}\frac{\partial^2}{\partial t^2}\mathbf{E}(\mathbf{r},t) = \frac{4\pi}{c^2}\frac{\partial^2}{\partial t^2}\mathbf{P}(\mathbf{r},t) \ . \tag{2}$$

In the following we consider standard quantum-well (QW) systems which are grown between larger bandgap buffer material to ensure efficient carrier confinement. In a semiconductor microcavity, the QWs are placed between Bragg mirrors which consist of quarter-wavelength dielectric layers, see Fig. 5.1. QW buffer layers and mirror layers are usually nonabsorbing and optically inactive for frequencies close to the QW band edge. Then the light field interacts nonresonantly with the QW buffer layers and the mirror layers and resonantly with the QWs. The nonresonant polarization can be calculated from a background susceptibility using

$$\mathbf{P}_B(\mathbf{r},t) = \chi_B(\mathbf{r})\, \mathbf{E}(\mathbf{r},t) \tag{3}$$

where χ_B represents the excitation-independent refractive index

$$n^2(\mathbf{r}) = 1 + 4\pi \chi_B(\mathbf{r}) \ . \tag{4}$$

In the simplest approximation, this refractive index is real and does not vary with frequency in the parameter range of interest. For the QWs the

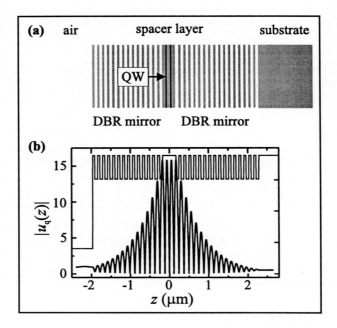

Fig. 5.1. Schematics of a planar semiconductor microcavity consisting of two distributed Bragg reflectors (DBR), spacer layers, and a quantum well (QW) in the cavity. The Bragg mirrors consist of $\lambda/4$ layers made of semiconductor material with different refractive index.

resonant interaction of the states close to the band edge is described by the QW polarization $\mathbf{P}_{QW}$ and the nonresonant interaction with other states is included in the background contribution $\mathbf{P}_B$, such that

$$\mathbf{P} = \mathbf{P}_{QW} + \mathbf{P}_B . \tag{5}$$

Hence, we can write Eq. (2) as,

$$\left[\frac{\partial^2}{\partial z^2} - \frac{n^2(z)}{c^2}\frac{\partial^2}{\partial t^2}\right]\mathbf{E}(z,t) = \frac{4\pi}{c^2}\frac{\partial^2}{\partial t^2}\mathbf{P}_{QW}(z,t) \tag{6}$$

where we again assumed light propagation in z direction, i.e. orthogonal to the layers of the QWs and Bragg mirrors.

The QW polarization $\mathbf{P}_{QW}(z,t)$ can be calculated from the microscopic semiconductor theory outlined in the previous chapters of these lecture notes. In the linear regime, it is convenient to introduce a Fourier transformed polarization,

$$\mathbf{P}_{QW}(z,t) = \int \frac{d\omega}{2\pi} e^{-i\omega t} \mathbf{P}_{QW}(z,\omega) , \tag{7}$$

and similarly a Fourier transformed optical field $\mathbf{E}(z,\omega)$. The spatial extension of the QW polarization $\mathbf{P}_{QW}(z,\omega)$ is restricted to the QW layers; the z-dependence of $\mathbf{P}_{QW}(z,\omega)$ is determined by the carrier confinement wave functions $\zeta(z)$ in the QW, compare Sec. 2.4. Hence, the QW polarization is given by

$$\mathbf{P}_{QW}(z,\omega) = \mathbf{P}_{QW}(\omega) \, |\zeta(z)|^2 \ . \tag{8}$$

The matrix element $\mathbf{P}_{QW}(\omega)$ contains the sum over all possible dipole transitions. In a QW Bloch basis with the in-plane carrier momentum $\mathbf{k}_\perp \equiv \mathbf{k}$, we use

$$\mathbf{P}_{QW}(\omega) = \frac{1}{A} \sum_{\mathbf{k}} \mu_{\mathbf{k}}^* p_{\mathbf{k}}(\omega) + \text{c.c.} \ . \tag{9}$$

If the QW polarization depends only linearly on the light field $p_{\mathbf{k}}(\omega)$ obeys the equation,

$$\hbar(\omega_{\mathbf{k}} - \omega - i\gamma) p_{\mathbf{k}}(\omega) = \mu_{\mathbf{k}} \, \mathbf{E}_{QW}(\omega) + \sum_{\mathbf{k}'} V_{|\mathbf{k}-\mathbf{k}'|} p_{\mathbf{k}'}(\omega) \ , \tag{10}$$

where $\hbar\omega_{\mathbf{k}}$ is the transition energy, compare Chap. 3. Eq. (10) is the Fourier transform of Eq. (3-89) for the case of a QW system. The driving term in Eq. (10) contains the effective field component that interacts with the QW,

$$\mathbf{E}_{QW}(\omega) = \int dz \, \mathbf{E}(z,\omega) \, |\xi(z)|^2 \ . \tag{11}$$

Since the light wave length is much larger than the thickness of typical QWs, the optical field is practically constant over the QW extension. Therefore we can take for $\mathbf{E}_{QW}(\omega)$ simply the optical field at the QW position. In the linear regime we can relate $\mathbf{P}_{QW}$ and $\mathbf{E}_{QW}$ by the QW susceptibility $\chi(\omega)$

$$\mathbf{P}_{QW}(\omega) = \chi(\omega) \, \mathbf{E}_{QW}(\omega) \ , \tag{12}$$

which contains a sum over all excitonic bound and continuum states.

With Eqs. (8), (11) and (12) a closed integral equation for the optical field can be obtained from the wave equation where the properties of the semiconductor QWs enter only through the independently calculated susceptibility $\chi(\omega)$ and the confinement wave function $\xi(z)$,

$$\left[\frac{\partial^2}{\partial z^2} + \frac{\omega^2}{c^2} n^2(z) \right] \mathbf{E}(z,\omega) = -\frac{4\pi\omega^2}{c^2} \, \chi(\omega) \, |\xi(z)|^2 \int dz' \, \mathbf{E}(z') \, |\xi(z')|^2 \ . \tag{13}$$

The main advantage of this equation is that it can be solved analytically for a single QW. Formulating this solution in terms of a transfer-matrix, also multiple QWs and QWs in a microcavity can be treated. In this approach the nonlocal linear QW response, radiative broadening, and coupling effects are fully included.

Since the QW width is small compared to the light wavelength, we can use

$$|\xi(z)|^2 = \delta(z - z_0) \ , \tag{14}$$

where z_0 is the QW position, on the RHS of Eq. (13). Then the wave equation simplifies to

$$\left[\frac{\partial^2}{\partial z^2} + \frac{\omega^2}{c^2}n^2(z)\right] \mathbf{E}(z,\omega) = -\frac{4\pi\omega^2}{c^2} \chi(\omega) \mathbf{E}(z_0,\omega)\delta(z - z_0) \ . \tag{15}$$

In the following we want to obtain a formal solution for light propagating normal to the interfaces of the heterostructure. First we consider the situation of an interface between two adjacent layers denoted by j and $j+1$. Placing the interface at $z = z_0$, we can write the field left (L) and right (R) of the interface as

$$\mathbf{E}_L(z,\omega) = \mathbf{L}_+ e^{iq_L z} + \mathbf{L}_- e^{-iq_L z} \tag{16}$$

and

$$\mathbf{E}_R(z,\omega) = \mathbf{R}_+ e^{iq_R z} + \mathbf{R}_- e^{-iq_R z} \ , \tag{17}$$

respectively. Here, we decomposed the total field into forward ($+$) and backward ($-$) travelling waves and denoted $q_{L/R} = \frac{\omega n_{L/R}}{c}$, where $n_{L/R}$ is the refractive index left or right of the interface.

The conditions

$$\mathbf{E}_L(z = z_0, \omega) = \mathbf{E}_R(z = z_0, \omega) \tag{18}$$

and

$$\frac{\partial}{\partial z}\mathbf{E}_L(z,\omega)|_{z_0} = \frac{\partial}{\partial z}\mathbf{E}_R(z,\omega)|_{z_0} \ , \tag{19}$$

which have to be satisfied for any normal incident field, yield

$$\begin{pmatrix} L_+ \\ L_- \end{pmatrix} = \hat{M} \begin{pmatrix} R_+ \\ R_- \end{pmatrix} \ , \tag{20}$$

where the *transfer matrix* $\hat{M}$ is given by

$$\hat{M} = \frac{1}{2} \begin{pmatrix} \left(1 + \frac{q_R}{q_L}\right) e^{iz_0(q_R - q_L)} & \left(1 - \frac{q_R}{q_L}\right) e^{-iz_0(q_R + q_L)} \\ \left(1 - \frac{q_R}{q_L}\right) e^{iz_0(q_R + q_L)} & \left(1 + \frac{q_R}{q_L}\right) e^{-iz_0(q_R - q_L)} \end{pmatrix} \ . \tag{21}$$

Similarly, we can construct a transfer matrix for a QW surrounded by buffer material. For this purpose we compute the propagation from the left buffer layer into the QW as,

$$\begin{pmatrix} E_+^{QW} \\ E_-^{QW} \end{pmatrix} = \hat{M}_1 \begin{pmatrix} L_+ \\ L_- \end{pmatrix} \tag{22}$$

and from the QW into the right buffer layer as

$$\begin{pmatrix} R_+ \\ R_- \end{pmatrix} = \hat{M}_2 \begin{pmatrix} E_+^{QW} \\ E_-^{QW} \end{pmatrix} , \qquad (23)$$

such that

$$\begin{pmatrix} R_+ \\ R_- \end{pmatrix} = \hat{M}_2 \, \hat{M}_1 \begin{pmatrix} L_+ \\ L_- \end{pmatrix} \equiv \hat{M}_{QW} \begin{pmatrix} L_+ \\ L_- \end{pmatrix} . \qquad (24)$$

Assuming that the buffer layers left and right of the QW have the same background refractive index n and using Eqs. (12) and (14) to characterize the QW we obtain

$$\hat{M}_{QW} = \begin{pmatrix} 1+Y & Ye^{-2iqz_0} \\ -Ye^{2iqz_0} & 1-Y \end{pmatrix} \qquad (25)$$

with

$$Y = \frac{iq}{2\varepsilon} \chi(\omega) , \qquad (26)$$

where $\varepsilon = n^2$. From a given transfer matrix $\hat{M}_{QW}$ the transmission $t = R_+/L_+$ and the reflection $r = L_-/L_+$ for an incident wave from the LHS can be determined directly. Solving

$$\begin{pmatrix} t \\ 0 \end{pmatrix} = \hat{M}_{QW} \begin{pmatrix} 1 \\ r \end{pmatrix} \qquad (27)$$

for r and t leads to

$$r(\omega) = -\frac{M_{21}}{M_{22}} = \frac{i \frac{q}{2\varepsilon} \chi(\omega)}{1 - i \frac{q}{2\varepsilon} \chi(\omega)} \, e^{2iqz_0} , \qquad (28)$$

$$t(\omega) = \frac{M_{11}M_{22} - M_{12}M_{21}}{M_{22}} = \frac{1}{1 - i \frac{q}{2\varepsilon} \chi(\omega)} , \qquad (29)$$

where M_{ij} are the matrix elements of $\hat{M}_{QW}$. The advantage of the transfer-matrix formulation is that it can be easily extended to multiple QWs at arbitrary positions. Since the transfer matrix determines the field coefficients on one side of the QW in terms of coefficients on the other side, a transfer matrix for the combined system of several QWs follows from successively multiplying the transfer matrices of the individual QWs,

$$\hat{M}_{MQW} = \hat{M}_{QW}^N \cdot \hat{M}_{QW}^{N-1} \cdot \ldots \cdot \hat{M}_{QW}^2 \cdot \hat{M}_{QW}^1 , \qquad (30)$$

where $\hat{M}_{QW}^1$ ($\hat{M}_{QW}^N$) is the transfer matrix of the outermost left (right) QW. Then from Eq. (27) reflection and transmission spectra of the combined system can be determined directly.

a) Analytical Solutions for 1s-Excitons

Restricting the analysis to 1s-excitons for this subsection only, we obtain from the two-dimensional Elliott formula

$$\chi_{1s} = -g \frac{|\mu_{cv}|^2}{\hbar\omega - E_{1s} + i\gamma} \tag{31}$$

with the 1s-exciton energy E_{1s} and the nonradiative homogeneous exciton broadening γ. The oscillator strength is determined by the dipole coupling matrix element μ_{cv} and the 1s wave function entering $g = |\phi_{1s}^{2D}(r = 0)|^2$. Using χ_{1s}, we can evaluate the transfer matrix for a single QW,

$$\hat{M}_{QW} = \frac{1}{\hbar\omega - E_{1s} + i\gamma} \tag{32}$$

$$\times \begin{pmatrix} [\hbar\omega - E_{1s} + i(\gamma - \Gamma)] & -i\Gamma\, e^{-2iqz_0} \\ i\Gamma\, e^{2iqz_0} & [\hbar\omega - E_{1s} + i(\gamma + \Gamma)] \end{pmatrix},$$

where a *radiative broadening*,

$$\Gamma = \frac{q}{2\varepsilon} g\, |d_{cv}|^2, \tag{33}$$

has been introduced. With Eqs. (32) we can determine the 1s-exciton reflection, transmission and absorption spectra of a single QW,

$$R(\omega) = |r(\omega)|^2 = \frac{\Gamma^2}{(\hbar\omega - E_{1s})^2 + (\gamma + \Gamma)^2}, \tag{34}$$

$$T(\omega) = |t(\omega)|^2 = \frac{(\hbar\omega - E_{1s})^2 + \gamma^2}{(\hbar\omega - E_{1s})^2 + (\gamma + \Gamma)^2}, \tag{35}$$

$$A(\omega) = 1 - R(\omega) - T(\omega) = \frac{2\gamma\Gamma}{(\hbar\omega - E_{1s})^2 + (\gamma + \Gamma)^2}. \tag{36}$$

The appearance of the additional radiative broadening Γ is a consequence of spatial boundary conditions that have to be satisfied by the solutions of the interacting light-exciton system.

Physically, the radiative broadening of the excitonic resonance results from the lack of momentum conservation in the QW growth direction. A light field, that is resonant with the exciton, can propagate in a three-dimensional semiconductor as a *polariton mode*. The propagation is only limited by the intrinsic semiconductor dephasing as well as material imperfections and the boundary of the crystal. For the light propagating through a QW, only the in-plane momentum has to be conserved. Hence the excited QW polarization can decay due to radiation emitted in the forward and backward directions. This leads to an additional decay channel which in good samples is the dominant one. For GaAs parameters, we obtain from Eq. (33) a radiative lifetime $T_{rad} =$

$\hbar/\Gamma \approx 13$ ps. We see from Eqs. (34)-(36) that small radiative coupling $\Gamma \ll \gamma$, the QW transmission approaches unity while reflection and absorption vanish. On the other hand, for vanishing dephasing γ, only the absorption vanishes.

Multiple QWs exhibit interesting optical coupling effects. (Stroucken et al. 1996; Andreani 1994; Citrin 1996) For two QWs analytical results can be obtained by multiplying two transfer matrices (32). Assuming $\lambda/2$ distance between the QWs (Bragg condition), the reflectivity is given by

$$R_{\lambda/2} = \frac{4\Gamma^2}{(\hbar\omega - E_{1s})^2 + (\gamma + 2\Gamma)^2} \ . \tag{37}$$

In comparison to the single QW, the radiative broadening is enhanced by a factor two and the reflectivity at the 1s-exciton resonance is increased by a factor $4\frac{(\gamma+\Gamma)^2}{(\gamma+2\Gamma)^2}$ due to in-phase coupling of the QW fields. For a $\lambda/4$ distance between the QWs (anti-Bragg condition), destructive interference reduces the QW reflectivity. The reflectivity spectrum,

$$R_{\lambda/4} = \frac{2\Gamma^2}{(\hbar\omega - E_{1s} - \Gamma)^2 + (\gamma + \Gamma)^2} \cdot \frac{2\Gamma^2}{(\hbar\omega - E_{1s} + \Gamma)^2 + (\gamma + \Gamma)^2} \ , \tag{38}$$

contains a product of two Lorentzians each of which is shifted by the radiative broadening Γ. Correspondingly, in the limit $\gamma < \Gamma$ a double-peak structure can be obtained.

b) Numerical Results for Single Quantum Well

The QW susceptibility containing all excitonic bound and Coulomb-enhanced continuum states can be obtained from a numerical solution of Eq. (10). As an example we consider the electron heavy-hole transition of the lowest subband for an 8 nm GaAs QW. In this case, the exciton binding energy is about $2.4E_B$ where E_B is the three-dimensional binding energy and $4E_B$ is the binding energy in perfectly two dimensional system.

The reflection, transmission and absorption spectra for a single QW with dephasing $\gamma = 0.05 E_B$ are shown in Fig. 5.2a. At the 1s-exciton resonance, the transmission is reduced and reflection as well as absorption is possible. When we compare the natural logarithm of the transmission with the imaginary part of the susceptibility in Fig. 5.2b, we see a small deviation due to radiative broadening. This deviation is enhanced for smaller dephasing or due to the radiative coupling between multiple QWs (Stroucken et al. 1996).

c) Excitons in a Microcavity

The linear treatment of the light propagation in QWs and multiple QWs can be readily extended to QWs in a semiconductor microcavity. The distributed

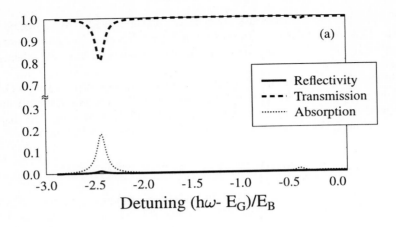
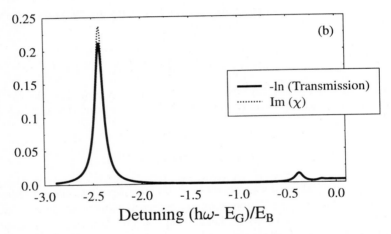

Fig. 5.2 (a). Computed reflection, transmission and absorption spectrum of a 8nm GaAs quantum well.
Fig. 5.2 (b). Comparison between the logarithm of the transmission and the imaginary part of the susceptibility. (Jahnke et al. 1997)

Bragg (DBR) mirrors consist of quarter-wavelength layers with alternating refractive index. For every surface between two mirror layers we use the transfer matrix, Eq. (21). By multiplying the matrices of the dielectric layers, a transfer matrix for both DBR mirrors of a microcavity can be obtained. The successive multiplication of the transfer matrices for the first DBR mirror, the QWs and the second DBR mirror leads to a transfer matrix for the microcavity.

The resulting semiclassical treatment of the light propagation, that incorporates a linear QW susceptibility, describes the coupling between the exciton

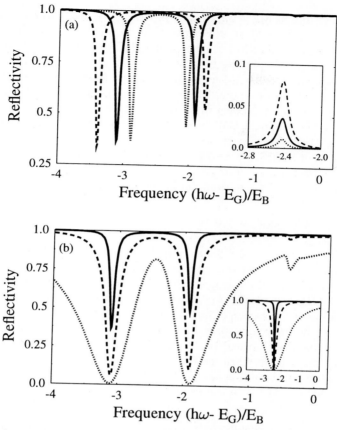

Fig. 5.3 (a). Reflection spectra of a microcavity with different number of quantum wells. (Dashed line = 4 QW, solid line = 2 QW, dotted line = 1 QW). The inset shows the corresponding spectra of the quantum-wells without the Bragg mirrors.
Fig. 5.3 (b). Reflection spectra of a microcavity with two quantum wells and different number of mirror layers: (left/right) solid line (14/16.6), dashed line (10/12.5), dotted line (6/8.5). The inset shows the cavity mode without the quantum wells. (Jahnke et al. 1997)

resonance and the single cavity mode. This *normal mode coupling* (NMC) results in a double peaked resonance structure in linear reflection, transmission and absorption spectra. Fig. 5.3 shows examples of linear normal-mode spectra which are calculated numerically from the susceptibility of the 8 nm QW shown in Fig. 5.2, and the microcavity transfer matrix. We consider a cavity with two Bragg mirrors of 99.6% reflectivity. The first mirror (exposed to air) and second mirror (on a GaAs substrate) contain 14 and 16.5 quarterwave pairs of GaAs ($n=3.61$) and AlAs ($n=2.95$), respectively. A $\frac{3}{2}\lambda$ GaAs spacer between the mirrors leads to two central anti-nodes of the static cavity field.

For the solid line in Fig. 5.3a, in each of the cavity anti-nodes a single 8 nm $In_{0.04}Ga_{0.96}As$ QW is located. For two QWs in every field anti-node (dashed line), the larger oscillator strength of the QW system leads to an increase of the normal-mode splitting by a factor $\sqrt{2}$. If only a single QW is placed in one of the two cavity anti-nodes (dotted line), the normal-mode splitting is reduced by a factor $1/\sqrt{2}$. The corresponding reflectivity of the QW(s) without microcavity is shown in the inset. In Fig. 5.3b we compare cavities with two QWs (a single QW in each of the two cavity anti-nodes) and a reduced number of mirror layers. For 6 and 8.5 quarterwave pair layers and a reflectivity of 88.3% and 86.0%, respectively, normal-mode coupling is still possible (dotted line). However, due to broadening of the cavity resonance (shown in the inset) the normal-mode peaks are strongly washed out.

d) Incoherent Exciton Saturation

After the discussion of the linear optical properties of QW inside and outside a semiconductor microcavity we now analyze a pump-probe situation. Assuming a pump pulse that generates carriers by exciting the system in the interband transition region, we have to compute the linear response to a weak probe field in the presence of incoherent electron-hole pairs.

For a sufficiently long time delay between the electron-hole-pair generation and the weak optical probe pulse, carrier-carrier and carrier-phonon scattering leads to equilibration of the carriers within their bands so that we can make the quasi-equilibrium approximation. For the polarization we then have to solve Eq. (3-81) in the form

$$\frac{dp_\mathbf{k}}{dt} = -i\omega_\mathbf{k} p_\mathbf{k} - i\Omega_\mathbf{k}(f_{e\mathbf{k}} + f_{h\mathbf{k}} - 1) - \sum_{\mathbf{k}'} \Lambda_{\mathbf{k}\mathbf{k}'} p_{\mathbf{k}'} \qquad (39)$$

with $\Lambda_{\mathbf{k}\mathbf{k}'}$ given by Eqs. (4-14) - (4-17), with $n_{\alpha\mathbf{k}}$ replaced by $f_{\alpha\mathbf{k}}$.

As an example we show numerical results for the saturation of the excitonic susceptibility in the presence of a free-carrier plasma. Fig. 5.4 (left) shows the computed exciton spectrum for a given carrier density ($10^{10} cm^{-2}$) and temperature (77 K). The solid line is obtained if all correlation terms in Eq. (39) are considered. For comparison, the dashed line shows the result if exchange contributions $\propto V_q V_{|\mathbf{k}-\mathbf{k}'+\mathbf{q}|}$ are neglected in the scattering terms $\Lambda_{\mathbf{k}\mathbf{k}'}$. Then the broadening increases by almost a factor of two. In the pure dephasing limit, where only the diagonal contributions are considered and the off-diagonal contributions in $\Lambda_{\mathbf{k}\mathbf{k}'}$ are neglected, the broadening is strongly overestimated (dotted line). Hence, we see clearly, how the off-diagonal dephasing compensates diagonal dephasing to a large extent.

In Fig. 5.4 (right), the full calculation is compared with the approximation of a constant damping rate γ. For a small carrier density ($10^{10} cm^{-2}$) the full calculation (thick solid line) can be fitted by a constant dephasing rate $\gamma = 0.05 E_B$ (thick dashed line). Increasing the carrier density to $10^{11} cm^{-2}$,

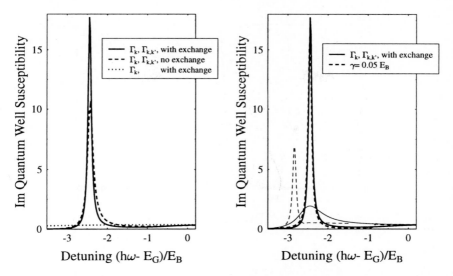

Fig 5.4 (left). Imaginary part of the quantum-well susceptibility showing the influence of the different contributions to the polarization decay. The solid line shows the result of the full calculation, the long dashed line is computed by dropping the exchange terms in the different contributions to Eqs. (4-13), and the dotted line shows the result obtained by neglecting the non-diagonal contribution, Eq. (4-17). **Fig. 5.4 (right).** Comparison of the full solution with a constant dephasing approximation for the densities 10^{10}cm^{-2} (thick lines) and 10^{11}cm^{-2} (thin lines). (Jahnke et al. 1996)

we obtain with constant damping and static screening the well-known but unphysical shift of the 1s-exciton whereas the full dephasing calculation does not exhibit this shift. For the higher carrier density and the same constant damping, the height of the 1s-exciton peak is only reduced by a factor of about 2.5 due to phase-space filling and screening. If the increased broadening is also taken into account within the full calculation, the height of the 1s-exciton peak is reduced almost by an order of magnitude. Figure 5.5 (top) shows the saturation of the 1s-exciton for increasing plasma density computed within the full dephasing model. These results are in good agreement with experimental results (Jahnke et al. 1996).

To compute the optical response for a microcavity containing QWs, we use the QW susceptibility within our transfer-matrix calculation for the microcavity design. As an example we consider a cavity with a $\frac{3}{2}\lambda$ GaAs spacer between GaAs/AlAs mirrors. For the top mirror (exposed to air) and bottom mirror (on a substrate) a reflectivity of 99.6% is obtained with 14 and 16.5 quarterwave pairs. A single 8 nm In$_{0.04}$Ga$_{0.96}$As QW is placed in each of the two cavity anti-nodes. The cavity wavelength is chosen to coincide with the 1s-exciton resonance of the QWs. The calculated microcavity trans-

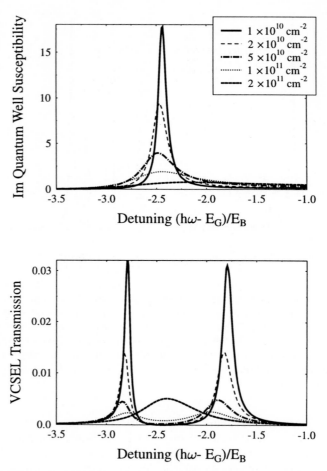

Fig. 5.5 (top). Computed saturation of the quantum-well exciton resonance for different incoherent carrier densities.

Fig. 5.5 (bottom). Corresponding reduction of the normal mode resonances of the quantum-well microcavity system. (Jahnke et al. 1996)

mission is shown in Fig. 5.5 (bottom). At low excitation levels we again see the double-peaked transmission characteristic for the normal mode coupling (NMC) between exciton and cavity resonance. For increasing bleaching of the 1s-exciton resonance with increasing carrier density, we find a strong reduction of the NMC peak-height with only small reduction of the NMC splitting. The reduced transmission and the increasing width of the individual NMC peaks indicate the strong broadening of the exciton resonance, whereas the small reduction of the splitting clearly reveals the minor reduction of the exciton oscillator strength within a large plasma density range. With increasing plasma density the renormalized band edge approaches the energetically stable 1s-exciton resonance. The rather abrupt replacement of the NMC doublet by a single transmission peak occurs when the exciton resonance is fully bleached and the cavity resonance becomes degenerate with the band edge.

5.2 Nonlinear Exciton Saturation

As the first step towards the treatment of the nonlinear light propagation problem in QWs and microcavities, we use the full semiconductor Bloch equations with the microscopic carrier relaxation and dephasing terms discussed in the previous chapter. In the following, we combine this model with a solution of the wave equation (6) for the optical field that describes the nonlinear field dynamics. For time-dependent linear studies or for a nonlinear field dependence of the polarization, it is convenient to solve the wave equation directly in the time-domain.

We assume a sequence of narrow QWs for which the time dependent polarization can be written as

$$P_{QW}(z,t) = \sum_i P_{QW,i}(t)\, \delta(z - z_i) \; , \tag{40}$$

where $P_{QW,i}$ is the polarization of the QW at position z_i. For this case, the analytical solution of the wave equation (6) can be obtained in Fourier space. Introducing the expansion

$$E(z,t) = \frac{1}{(2\pi)^2} \int dq\, d\omega\, e^{i(qz-\omega t)} E(q,\omega) \; , \tag{41}$$

the Fourier coefficients obey the equation

$$E(q,\omega) = \frac{4\pi\, \omega^2}{c^2 q^2 - \omega^2 n^2} \sum_i P_{QW,i}(\omega)\, e^{-iqz_i} \; . \tag{42}$$

With the residue theorem, this leads to the space-time domain solution

$$E(z,t) = -\frac{2\pi}{nc} \sum_i \frac{\partial}{\partial t} P_{QW,i}(t - \frac{n|z - z_i|}{c}) + C^{\pm}\, E_0^{\pm}(t \mp \frac{z}{c}) \; . \tag{43}$$

Since we need the general solution of the inhomogeneous wave equation (z) we have added the general solution of the homogeneous wave equation, $C^{\pm}\, E_0^{\pm}(t \mp \frac{z}{c})$ describing free incident waves.

As a first example, we discuss the light propagation through a single QW at $z = 0$. An incident wave is applied on the QW in the forward direction choosing $C^+ = 1$ and $C^- = 0$. The inhomogeneous solution describes the QW emission as forward and backward traveling waves for $z > 0$ and $z < 0$, respectively, as can be seen from

$$\frac{\partial}{\partial t} P_{QW}(t - \frac{|z - z_i|}{c}) = \begin{cases} \frac{\partial}{\partial t} P_{QW}(t - \frac{z}{c}) & z > 0 \\ \frac{\partial}{\partial t} P_{QW}(t + \frac{z}{c}) & z < 0 \end{cases}. \quad (44)$$

The transmitted and reflected fields, E^+ and E^-, can be found by taking the forward propagating component of the solution for $z > 0$ and the backward propagating part of the solution for $z < 0$. Together with the field at the QW position we obtain

$$E^+(t - \frac{z}{c}) = E_0^+(t - \frac{z}{c}) - \frac{2\pi}{nc} \frac{\partial}{\partial t} P_{QW}(t - \frac{z}{c}), \quad (45)$$

$$E^-(t + \frac{z}{c}) = -\frac{2\pi}{nc} \frac{\partial}{\partial t} P_{QW}(t + \frac{z}{c}), \quad (46)$$

$$E_{QW}(t) = E_0(t) - \frac{2\pi}{nc} \frac{\partial}{\partial t} P_{QW}(t). \quad (47)$$

For the transmitted and reflected field components of a system of several QWs Eq. (43) yields

$$E^+(t - \frac{z}{c}) = E_0^+(t - \frac{z}{c}) - \frac{2\pi}{nc} \sum_{i=1}^{N} \frac{\partial}{\partial t} P_{QW,i}(t - \frac{z - z_i}{c}), \quad (48)$$

$$E^-(t + \frac{z}{c}) = -\frac{2\pi}{nc} \sum_{i=1}^{N} \frac{\partial}{\partial t} P_{QW,i}(t + \frac{z - z_i}{c}). \quad (49)$$

a) Numerical Results for Single Quantum Well

As an application we first investigate the saturation of the excitonic resonance in a single QW. For this purpose we show in Fig. 5.6a the time evolution of the QW polarization for propagation of a 100 fs laser pulse through a single 8 nm GaAs QW. Using a small Rabi energy $\Omega_R = \mu_{cv} E$ of an externally applied pulse, the solution remains in the linear regime and the polarization decay is governed by the background damping. The temporal oscillations in the polarization are quantum beats between the excitonic resonances, most prominently the 1s-resonance and the higher states. At elevated pulse energies, enhanced polarization decay is obtained from increased efficiency of carrier and polarization scattering. This is the phenomenon of *excitation induced dephasing*. The calculated spectra of the transmitted pulses are shown in Fig. 5.6b. Similar to the case of quasi-equilibrium excitation, bleaching of the exciton transmission occurs basically without any shift of the 1s-exciton resonance.

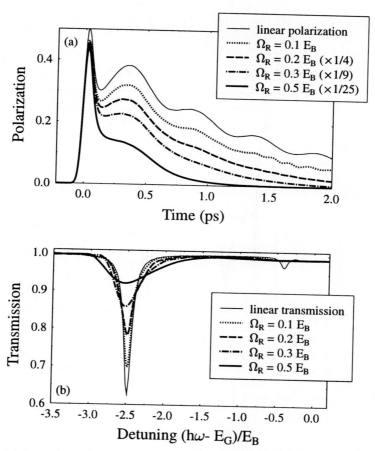

Fig. 5.6 (a). Time dependent polarization in a quantum well after resonant exciton excitation with a 100 fs laser pulse of different Rabi energy $\Omega_R = \mu E$.
Fig. 5.6 (b). Corresponding quantum-well transmission spectra. (Jahnke et al. 1997)

b) Microcavity Results

Using the nonlinear theory, we can study the saturation of the excitonic normal-mode coupling (Jahnke et al. 1997). Calculations of the time-resolved reflected signal for increasing intensity of the applied 100 fs pulse are shown in Fig. 5.7. We see that the time dependent polarization exhibits oscillations which are the real-time counterpart of the appearance of the normal-mode resonance doublet in the frequency domain (inset to Fig. 5.7). The modulation depth of the polarization oscillations decreases with increasing intensity, whereas the oscillation period is basically excitation independent. These time domain results are then consistent with results such as those in Fig. 5.5 showing the excitation dependent saturation of the normal-mode resonances without loss of excitonic oscillator strength.

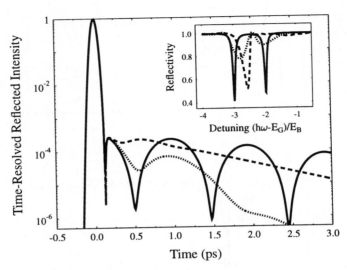

Fig. 5.7. Reflected intensity as function of time after excitation of a microcavity with a 100 fs pulse and different Rabi energies $\Omega_R = 0.01\varepsilon_R$ (solid line), $\Omega_R = \varepsilon_R$ (dotted line) and $\Omega_R = 2\varepsilon_R$ (dashed line). The inset shows the corresponding reflectivity spectra. [Jahnke et al. (1997)]

5.3 Microcavity Luminescence

So far, we have used semiclassical theory, where the medium is treated quantum mechanically but the electromagnetic radiation is modelled as a classical light field. This approach is usually well justified as long as the classical fields exceed the vacuum fluctuations, which are almost always much smaller than the weakest probe beam. However, photoluminescence is an important phenomenon which cannot be explained semiclassically. The results of the previous sections, e.g. Figs. 5.6 and 5.7, show that without external driving field the polarization and the coherent microcavity field $E = \langle E \rangle$ decay away typically on a ps time scale after the excitation pulse. However, in many cases a substantial number of incoherent electrons and holes remains excited in the system. The system can then reach its ground state via non-radiative electron-hole recombination or radiatively through spontaneous emission leading to photoluminescence out of the quantum well.

A quantum treatment of light is required to describe photoluminescence because the field has nonclassical properties, e.g., $\langle E \rangle = 0$ but $\langle EE \rangle \neq 0$. The fully quantum mechanical analysis of the interacting photon-semiconductor electron-hole system poses a considerable challenge to current theories. In the semiclassical calculations, the major difficulties arise from the consistent inclusion of carrier-carrier Coulomb interaction effects. In addition to that we now need a fully quantum mechanical theory for the interacting carrier-photon system. A detailed description of the semiconductor luminescence quantum theory is beyond the scope of these lecture notes and will be published elsewhere (Kira et al., to be published).

Here, we merely present the basic results and discuss the intuitive physics behind the results. To avoid a mixture of coherent and incoherent effects, which is an interesting subject on its own, we focus on the theoretical analysis of photoluminescence experiments where carriers are non-resonantly generated in the QW by optical excitation high above the semiconductor bandedge. Hence, there is no coherent field or intraband polarization generated in the vicinity of the exciton resonance and we can use

$$p_\mathbf{k} = 0, \qquad \langle d_q(t) \rangle = 0 \;, \tag{50}$$

where d_q is the destruction operator of a photon in the mode q. As discussed earlier, an initially generated distribution of electrons and holes rapidly relaxes into quasi-equilibrium Fermi-Dirac distributions within the respective bands. Hence, we only need dynamic equations for the photon operators and the operator for the semiconductor interband polarization. Since we are interested in the photoluminescence intensity spectrum, we want to compute the quantity $\langle d_q^+ d_q \rangle(t)$. Using the light-matter coupling Hamiltonian

$$H_{l-m} \propto \sum \left(A_{q,\mathbf{k}} \, d_q^+ a_{\mathbf{k+q}} b_{-\mathbf{k}} + h.c. \right) \;, \tag{51}$$

we obtain from the Heisenberg equation of motion

$$i\hbar \frac{\partial}{\partial t} \langle d_q^+ d_{q'} \rangle = \hbar \left(\omega_{q'} - \omega_q \right) \langle d_q^+ d_{q'} \rangle$$
$$+ iF_q \sum_{\mathbf{k}'} \mu_{cv} \langle a_{\mathbf{k}'}^+ b_{-\mathbf{k}'}^+ d_{q'} \rangle + iF_{q'} \sum_{\mathbf{k}'} \mu_{cv}^* \langle d_q^+ b_{-\mathbf{k}'} a_{\mathbf{k}'} \rangle \;, \tag{52}$$

where F_q is the effective mode strength at the QW position. Eq. (52) shows that the expectation value $\langle d_q^+ d_q \rangle$ is driven by terms containing light-matter correlations like $\langle d_q^+ a_\mathbf{k} b_{-\mathbf{k}} \rangle$. The dynamic equations for these correlations include Coulombic many-body contributions which once again lead to the hierarchy problem of continuous coupling to correlation functions of higher order. To obtain a closed set of equations we therefore have to use the truncation schemes discussed in the previous chapter of these lecture notes. At the Hartree-Fock level we obtain the coupled equations

$$i\hbar \frac{\partial}{\partial t} \langle d_q^+ a_\mathbf{k} b_{-\mathbf{k}} \rangle = \left(\frac{\hbar^2 k^2}{2m_r} - \hbar\omega_q + \varepsilon_g - \sum_{\mathbf{k}'} V_{\mathbf{k}-\mathbf{k}'} (n_{e\mathbf{k}'} + n_{h\mathbf{k}'}) \right) \langle d_q^+ a_\mathbf{k} b_{-\mathbf{k}} \rangle$$
$$+ (n_{e\mathbf{k}} + n_{h\mathbf{k}} - 1) \, \Omega(\mathbf{k}, q) + n_{e\mathbf{k}} n_{h\mathbf{k}} \Omega_{SE}(\mathbf{k}, q) \;, \tag{53}$$

and

$$i\hbar\frac{\partial}{\partial t}n_{\alpha\mathbf{k}} = -2i\sum_q \text{Im}\left[\mu_{cv}^\star iF_q\langle d_q^+\hat{p}_\mathbf{k}\rangle\right], \quad \alpha = e, h \,. \tag{54}$$

Equations (52)-(54) give a closed set of semiconductor luminescence equations with the renormalized stimulated emission/absorption term

$$\Omega(\mathbf{k},q) = \mu_{cv}\left(\sum_{q'}iF_{q'}\langle d_{q'}^+ d_{q'}\rangle - \sum_{\mathbf{k}'}\mu_{cv}^\star\langle d_q^+ b_{-\mathbf{k}'}a_{\mathbf{k}'}\rangle\right)$$
$$+ \sum_{\mathbf{k}'}V_{\mathbf{k}-\mathbf{k}'}\langle d_q^+ a_{\mathbf{k}'}b_{-\mathbf{k}'}\rangle \tag{55}$$

similar to the renormalized Rabi energy of a classical field, Eq. (3.85). In Eq. (53) the rate of spontaneous emission is given by

$$\Omega_{SE}(\mathbf{k},q) = iF_q\mu_{cv} \,. \tag{56}$$

The correlation $\langle d_q^+ a_\mathbf{k} b_{-\mathbf{k}}\rangle$ gives the amplitude of a process where an electron-hole pair, having zero center of mass momentum, recombines by emitting a photon with vanishing in-plane momentum. Even if the field-particle and the field-field correlations are initially taken to be zero, correlations start to build up because of the term $n_{e\mathbf{k}}n_{h\mathbf{k}}\Omega_{SE}(\mathbf{k},q)$ in Eq. (53). This driving term is directly associated with spontaneous emission triggering the recombination process. According to the factor $n_{e\mathbf{k}}n_{h\mathbf{k}}$, the spontaneous recombination takes place only if an electron and hole are present simultaneously. The stimulated emission/absorption $\Omega(\mathbf{k},q)$ strongly influences the photoluminescence spectrum, such that the resulting photoluminescence reflects the dynamic interplay of the field-field and field-particle correlations affected by the elementary processes of spontaneous emission and stimulated emission or absorption.

The semiconductor luminescence equations partially resemble the semiconductor Bloch equations describing the interaction of classical fields with the semiconductor. The semiclassical calculation used in the previous sections of this chapter includes screening and dephasing due to carrier-carrier scattering and polarization scattering, which are beyond the Hartree-Fock approximation. Such a fully microscopic calculation of photoluminescence remains a major challenge for the future. However, since the full quantum theory can be reduced to semiclassical calculation, these effects can be described relatively well, at least for incoherent excitation, using the screened Hartree-Fock approximation (see Sec. 4.4) and a dephasing rate which is extracted from an independent quantum kinetic calculation. This simplifies the quantum calculations considerably.

a) Numerical Results for Single Quantum Well

As a first application we compute the steady-state luminescence spectra of a single 8 nm QW for various carrier densities. The results are shown in Fig. 5.8a in direct comparison with the corresponding absorption spectra, Fig. 5.8b. We see that the photoluminescence spectra have their maximum close to the excitonic absorption peak. Furthermore, the QW luminescence stays peaked for much higher carrier densities than the excitonic absorption. Even for relatively high densities, where the excitonic resonance has vanished from the absorption spectra, the semiconductor luminescence remains peaked, however, it gradually becomes asymmetric.

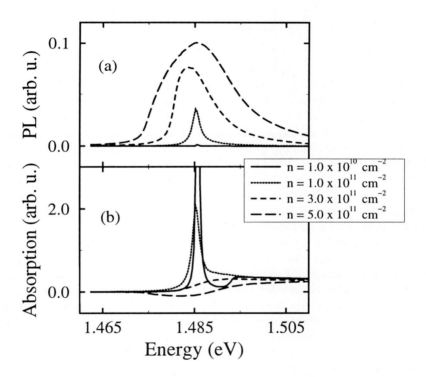

Fig. 5.8. Computed quantum-well luminescence (a) and absorption spectra (b) for different electron-hole plasma densities (Kira et al. 1998)

b) Microcavity Results

Next we compute the luminescence for QW inside a microcavity. In Fig. 5.9 we show examples of calculations for different detunings Δ between cav-

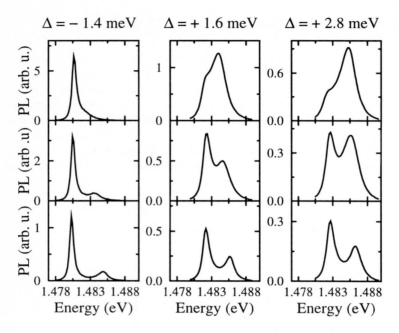

Fig. 5.9. Computed luminescence spectra for different detunings Δ between cavity and exciton resonance. The carrier densities are 2.1, 1.7 and 1.25×10^{11} cm^{-2} from top to bottom. (Kira et al. 1998)

ity mode and exciton resonance, clearly demonstrating double peaked emission. For different excitiation levels the relative height of the emission peaks changes. In particular, we note that for positive detunings ($\Delta > 0$), the high energy peak gradually overtakes the low energy peak. After its experimental observation this nonlinear behavior has led to erroneous speculations concerning excitonic condensation effects. However, in reality we can understand the nonlinear microcavity luminescence under incoherent excitation conditions in detail at the level of theory presented here simply by analyzing the different microscopic contributions to the spontaneous emission in a microcavity (Kira et al. 1997). These investigations show that the nonlinear luminescence properties of QW in microcavities under incoherent pumping conditions result from the interplay between phase-space, renormalization and light-coupling contributions in the interacting electron-hole-photon system.

Generally, the fully quantum mechanical theory can serve as starting point for even more elaborate treatments, e.g. of photon correlation effects, dynamic luminescence properties and the intricate interplay between coherent and incoherent excitation and emission properties. Work along these lines is in progress as well as attempts to include additional effects, such as the influence of phonon coupling, admixture of radiative and guided/evanescent modes as well as scattering of carriers by static disorder.

References

Light coupling effects in multi-quantum well systems are analyzed e.g. in
Stroucken, T., A. Knorr, P. Thomas, and S.W. Koch, (1996), Phys. Rev. **B 53**, 2026.
Andreani, L.C., (1994), Phys. Lett. **A 192**, 99 and references.
Citrin, D.S., (1996), Phys. Rev. **B 54**, 1 and references.

Normal mode coupling in semiconductor microcavity systems was first reported in
Weisbuch, C., M. Nishioka, A. Ishikawa, and Y. Arakawa, (1992), Phys. Rev. Lett. **69**, 3314.

Many more references are given in the other articles of this lecture note volume.

The microscopic theory at the semiclassical level is developed in
Jahnke, F., M. Kira, and S.W. Koch, (1997), Z. Physik **B 104**, 559.
Jahnke, F., et al., (1996), Phys. Rev. Lett. **77**, 5257.

For the fully quantum mechanical theory see:
Kira, M. et al., (1997), Phys. Rev. Lett. **77**, 5170
and Kira, M., F. Jahnke and S.W. Koch, to be published.

An Introduction to Photonic Crystals

J.D. Joannopoulos

Massachusetts Institute of Physics, Department of Physics, Cambridge, Massachusetts 02139, USA

LECTURE I

1 Motivation

For the past 50 years, semiconductor physics has played a vital role in almost every aspect of modern technology. Advances in this field have allowed scientists to tailor the conducting properties of certain materials and have initiated the transistor revolution in electronics. New research suggests that we may now be able to achieve similar accomplishments with light. The key lies in the use of a new class of materials called *photonic crystals* (Joannopoulos et al. 1995).

The underlying concept behind these materials stems from early work by Yablonovitch (1987) and John (1987). The basic idea is to design materials that affect the properties of photons in much the same way that solids affect the properties of electrons. From a practical point of view, there are many instances where one would like to replace the electron with the photon in technological applications. A photon is faster than an electron and suffers fewer losses because it interacts much more weakly than an electron. Of course, this "weak" interaction is also responsible for the difficulties which arise in trying to control light.

In a semiconductor or metal, the electronic structure is determined by the nature of the atomic potential and the periodicity of the solid. For photon modes, it is the dielectric response of a material that determines their properties. Thus let us examine what happens if we have a periodic arrangement of macroscopic homogeneous (for simplicity) dielectric media, i.e. a "photonic crystal". There will be many parallels between our discussion and the language of elementary quantum mechanics.

2 The Master Equation

The equations that govern all of macroscopic electromagnetism including the propagation of light in a photonic crystal are of course Maxwell's equations:

$$\nabla \cdot B = 0 \tag{1}$$
$$\nabla \cdot D = 4\pi\rho \tag{2}$$
$$\nabla \times E + \frac{1}{c}\frac{\partial B}{\partial t} = 0 \tag{3}$$
$$\nabla \times H - \frac{1}{c}\frac{\partial D}{\partial t} = \frac{4\pi J}{c} \tag{4}$$

In the absence of free charges and currents, $\rho = J = 0$. Next we consider the constitutive equations which relate D to E and B to H. For example, quite generally:

$$D_i = \sum_j \epsilon_{ij}(r,\omega)E_j + \sum_{j,k} \chi_{ijk} E_j E_k + \vartheta(E^3) + \cdots \tag{5}$$

But for many dielectric materials, it is a good approximation to assume:

1. Non-linear terms are negligible, $\Rightarrow \chi = 0$;
2. Isotropic macroscopic media, $\Rightarrow \epsilon$ is a scalar;
3. No losses, $\Rightarrow \epsilon$ is purely real;
4. Weak ω dependence of ϵ. Instead we simply choose the value of ϵ appropriate to the frequency of interest, i.e. the ω-dependence of ϵ is slowly varying.

With those assumptions, $D(r) = \epsilon(r)E(r)$ & $H(r) = B(r)$ $[\mu = 1]$, so that

$$\nabla \cdot H(r,t) = 0 \tag{6}$$
$$\nabla \cdot \epsilon(r)E(r,t) = 0 \tag{7}$$
$$\nabla \times E(r,t) + \frac{1}{c}\frac{\partial H(r,t)}{\partial t} = 0 \tag{8}$$
$$\nabla \times H(r,t) - \frac{\epsilon(r)}{c}\frac{\partial E(r,t)}{\partial t} = 0 \tag{9}$$

The last two equations above are reminiscent of coupled time-dependent Schrödinger-like equations. So as in the case of the Schrödinger equation, let us proceed to derive a time-independent equation that describes the stationary solutions. Since Maxwell's equations are linear, any solution can be expanded in terms of the stationary states. We thus seek an equation that describes solutions of the form:

$$H(r,t) = H(r)e^{i\omega t} \quad E(r,t) = E(r)e^{i\omega t} \tag{10}$$

Substitution of (10) into (6) and (7) gives at once

$$\nabla \cdot H(r) = \nabla \cdot D(r) = 0. \tag{11}$$

Thus the fields are built up of waves that are transverse. For example, if $H(r) = \sum_q h_q e^{iq\cdot r}$, then $\nabla \cdot H = 0 \Rightarrow \sum_q iq \cdot h_q e^{iq\cdot r} = 0, \forall r \Rightarrow q \cdot h_q = 0 \Rightarrow q \perp h_q$. Substitution of (10) into (8) and (9) gives

$$\nabla \times E(r) + \frac{i\omega}{c} H(r) = 0 \quad \nabla \times H(r) - \frac{i\omega}{c}\epsilon(r) E(r) = 0 \qquad (12)$$

If we now take $\nabla \times$ of the 2nd equation in (12) and substitute for the 1st, we get,

$$\nabla \times \left[\frac{1}{\epsilon(r)} \nabla \times H(r)\right] = \left(\frac{\omega}{c}\right)^2 H(r) \quad \text{with } \nabla \cdot H = 0 \qquad (13)$$

This is the Master Equation. Once we know $H(r)$ we can get $E(r)$ (or $D(r)$) from equation (12) via

$$E(r) = \left(-\frac{ic}{\omega\epsilon(r)}\right) \nabla \times H(r) \qquad (14)$$

Note that (13) is just an eigenvalue problem! Moreover, the operator $\Theta \equiv \nabla \times \frac{1}{\epsilon} \nabla \times$ is a linear Hermitian operator and thus reminiscent of the Hamiltonian for electrons. Thus all the beautiful properties of eigenfunctions of the Hamiltonian follow for the solutions $H(r)$. That is:

1. they have real eigenvalues (also for Θ we can show they are all positive);
2. they can be made orthogonal;
3. they can be catalogued by their symmetry properties;
4. they can be obtained by a variational principle.

At this point the reader may wonder why we have put everything in terms of the magnetic field rather than the electric field. The corresponding eigenvalue problem for the electric field is a generalized Hermitian problem rather than the simple Hermitian problem obtained for the magnetic field. The source of the asymmetry between E and B was our approximation of constant magnetic permeability but varying dielectric constant, but this is a very good approximation in most materials.

2.1 Translational symmetries

Let us now examine the effects of discrete translational symmetry (i.e. a crystal) on the eigenfunctions of Θ. The fact that we have a crystal (an infinite discrete, periodic system) can be expressed by

$$\epsilon(r) = \epsilon(r + R) \quad \forall R \in \{R\} \qquad (15)$$

The $\{R\}$ are called lattice vectors. For a simple 1D system of dielectric spheres in air separated by distance a, the $\{R\} = \{ma\}$ with $m = 0, \pm 1, \pm 2$, etc. The vector a is called a primitive lattice vector and defines a primitive unit as we have already alluded to in the case of transversality of our fields, to think in terms of expansions of waves, i.e. $\{e^{iq \cdot r}\}$. Thus, let us consider the consequences of periodicity on the Fourier transform of $\epsilon(r)$:

$$\epsilon(r) = \sum_q \epsilon_q e^{iq \cdot r} = \epsilon(r+R) = e^{iq \cdot R} \sum_q \epsilon_q e^{iq \cdot r} = e^{iq \cdot R} \epsilon(r) \qquad (16)$$

thus,

$$e^{iq \cdot R} = 1 \quad \forall \{R\} \qquad (17)$$

This severely restricts the set of q for which $\epsilon_q \neq 0$! It turns out that to satisfy (17) $\forall R$, q must be of the form

$$q = m_1 b_1 + m_2 b_2 + m_3 b_3; \quad \{m_j\} \text{ integer} \qquad (18)$$

with

$$b_1 = 2\pi \frac{a_2 \times a_3}{a_1 \cdot a_2 \times a_3} \quad \text{and} \quad b_i \cdot a_j = 2\pi \delta_{ij} \qquad (19)$$

where

$$R = m_1 a_1 + m_2 a_2 + m_3 a_3; \quad \{m_i\} \text{ integer} \qquad (20)$$

The set of $\sum_j m_j b_j \equiv G$ are called reciprocal lattice vectors because they define a crystalline lattice in reciprocal space. Thus any periodic function satisfies

$$f(r) = \sum_G e^{iG \cdot r} f_G \qquad (21)$$

We shall discover shortly that just as in real space, we can describe a crystal simply by looking at its primitive cell, in reciprocal space we can describe all the solutions of Θ by looking into a primitive cell of the reciprocal lattice spanned by b_i.

Lets us now return to the problem at hand, what are the effects of periodicity on the eigenfunctions $H(r)$ of Θ?

Given that $\epsilon(r) = \epsilon(r+R) \quad \forall R$, let us define $O_R = e^{-R \cdot \nabla}$ such that $O_R \epsilon(R) = \epsilon(r-R)$, then Θ commutes with the discrete translation operator O_R

$$[O_R, \Theta] = 0 \quad \text{with} \quad O_R \equiv e^{-R \cdot \nabla} \quad \text{and} \quad O_R f(r) = f(r-R) \qquad (22)$$

Thus, as we recall from our Q.M., the eigenfunctions of Θ can be chosen to be simultaneous eigenfunctions of O_R and the eigenfunctions of O_R are much easier to deal with. So let's build on this idea.

Let ϕ_k be an eigenstate of O_R with eigenvalue λ_k, i.e.

$$O_R \phi_k = \lambda_k \phi_k \qquad (23)$$

Now since $[O_R, \Theta] = 0$, we have,

$$O_R(\Theta\phi_k) = \Theta(O_R\phi_k) = \lambda_k(\Theta\phi_k) \tag{24}$$

Thus $\Theta\phi_k$ is also an eigenvalue of O_R with the same eigenvalue λ_k. If there are degenerate states ϕ_k^β with eigenvalue λ_k, then

$$\Theta\phi_k = \sum_\beta C_{k,\beta}\phi_k^\beta \tag{25}$$

Thus Θ can be diagonalized in the subspace ϕ_k^β and the eigenfunctions $H(r)$ can be labeled by k and satisfy:

$$H_k(r) = \sum_\beta C'_{k,\beta}\phi_k^\beta \ \ \text{with} \ \ O_R\phi_k^\beta = \lambda_k\phi_k^\beta \tag{26}$$

Thus any eigenstates of Θ (even degenerate ones) can be chosen to satisfy

$$O_R H_k(r) = \lambda_k H_k(r) = H_k(r-R) \tag{27}$$

Now what is λ_k? A convenient set of complete orthogonal eigenfunctions of O_R are simply the set $e^{ik\cdot r}$. Thus

$$O_R e^{ik\cdot r} = e^{-ik\cdot R} e^{ik\cdot r} \ \ \text{with} \ \ \lambda_k = e^{-ik\cdot R} \tag{28}$$

(Note that the subspace ϕ_k^β in this case is just $e^{i(k+G)\cdot r}\forall G$.) Call this subspace $S_{\mathbf{k}}$ for each $\mathbf{k}$. The vectors $\mathbf{k}$ can be taken to lie within the first Brillouin zone of the reciprocal lattice, which is just one maximal set of vectors with no two differing by a reciprocal lattice vector (since $S_{\mathbf{k}} = S_{\mathbf{k+G}}$ for a reciprocal lattice vector $\mathbf{G}$).

In order to diagonalize Θ and find the modes of the master equation, we only need to diagonalize within these countably infinite subspaces of plane waves $S_{\mathbf{k}}$. The resulting eigenstates will be combinations of plane waves $e^{i(k+G)\cdot r}$, and will not be plane waves themselves but instead of the form

$$H_k(r) = e^{ik\cdot r}u_k(r) \tag{29}$$

where $u_k(r)$ is periodic in the real space lattice. Substitution of this form into the master equation gives the reduced master equation

$$(ik+\nabla)\times\frac{1}{\epsilon(r)}(ik+\nabla)u_k(r) = \frac{\omega(k)^2}{c^2}u_k(r). \tag{30}$$

This equation can be solved numerically for all $\mathbf{k}$ in the first Brillouin zone to give the photonic bands, $\omega_n(k)$, of the crystal.

2.2 Rotational symmetries

The region of the Brillouin zone which needs to be studied can be further reduced by other discrete symmetries. Discrete rotational symmetries of the photonic crystal are also symmetries of the reciprocal lattice, so that only an irreducible sector of the B.Z. need be studied. Let's look explicitly at some rotation operator O_A corresponding to the rotation $A \in O(3)$ (proper and improper rotations allowed). Since A is a symmetry element of the photonic crystal, $[O_A, \Theta] = 0$. If H_k is an eigenstate of Θ, then it follows from

$$\Theta(O_A H_k) = O_A(\Theta H_k) = \frac{\omega_k^2}{c^2}(O_A H_k) \tag{31}$$

that $O_A H_k$ is also an eigenstate of Θ. Now we show $O_A H_k$ is the Bloch state H_{Ak}. If O_R is the translation operator by a lattice vector R, $O_R O_A = O_A O_{A^{-1}R}$ and

$$O_R(O_A H_k) = O_A O_{A^{-1}R} H_k = O_A e^{-ik \cdot A^{-1} R} H_k = e^{-iAk \cdot R}(O_A H_k). \tag{32}$$

It follows immediately that $\omega_n(k) = \omega_n(Ak)$, so that the bands need only be found within the B.Z. reduced modulo A. This defines the so-called irreducible B.Z.

Let us now examine mirror symmetry in a little more detail. In a two-dimensional periodic photonic crystal (i.e., one with continuous translational symmetry in the z direction, or no z direction at all), mirror symmetry through the plane $z = 0$ leads to a useful separation of modes into two symmetry classes. Let O_M be the operator for mirror reflections M through the plane. Since $O_M{}^2 = 1$,

$$O_M H_k(r) = e^{i\phi} H_{Mk}. \tag{33}$$

If $Mk = k$ (k is in the plane), the eigenvalues of O_M are ± 1. So $MH_k(r) = \pm H_k(r)$. From Maxwell's equations $ME_k(r) = \pm E_k(r)$. Since E transforms like a vector and H like a pseudovector, there are only two classes: even ($\lambda = +1$), with nonzero E_x, E_y, H_z, and odd, with nonzero E_z, H_x, H_y. We assumed in the above that H_k was nondegenerate, but in the degenerate case combinations of states can be chosen to satisfy the classification. Thus for all two-dimensional photonic crystals at $k_z = 0$, all modes are either even (TE) or odd (TM), with electric and magnetic fields transverse to the mirror plane normal, respectively.

3 Scale Invariance

Maxwell's equations for macroscopic media have no fundamental length scale, and consequently neither does the master equation. Suppose we know the modes for dielectric function $\epsilon(r)$. Then for dielectric function $\epsilon'(r) = \epsilon(r/s)$ the master equation is

$$\nabla \times \left(\frac{1}{\epsilon(r/s)} \nabla \times H'(r) \right) = \frac{\omega'^2}{c^2} H'(r). \tag{34}$$

But if $H_n(r)$ are the solutions of the original equation with frequencies ω_n, then the solutions of the primed equation are just $H'_n(r) = H_n(r/s)$ with $\omega'_n = \omega_n/s$.

This scaling is of considerable practical use. For instance, in order to understand how a structure behaves at micron length scales, the structure can be built at millimeter scales and its bands measured, then scaled. Calculations need only be performed at one length scale in order to understand the band structure at all scales. Losses, nonlinearities, and other phenomena ignored in these approximations may break the scale invariance slightly, but it remains a useful theoretical and experimental property.

LECTURE II

4 Two-dimensional Results

In the last lecture we developed a simple formalism for modes in photonic crystals by analogy with quantum mechanics. The resulting equations can be analyzed by the same techniques used to analyze electron bands in periodic potentials (e.g. Meade et al. 1993). So far our methods are exact within the assumptions listed above and we expect good agreement with experiment. In this lecture we will try to develop some intuition about how the photonic bands are determined by the periodic dielectric structure.

Figure 1 shows the measured transmission spectrum (Robertson et al. 1992) of a macroscopic crystal of dielectric rods in air with rod diameters ~ 1 mm, and inter-rod spacing ~ 2 mm. There are about 175 rods in the experimental setup. We are interested in the propagation properties of light through this crystal in the plane $k_z = 0$. Recall that such modes can be classified as TE (electric field in plane) or TM (magnetic field in plane), depending on their behavior (even or odd) under reflection in the $x-y$ plane. The spectrum was obtained using COMETS (Coherent Microwave Transient Spectroscopy). Note that some theoretically predicted photonic bands do not show up in the experimental data. These bands are actually there in the system but cannot be coupled into by this particular experiment.

In Fig. 2 we show the complete photonic band structure. Note that for the TE modes there are solutions at every frequency ω, while for the TM modes there is a range of ω values for which no TM modes exist. This is a photonic band gap. In order to understand why there is a gap for TM modes but not for TE modes, we develop a variational form of the master equation.

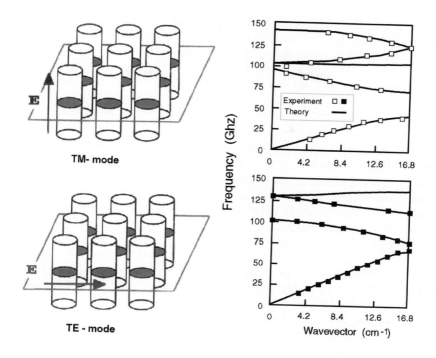

Fig. 1. TM (top) and TE (bottom) photon bands along (10) for a square lattice of alumina rods in air.

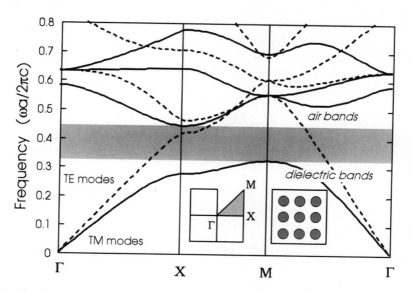

Fig. 2. Photon bandstructure for a square lattice of dielectric rods in air along various symmetry directions. Note the large gap for the TM modes.

Theorem: The eigenmodes of the master equation (13) are stationary points of the functional

$$E_f(\mathbf{H}) = \int d^3\mathbf{r}\, \frac{1}{\epsilon(\mathbf{r})} |\nabla \times \mathbf{H}(\mathbf{r})|^2 \tag{35}$$

within the space of functions $\mathbf{H}(\mathbf{r})$ of fixed norm

$$\int d^3\mathbf{r}\, H^2(\mathbf{r}) = C. \tag{36}$$

The functional is minimized by the lowest-frequency mode, and successively higher modes are minima in the subspace orthogonal to the previous modes. This is the analogue of the functional $\int V|\psi|^2 + \hbar^2(\nabla\psi)^2/2m$ which reproduces Schrödinger's equation. Since $\nabla \times \mathbf{H} \sim \mathbf{D}$, it is convenient to think of the displacement field as tending to concentrate in regions of high dielectric constant (small $1/\epsilon$).

Now we can try to understand the formation of gaps. Consider the second-lowest band of modes, which must be orthogonal to the first-lowest band and hence have a node in D in the high-dielectric region. This allows less concentration of D in the high-dielectric regions so that the second-lowest band does indeed have higher frequency.

So far the argument holds for both TE and TM modes. The essential difference between the two cases comes from the vector nature of D. For a TE mode, the displacement field lines lie in the plane and necessarily cross low-dielectric regions, while for TM modes the displacement field lines are along z. Thus for TM modes there is a much larger difference between the first and second band, since in the first band D lies almost entirely within the high-dielectric regions, as illustrated in Fig. 3. For TE modes there is a small difference in frequency between the first and second band but not enough to cause an omnidirectional gap.

This can be made quantitative by introducing the ratio

$$f = \frac{\int_{\text{dielectric}} d^3\mathbf{r}\, \mathbf{E}^*(\mathbf{r}) \cdot \mathbf{D}(\mathbf{r})}{\int_{\text{total}} d^3\mathbf{r}\, \mathbf{E}^*(\mathbf{r}) \cdot \mathbf{D}(\mathbf{r})} \tag{37}$$

which determines what fraction of the electromagnetic energy is contained in the dielectric regions. For dielectric rods in air, some computed values at X are

	f	
	band 1	band 2
TM	0.8	0.3
TE	0.2	0.1

Note the large difference in f between the first and second TM bands, which corresponds to the existence of an omnidirectional gap. The above vector argument suggested that TM gaps are formed for connected air regions. For connected dielectric regions, we expect a TE gap, by the same reasoning.

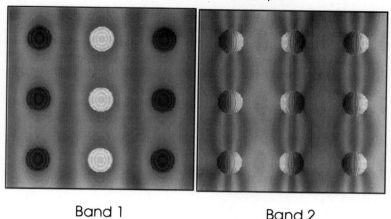

Fig. 3. Displacement field at the X-point for TM bands 1 and 2. Black and white represent large negative and positive values of D, respectively. Grey regions represent small amplitudes of the field.

Indeed, values of f for connected dielectric regions (air columns in dielectric) at X are:

	f band 1	band 2
TM	0.8	0.7
TE	0.7	0.2

As predicted, the structure has a TE gap (see e.g. Joannopoulos et al. 1995).

How can we design a structure with a gap for both TE and TM polarizations? One approach is to make a structure with thin connecting dielectric veins. A structure which works well consists of a triangular lattice of air columns in dielectric. If the diameters of the air columns are large, this structure will have a gap for all polarizations. The gap-width-to-midgap-frequency ratio can be made as large as $17\% = \delta\omega/\omega_0$, as shown in Fig. 4.

Why is it useful to quote the width of a gap in terms of this ratio? The answer goes back to the scaling properties of Maxwell's equations discussed in the first lecture. Once we know there is a gap for a certain structure, we can construct the structure at any length scale with the same fractional gap width. For instance, we can build a model structure at the microwave scale, measure its properties, and then scale everything up to optical frequencies, where measurements are more difficult.

You may have noticed that we plot bands along symmetry lines. The reason is that usually if a material has a gap it will have a gap on one of the Brillouin zone surfaces which are of higher symmetry than a generic

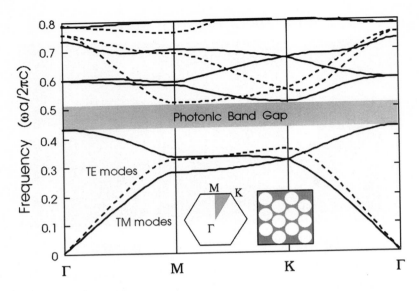

Fig. 4. Photon band structure for a triangular lattice of air columns in dielectric. Note a complete photonic band gap for both polarizations of light.

interior point. There are exceptional structures with gaps at low-symmetry points, but these are rare, as suggested by the following argument. Consider calculating the band structure by diagonalizing within the first Brillouin zone. The strongest matrix elements will be between states differing by a reciprocal lattice vector, and all such states occur on high-symmetry planes. (This is the same argument used to show that electron band gaps in a weak periodic potential develop first at the zone faces.)

5 Three-dimensional Crystals and Defects

It is also possible to find gaps in fully three-dimensional structures. In fact, the diamond structure of air spheres in high dielectric (e.g. $\epsilon = 13$) can be made to have a huge 29% omnidirectional band gap (Ho et al. 1990). Unfortunately it has not yet proved possible to make this structure in the laboratory. Diamond has six air channels running through the material. If we settle for three air channels through the material, each at 35°, the resulting structure (shown in Fig. 5) has a 17% band gap. This structure is known affectionately as "Yablonovite" after its inventor (Yablonovitch et al. 1991a). It is the first photonic crystal fabricated and measured to have a complete photonic band gap.

Now that we have a basic understanding of perfect photonic crystals, let us turn our attention to the properties of defects. A 0-dimensional defect

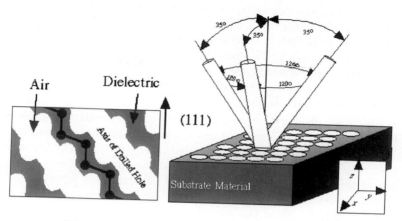

Fig. 5. The method for fabrication of *Yablonovite*.

(point defect) can support a trapped photonic mode inside a photonic band gap. Because frequencies inside the gap cannot penetrate the bulk of the crystal, they are confined to the defect. The defect must be large enough, however, to support a mode in the gap. Figure 6 shows how trapped defect modes appear in the spectrum with increasing defect size. The experimental results are from Yablonovitch et al. (1991b). In this plot, the air defect mode appears at a much smaller size than the dielectric defect because the air defect changes the topology of the structure by cutting a dielectric vein.

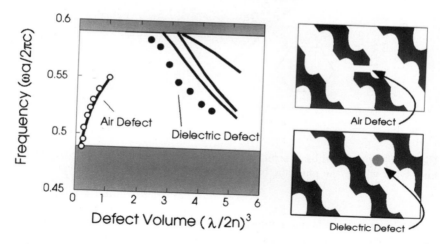

Fig. 6. Plots of air and dielectric defect frequencies within the band gap of Yablonovite as the defect size varies. Solid lines are theory and dots are experiment.

The new defect modes are analogous to modes in doped semiconductors: the air defect acts like a repulsive potential, and the dielectric defect like an attractive potential. Defects which are extended in one dimension may have a number of interesting applications and will be discussed in detail in the last lecture. Surface modes which are extended in two dimensions arise naturally at the boundaries of three-dimensional crystals (the edge of a crystal is a "defect" in the periodic lattice).

In Fig. 7 we display the photon states associated with a surface of the Yablonovite crystal. The modes of this system can be grouped into four categories: 1. *EE* (extended in air, extended in crystal); 2. *ED* (extended in air, decaying in crystal) which correspond to vacuum modes reflected at the surface of the crystal; 3. *DE* (decaying in air, extended in crystal) which are crystal modes not satisfying the vacuum dispersion relation, so that they are trapped in the bulk of the crystal; 4. *DD* (decaying in air, decaying in crystal) which are trapped at the surface.

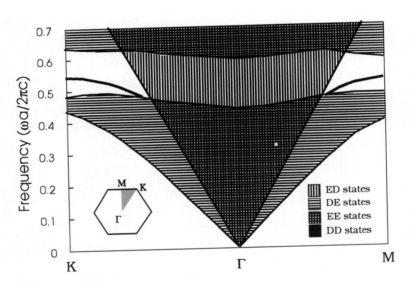

Fig. 7. The projected band structure of the (111) surface of Yablonovite. The black lines lying within the photonic band gap represent bona fide surface states.

A sharp surface is determined by its inclination (the angle of the surface) and its termination (a number varying from 0 to 1 which describes where the surface cuts the crystal unit cell). An important result is that if a crystal has a photonic band gap, *then for any inclination there exists some termination which yields a surface mode* (a *DD* mode in the earlier classification).

LECTURE III

In this lecture we study in more detail the properties of the defects: point and line defects, in a photonic crystal. These defects not only display interesting physical phenomena, but also have potential for use in various applications of the optoelectronic industry.

We shall use a simple 2D lattice as our working example. The perfect crystal is made up of a square lattice of dielectric rods. As we have discussed previously, such structure possesses a photonic bandgap for the TM polarization.

6 Point Defects

A point defect can be introduced into the otherwise perfect lattice by changing the size of one of the dielectric rods. Such a defect can localize an electromagnetic wave, much in the same way a donor or acceptor atom localizes electronic wavefunction. Following this analogy, there are also two types of the point defects in a photonic crystal: an "air" defect, created by using a rod that is smaller than the rest, and a "dielectric" defect, created by using a rod that is larger than the rest, as shown in Fig. 8.

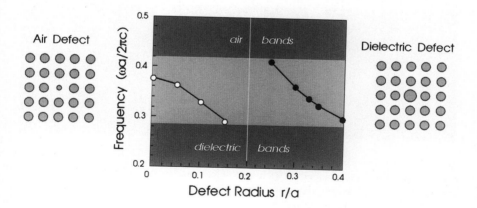

Fig. 8. Plots of air and dielectric defect frequencies within the band gap of the square lattice of rods in air as the defect size varies.

The frequencies of the defect mode can be tuned by altering the radius of the defect. In the case of an "air" defect, we begin with a perfect crystal

— where every rod has a radius of $0.20a$ — and gradually reduce the radius of a single rod. Initially, the perturbation is too small to localize a mode in the crystal. When the radius reaches $0.15a$, a resonant mode appears in the vicinity of the defect. Since the defect involves removing dielectric material in the crystal, the mode appears at a frequency close to the lower edge of the band gap. As the radius of the rod is further reduced, the frequency of the resonance sweeps upward across the gap, and eventually reaches $f = 0.38c/a$ when the rod is completely removed.

The frequency of the "dielectric" defect mode can be tuned as well by changing the size of the defect. Again, starting from a perfect crystal, we gradually increase the radius of a rod. When the radius reaches $0.25a$, two doubly degenerate modes appear at the top the gap, since the defect involves adding material. The modes sweep downward across the gap as the radius increases, eventually disappearing into the continuum below the gap when the radius becomes larger than $0.40a$.

In addition to mode frequencies, the symmetry of the mode can be altered as well. While the "air" defect creates a monopole mode, the "dielectric" defect can create doubly degenerate dipole modes (as illustrated in Fig. 9). More complex modal patterns are possible as the size of the defect is increased.

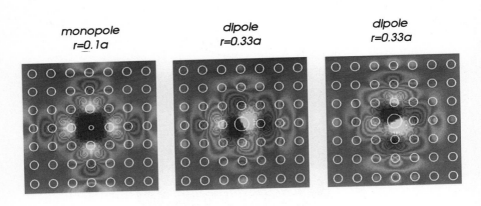

Fig. 9. Electric fields associated with two defect modes shown in Fig. 8. Note that the dipole mode is doubly degenerate.

The ability to change the frequency and the symmetry of the defect modes can have profound consequences in the control of spontaneous emission. The rate of spontaneous emission is related to the density of states of available modes and the transition matrix element between the initial and the final states.

$$R = \frac{2\pi}{\hbar} | <f|H_{\text{int}}|i> |^2 \rho(\omega_{\text{fi}}) \tag{38}$$

A defect in a photonic crystal offers control to both the density of states and the matrix elements (Joannopoulos et al. 1997).

The effects of perfect photonic crystal lattice and defect structure on the density of available modes is illustrated in Fig. 10. In free space, the density of states goes as ω^2, which determines the "natural" rate emission. In a photonic bandgap of a perfect photonic crystal, however, no modes are available for the transition and the spontaneous emission can be suppressed (Yablonovitch 1987). On the other hand, when a defect mode is introduced in the gap, the DOS at the resonance frequencies is greatly increased, leading to enhancement of spontaneous emission.

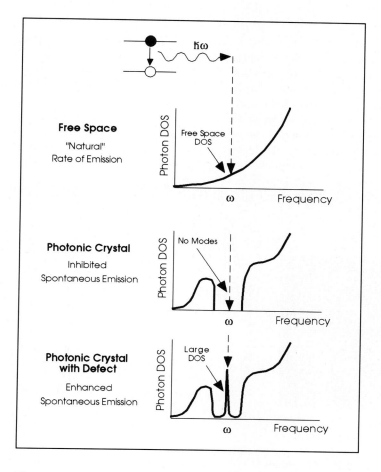

Fig. 10. Control of spontaneous emission with photonic crystals.

In addition to altering the density of states, photonic crystals offer control over the symmetry of the photon wave function and could dramatically influence the selection rule. In the case of a light-emitting transition in free space, the photon emitted typically has s-like symmetry, transitions between s-like and p-like atomic states are allowed, while transition between two s-like or two p-like states are forbidden. In a photonic crystal, on the other hand, the photon can have p-like symmetry, the selection rule is then completely different. The transition between s-like and p-like states is now forbidden, while the transition between two s-like or two p-like states is allowed.

The ability to control spontaneous emission using a photonic crystal could have a huge impact on laser design. Lasing conditions in materials are associated with relative rates of transitions between different levels. The capability to affect these transition rates could lead to, for example, lasing at wavelengths where no ordinary laser is currently available. It could also lead to higher efficiency or lower threshold for available laser systems.

As a major step towards building such photonic crystal cavities, an airbridge microcavity has recently been fabricated and successfully tested by Foresi et al (1997). The results are shown in Fig. 11. The localization here relies on a 1D photonic bandgap along the waveguide and total internal reflection, i.e. index confinement, along the other two directions. To maximize the index difference between the cavity and the surrounded media, part of the substrate underneath the cavity is oxidized, resulting in a monorail-type geometry. Fabrication of such structures represents a major achievement in microlithography. It opens the way for incorporating a laser cavity onto a waveguide and could lead to novel micro optoelectronic devices, such as optical switches and in-plane microlasers.

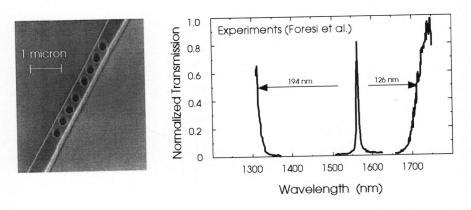

Fig. 11. SEM and measured transmission for a single mode photonic band gap microcavity designed to operate near 1.56 microns.

7 Line Defects

We now turn our attention to line defects in a photonic crystal. A line defect can be created, for example, by removing one row in an otherwise perfect array of dielectric rods. Such a defect leads to a band of states in the gap which behaves like a single-mode waveguide band. The electric field for any state in this band turns out to be highly localized along the defect and decays exponentially away from it (see e.g. Joannopoulos et al. 1995).

Thus the photonic crystal waveguide is unique in that it can guide optical light in an air region surrounded by a higher dielectric region. This is because it operates on an entirely different guiding mechanism. A conventional waveguide, based on the principle of total-internal reflection, can only guide light in a high-dielectric strip surrounded by low dielectric media.

The operation of a conventional waveguide, based on the principle of total internal reflection, is restricted by radiation losses to moderate curvature bends. In fact, when light is steered around a corner in such a guide, the radius of curvature must well exceed the wavelength the light even for high dielectric contrasts in order to avoid large losses at the corners. When a bend is introduced into a photonic crystal waveguide, on the other hand, no power will be radiated out of the guide as it travels around the bend, because there are no extended modes into which the propagating mode can couple. Light will either be transmitted or reflected; only back reflection will hinder perfect transmission.

To study the transmission and reflection properties of photonic crystal waveguide bends, we simulate the propagation of an EM wave as shown in Fig. 12. In the simulation, a dipole located at the entrance of the waveguide creates a pulse with a Gaussian envelope in time. The field amplitude is monitored inside the guide at two points, one before the bend (point A) and one after (point B) as indicated in the panel. Although most of the light that reaches the edge of the computational cell is absorbed by the boundaries, some light gets reflected back from the ends of the waveguide. By using a sizable computational cell of 100×120 lattice constants and by positioning each monitor point appropriately, we can distinguish and separate all the different pulses propagating in the cell; the useful pulses, such as the input pulse and the pulses reflected by and transmitted through the bend, and the parasite pulses which are reflected from the edges of the cell. These pulses are clearly shown in Fig. 13.

In the specific simulation considered here, six pulses are sent down the guide, covering different ranges of frequencies as indicated in Fig. 14. The pulses are then Fourier transformed to obtain the reflection and the transmission coefficients at each frequency. The transmission and reflection coefficients do add up to unity for every frequency in the gap, which confirms that there is no observable radiation loss, in spite of the close proximity of the waveguide to the edge of the computational cell. The transmission drops sharply to zero below the cutoff frequency of the guided mode. The transmis-

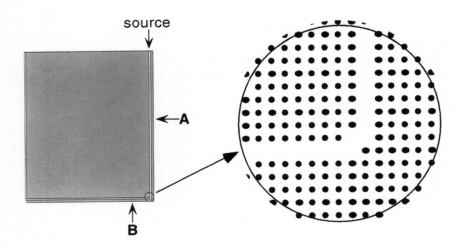

Fig. 12. Schematic view of computational cell used to study transmission around sharp waveguide bends in a photonic crystal.

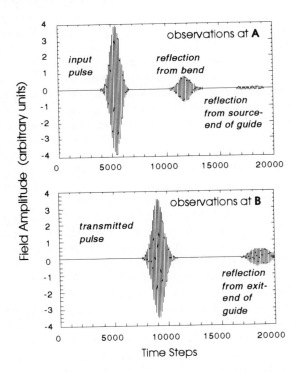

Fig. 13. Field amplitude recorded at **A** and **B**, as a function of time.

sion for frequencies $\omega < 0.392 \ 2\pi c/a$ is larger than 95%, and reaches 100% when $\omega = 0.353 \ 2\pi c/a$. The field pattern of the propagating mode can be observed by a CW excitation of the guided mode. We show in the right panel of Fig. 14. the electric field pattern for the case where $\omega = 0.353 \ 2\pi c/a$. The mode is completely confined inside the guide, and the light wave travels smoothly around the sharp bend, even though the radius of curvature of the bend is smaller than the wavelength of the light.

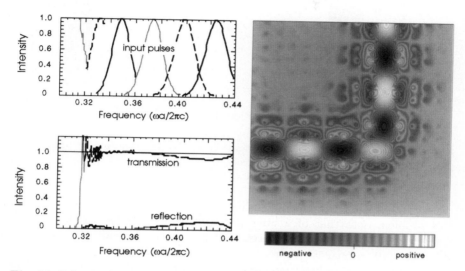

Fig. 14. Left panel: spectral profile of six input pulses and the associated computed transmission and reflection coefficients. Right panel: electric field pattern in the vicinity of the bend for frequency $a = 0.353(2\pi c/a)$.

Both the high transmission through the bend and the oscillatory behavior of the transmission spectrum can be explained by a simple model. The bent photonic crystal waveguide structure can be viewed as separate waveguide sections, one in the (01) direction and one in the (10) direction, connected by a short waveguide in the (11) direction. For a given frequency ω, there is a single wavevector $k(\omega)$ for the guided modes in any particular waveguide section. We label these wavevectors $k_1(\omega)$ for propagation along the (01) or (10) direction, and $k_2(\omega)$ for the (11) direction. These wavevectors are given by the dispersion relations shown in Fig. 15.

The transmission through the bend is modeled as a simple one-dimensional scattering process in which the mode propagating with wavevector k_1 is scattered into the mode with wavevector k_2, then back into the mode with wavevector k_1. At the interface, continuity of the field and of its derivative is required, as one would do in the case of a plane EM wave normally incident on a boundary between materials with different refractive indices. By com-

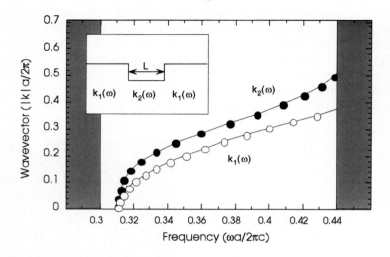

Fig. 15. Dispersion relations $k_1(\omega)$ and $k_2(\omega)$ for propagation of light along (10)/(01) and (11) directions, respectively.

plete analogy with the one-dimensional Schrödinger equation, we can map this problem onto that of a wave propagating in a "dielectric potential". This potential consists of three constant pieces, corresponding to the (01), (11), and (10) propagation directions, respectively, as shown in the inset of Fig. 15. This model differs from the usual one-dimensional scattering problem in that the depth of the well, determined by the difference $|k_1(\omega)|^2 - |k_2(\omega)|^2$, now depends on the frequency of the traveling wave.

The reflection coefficient is then given by

$$R(\omega) = \left[1 + \left(\frac{2k_1\omega)k_2(\omega)}{(k_1^2(\omega) - k_2^2(\omega))\sin(k_2(\omega)L)}\right)^2\right]^{-1} \quad (39)$$

The sole parameter in determining the reflection coefficient is the length L of the well (or of the bend). To set this parameter, we can select a single point from the computational results shown earlier in Fig. 14. We choose the point at $\omega = 0.353\ 2\pi c/a$, where the reflection coefficient is zero. Our choice of solution is $L = 1.33\sqrt{2}a$, which is the one closest to the physical length of the (11) portion of the waveguide.

The validity of this model is demonstrated by varying the length of the (11) waveguide section and compare the reflection coefficients computed from the numerical simulations to those obtained from the analytic expression. The value $L = 1.33\sqrt{2}a$ found above is used to set the parameter L in each case. As we vary the bend length by integer multiples of $\sqrt{2}a$, the effective length L should also change by the same amount, giving $L = 0.33\sqrt{2}a$, $L = 1.33\sqrt{2}a$, $L = 2.33\sqrt{2}a$, and $L = 3.33\sqrt{2}a$ as illustrated in Fig. 16. Good agreement is found between the one-dimensional scattering model (solid line) and the numerical simulations (diamonds). The model correctly predicts the frequencies where the reflection coefficients vanishes, as well as the general quantitative features of the transmission spectrum.

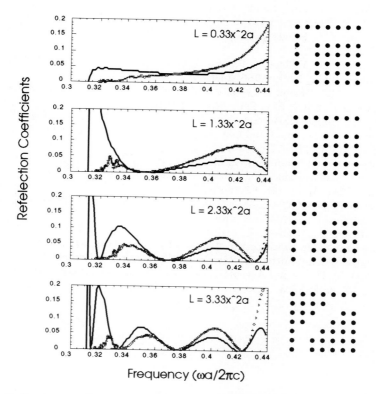

Fig. 16. Reflection coefficients computed from numerical simulations (diamonds) and from 1D scattering theory (solid line) for four different bend geometries.

One notes that the 90° bend with zero radius of curvature, as shown in the top panel of Fig. 16, is not described in this model by a uniformly constant potential, but by a potential with an effective length $L = 0.33\sqrt{2}a$. This length is extrapolated from the bends with longer (11) sections. The model accurately predicts the existence of reflection from the bend, with transmission exceeding 95% for guided modes below $\omega = 0.403 \; 2\pi c/a$. This behavior is in marked contrast to that of a conventional dielectric waveguide with a sharp 90° bend. Power transmission reaches at most 30% even for a guide with a refractive index contrast of 3.5 to 1 with its surroundings, due to large radiation loss at the corner.

References

Foresi J., Villeneuve P., Ferrera J., Thoen E., Steinmeyer G., Fan S., Joannopoulos J., Kimerling L., Smith H., Ippen E. (1997): Photonic-bandgap Microcavities in Optical Waveguides. Nature **390**, 143–145

Ho K.M., Chan C., Soukoulis C. (1990): Existence of Photonic Gaps in Periodic Dielectric Structures. Phys. Rev. Lett. **65**, 3152–3155

Joannopoulos J., Meade R., Winn J. (1995): Photonic Crystals. Princeton Press, Princeton, N.J.

Joannopoulos J., Villeneuve P., Fan S. (1997): Photonic Crystals: Putting a New Twist on Light. Nature **386**, 143–149

John S. (1987): Strong Localization of Photons in Certain Disordered Dielectric Suprlattices. Phys. Rev. Lett. **58**, 2486–2489

Meade R., Rappe A., Brommer K., Joannopoulos J. (1993): Accurate Theoretical Analysis of Photonic Band Gap Materials. Phys. Rev. B**48**, 8434–8437 Erratum: Phys. Rev. B**55**, 15942 (1997)

Robertson W., Arjavalingam G., Meade R., Brommer K., Rappe A., Joannopoulos J. (1992): Measurement of Photonic Band Structure in a 2D Periodic Dielectric Array. Phys. Rev. Lett. **68**, 2023–2026

Yablonovitch E. (1987): Inhibited Spontaneous Emission in Solid State Physics and Electronics. Phys. Rev. Lett. **58**, 2059–2062

Yablonovitch E., Gmitter T., Leung K. (1991a): Photonic Bandstructure: The fcc Case Employing Nonspherical Atoms. Phys. Rev. Lett. **67**, 2295–2298

Yablonovitch E., Gmitter T., Meade R., Brommer K., Rappe A., Joannopoulos J. (1991b): Donor and Acceptor Modes in Photonic Band Structures. Phys. Rev. Lett. **67**, 3380–3383

Linear Optical Properties of Semiconductor Microcavities with Embedded Quantum Wells

Vincenzo Savona

Institut de Physique Théorique, Ecole Polytechnique Fédérale,
CH-1015 Lausanne, Switzerland

Abstract. An overview of the theory of the linear optical response of planar semiconductor microcavities with embedded quantum wells is presented. In particular, the optical properties close to the excitonic transition in the strong coupling regime are addressed and the formalism of exciton polaritons is used. First, the transfer matrix formalism is introduced in order to solve Maxwell equations for the Fabry-Pérot microcavity with distributed Bragg reflectors and to study the cavity mode features. Then, the coupling to a quantum well excitonic resonance is included within the semiclassical formalism for the optical response. The main qualitative and quantitative features of microcavity polaritons are illustrated through several calculated optical spectra and, afterwards, a more formal description of the polariton modes is provided. Finally, the problem of the full quantum description of the exciton photon coupling is briefly addressed. The quasimode formalism is introduced and, as an example of application, a simple model for microcavity photoluminescence under nonresonant continuous wave excitation is presented.

1 Introduction

Semiconductor microcavities with embedded heterostructures exhibiting an excitonic resonance have been studied for the first time [Rarity (1996)] ten years ago. The first studies, both theoretical and experimental, were stimulated by the results obtained in the domain of atomic spectroscopy for atom–cavity systems. In particular, two main objectives were the measurement of a modification – either enhancement or inhibition – of the excitonic spontaneous emission rate and the observation of the vacuum field Rabi splitting in the strong coupling regime. These two objectives have been achieved in the early nineties [Yokoyama (1990), Weisbuch (1992)] and, since then, the field has grown at a rather unexpected rate [Burstein (1995)]. The reason of this growth is twofold. From an application point of view, microcavities have been considered a very promising benchwork for the fabrication of novel light emitting devices [Ebeling (1993), Benisty (1998)]. On the other hand, people working on more fundamental issues have realized that microcavities could provide a great help in understanding the physics of semiconductor heterostructures in the range of energies close to the band gap. It is for this reason that recently all the main steps in the investigation of the optical properties of semiconductor heterostructures are being applied to microcavities. In the concluding chapter of these notes we will mention some of the most recent theoretical and experimental works that have provided evidence for new interesting phenomena and better understanding in the physics of semiconductors.

These notes are intended to review the linear optical properties of multi-layered planar dielectric structures embedding quantum wells with excitonic resonances (The theory of nonlinear optical properties of microcavities is covered in detail in the lectures by Prof. S. W. Koch in this same volume). The physical system that we will focus on is the semiconductor microcavity with embedded quantum wells, which has been object of research in the past ten years. We will adopt Maxwell equations to describe the electromagnetic properties of dielectric multilayered structures and in particular the monochromatic plane wave propagation. The dielectric media will be considered as homogeneous and non absorbing, thus described by a real dielectric constant. The solution of Maxwell equations will be obtained by means of the transfer matrix formalism which is the most useful for these kind of structures. The transfer matrix formalism will be introduced in the first part of the notes. The second part will present the derivation of the optical response of a Fabry-Pérot (FP) resonator with arbitrary mirror structure. The Airy's formula describing the FP complex transmission and reflection coefficient will be obtained. The two special cases of a constant mirror reflectivity and of a Distributed Bragg Reflector will be considered. In both cases the FP transmission and reflection spectra are presented and the dispersion of the FP modes is derived. The electromagnetic properties of an exciton level of a quantum well in the linear regime are described by a resonant linear susceptibility. We will derive the transfer matrix of a quantum well starting from the linear susceptibility. Then, we will again calculate the optical response of a FP embedding one quantum well. The treatment allows to derive the dispersion of the optical resonances of the system. These resonances are interpreted as coupled modes originating from the exciton and the cavity mode, namely they are the polariton modes of the system. The polariton properties are consequently studied and simple expression for the relevant quantities are derived. The last chapter is devoted to a brief introduction to the quantum theory of microcavity polaritons. We define the microcavity photon and exciton operators and present the polariton Hamiltonian. We discuss in detail the so called *quasimode* approximation, which greatly simplifies the problem of the coupling to the electromagnetic field outside the cavity. By means of this approximation we present a simple model for microcavity polariton photoluminescence which succeeds in explaining the existing experimental findings.

2 Maxwell equations and the transfer matrix approach

We consider a structure made of a stack of different layers of given thicknesses and infinite lateral extension. The whole structure is thus planar and translational invariant along the plane. The axis perpendicular to the plane is here indicated as the **z** axis. The layers are assumed to be made of homogeneous materials with a uniform, frequency independent dielectric constant which differs from layer to layer. A sketch of a sample structure is shown in Fig. 1.

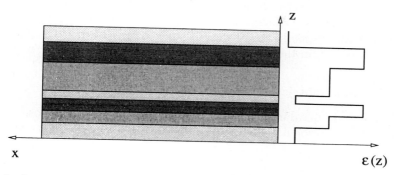

Fig. 1. An example of a multilayered dielectric structure. The corresponding dielectric constant profile $\epsilon(z)$ is plotted on the right.

Assuming a monochromatic field at frequency ω, in absence of charge or current density, the Maxwell equation for the electric field reads [Jackson (1975)]

$$\nabla^2 \mathbf{E}(\mathbf{r}, z) + \frac{\omega^2}{c^2} \epsilon(z) \mathbf{E}(\mathbf{r}, z) = 0 \; , \tag{1}$$

where $\mathbf{r}$ is the in–plane position vector and $\epsilon(z)$ is the dielectric constant profile. According to our assumptions, the function $\epsilon(z)$ is piecewise constant. In addition, it is assumed to vary only within a finite z interval, which means that the thickness of our multilayered structure is finite. Because of the in–plane translational invariance, the solutions of (1) are plane waves along the in–plane direction. For each given in–plane wave vector $\mathbf{k}_\parallel$ and polarization thus, we write

$$\mathbf{E}_{\mathbf{k}_\parallel}(\mathbf{r}, z) = \boldsymbol{\epsilon}_{\mathbf{k}_\parallel} \, U_{\mathbf{k}_\parallel,\omega}(z) e^{i\mathbf{k}_\parallel \cdot \mathbf{r}} \; , \tag{2}$$

where $\boldsymbol{\epsilon}_{\mathbf{k}_\parallel}$ is the polarization vector. Replacing into (1) we get a one dimensional problem for the mode function $U_{\mathbf{k}_\parallel,\omega}(z)$:

$$\frac{d^2 U_{\mathbf{k}_\parallel,\omega}(z)}{dz^2} + \left(\frac{\omega^2}{c^2} \epsilon(z) - k_\parallel^2 \right) U_{\mathbf{k}_\parallel,\omega}(z) = 0 \; . \tag{3}$$

This equation may be separately solved in each homogeneous layer. For a layer with dielectric constant ϵ we have

$$U_{\mathbf{k}_\|,\omega}(z) = E_l(\mathbf{k}_\|)e^{-ik_z z} + E_r(\mathbf{k}_\|)e^{ik_z z} , \qquad (4)$$

$$k_z = \sqrt{\frac{\omega^2}{c^2}\epsilon - k_\|^2} . \qquad (5)$$

The solution represents two monochromatic waves traveling in opposite directions. It turns out from (5) that propagating waves exist only for $(\omega^2/c^2)\epsilon > k_\|^2$, otherwise the solution is an evanescent wave along $\mathbf{z}$. The quantities E_l and E_r are complex coefficients which have to be determined by imposing Maxwell boundary conditions at each interface between two layers. This task is very simple within the transfer matrix approach.

2.1 Transfer matrices

In relation to the one dimensional problem (3), we define for each position z in space a two dimensional vector, with components given by the two coefficients in (4), as

$$\begin{bmatrix} E_r \\ E_l \end{bmatrix} . \qquad (6)$$

We drop the $\mathbf{k}_\|$-dependence of the E_l and E_r coefficients, since the problem is separate in $\mathbf{k}_\|$-space. For an arbitrary structure, one can write the field in the form (4) for two points z_1 and z_2 at the two boundaries of the structure, as illustrated in Fig. 2. Maxwell boundary conditions across the structure

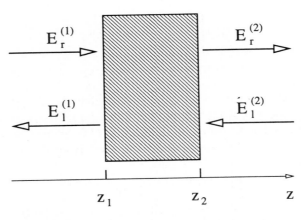

Fig. 2. Fields propagating on the two sides of a planar structure.

will result in a linear relation between the coefficients in z_1 and z_2, which we write as

$$\begin{bmatrix} E_r^{(2)} \\ E_l^{(2)} \end{bmatrix} = \begin{bmatrix} M_{11} & M_{12} \\ M_{21} & M_{22} \end{bmatrix} \begin{bmatrix} E_r^{(1)} \\ E_l^{(1)} \end{bmatrix} . \qquad (7)$$

The matrix M thus defined is the *transfer matrix* of the structure we are considering.

The most important property of transfer matrices is also a very intuitive one: they can be composed to obtain transfer matrices of larger structures. This means that, given two structures characterized by the matrices M_1 and M_2, following each other in the spatial order from left to right, the transfer matrix of the overall structure is simply $M = M_2 M_1$. This property is what makes transfer matrices so powerful. In fact, starting from the matrices for the simplest elements, namely the homogeneous layer of given thickness and the simple interface, one can simply derive the wave propagation for arbitrarily complex planar structures.

We give here, without proof, these elementary transfer matrices. It is straightforward to show that the transfer matrix corresponding to the propagation from z_1 to z_2 in a homogeneous medium is given by

$$M_{\text{hom}} = \begin{bmatrix} e^{ik_z(z_2-z_1)} & 0 \\ 0 & e^{-ik_z(z_2-z_1)} \end{bmatrix} . \tag{8}$$

The transfer matrix for an interface at position z_0 between two dielectric layers is defined as the matrix which relates the vectors of type (6) on the two sides of the interface, namely at $z \to z_0^+$ and $z \to z_0^-$. It is different for the two different polarizations TE (Transverse Electric) and TM (Transverse Magnetic)[1], which are illustrated in Fig. 3 and Fig. 4.

The transfer matrix for an interface and for TE polarization is given by

$$M_{TE} = \begin{bmatrix} \dfrac{k_z^{(2)} + k_z^{(1)}}{2k_z^{(2)}} & \dfrac{k_z^{(2)} - k_z^{(1)}}{2k_z^{(2)}} \\[2ex] \dfrac{k_z^{(2)} - k_z^{(1)}}{2k_z^{(2)}} & \dfrac{k_z^{(2)} + k_z^{(1)}}{2k_z^{(2)}} \end{bmatrix} , \tag{9}$$

while that for the TM polarization reads

$$M_{TM} = \begin{bmatrix} \dfrac{n_2^2 k_z^{(1)} + n_1^2 k_z^{(2)}}{2n_1 n_2 k_z^{(2)}} & \dfrac{n_2^2 k_z^{(1)} - n_1^2 k_z^{(2)}}{2n_1 n_2 k_z^{(2)}} \\[2ex] \dfrac{n_2^2 k_z^{(1)} - n_1^2 k_z^{(2)}}{2n_1 n_2 k_z^{(2)}} & \dfrac{n_2^2 k_z^{(1)} + n_1^2 k_z^{(2)}}{2n_1 n_2 k_z^{(2)}} \end{bmatrix} . \tag{10}$$

Here

$$k_z^{(j)} = \sqrt{\frac{\omega^2}{c^2}\epsilon_j - k_\parallel^2} , \tag{11}$$

---

[1] These two polarizations are also indicated by s and p polarizations, respectively. See e.g. [Jackson (1975)]

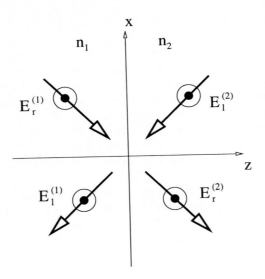

Fig. 3. Propagation of TE-polarized waves at a dielectric interface. The electric field polarization points outside from the page.

where $j = 1, 2$ indicate the left and right side material respectively, and $n_j = \sqrt{\epsilon_j}$. These transfer matrices may be derived by a straightforward application of Maxwell boundary conditions.

From a practical point of view, the three expressions (8), (9), (10) together with the composition law is all one needs to employ the transfer matrix approach by her own. It is enough to put the input field on the left side of a structure and one gets the output field on the right side. There are however some properties which will allow us to perform analytical calculations on transfer matrices.

2.2 Transfer matrices under time and space inversion

The Maxwell boundary conditions must be invariant under time reversal. This means that the complex coefficients of the transfer matrix do not change if we reverse the time evolution, provided we always stick to the convention that the first component of the 2-D vector is the right-propagating wave.

To derive the action of time reversal on the 2-D field vectors we should recall that the amplitude of the electric field is given by the real part of its representation in terms of complex exponentials:

$$E(\mathbf{r}, z, t) = \mathcal{R}e\left[\left(E_l e^{-ik_z z} + E_r e^{ik_z z}\right) e^{i\mathbf{k}_\parallel \cdot \mathbf{r}} e^{-i\omega t}\right] . \tag{12}$$

It is then easy to verify that the time reversal operator $\hat{T}$ acts as

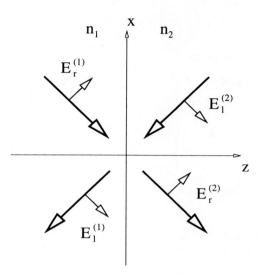

Fig. 4. Propagation of TM-polarized waves at a dielectric interface. The polarization vectors are indicated by the smaller arrows.

$$\hat{T}E(\mathbf{r}, z, t) = E(\mathbf{r}, z, -t)$$
$$= \mathcal{R}e\left[\left(E_r^* e^{-ik_z z} + E_l^* e^{ik_z z}\right) e^{-i\mathbf{k}_\| \cdot \mathbf{r}} e^{-i\omega t}\right]. \quad (13)$$

Thus, in addition to the sign reversal $\mathbf{k}_\| \to -\mathbf{k}_\|$, $\hat{T}$ acts on the 2-D vectors $[E_r, E_l]$ as

$$\hat{T}\begin{bmatrix} E_r \\ E_l \end{bmatrix} = \begin{bmatrix} E_l^* \\ E_r^* \end{bmatrix}. \quad (14)$$

The time reversal invariance allows us to relate the four elements of a general transfer matrix to the complex reflection and transmission coefficients of the corresponding structure. Consider the situation in which a unitary wave is incoming from the left, a wave of amplitude r is reflected in the opposite direction and a wave of amplitude t is transmitted from the right boundary of the structure, as shown in Fig. 5. Then

$$\begin{bmatrix} t \\ 0 \end{bmatrix} = \begin{bmatrix} M_{11} & M_{12} \\ M_{21} & M_{22} \end{bmatrix} \begin{bmatrix} 1 \\ r \end{bmatrix}, \quad (15)$$

which gives

$$r = -\frac{M_{21}}{M_{22}}, \quad t = \frac{\det(M)}{M_{22}}. \quad (16)$$

Now we apply the time reversal operator to the relation (15), obtaining

$$\begin{bmatrix} 0 \\ t^* \end{bmatrix} = \begin{bmatrix} M_{11} & M_{12} \\ M_{21} & M_{22} \end{bmatrix} \begin{bmatrix} r^* \\ 1 \end{bmatrix}, \quad (17)$$

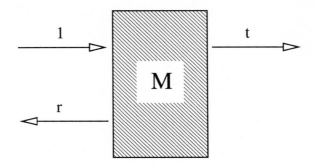

Fig. 5. Waves originating from a wave of unitary amplitude incoming from the left.

where it has been assumed that the transfer matrix is invariant under time reversal, namely $\hat{T}M\hat{T}^{-1} = M$. Then two other relations are obtained:

$$r^* = -\frac{M_{12}}{M_{11}} \quad , \quad t^* = \frac{\det(M)}{M_{11}} \quad . \tag{18}$$

We need a further step which consists in finding the determinant of the transfer matrix. To do this, we define the reflectance $R = |r|^2$ and the transmittance $T = |t|^2/\alpha_{12}$, where

$$\alpha_{12} = \frac{\mathcal{R}e(k_{1z})}{\mathcal{R}e(k_{2z})} \tag{19}$$

for TE polarization and

$$\alpha_{12} = \frac{\mathcal{R}e(k_{1z})}{\mathcal{R}e(k_{2z})} \frac{n_2^2}{n_1^2} \tag{20}$$

for TM polarization. Here, n_1, n_2 are the refractive indices of the left and right side materials respectively. Then, by using the relation $R + T = 1$ together with (16) and (18), very simple algebra brings to the result

$$\det(M) = \alpha_{12} \quad . \tag{21}$$

This result is very general and is a direct consequence of energy conservation. In particular, when the refraction indices on the left and right sides of an arbitrary planar structure are equal, the transfer matrix is unimodular. Once the determinant of a transfer matrix is known, Eqs. (16) and (18) can be solved for the matrix elements M_{ij}. It turns out that the general expression of a transfer matrix in terms of the reflection and transmission coefficients is

$$M = \alpha_{12}^2 \begin{bmatrix} \dfrac{1}{t^*} & -\dfrac{r^*}{t^*} \\ -\dfrac{r}{t} & \dfrac{1}{t} \end{bmatrix} \quad . \tag{22}$$

The corresponding inverse matrix which expresses the left field as a function of the right field is

$$M^{-1} = \frac{1}{\alpha_{12}^2} \begin{bmatrix} \frac{1}{t} & \frac{r^*}{t^*} \\ \frac{r}{t} & \frac{1}{t^*} \end{bmatrix}. \qquad (23)$$

These expressions are generally valid for any planar multilayered structure without absorption. The general form of a transfer matrix of a symmetric structure which includes absorbing elements has been discussed by Citrin [Citrin (1993)] and will not be reported here.

Another important simmetry operation is the space inversion along the z direction. Of course, in general a planar multilayered structure is *not* invariant under such a simmetry operation. However, we are going to use the space inversion operator when describing the Fabry-Pérot resonator. By applying the inversion of the z coordinate on the electric field as expressed in (12), we obtain the obvious result that the left and right traveling waves are simply exchanged

$$\hat{\Pi}_z \begin{bmatrix} E_r \\ E_l \end{bmatrix} = \begin{bmatrix} E_l \\ E_r \end{bmatrix}. \qquad (24)$$

The transfer matrix $\tilde{M}$ of the spatially reflected structure can be expressed in terms of the matrix elements of the originary structure. In particular, if M has been defined in terms of the reflection and transmission coefficients, like in the first sketch of Fig. 6, then the spatially reflected structure will have the same r and t for light incoming from the right, as in the second sketch of Fig. 6. Then, $\tilde{M}$ can be derived from the relation

$$\begin{bmatrix} r \\ 1 \end{bmatrix} = \begin{bmatrix} \tilde{M}_{11} & \tilde{M}_{12} \\ \tilde{M}_{21} & \tilde{M}_{22} \end{bmatrix} \begin{bmatrix} 0 \\ t \end{bmatrix}, \qquad (25)$$

which provides two equations for the four matrix elements. The two remaining constraints simply come from the requirement that the most general transfer matrix must be of the form (22). We give directly the final result which is

$$\tilde{M} = \frac{1}{\alpha_{12}^2} \begin{bmatrix} \frac{1}{t^*} & \frac{r}{t} \\ \frac{r^*}{t^*} & \frac{1}{t} \end{bmatrix}, \qquad (26)$$

where we assume that the refraction indices on the left and right sides are always n_1 and n_2 (practically, we have flipped the structure but left the same materials on the left and on the right of it).

Expr. (26) is useful because it allows to write the transfer matrix of a structure once we know the coefficients r and t corresponding to light incoming from the right. This expression will be used in the next section to derive

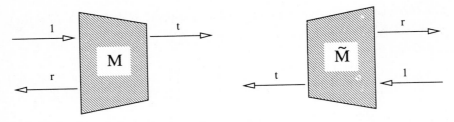

Fig. 6. The same structure as in Fig. 5 and the corresponding spatially structure.

the properties of a Fabry-Pérot resonator, where we will assume to know the reflection and transmission coefficients of the two mirrors for light *inside* the Fabry-Pérot.

We remark that the spatially reflected matrix here obtained is substantially different from the inverse matrix (23). The spatially reflected matrix allows to define a structure in terms of the reflection and transmission coefficient for light incoming onto its right surface. In fact, this matrix gives the "output" fields (those on the right side) when applied to the "input" fields (those on the left side) for a structure which is the mirror image of the originary structure. On the other hand, the inverse matrix simply allows to derive the "input" fields once the "output" fields are known for the given structure.

3 The Fabry-Pérot resonator

We are now going to describe the properties of the simplest light confining device: the Fabry-Pérot resonator. The Fabry-Pérot resonator is a planar structure made of two parallel mirrors. The mirrors can be of any kind, so in our treatment we will consider two generic mirrors described by their reflection and transmission coefficients. The Fabry-Pérot resonator is illustrated in Fig. 7. The central body between the two mirrors has refraction index n_c and is called *spacer*. For the moment we do not make any assumption about the thickness L_c of the spacer. The refraction indices for the left and right materials are n_1 and n_2 respectively. As illustrated in Fig. 7, we consider a plane wave of unit amplitude incoming from the left of the structure and consequently define the reflection and transmission coefficients r and t of the whole structure. The complex quantities a and b indicate the amplitudes of the two waves propagating in the spacer region. In the most general case, the two mirrors are different and we have denoted their reflection and transmission coefficients by r_1, t_1 and r_2, t_2 respectively. It is important to remark that these coefficients are defined, for each mirror, for light incoming from the spacer region.

We look for the transfer matrix of the whole structure. This is simply obtained by applying the knowledge gathered in the previous section. In particular, we have to compose three transfer matrices corresponding to the left

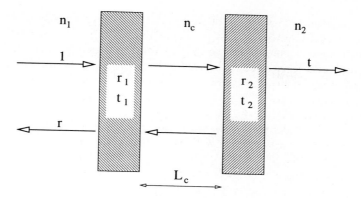

Fig. 7. The sketch of a Fabry-Pérot resonator with two generic mirrors.

mirror, the spacer and the right mirror, called M_1, M_s and M_2 respectively. They are given by

$$M_1 = \alpha_{1c}^2 \begin{bmatrix} \dfrac{1}{t_1^*} & \dfrac{r_1}{t_1} \\ \dfrac{r_1^*}{t_1^*} & \dfrac{1}{t_1} \end{bmatrix}, \tag{27}$$

$$M_s = \begin{bmatrix} e^{ik_z L_c} & 0 \\ 0 & e^{-ik_z L_c} \end{bmatrix}, \tag{28}$$

$$M_2 = \dfrac{1}{\alpha_{2c}^2} \begin{bmatrix} \dfrac{1}{t_2^*} & -\dfrac{r_2^*}{t_2^*} \\ -\dfrac{r_2}{t_2} & \dfrac{1}{t_2} \end{bmatrix}. \tag{29}$$

A few important remarks follow. The matrix M_2 is just given by the expression (22) because r_2 and t_2 are defined for light coming from the left, as in our convention. On the other hand, the matrix M_1 derives from expression (26), since the coefficients r_1 and t_1 are defined for light coming from the right of the mirror.

The transfer matrix of the Fabry-Pérot structure is

$$\begin{aligned} M_{\rm FP} &= M_2 \cdot M_s \cdot M_1 \\ &= \alpha_{12}^2 \begin{bmatrix} \dfrac{e^{ik_z L_c} - r_1^* r_2^* e^{-ik_z L_c}}{t_1^* t_2^*} & \dfrac{r_1 e^{ik_z L_c} - r_2^* e^{-ik_z L_c}}{t_1 t_2^*} \\ \dfrac{r_1^* e^{-ik_z L_c} - r_2 e^{ik_z L_c}}{t_1^* t_2} & \dfrac{e^{-ik_z L_c} - r_1 r_2 e^{ik_z L_c}}{t_1 t_2} \end{bmatrix}. \end{aligned} \tag{30}$$

Let's make a few considerations about this large expression. First, one should remark that M_{FP} is of the general form (22), as expected. In fact, the two diagonal elements are conjugate to each other, as well as the two off-diagonal ones. Using (22) we can also derive in a straightforward way the transmission and reflection coefficients of the Fabry-Pérot, which are given by

$$t_{\text{FP}} = \frac{1}{[M_{\text{FP}}]_{22}}$$
$$= \frac{t_1 t_2 e^{ik_z L_c}}{1 - r_1 r_2 e^{2ik_z L_c}}, \quad (31)$$

$$r_{\text{FP}} = t_{\text{FP}}[M_{\text{FP}}]_{12}$$
$$= -\frac{t_2}{t_2^*} \frac{r_2^* - r_1 e^{2ik_z L_c}}{1 - r_1 r_2 e^{2ik_z L_c}}. \quad (32)$$

These expressions are the most general ones for a Fabry-Pérot resonator, once the properties of the mirrors are known. In general, the reflection and transmission coefficients of the two mirrors are complex quantities that depend on frequency and in-plane wave vector. We consider two particular cases in the following, namely the metallic mirror case and the distributed Bragg reflector case.

3.1 Fabry-Pérot with metallic mirrors

We take the square moduli of (31) and (32) and obtain the reflectance and transmittance of the Fabry-Pérot structure. Some straightforward algebra leads to the expressions

$$T_{\text{FP}} = \frac{|t_{\text{FP}}|^2}{\alpha_{12}}$$
$$= \frac{1}{\alpha_{12}} \frac{|t_1 t_2|^2}{(1 - |r_1 r_2|)^2 + 4|r_1 r_2| \sin^2(k_z L_c + \phi/2)}, \quad (33)$$

$$R_{\text{FP}} = |r_{\text{FP}}|^2$$
$$= \frac{(|r_2| - |r_1|)^2 + 4|r_1 r_2| \sin^2(k_z L_c + \phi/2)}{(1 - |r_1 r_2|)^2 + 4|r_1 r_2| \sin^2(k_z L_c + \phi/2)}, \quad (34)$$

where $\phi = \arg(r_1) + \arg(r_2)$ is the sum of the phases of the two reflection coefficients. These expressions are the so called *Airy's formulae* of the Fabry-Pérot structure [Born and Wolf (1993)]. They are valid in general for complex, frequency dependent values of the reflection and transmission coefficients of the mirrors. We notice that the quantity $\delta = 2k_z L_c + \phi$, which is the phase change of light undergoing a round trip inside the resonator, enters the argument of the sine function in expressions (33) and (34). Keep this in mind until we will derive the same expressions for the distributed Bragg reflector case.

We are not going to provide here a theory for the reflection and transmission of radiation at a boundary with a metallic medium (See e.g. J. D. Jackson, *Classical Electrodynamics* [Jackson (1975)], for a full account). In general, the reflection coefficient at a metallic boundary is a complex, frequency dependent value. In the limit of a perfectly reflecting metallic surface, considering TE polarization, the constraint of zero electric field in the body of the metal implies that the incident plus reflected field must vanish at the boundary, giving $r = -1$. However, a thin metallic mirror can both transmit and absorb light. Thus, in general, the modulus of the reflection coefficient is strictly less than one (the situation we are actually interested in). In this case, however, in order to preserve boundary conditions the reflection and transmission coefficients must be complex valued. The important simplification that we introduce to describe our idealized situation, is the neglection of the frequency and polarization dependence of r_1, r_2, t_1 and t_2. This approximation will however allow us to understand the main properties of Fabry-Pérot resonators and to establish a useful analogy with the case of Bragg reflectors considered in the following.

In order to have a qualitative and quantitative insight in the properties of such structure, we plot in Fig. 8 the reflectance and transmittance of the Fabry-Pérot as a function of the adimensional parameter $k_z L_c$, for three different choices of the reflection coefficients of the mirrors. In this plot for simplicity we assume real reflection coefficients, thus neglecting the phase ϕ. We recall that we are neglecting any frequency dependence of r_1 and r_2. As a consequence, all the frequency dependent features of the spectra in Fig. 8 are due to the multiple interference induced by the resonator.

The main feature of the spectra in Fig. 8 is the presence of resonant peaks at regular intervals. This is the "filtering" effect of the Fabry-Pérot, which results in a series of resonances regularly spaced. The position of the peaks is given by the relation $k_z L_c = N\pi$, together with the expression (5) for k_z. In the most general case with a finite phase ϕ depending on ω the relation becomes

$$k_z L_c + \phi(\omega) = N\pi \qquad (35)$$

and the phase is generally responsible for a change in the shape and position of the peaks. In our particular case, we observe an increase of the width of the peaks and a decrease of their intensities for decreasing r_2. Of course, the narrower the peaks, the better the frequency selectivity of the resonator. We can calculate the cavity full width at half maximum (FWHM) within the assumption of narrow peaks, by taking the first order development of the sine function in the demoninator of (33). Developing close to the N-th resonance, we obtain

$$T_{\text{FP}} \simeq \frac{1}{\alpha_{12}} \frac{|t_1 t_2|^2}{(1-|r_1 r_2|)^2 + 4|r_1 r_2|(k_z L_c - N\pi)^2} \,. \qquad (36)$$

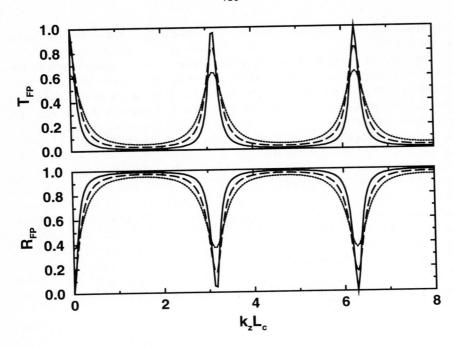

Fig. 8. T_{FP} (upper plot) and R_{FP} (lower plot) for different real values of r_1 and r_2. The first mirror has $r_1^2 = 0.8$ in the three curves. Full line: $r_2^2 = 0.8$. Dashed line: $r_2^2 = 0.6$. Dotted line: $r_2^2 = 0.4$.

For normal incidence, $k_z = \omega n_c/c$. Then, after some algebra, we obtain

$$T_{\text{FP}} \simeq \frac{1}{\alpha_{12}} \frac{v^2|t_1 t_2|^2/(4|r_1 r_2|L_c^2)}{\frac{(1-|r_1 r_2|)^2 v^2}{4|r_1 r_2|L_c^2} + (\omega - \omega_N)^2} \ , \tag{37}$$

where $v = c/n_c$ and $\omega_N = N\pi v/L_c$. This is a Lorentzian lineshape and the FWHM is given by

$$2\gamma_c = \frac{1-|r_1 r_2|}{\sqrt{|r_1 r_2|}} \frac{v}{L_c} \ . \tag{38}$$

The selectivity of the resonator can be quantified in terms of the linewidth and the separation between peaks. We define the *finesse* of the Fabry-Pérot resonator as the ratio between the peak separation and the linewidth. In our case we obtain

$$\mathcal{F} = \frac{\pi v}{2\gamma_c L_c} = \pi \frac{\sqrt{|r_1 r_2|}}{1-|r_1 r_2|} \ . \tag{39}$$

The last information we can derive from our simple Fabry-Pérot with constant mirrors is the dispersion of the Fabry-Pérot modes. We should not forget that the light propagating in our structure may have a finite in–plane wave vector $\mathbf{k}_\parallel$ and thus a finite propagation angle with respect to the normal direction. The dispersion of the cavity modes is the $\mathbf{k}_\parallel$ dependence of the resonance frequencies of the Fabry-Pérot. We derive the dispersion from (5) together with the resonance condition $k_z L_c = N\pi$ and obtain:

$$\left(\frac{\omega^2}{c^2} n_c^2 - k_\parallel^2\right)^{\frac{1}{2}} L_c = N\pi \ . \tag{40}$$

The dispersion of the lowest Fabry-Pérot modes is shown in Fig. 9

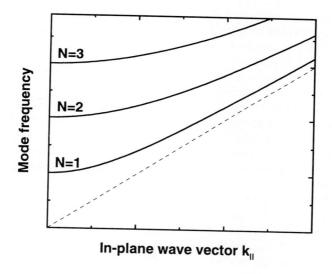

Fig. 9. The dispersion curves of a Fabry-Pérot resonator. The modes with $N = 1, 2, 3$ are plotted. The dashed line is the two-dimensional free photon dispersion $\omega = ck_\parallel / n_c$.

Let's keep in mind this plot too because we will need it when describing the microcavity polariton dispersion curves. For the moment, we remark that the spacing between the modes is a function of $\mathbf{k}_\parallel$ and that all the modes have the free photon dispersion line $\omega = vk_\parallel$ as an asymptote for large values of $k_\parallel$.

3.2 Fabry-Pérot with distributed Bragg reflectors

So far we have described the properties of a simplified Fabry-Pérot structure with equal mirrors having frequency independent reflection coefficient. Semiconductor microcavities are essentially Fabry-Pérot resonators with a peculiar mirror structure [Burstein (1995)]. The mirrors, called Distributed Bragg Reflectors (DBRs) [McLeod (1986), Benisty (1998)], are stacks of semiconductor layers with two alternating refraction indices, as in Fig. 10. The two indices, that we will call n_1 and n_2 with $n_1 > n_2$, and the thickness of the two layers L_1 and L_2 are chosen in order for the two layers to have the same optical thickness $L_1/n_1 = L_2/n_2 = \lambda/4$. A DBR designed in such a way presents a wavelength interval centered at λ in which the square modulus of the reflection coefficient at normal incidence is very close to one, provided the number of pairs is sufficiently high. In addition, the phase of the reflection coefficient within this region, called "stop band", behaves linearly as a function of the frequency. The modulus and the phase of the reflection coefficient of a typical DBR are plotted in Fig. 11 for light at normal incidence. The DBR thus behaves as a very good mirror, within a given frequency window. This property is preserved also for different incident angles. To show this, we plot in Fig. 12 the reflectance of a DBR for its central frequency $\omega_m = v\pi/\lambda$ and TE polarization as a function of the normalized in–plane wave vector $\mathbf{k}_\parallel/k_0$, where $k_0 = \omega_m/v$ and v is the light velocity in the medium of the incoming wave. We remark that in both plots, the reflectance outside the stop band shows a highly oscillatory behaviour. In Fig. 12, in particular, the reflectance is characterized by a number of peaks corresponding to the number of pairs of layers in the DBR. These peaks correspond to side resonances due to the multiple interference in the structure. These resonances have a very strong influence on the physics of any semiconductor microcavity device, as we will see in detail later.

The reflection coefficient of a DBR can be calculated using the transfer matrix formalism. Here we report (without derivation) a useful parametrization of the reflection coefficient $r(\omega)$ at normal incidence, which is valid inside the stop band and for a sufficiently high number of pairs of layers. The refraction indices on the left and right side of the mirror are indicated as n_l and n_r, respectively[2]. We assume that the first layer on the left side of the mirror (the side whose reflectivity we are interested in) has the low refraction index n_2 and that the index of the left material n_l is larger than n_2[3].

---

[2] The reader will remark the sudden but necessary change in notation that has been introduced. From now on, n_1 and n_2 will denote the refraction indices of the DBR layers. The former n_1 and n_2, indicating the indices on the left and right side of a structure, will now be denoted by n_l and n_r.

[3] The other possible choice is n_1 as a starting layer and $n_l < n_2$. These two arrangements maximize the reflectance in the stop band region. The important prescription for the design of a DBR is to avoid "steps", namely three successive layers with increasing or decreasing refraction indices. With this prescription at

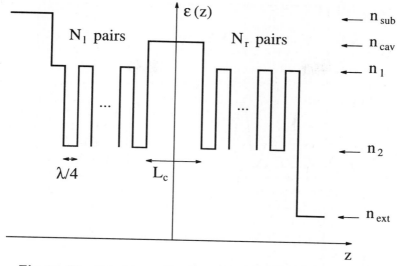

Fig. 10. The dielectric profile of a typical semiconductor microcavity.

Moreover, we will always assume that the DBR has an even number of layers $2N$. The reflectance $R = |r(\omega)|^2$ then is approximately constant and given by the value at the center of the stop band $\omega = \omega_m$. The approximate result, obtained neglecting the quantity $(n_2/n_1)^{2N}$ compared to $(n_1/n_2)^{2N}$ is

$$R = 1 - 4\frac{n_r}{n_l}\left(\frac{n_2}{n_1}\right)^{2N} \quad . \quad (n_1 > n_2) \tag{41}$$

The phase of $r(\omega)$ has approximately a linear frequency dependence inside the stop band, as seen in Fig. 11. Within the same approximations introduced for the previous expression, the phase is given by

$$\phi_r(\omega) = \frac{n_l L_{\text{DBR}}}{c}(\omega - \omega_m) \; , \tag{42}$$

where L_{DBR} represents an effective thickness given by

$$L_{\text{DBR}} = \frac{\lambda}{2}\frac{n_1 n_2}{n_l(n_1 - n_2)} \quad (n_1 > n_2) \; . \tag{43}$$

Here, λ is the chosen optical thickness (the mirror layers have optical thickness $\lambda/4$). We remark that the quantity L_{DBR} does not depend on the number of pairs. Actually, we should remember that the above expressions are valid

---

hand, one easily realizes that the two cases here mentioned are the only relevant ones. See Ref. [McLeod (1986)] for details.

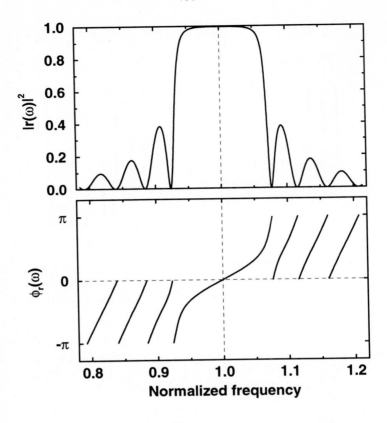

Fig. 11. (a) The reflectance of a 20-pair-DBR with refraction indices $n_1 = 3.0$ and $n_2 = 3.6$ is plotted. (b) The corresponding phase of the complex reflection coefficient.

for sufficiently large number of pairs (not too close to one). In this situation, L_{DBR} is not a real penetration depth for the electromagnetic field but, rather, an effective optical length that expresses the phase change of the electromagnetic wave upon reflection. We will see in a while how this parameter influences the properties of a Fabry-Pérot. The analytical approximations (41), (42) and (43), as well as the other considerations that we have made in relation to the DBR structures, can be derived in a straightforward but tedious way from the transfer matrix formalism. Here we do not report such derivation, since we are interested only in the practical aspects related to microcavities, and refer to the book by H. A. Mc Leod [McLeod (1986)] for the readers who want a deeper insight on this problem. In what follows we

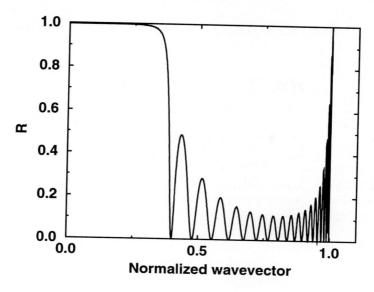

Fig. 12. The reflectance of a 20-pair-DBR plotted at the mirror resonance ω_m, for TE polarization, as a function of the normalized in–plane wave vector $k_\parallel/k_0$.

will always assume that the DBR reflectivity can be parametrized using the three expressions above.

A Fabry-Pérot resonator can be built using two DBRs. The expressions already introduced for the reflection and transmission coefficient (32) and (31) are still valid, as well as those for the reflectance and transmittance (34) and (33).

At this point we can consider the role of the cavity thickness L_c and understand what makes the difference between a *microcavity* and a "macroscopic" Fabry-Pérot resonator. When the thickness of the spacer L_c is large with respect to the wavelength λ, the mirror frequency ω_m will be close to some ω_N with N large. Since the spacing between successive ω_N values is large with respect to ω_m, several cavity modes will appear, closely spaced, within the mirror stop band. A microcavity instead, is a Fabry-Pérot resonator whose spacer has a thickness equal to a small multiple of $\lambda/2$. In such a system, typically only one cavity resonance occurs within the stop band, corresponding to one of the lowest resonances of the spacer. The resulting electromagnetic mode in this case is a narrow isolated peak, as we will see later on. From now on we will be interested in the microcavity system. We will use the terms λ-cavity or $\lambda/2$-cavity to denote a microcavity with $L_c = \lambda$ or $L_c = \lambda/2$ respectively, while ω_c will denote the spacer resonance that falls inside the stop band region of the mirrors. Incidentally, we remark that the phase of the reflection coefficient of the DBRs at the stop band center is zero, as appears from (42). This means that the Fabry-Pérot mode inside the spacer will be a plane wave with antinodes at the mirror boundaries. Usually, as we will see in the next section, microcavities are designed in order to maximize the amplitude of the field somewhere inside the spacer. The $\lambda/2$-cavity, thus, is not suitable in the present case since it presents a node at the center of the spacer. $\lambda/2$-cavities can be built if the other prescription for the mirror design is chosen (low spacer index and high index for the first mirror layer), since in that case the phase of the reflection coefficient at the mirror boundary is π at resonance [McLeod (1986)].

Again, by replacing (41), (42) and (43) into (33) and by developing the sine function around the cavity resonance, an expression analogous to (37) is obtained, and the corresponding cavity mode linewidth can be derived:

$$2\gamma_c = \frac{1-R}{\sqrt{R}} \frac{c}{n_c(L_c + L_{\mathrm{DBR}})} \ . \tag{44}$$

We see that the mode linewidth is in general narrower than the one of a Fabry-Pérot with metallic mirrors of equal reflectance. This effect is a direct consequence of the frequency dependence of the phase of $r(\omega)$. In fact, at normal incidence, an expression analogous to (35) for the condition of constructive interference in a round trip holds. For normal incidence this expression reads

$$\frac{n_c}{c}[(\omega-\omega_c)L_c + (\omega-\omega_m)L_{\mathrm{DBR}}] = N\pi \ , \tag{45}$$

where N is an arbitrary integer. In Eq. (45) the term proportional to L_{DBR} originates from the phase change upon reflection $\phi(\omega)$ appearing in (35). In the special case where the mirrors are designed with a resonant frequency ω_m equal to the spacer resonance ω_c, Eq. (45) reads

$$\frac{n_c}{c}(\omega-\omega_c)(L_c + L_{\mathrm{DBR}}) = N\pi \ . \tag{46}$$

Such expression corresponds to the resonance condition of an ideal Fabry-Pérot with zero phase change at the mirrors and an effective thickness $L_c + L_{\mathrm{DBR}}$. This intuitively explains the narrowing of the Fabry-Pérot line with respect to the metallic mirrors case: the phase change on a round trip varies more rapidly as a function of frequency, thus the resonance condition is satisfied within a narrower frequency window. From Eq. (45) it also turns out that, in the general case when ω_c does not correspond to an integer multiple of ω_m, the resonance condition is met for a frequency ω different from both ω_m and ω_c. Moreover, in typical DBRs made of III-V semiconductors, L_{DBR} is about one order of magnitude larger than the microcavity thickness L_c. In this case, and for small mirror-cavity detuning, the Fabry-Pérot resonance frequency is practically determined by the mirror resonance. This is a very important property that has to be considered when designing a semiconductor microcavity. From now on, we will always assume a cavity design such that $\omega_c = \omega_m$ at normal incidence, and will denote with ω_c both the cavity and the mirror resonance frequency. From a practical point of view, this choice minimizes the cavity mode linewidth because the cavity mode frequency falls exactly at the center of the stop band region where the mirror reflectance, and consequently the light confinement, is maximum.

Fig. 13 shows the plot of the microcavity reflectance for a typical λ-cavity and normal incidence. The high reflectance window around $\omega = \omega_c$ originates from the stop band of the two Bragg mirrors. The cavity mode appears as a very narrow peak at the center of this region. Outside the stop band, a rather pronounced peak structure appears. This structure arises from the interplay between the oscillations in the reflectance of the two mirrors outside the stop band, as seen in Fig. 11. It can be given various physical interpretation. In particular, the minima in reflectivity outside the stop band can be interpreted as confined modes other than the main cavity mode, arising from the peaks in the DBR reflectivity outside the stop band. Consistently with the previous remarks, these modes present a linewidth larger with respect to the main cavity mode, because the mirror reflectivity is much smaller than at the stop band center, as seen in Fig. 11. These cavity modes exist for all DBR microcavities. They are called *leaky modes*, because the electromagnetic field can "leak" outside the structure more efficiently at the corresponding frequencies. The leaky modes exist for all values of the in–plane wave vector and present a dispersion similar to that of the main cavity mode. From Fig. 12 we can deduce that, at the cavity resonance ω_c, leaky modes exist for several values of $k_\parallel$ corresponding to the maxima in the mirror reflectance. Their number is equal to the number of pairs of layers N of the DBR on the right side (the "substrate" side, because we have chosen $n_r = 3.5$). This is illustrated in Fig. 14, where the dispersion of the microcavity resonances for a typical structure is plotted. In this particular case with $n_l = 1$ (air) and $n_r \geq n_c$ (substrate), the leaky modes at $\omega = \omega_c$ can only leak into the substrate because of the total internal reflection on the air side. Leaky modes are a very important feature of microcavities and play a crucial role in the design of microcavity light emitting devices. In fact, they act as very efficient channels for emission into the substrate, where the light is absorbed and consequently wasted. They thus constitute one of the main limitation to the overall efficiency of a light emitting device. We will come back on this point when discussing the radiative linewidth of microcavity polaritons.

Another interesting interpretation that has been given to the modes of a Fabry-Pérot resonator with DBRs is that of impurity levels in a one-dimensional *photonic band gap* structure [Stanley (1993)]. Let us imagine a single DBR made of $N_l + N_r$ pairs of $\lambda/4$ layers. For a sufficiently large number of layers, this is a periodic structure. Consequently the electromagnetic modes, which obey the one-dimensional wave equation (3), present a band

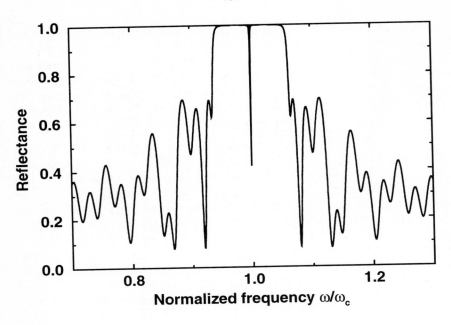

Fig. 13. The reflectance of λ-microcavity with DBRs as a function of the normalized frequency. The indices are $n_c = n_1 = n_r = 3.5$, $n_2 = 3$ and $n_l = 1$, while the number of pairs is 13 for the left DBR and 21.5 for the right DBR. The spacer thickness is $L_c \simeq 236.3$ nm, corresponding to $\hbar\omega_c = 1.5$ eV.

structure with a gap, in analogy with the electronic states of an insulating crystal. The gap exactly corresponds to the stop band of the whole DBR. Now we may increase the thickness of the lower index layer in the $N_l + 1$-th pair. This variation can be seen as the presence of an impurity in the periodic structure. Stanley et al. [Stanley (1993)] have shown that, starting from $\lambda/4$, the lowest mode in the "conduction" band of the structure splits towards lower energies, like a donor level in an insulating crystal. When the thickness is increased up to $\lambda/2$, this impurity level shifts to the midgap position. This is exactly the situation of a $\lambda/2$-cavity with two DBRs. A further increase of the thickness of the layer up to $3/4$ λ will bring the impurity level to the top of the "valence" band. Now the structure is again equivalent to a single DBR with $N_l + N_r$ pairs, because the $3/4$ λ layer induces the same variation to the optical phase of a propagating wave as a $\lambda/4$ layer. Still increasing the thickness of the central layer, another impurity level splits from the upper band and shifts to midgap for a layer thickness equal to λ. This is the situation of a λ cavity with DBRs. And so on...

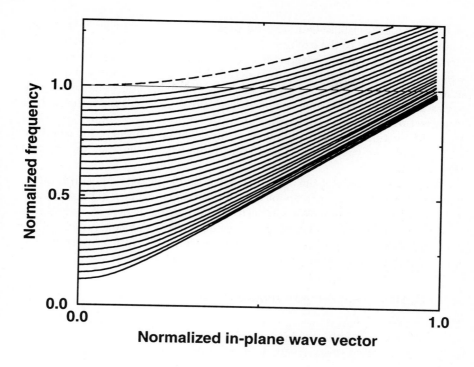

Fig. 14. The dispersion of the main cavity mode and of the leaky modes of a typical microcavity with DBRs. The dashed line is the main cavity mode while the full lines are the leaky modes at frequencies below the main mode. The leaky modes existing above the main cavity mode are not plotted. The thin horizontal line corresponds to $\omega = \omega_c$.

We end this digression about Fabry-Pérot resonators and refer to the fundamental literature [Jackson (1975), McLeod (1986), Born and Wolf (1993)] for a more specific treatment. In the next section we will address the problem of a quantum well, having an optically active exciton level, embedded in a semiconductor microcavity.

4 Optical response of a microcavity embedded quantum well

We are now going to study the linear optical response of quantum wells embedded in a semiconductor microcavity, close to the excitonic resonance. We will focus on the linear optical response to a plane electromagnetic wave, like in the previous sections. Our starting point will be the transfer matrix of a quantum well close to the excitonic transition. The explicit form of this transfer matrix will be given, without derivation. The derivation of the quantum well transfer matrix would require a detailed treatment of the exciton states in a quantum well and of their coupling to electromagnetic radiation. These topics are beyond the scope of the present notes and will be addressed by other lecturers (Prof. S. W. Koch) during this school. We will limit ourselves to a brief introduction that will mention the essential aspects of the problem, and we will take the quantum well transfer matrix as the starting point of this section.

After a description of the optical response of microcavity embedded quantum wells, we will introduce the picture of polaritons as mixed exciton-photon resonances of the system. We will derive the polariton dispersion relation directly from the transfer matrix of the system. The dispersion curves will be reviewed and some considerations about the main physical aspects will be made. Finally, a few useful approximations will be derived.

4.1 Optical response of a quantum well exciton

Semiconductor quantum wells, like bulk semiconductors, are characterized by exciton states below the conduction band edge. The optical transition related to quantum well excitons is characterized by a given oscillator strength per unit surface [Andreani (1995)]. The selection rules for the optical matrix element depend on the particular exciton state, the simmetry of the underlying crystal and the growth axis. Here we are not interested in the details of the coupling. The only important selection rule, which holds independently of the material parameters, is the one originating from the translational simmetry of the quantum well along its plane. Due to this simmetry, the in–plane wave vector $\mathbf{k}_\|$ is a good quantum number for exciton states. Consequently, since the same simmetry holds for the electromagnetic field, the in–plane wave vector is conserved in the exciton-photon coupling. Thus, an electromagnetic plane wave with a given $\mathbf{k}_\|$ will only be coupled to exciton states having the same $\mathbf{k}_\|$. On the other hand, an exciton state with a given $\mathbf{k}_\|$ is coupled to all the electromagnetic waves with the same $\mathbf{k}_\|$ and all the possible values of the orthogonal component of the wave vector k_z. This situation is substantially different from the bulk semiconductor case, in which both excitons and electromagnetic field obey three-dimensional spatial invariance and a one-to-one coupling scheme holds. This peculiarity of quantum wells is at the origin of the finite radiative recombination rate of quantum well excitons [Agranovich (1966), Andreani (1991)].

We write the electromagnetic polarization $\mathbf{P}(\omega, \mathbf{k}_\|, z)$ in the most general way as a linear function of the electric field $\mathbf{E}(\omega, \mathbf{k}_\|, z)$, as

$$\mathbf{P}(\omega, \mathbf{k}_\|, z) = \int_{-\infty}^{+\infty} dz' \chi(z, z', \mathbf{k}_\|, \omega) \mathbf{E}(\omega \mathbf{k}_\|, z') \; , \tag{47}$$

where the fields are expressed in reciprocal $\mathbf{k}_\parallel$ space and real space along the z direction, and $\chi(z, z', \mathbf{k}_\parallel, \omega)$ is the linear susceptibility. From this expression, the $\mathbf{k}_\parallel$ selection rule clearly appears. On the other hand, the coupling is nonlocal along the z direction, because of the absence of translational simmetry in that direction. The linear susceptibility of the excitonic transitions in a quantum well is written as [Tassone (1990)]

$$\chi_{\text{QW}}(z, z', \mathbf{k}_\parallel, \omega) = \frac{1}{\hbar} \sum_n \frac{\boldsymbol{\mu}_{\text{cv}} \otimes \boldsymbol{\mu}_{\text{cv}}}{\omega - \omega_n(\mathbf{k}_\parallel) + i\delta} |F_n(0)|^2 \rho(z)\rho(z') , \qquad (48)$$

with $\delta \to 0^+$. The sum in (48) runs over the different exciton levels for a given $\mathbf{k}_\parallel$ and $\hbar\omega_n(\mathbf{k}_\parallel)$ is the corresponding exciton energy with spatial dispersion ($\mathbf{k}_\parallel$-dependence). The quantities $\boldsymbol{\mu}_{\text{cv}}$, $F_n(\boldsymbol{\rho})$ and $\rho(z)$ are related to the description of the exciton states in the envelope function approximation [Knox (1963), Bastard (1989), Andreani (1995)]. The vector $\boldsymbol{\mu}_{\text{cv}}$ is the dipole matrix element between conduction and valence Bloch states; $F_n(\boldsymbol{\rho})$ is the envelope function for the electron-hole relative motion along the plane; $\rho(z) = f_e(z)f_h(z)$ is the product of the electron and hole envelope functions for the motion along the z direction. This latter quantity is called the exciton confinement function. We remark that in general the susceptibility (48) is a tensor resulting from the tensor product $\boldsymbol{\mu}_{\text{cv}} \otimes \boldsymbol{\mu}_{\text{cv}}$. This tensor product, together with the envelope functions and the exciton dispersion, contains all the information about the particular material and design of the quantum well.

The optical response of the quantum well is given by the excitonic susceptibility (48) plus a background dielectric constant that accounts for all the other optical transitions of the system. Thus, we may write the total quantum well susceptibility as

$$\chi(z, z', \mathbf{k}_\parallel, \omega) = \frac{\epsilon_\infty - 1}{4\pi} \delta(z - z') + \chi_{\text{QW}}(z, z', \mathbf{k}_\parallel, \omega) , \qquad (49)$$

where the Dirac delta function expresses the locality of the background dielectric constant. In practice, the confinement function $\rho(z)$ is nonzero in the quantum well region and vanishes into the barriers. By neglecting the barrier penetration of the confinement function[4], we may assume that the quantum well region is described by the susceptibility (49), while the barrier is taken as an homogeneous medium with dielectric constant ϵ_∞. We neglect the mismatch between the quantum well and the barrier background dielectric constants: this can be done without any loss of generality and the dielectric mismatch can be included at a later stage by means of transfer matrices.

Once the exciton susceptibility is known, Maxwell equations can be solved and the quantum well reflection and transmission coefficients can be derived.

---

[4] This is not valid for shallow quantum wells neither for very narrow quantum wells where the quantum well thickness is much smaller than the exciton Bohr radius a_B^* [Bastard (1989)]

This derivation is very clearly presented in the literature [Tassone (1990), Tassone (1992)] and thus will be omitted in the present notes. We give directly the form of these coefficients, assuming one exciton level in a quantum well grown along the (100) axis of a material with cubic simmetry. This is the most common case of GaAs/AsGaAs quantum wells. In this case, exciton levels originate both from the light and heavy hole valence subbands and can have three different polarizations: T (transverse with respect to $\mathbf{k}_\parallel$), L (longitudinal) and Z (perpendicular to the quantum well plane). The details about the exciton selection rules in these materials are neatly presented in [Andreani (1995)]. Here we restrict ourselves to the two lowest exciton levels in GaAs/AsGaAs quantum wells, corresponding to T- and L- heavy hole excitons. These two exciton levels obey a very simple selection rule: the transverse exciton couples to TE modes only, while the longitudinal exciton couples to TM modes only. With this in mind, we may report the reflection and transmission coefficients for a quantum well, as derived in [Tassone (1990), Tassone (1992)], which read

$$r_{\rm QW}(\mathbf{k}_\parallel,\omega) = -\mathrm{i}e^{\mathrm{i}k_z L_{\rm QW}} \frac{\bar{\varGamma}_\alpha(\mathbf{k}_\parallel,\omega)}{\omega - \bar{\omega}_\alpha(\mathbf{k}_\parallel,\omega) + \mathrm{i}\bar{\varGamma}_\alpha(\mathbf{k}_\parallel,\omega)} \ , \tag{50}$$

$$t_{\rm QW}(\mathbf{k}_\parallel,\omega) = e^{\mathrm{i}k_z L_{\rm QW}} \frac{\omega - \bar{\omega}_\alpha(\mathbf{k}_\parallel,\omega)}{\omega - \bar{\omega}_\alpha(\mathbf{k}_\parallel,\omega) + \mathrm{i}\bar{\varGamma}_\alpha(\mathbf{k}_\parallel,\omega)} \ , \tag{51}$$

where $\alpha = T, L$ labels the two different polarizations. The quantities $\bar{\varGamma}_\alpha(\mathbf{k}_\parallel,\omega)$ and $\bar{\omega}_\alpha(\mathbf{k}_\parallel,\omega)$ are defined as

$$\bar{\omega}_{\rm T}(\mathbf{k}_\parallel,\omega) = \omega(\mathbf{k}_\parallel) - 4\pi \frac{\mu_{\rm cv}^2 |F(0)|^2}{\epsilon_\infty \hbar} k_0^2 P(k_z) \ , \tag{52}$$

$$\bar{\varGamma}_{\rm T}(\mathbf{k}_\parallel,\omega) = 2\pi \frac{\mu_{\rm cv}^2 |F(0)|^2}{\epsilon_\infty \hbar} \frac{k_0^2}{k_z} Q^2(k_z) \ , \tag{53}$$

$$\bar{\omega}_{\rm L}(\mathbf{k}_\parallel,\omega) = \omega(\mathbf{k}_\parallel) - 4\pi \frac{\mu_{\rm cv}^2 |F(0)|^2}{\epsilon_\infty \hbar} k_z^2 P(k_z) \ , \tag{54}$$

$$\bar{\varGamma}_{\rm L}(\mathbf{k}_\parallel,\omega) = 2\pi \frac{\mu_{\rm cv}^2 |F(0)|^2}{\epsilon_\infty \hbar} k_z Q^2(k_z) \ , \tag{55}$$

where $k_0^2 = (\omega^2/c^2)\epsilon_\infty$. These quantities are the real ($\bar{\omega}$) and imaginary ($\bar{\varGamma}$) parts of the correction to the exciton energy due to the coupling to the electromagnetic field. The meaning of a complex, ω-dependent energy correction would require an extensive discussion which is beyond the scope of the present notes. As a matter of fact, the imaginary part to the exciton energy represents the intrinsic radiative probability per unit time that the exciton acquires when coupled to the one-dimensional photon continuum labeled by k_z. In the same way, the second term on the right side of (52) and (54) is a radiative shift of the exciton energy. The fact that these corrections are ω-dependent is better understood within a quantum formalism. In particular, by replacing the exciton dispersion $\omega = \omega(\mathbf{k}_{\|})$, one obtains the result of the second order perturbation theory for the exciton-photon coupling [Andreani (1991), Andreani (1994), Citrin (1994), Andreani (1995)], called *exciton pole approximation*. In the present semiclassical treatment, the system is probed with a monochromatic beam of frequency ω and the ω-dependence of the above quantities directly follows from Maxwell equations. The functions $P(k_z)$ and $Q(k_z)$ are superposition integrals of the exciton wave function and the electromagnetic wave, defined as

$$P(k_z) = -\frac{1}{2k_z} \int_{-\frac{L_{QW}}{2}}^{+\frac{L_{QW}}{2}} \int_{-\frac{L_{QW}}{2}}^{+\frac{L_{QW}}{2}} \sin(k_z|z-z'|)\rho(z)\rho(z')\,\mathrm{d}z\mathrm{d}z' \ , \quad (56)$$

$$Q(k_z) = \int_{-\frac{L_{QW}}{2}}^{+\frac{L_{QW}}{2}} \cos(k_z z)\rho(z)\,\mathrm{d}z \ . \quad (57)$$

These integrals, as well as the expressions (52), (53), (54) and (55), are valid for $k_{\|} < k_0$ namely in the *radiative region*, in which the solutions of the electromagnetic field are propagating waves (the situation we are interested in). Analogous expressions hold in the nonradiative region $k_{\|} > k_0$ where only surface modes (evanescent waves along the z direction) can exist [Tassone (1990), Tassone (1992)]. Since the quantum well thickness L_{QW} is usually much smaller than the wavelength, the long wavelength approximation can be introduced for these integrals. Then, $P(k_z) \simeq 0$ and $Q(k_z) \simeq 1$ to order $(k_z L_{QW})^2$. In what follows we will always assume this approximation to hold.

All the previous considerations about the quantum well semiclassical response are supported by a vast literature (See e.g. Ref. [Andreani (1995)] and references therein for a very clear account of the optical properties of excitons in semiconductors) to which the reader can refer for the details of the formalism.

By means of the results of Section 2 it is now easy to derive the transfer matrix for a quantum well starting from its reflectivity and transmission (50) and (51). The result is reported here:

$$M_{QW} = \begin{bmatrix} \dfrac{\Delta_\alpha - \mathrm{i}\bar{\Gamma}_\alpha}{\Delta_\alpha} e^{\mathrm{i}k_z L_{QW}} & -\dfrac{\mathrm{i}\bar{\Gamma}_\alpha}{\Delta_\alpha} \\ \dfrac{\mathrm{i}\bar{\Gamma}_\alpha}{\Delta_\alpha} & \dfrac{\Delta_\alpha + \mathrm{i}\bar{\Gamma}_\alpha}{\Delta_\alpha} e^{-\mathrm{i}k_z L_{QW}} \end{bmatrix}, \quad (58)$$

where $\Delta_\alpha = \omega - \bar{\omega}_\alpha(\omega, \mathbf{k}_{\|})$.

4.2 Microcavity polariton spectra

We illustrate here the basic phenomenology of the microcavity polariton optical response, in order to test the method we have introduced and to give a first insight into the behaviour of our system. We will consider a λ–cavity with two different DBRs and a single quantum well embedded at the center of the spacer. In particular, the structure of the microcavity will be the same as in Fig. 10 with the same parameters as those used to obtain the plot in Fig. 13. We will consider spectra at normal incidence. The quantum well parameters relevant to this model are three: the exciton energy at $k_\parallel = 0$, the exciton oscillator strength per unit area and the phenomenological exciton homogeneous broadening. The exciton energy is chosen to be $\omega_e = 1.5$ eV ($\omega(\mathbf{k}_\parallel) = \omega_e + \hbar/(2M_e)k_\parallel^2$) in order to be resonant to the cavity mode. The exciton oscillator strength per unit area is proportional to the constant

$$\Gamma_0 = 2\pi k_0 \frac{\mu_{cv}^2 |F(0)|^2}{\epsilon_\infty \hbar} , \qquad (59)$$

that appears also in (53) and (55) and represents the exciton radiative rate at $\mathbf{k}_\parallel = 0$ [Tassone (1990)]. We will adopt this quantity as a measure of the exciton-photon coupling in a quantum well. We use the parameter typical of a 100 Å GaAs-AlAs quantum well $\hbar \Gamma_0 = 32$ μeV. The phenomenological exciton homogeneous broadening γ_e is added as an imaginary part to the bare exciton energy in (50) and (51): $\omega_e \to \omega_e - i\gamma_e$. We leave the important remarks about this replacement to the next section and consider for now the parameter γ_e as an additional phenomenological broadening, assigning the value $\gamma_e = 1$ meV.

Fig. 15 shows the microcavity reflectance, as calculated applying the transfer matrix formalism that we have learned so far. Comparison with Fig. 13 reveals the presence of two peaks in place of the originary cavity mode. The remaining features of the spectrum (stop band, side lobes and so on), are unchanged. These two peaks are the manifestation of the *strong coupling regime* between the exciton and the electromagnetic mode of the cavity. An enlargement of this feature in the spectrum is shown in Fig. 16, where the transmittance and absorption spectra (obtained as $A = 1 - R - T$) are also shown. The two peaks are symmetrically shifted with respect to the resonance energy and the energy splitting is $\hbar\Omega_R \simeq 4$ meV. This splitting is the analogous for this system of the *vacuum field Rabi splitting* that has been known since long in atomic physics [Meystre (1992), Haroche (1992), Kaluzny, 1983, Sanchez-Mondragon (1983), Thompson (1992), Hood (1998)]. From this point of view, due to the in–plane translational simmetry, our structure is the one-dimensional analog of a pointlike atom in a 3-D optical cavity. We would like to point out, however, that this analogy holds only in the restricted framework of the linear regime. In particular, within the single excitation regime (one photon and one atom/exciton in the system) the Jaynes–Cummings Hamiltonian [Meystre (1992)] and the Polariton Hamiltonian are strictly the same. As soon as the number of excitations in the system is large[5] the nonlinearities of the two systems are totally different and require different specific theoretical treatment [Meystre (1992), Hood (1998), Jahnke (1997)].

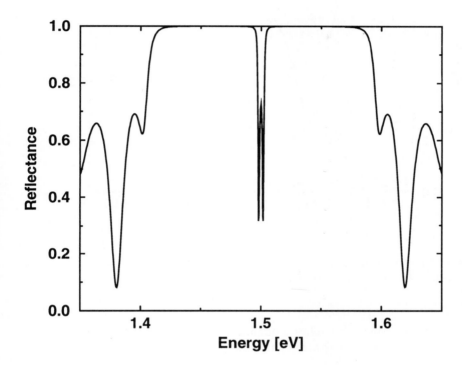

Fig. 15. The reflectance of a microcavity with one quantum well embedded at the center of the spacer. The two energy split peaks appear clearly at the center of the stop band.

The Rabi splitting can also be interpreted in terms of normal mode coupling between the exciton and the cavity modes. In this case, the one-to-one selection rule in the coupling between the exciton level and the quasi-discrete cavity mode traces back to the polariton picture in a bulk semiconductor. Our system is thus the 2-D analog of a bulk semiconductor and the Rabi splitting corresponds to the polariton splitting [Andreani (1995)]. As a matter of fact,

---

[5] for an atomic cavity this means more than one atom or one photon in the cavity; for the excitonic system nonlinearities appear when the exciton density approaches the saturation value

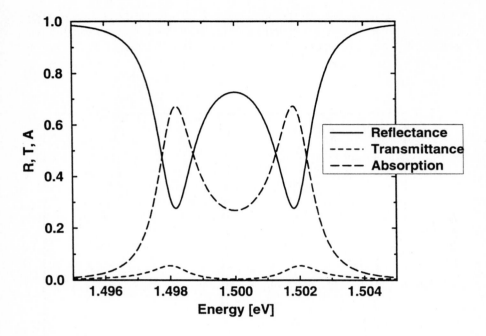

Fig. 16. The details of the microcavity reflectance, transmittance and absorption spectra. An energy splitting of about 4 meV characterizes the three curves.

the polariton picture implies a quantum mechanical coupling, the polariton modes being the eigenstates of the coupled system [Hopfield (1958)]. However, it is well known that in the present case the quantum and semiclassical treatments give *exactly* the same results, because the exciton-photon coupling is quadratic in the two fields [Savona (1995), Savona (1996a)]. Thus, at least for the moment, the reader should feel free to use both pictures indifferently when describing the microcavity system. In the next section we will show how the microcavity polariton dispersion can be rigorously extracted from the present semiclassical treatment. In the end, we must keep in mind that we are just dealing with the system of two coupled harmonic oscillators.

The existence of a finite Rabi splitting depends on all the parameters of the system: exciton oscillator strength, mirror reflectance, exciton broadening. In Fig. 17 we show how the reflectance spectrum is modified for decreasing Γ_0 (an unphysical situation because the single quantum well oscillator strength is a fixed property of the material). We see that the splitting disappears for a finite value of Γ_0 and the system makes a transition to the *weak coupling regime*. We will derive in the next section the dependence of the Rabi splitting on all the system parameters. We remark however that in the linear response formalism the strong coupling is defined as the appearance of a two peak structure in the optical quantities. As a matter of fact, the energy splittings in R, T and A are in principle quantitatively different [Savona (1995)]. Only in the limit of a large splitting, compared to the linewidths, the splitting is approximately the same for the three spectra. Otherwise, it is even possible that the splitting is present in some of the optical constants and absent in the others. The distinction between weak and strong coupling regime may thus be ambiguous when close to the transition between the two. This ambiguity is still increased if we consider that the broadening mechanisms in reality (typically exciton-phonon coupling [Schulteis (1986)], disorder [Glutsch (1994), Glutsch (1996), Zimmermann (1997)] and excitation induced dephasing [Lindberg (1988), Jahnke (1996)]) do not act in such a simple way as our phenomenological parameter γ_e. Even though a situation where Rabi splitting is only present for some of the optical constants has never been observed experimentally, the reader should keep in mind that strong coupling is well defined only when the splitting is considerably larger than the peak linewidths.

The two coupled oscillators picture is confirmed by the analysis of the spectra for varying exciton-cavity detuning. In Fig. 18 we plot the reflectance spectra of our system for different values of the detuning $\delta = \omega_c - \omega_e$, keeping ω_e fixed and varying ω_c (dashed lines). We see that the polariton peaks show an anticrossing behaviour typical of the two coupled oscillators.

The first experimental confirmation of this phenomenon came in 1992 in the work by Weisbuch *et al.* [Weisbuch (1992)] who fabricated and studied the first microcavity system exhibiting strong coupling regime. Since then, the domain has grown considerably and many properties of the strong coupling regime have been investigated [Rarity (1996)].

4.3 Microcavity polariton dispersion relations

In this section we will perform a more formal analysis of the microcavity embedded quantum well system and derive the polariton dispersion relations from its transfer matrix.

The exciton-polariton is the mixed exciton-radiation mode that sets up in an insulating crystal close to the exciton transition. In bulk semiconductors, because of the translational invariance, each polariton state is a quantum superposition of one exciton and one photon mode with a given wave vector [Hopfield (1958)]. The polaritons in bulk are thus stationary states. The optical transition between the polariton and the ground state of the crystal can be probed because of the finite size of the sample that allows coupling to the external electromagnetic field through the sample boundaries [Andreani (1994), Sumi (1976), Ulbrich (1979), Weisbuch (1977),

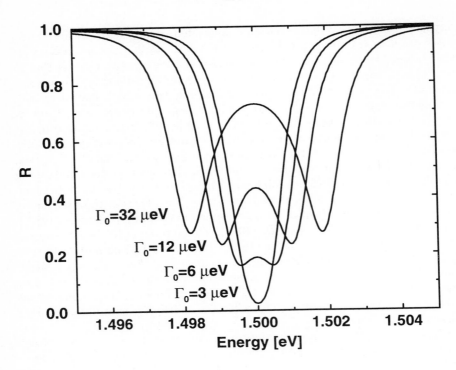

Fig. 17. The transition from strong to weak coupling for decreasing Γ_0 is illustrated.

Weisbuch (1979), Weisbuch (1982)]. In a quantum well the translational simmetry along the growth direction is broken for the exciton degrees of freedom. Consequently, one exciton with given in–plane wave vector $\mathbf{k}_\parallel$ is coupled to a continuum of photon modes with the same $\mathbf{k}_\parallel$ and all the possible values of the remaining component k_z. This gives rise to a situation where the exciton is no longer a stationary state and has a finite recombination probability per unit time [Agranovich (1966), Andreani (1991)]. If we think of the system in terms of eigenstates of the exciton-photon coupling Hamiltonian, these eigenstates form a continuum for each value of $\mathbf{k}_\parallel$. Polaritons are resonances in this continuum, just like Fano resonances [Fano (1961)]. In this sense, polariton resonances are rigorously characterized by poles of the quantum mechanical exciton propagator (Green function) in the complex energy plane [Cohen-Tannoudji (1988)]. The real and imaginary part of the poles represent the "peak" position of the resonance and its linewidth respectively. This viewpoint and the calculation of the exciton-polariton dispersion in quantum wells have been developed by several authors in the literature

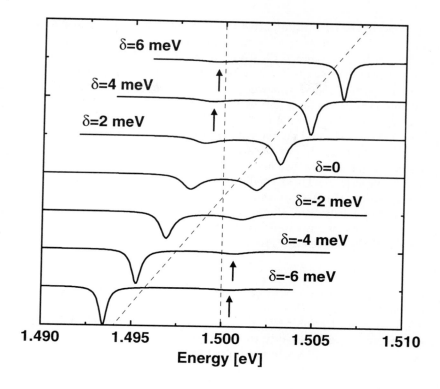

Fig. 18. The reflectance spectra for different values of the detuning δ. The arrows indicate the peak positions when the peaks are only slightly pronounced. The two dashed lines indicate the uncoupled exciton and cavity mode energies.

[Jorda (1993), Citrin (1994)]. When a quantum well in embedded in a microcavity, the situation is not much different. In fact, because of the finite mirror transmittance, the photon modes for a given $\mathbf{k}_\parallel$ still form a continuum, even though this continuum is structured into narrow peaks corresponding to the Fabry-Pérot modes. The quantum mechanical approach to the exciton-polariton dispersion is thus still valid and the polariton dispersion in a semiconductor microcavity has been calculated recently [Savona (1996a)].

What is the relation between this picture and the present semiclassical treatment of the optical response? As we already mentioned, the two approaches are equivalent. Here, by "equivalent" we mean that all the physical quantities that are derived with one of the two approaches can also be derived with the other. Then, of course, the two approaches are complementary to each other. In fact, while the semiclassical response naturally provides the optical response to an electromagnetic wave, within the quantum formalism the concepts of exciton radiative rate, photoluminescence and dynamics find a natural interpretation. The quantum well polariton dispersion can be obtained within the semiclassical formalism if one identifies the polaritons as resonances in the scattering process of an electromagnetic wave. Then, these resonances are hidden somewhere in the transfer matrix. The polariton dispersion has already been derived from the semiclassical treatment in the case of quantum wells [Tassone (1990)], multiple quantum wells and superlattices [Andreani (1994)] and microcavity embedded quantum wells [Savona (1995)].

The resonances in our system can be derived from the poles of reflection and transmission coefficients, these latter being the electromagnetic scattering amplitudes. These coefficients are expressed in terms of the transfer matrix as in (16). We are thus going to calculate the transfer matrix of the system previously studied: one quantum well at the center of a λ-cavity with two DBRs. We assume vanishing thickness for the quantum well. The transfer matrix of the whole structure is defined as

$$M_{MC} = M_2 \cdot M_s \cdot M_{QW} \cdot M_s \cdot M_1 , \qquad (60)$$

where M_1, M_2 are given by (27) and (29), M_s is the propagation in half the spacer length and is thus given by (28) where L_c is replaced by $L_c/2$, and M_{QW} is given by (58). We are going to skip the cumbersome steps of the derivation and give directly the final result for the matrix element $[M_{MC}]_{22}$ whose zeros correspond to the poles of the transmission:

$$[M_{MC}]_{22} = \frac{e^{-ik_z L_c}}{t_1 t_2}[i\bar{\Gamma}_\alpha(1 + r_1 e^{ik_z L_c})(1 + r_2 e^{ik_z L_c})]$$
$$+ \Delta_\alpha(1 - r_2 r_2 e^{2ik_z L_c})] . \qquad (61)$$

In the thin quantum well limit the two integrals (56) and (57) are zero and one respectively, and the polariton dispersion relation can be finally written as

$$\omega - \omega(\mathbf{k}_\parallel) + i\bar{\Gamma}_\alpha \frac{(1 + r_1 e^{ik_z L_c})(1 + r_2 e^{ik_z L_c})}{1 - r_2 r_2 e^{2ik_z L_c}} = 0 , \qquad (62)$$

where we should not forget that the mirror reflectivities $r_{1,2}$ are actually function of ω and $\mathbf{k}_\parallel$. Eq. (62) is exactly the same as in Ref. [Savona (1996a)] where it has been obtained by means of a quantum approach.

We briefly discuss the dispersion relation as obtained numerically from the solution of (62) on the complex energy plane. In Fig. 19 the real part of the solution, corresponding to the polariton energy, is plotted as a function of $k_\parallel$. The parameters are the same as in the previous evaluation of the polariton spectra. We must compare this figure with the plot of the main cavity modes in Fig. 9 and, particularly, with the plot of the leaky mode dispersion in Fig. 14. The first remark concerns the small wave vector region. Here, the exciton and cavity modes are resonant, as indicated by the dashed lines, and the resulting polariton modes exhibit an anticrossing behaviour, as expected in the strong coupling regime. In this region the two polariton resonances,

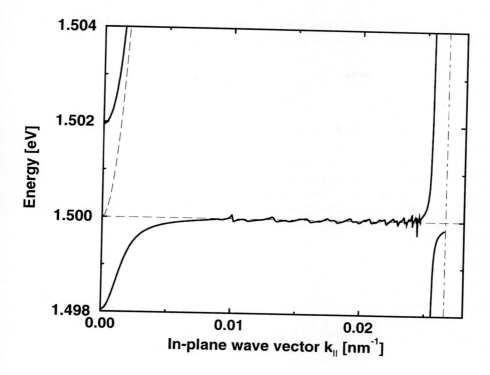

Fig. 19. The energy dispersion of the microcavity polaritons calculated numerically from (62) for TE polarization. The dashed lines indicate the uncoupled exciton and cavity modes, while the dashed-dotted line is the boundary of the radiative region $n_c \omega/c = k_\parallel$.

from a quantum mechanical viewpoint, have maximum admixture of exciton and photon. Outside this region, instead, the two modes approach the uncoupled exciton and cavity modes. However, the lower polariton branch shows a wiggling for wave vectors larger than about 0.01 nm^{-1}. These features correspond to the interaction of the exciton level with the leaky modes of the microcavity structure. In each point where the exciton dispersion crosses one leaky mode (see Fig. 14), the exciton light coupling gives rise to a radiative shift analogous to the one corresponding to the Rabi splitting at $k_\parallel = 0$. Close to the boundary of the radiative region, denoted by the dashed-dotted line in Fig. 19, another anticrossing is present. This is due to the interaction of the exciton with a quasi-guided cavity mode. In fact, in the region $n_c \omega/c > k_\parallel > n_2 \omega/c$ the electromagnetic field can propagate through the layers with index n_c or n_1 but is an evanescent wave in the layers with index

n_2. Consequently, the modes are still radiative but the probability of leaking into the substrate becomes very small. Eventually, one guided mode arises in this region and couples strongly with the exciton, because of its vanishing linewidth, giving rise to the observed feature. The calculation has been carried out only within the radiative region, where expressions (52)–(55) apply. The nonradiative or surface polariton modes [Tassone (1992)] have not been considered. In Fig. 20 we plot the imaginary part of the solutions of (62) corresponding to the polariton radiative linewidth. We see that the two linewidths are strictly equal at resonance. We will show in the next section that they both have half the cavity mode linewidth[6]. This is a general feature of the strong coupling at resonance and is peculiar of the exciton photon coupling. In fact, the nonradiative broadening mechanisms in general act differently on the two polariton branches and give rise to different broadenings [Fisher (1996), Whittaker (1998), Savona (1997a), Savona (1997b), Ell (1998)]. The upper polariton linewidth coincides with the bare cavity mode linewidth far from resonance. The abrupt change in slope corresponds to the onset of total internal reflection on the air side. The lower branch radiative linewidth has a maximum at resonance, where the radiative rate is generally enhanced with respect to the bare exciton radiative rate. Then the radiative rate shows several peaks in correspondence to the interaction with the leaky modes. This is a very important result. The leaky modes are strongly radiative, and the radiation is totally emitted into the substrate [Savona (1996a)]. In addition, leaky modes cover a large portion of the radiative phase space, because their 2-D density is proportional to $k_{\|}$. This is one of the major limitations to the design of light emitting devices, since most of the radiation is lost in the substrate where it gets absorbed. The other limitation to the internal efficiency is due to the guided modes (lying in the nonradiative region) that become radiative in structures with finite lateral size. Minimizing these two effects constitutes the key for achieving the best performance from a light emitting device [Benisty (1998)].

4.4 Useful approximations

The general expression for the dispersion of the quantum well polaritons in an arbitrary microcavity has already been presented and solved numerically in the previous section. However it is useful for many applications to consider an approximate form of the dispersion which is derived analytically. It has the advantage of allowing a simple and intuitive physical interpretation of the behavior of quantum well polaritons as a function of the parameters of the system.

---

[6] In Expr. (62) and in what follows we have introduced no phenomenological exciton broadening. The calculation thus gives the radiative contribution to the polariton broadening.

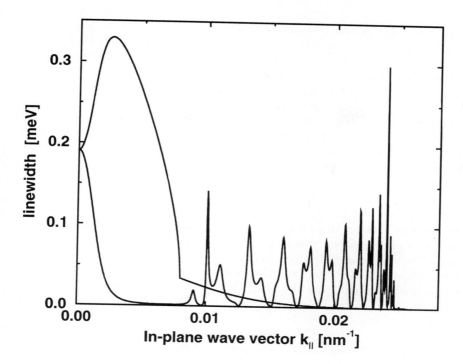

Fig. 20. The radiative linewidth of the microcavity polaritons calculated numerically from (62) for TE polarization.

Our starting expression is the polariton dispersion (62). We make three assumptions. First, we consider values of the frequency ω close to the resonance ω_c compared to the stop band width. Second, we neglect higher order cavity modes, since the energy separation between successive orders is much larger than the Rabi splitting. Third, we assume high mirror reflectance, so that the approximate expressions (41), (42) and (43) can be used. Replacing these expressions into (62) for $\mathbf{k}_\parallel = 0$ and performing straightforward algebra, we end up with the simple equation

$$(\omega - \omega_e + i\gamma_e)(\omega - \omega_c + i\gamma_c) = V^2 , \qquad (63)$$

where γ_c and V are given by the expressions

$$\gamma_c = \frac{c}{n_c(L_c + L_{\text{DBR}})} \frac{(1 - r_1 r_2)(1 + r_1)(1 + r_2)}{r_1(1 + r_2)^2 + r_2(1 + r_1)^2} , \qquad (64)$$

$$V^2 = \frac{c\Gamma_0}{n_c(L_c + L_{\text{DBR}})} \frac{(1 + r_1)^2(1 + r_2)^2}{r_1(1 + r_2)^2 + r_2(1 + r_1)^2} , \qquad (65)$$

and $r_{1,2} = \sqrt{R_{1,2}}$. Eq. (63) is simply the secular equation for the coupling between two damped harmonic oscillators, where the exciton damping constant γ_e is added phenomenologically, as in Section 4.2. This is no surprise, since the system is made of two fields, the exciton and the electromagnetic one, that are bosonic in nature (thus harmonic oscillators) and are linearly coupled. Nevertheless, equations (63), (64) and (65) are of great utility because they allow to roughly predict the characteristics of a semiconductor microcavity in a very straightforward way. In particular, by solving (63) in the resonant situation $\omega_e = \omega_c$, the real part of the two solutions allow to express the Rabi splitting as

$$\Omega_R = 2\sqrt{V^2 - \frac{1}{4}(\gamma_c - \gamma_e)^2} \ . \qquad (66)$$

We see immediately that this expression provides a precise criterium for the existence of the strong coupling regime, namely $4V^2 > (\gamma_e - \gamma_c)^2$. There are however several objections that may be raised. First, the quantity Ω_R here obtained is the energy splitting of the polariton resonances. This quantity is still different from the energy splitting in the spectra of the three optical quantities R, T and A. Thus it is possible that a strong coupling regime is predicted by (66) but the two peak feature does not appear in any of the optical spectra. This concept is discussed in detail in the work by Savona et al. [Savona (1995)], where approximate expressions for the splitting in the optical spectra are also derived under the same approximations that led here to (66). One striking example of the difference between (66) and the energy splitting in the optical spectra can be found in the hypothetical situation of very large γ_e and γ_c with $\gamma_e \simeq \gamma_c$. In this case, the difference $\gamma_c - \gamma_e$ in (66) is vanishing and a finite Rabi splitting occurs. However, the linewidths are so large that the two peaks in the spectra will be completely blurred and a single peaked feature will appear instead. Even though this example does not imply any kind of physical paradox, as a matter of fact the description of the exciton nonradiative broadening here adopted has revealed to be oversimplified for practically all the main exciton broadening (dephasing) mechanisms, as already mentioned above. We will spend a few words on this problem in the last section of this chapter. We recall that the strong coupling regime in microcavities presents a strong analogy with the polariton splitting in bulk systems. In fact, being both photons and excitons confined in two dimensions, the quantum well embedded in a microcavity is the two-dimensional analogous of a bulk semiconductor, with the only difference that the two modes have a finite energy broadening. Another analogy exists with the normal mode splitting which occurs in atomic physics when an atomic level interacts with the electromagnetic mode of a cavity [Carmichael (1989), Meystre (1992), Zhu (1990)]. Expression (66), in fact, had already been derived in that framework [Agarwal (1984)], and the term vacuum field Rabi splitting has been borrowed from atomic spectroscopy. Within our model, the other situation that can occur is the weak coupling regime, characterized by $4V^2 \leq (\gamma_e - \gamma_c)^2$. In this situation, two degenerate polariton resonances exist, which however present different energy broadenings according to the solutions of (63). The weak coupling regime does not preserve the analogy with the three-dimensional polaritons and is somewhat closer to the case of bare quantum well polaritons. In fact, the two degenerate resonances correspond to the exciton which either radiatively decays into the photon continuum or is dissipated by its nonradiative damping mechanism. In particular, the exciton spontaneous emission probability is enhanced with respect to the bare quantum well case, because of the resonant interaction with the cavity mode. In the weak coupling regime, the exciton radiative decay can be described as usual in terms of an irreversible exponential decay. In this case, the decay rate is the one obtained by the standard second order perturbation theory (Fermi golden rule). The first observation of enhanced spontaneous emission rate for a quantum well embedded in a microcavity in the weak coupling regime, dates to the work of Yokoyama *et al.* [Yokoyama (1990)], two years before the observation of the Rabi splitting by Weisbuch *et al.* [Weisbuch (1992)]. When the coupling is very weak, the decay probability here calculated approaches that of a bare quantum well [Savona (1995)].

In order to illustrate the transition from weak to strong coupling regime, we plot in Fig. 21 the real and imaginary part of the solutions of (63) as a function of the mirror reflectance (assuming in this case two equal mirrors). Varying the mirror reflectance corresponds to varying at the same time the effective coupling constant (65)[7] and the cavity mode linewidth (64). The plot is calculated at resonance and for $\gamma_e = 0$. We see that in the weak coupling regime the imaginary parts of the solutions are not degenerate. In particular, in the limit of vanishing reflectance, one of the two plots approaches the bare exciton radiative linewidth $\Gamma_0 = 0.032$ meV. This is important because some times in the literature the two coupled oscillator formula (63) is misused [Abram (1996)]. The error consists in considering the bare exciton radiative broadening (the radiative decay time of 13 ps in GaAs 100 Å quantum wells [Andreani (1991), Agranovich (1966)]) as a separate broadening mechanism that has to be included in γ_e. This is an erroneous approach as the present derivation clearly shows: the bare exciton radiative rate is naturally included in the treatment via the exciton-radiation coupling. In the planar system in fact, because of the $\mathbf{k}_{\parallel}$ selection rule, one exciton with a given $\mathbf{k}_{\parallel}$ has the cavity mode at the same $\mathbf{k}_{\parallel}$ as the only radiative recombination channel. The confusion comes from an erroneous analogy with the corresponding systems in atomic physics. In typical experiments on atoms, the cavity geometry is very peculiar [Haroche (1992)]. It is made of two spherical mirrors facing

---

[7] Important remark: varying the mirror reflectance does not modify the true exciton photon coupling given by the dipole matrix element. It just modifies the density of photon modes interacting with the exciton and, consequently, the effective coupling constant V^2 entering in (63).

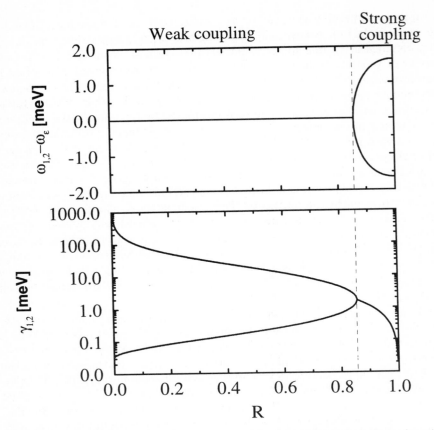

Fig. 21. The Rabi splitting and the polariton broadenings are plotted as a function of the DBR reflectance R. The transition between weak and strong coupling regime around $R = 0.85$ is evident. For $R \to 0$ one energy broadening tends to infinity (an unphysical solution) while the other approaches the bare quantum well spontaneous emission rate.

each other. Because of this configuration, a "stop band" with one or more sharp photon modes is formed in the three dimensional photon **k**–space. This region corresponds to the solid angle spanned by the mirrors with respect to the cavity center. An atom can undergo radiative decay by emitting photons in all directions. When put inside a cavity, in particular, it can emit a photon into the cavity mode, in strict analogy with exciton recombination in planar microcavities. But an atom in a cavity can also radiate in free space in the directions that are not affected by the cavity confinement. To these radiative processes, an additional independent radiative rate corresponds. Since the cavity confinement involves only a small portion of the overall emission angle, a good approximation consists in giving this additional radiative rate the corresponding free space value. This picture breaks down in the microcavity system where the $\mathbf{k}_\parallel$ selection rule holds. Consequently, all the allowed radiative processes are restricted to the exciton coupling with the single cavity mode and no additional excitonic radiative rate should be included in a theoretical treatment of microcavity exciton–polaritons.

The two oscillator formula turns out to be predictive for the polariton energies, particularly in the very strong coupling regime where the Rabi splitting is much larger than the polariton linewidth. Unfortunately, the same model completely fails to predict the polariton linewidths as measured in any optical experiment. The problem is related to the nonradiative broadening mechanisms that should be described more accurately by means of specific models. We briefly review this problem and the most recent achievements in the next section.

4.5 Polariton broadening

The broadening mechanisms acting on a quantum well exciton level are mainly of three kinds: exciton-phonon interaction, disorder and excitation induced dephasing. These three effects are not, in principle, independent of each other. We might remark, as an example, that the exciton–phonon scattering rates [Zimmermann (1997)] and the excitation induced exciton dephasing rates [Brinkmann (1998), Brinkmann (1996)] depend on the exciton wave function which is strongly affected by the presence of disorder. Needless to say, the problem of exciton broadening is a very complex one and the present approach, which consisted in including a phenomenological linewidth γ_e to account for the nonradiative exciton broadening, is by far oversimplified. We should mention that, in general, the main features of the microcavity polariton spectra in the strong coupling regime are not much affected by the details of the exciton broadening, provided this latter is much smaller than the Rabi splitting at resonance (the regime that we call very strong coupling). So, for example, the peak position versus detuning is very well described by the simple assumptions made so far. There are however some smaller details of the polariton spectra which require a more detailed description of the broadening mechanisms. We are going to briefly review the three main broadening mechanisms and the related predictions that have been made in recent years. We point out that we will always be making the assumption of independent broadening mechanisms. In other words we will consider separately each broadening mechanism as acting onto an otherwise noninteracting exciton state. This situation, as mentioned above, is never realized and must be considered as an approximation whose validity has to be tested from case to case.

We start with the broadening due to disorder. Disorder in high quality quantum wells mainly originates from alloy fluctuations and interface roughness. According to the works by Zimmermann [Zimmermann (1992)], disorder mainly affects the exciton in–plane center-of-mass degrees of freedom. The relative electron-hole motion and the motion along z are only affected to higher order in the ratio between the average fluctuation of the confinement energy and the exciton binding energy. The exciton lineshape in a disordered quantum well has been studied both theoretically and experimentally [Schnabel (1992), Glutsch (1994), Glutsch (1996)]. In the case of microcavity polariton, the general idea is that the exciton-photon coupling competes with the $k_\parallel$-nonconserving interaction of the exciton with a disordered potential. This led to the assumption that, in order to describe the optical response, the two interactions had to be trated on equivalent bases. Although a full treatment of the two interactions [Savona (1997a)] correctly reproduced the measured polariton linewidths as a function of detuning [Fisher (1996)], later it became clear that an important approximation led much more simply to the same results. The idea is to assume the lineshape of a bare quantum well exciton to be given. This lineshape may have been previously calculated by means of a microscopical model [Glutsch (1994)], separately measured [Ell (1998)], or defined phenomenologically [Kavokin (1998)]. Then, the semiclassical theory of the quantum well optical response is used, like in these notes, where the matrix element of a quantum well is convoluted with the given exciton lineshape. In other words, the exciton response is assumed to originate from a set of energy levels whose optical density is exactly given by the measured or calculated bare exciton lineshape. This approach is called Linear Dispersion Theory, or LDT for short. Whittaker [Whittaker (1997), Whittaker (1998)] has compared the result of LDT with the result of an exact numerical calculation of the microcavity polariton broadening and has shown that the differences are hardly distinguishable for realistic exciton and cavity parameters. Afterwards, the validity of the linear dispersion theory has been experimentally checked [Ell (1998)] by separately measuring the bare exciton lineshape and the polariton spectra on the same sample. Now it is well established that LDT can be used to calculate microcavity polariton spectra whenever the disorder broadening is dominant on the other broadening mechanisms. We must mention the works by Andreani [Andreani (1998)] and by Kavokin [Kavokin (1998)] where an exhaustive theoretical analysis of LDT and its implications on the optical response is carried out.

The physical reasons for which linear dispersion theory works well are not simple to explain. A handwaving argument might be the following. When the intracavity radiation interacts with the quantum well exciton, the amount of light scattered to different in–plane wave vectors due to the disordered exciton wave function is negligible with respect to the reflected and transmitted beams. Then, if we just need to describe the transmission, reflection and absorption of the microcavity[8], we may safely disregard any $k_\parallel$-nonconserving

---

[8] Other effects, such as the resonant light scattering into directions other than the transmitted or the reflected one, require the inclusion in the exciton–polariton

scattering event. The excitonic lineshape induced by the presence of disorder needs however to be included within the semiclassical treatment. This is done by including the disordered exciton lineshape at the level of the linear optical response function. This picture essentially works because the disorder is usually weak and particularly because the disorder is a static perturbation involving elastic scattering processes only. The same argument in principle does not hold for the two other sources of broadening, namely exciton-phonon scattering and exciton-exciton Coulomb scattering. In both cases, in fact, the exciton exchanges energy with the phonon or with the other exciton taking part in the scattering process. No definitive results exist for these two problems. In the case of exciton-phonon interaction, a theoretical model has been developed [Savona (1997b)] in which the broadening due to exciton-phonon scattering is calculated within Born approximation (second order perturbation theory) starting from the microcavity polariton states. In other words, it is assumed that the polariton states are those which undergo scattering with phonons. This is a natural assumption if the exciton–phonon scattering rates are much smaller than the exciton–photon coupling, which is the case for most of the strong coupling situations. Then, it can be shown that the peculiar dispersion curve of the lower polariton branch is responsible for a strong suppression of the lower polariton broadening mechanism. This is due to the steepness of the polariton dispersion near $\mathbf{k}_\parallel = 0$ (See Fig. 19) that makes the phase space allowed for polariton-phonon scattering very small. Consequently, for realistic values of the exciton and phonon parameters, it has been shown that the linewidth of the lower polariton at resonance is very small compared to that of the upper polariton, this latter being comparable to the bare exciton homogeneous broadening. Moreover, the linewidth linearly increases with temperature for both polaritons, but the two slopes are very different: for typical exciton and cavity parameters, to the lower polariton linewidth a slope of less than $0.5~\mu\mathrm{eV/cm}^{-1}$ corresponds, while a value larger by one order of magnitude (comparable with the bare quantum well case [Schulteis (1986)]) is found for the upper polariton. Although this model is simple and intuitive in itself and agrees with the first results on microcavity samples of exceptionally good quality[9] [Stanley (1997)], the problem needs further investigation and the question whether linear dispersion theory works independently of the broadening mechanism still lacks of a clear answer.

The problem of density induced polariton broadening is complex and here we just mention the most representative experimental and theoretical results. Measurements of the cavity polariton spectra under high excitation condi-

---

formalism of the light scattering onto the disordered quantum well exciton at least to first order in the scattering matrix element [Citrin (1996a), Citrin (1996b)].

[9] Generally, the homogenous broadening due to phonons is blurred by the disorder induced broadening at low temperatures. Thus, in order to quantify the homogeneous broadening at low temperatures in a linear optical measurement, a sample with an extremely low inhomogeneous broadening is required.

tions have been performed under nonresonant CW excitation [Houdré (1995)] and under resonant excitation by ultrashort pulses [Jahnke (1996), Rhee (1996), Lyngnes (1997)]. In both resonant and nonresonant excitation experiments, the cavity polariton peaks broaden for increasing excitation density. At the same time, the Rabi splitting decreases and eventually disappears. All these experimental results are satisfactorily explained by the semiconductor Maxwell-Bloch equations including second order Coulomb scattering terms [Haug (1994), Jahnke (1997)]. To the author's knowledge, no experiments have been performed in which the polariton spectra are measured under selective resonant excitation of one polariton peak only. Under resonant excitation and moderate excitation densities, the important broadening mechanism is the exciton-exciton Coulomb scattering [Ciuti (1998)]. Translated into the microcavity polariton system, due to the peculiar dispersion of the lower polariton, we may expect that the polariton–polariton scattering is strongly suppressed at the bottom of the lower polariton branch, in analogy with broadening due to exciton-phonon scattering. A first experimental evidence of this effect has recently been reported [Baumberg (1998)]. In this measurement, however, the observed linewidth suppression is very small compared to the theoretical expectation. This first experimental result should thus stimulate further detailed measurements of this phenomenon.

5 The quantum treatment and the quasimode approximation

The semiclassical treatment of the optical response of excitons in semiconductors, and in particular of microcavity polaritons, covers a large portion of the physics of these systems. It allows to derive quantities like the polariton dispersion and radiative rates which are usually assumed to be contained only in a full quantum treatment. In addition, a semiclassical formalism that includes Coulomb interaction between carriers to a more general extent than the simple electron-hole bound states, namely the semiconductor Maxwell-Bloch equations [Haug (1994), Lindberg (1988)], have had great success in the past ten years in describing the nonlinear phenomena in the optical response under high excitation regime. However, there are some physical phenomena that go beyond the semiclassical treatment and require a full quantum approach to be modeled. The most important of these phenomena is of course the photoluminescence. In a photoluminescence experiment, the system is initially in some excited configuration induced by means of an external perturbation (laser excitation or carrier transport) and then decays to its ground state giving rise to spontaneous or stimulated emission of light. This process is accompanied by relaxation through Coulomb interaction within the electron-hole system and through the interaction with other excitations, like phonons or impurities. It is clear that the situation depicted here requires a full quantum treatment of the exciton and photon fields.

In what follows we briefly recall the second quantization of the one-dimensional electromagnetic field described by Eq. (3) and write the Hamiltonian for the linear exciton-photon coupling. Then, we introduce a very useful approximation frequently used in the quantum models involving confined electromagnetic field: the quasimode approximation. By means of this approximation, one can treat the cavity mode as a discrete energy level and include the finite transmission of the cavity mirrors as an imaginary part in the time evolution of the cavity mode operator.

5.1 Second quantization of the exciton and cavity modes

The exciton field can be expressed in second quantization through the Bose operators $\hat{A}_{\mathbf{k}_\parallel}$ and $\hat{A}_{\mathbf{k}_\parallel}^\dagger$ obeying the commutation rule $[\hat{A}_{\mathbf{k}_\parallel}, \hat{A}_{\mathbf{k}'_\parallel}^\dagger] = \delta_{\mathbf{k}_\parallel, \mathbf{k}'_\parallel}$ [10]. The exciton Hamiltonian is thus

$$H_{\text{exc}} = \sum_{\mathbf{k}_\parallel} \hbar\omega(\mathbf{k}_\parallel) \hat{A}_{\mathbf{k}_\parallel}^\dagger \hat{A}_{\mathbf{k}_\parallel} \ . \tag{67}$$

In order to express the electromagnetic field in second quantization, we must choose an orthonormal set of modes as a basis for the canonical quantization. We take the modes already used in Chapter 2 for the Fabry-Pérot, that we depict in Fig. 22. The modes are

$$U_{\mathbf{k}_\parallel,\omega}(z) = \sqrt{\frac{\omega n_c^2}{k_z^{(l)} c^2}} \begin{cases} e^{ik_z^{(l)} z} + R(\mathbf{k}_\parallel,\omega) e^{-ik_z^{(l)} z} & z \in I \\ I(\mathbf{k}_\parallel,\omega) e^{ik_z z} + J(\mathbf{k}_\parallel,\omega) e^{-ik_z z} & z \in II \\ T(\mathbf{k}_\parallel,\omega) e^{ik_z^{(r)} z} & z \in III \end{cases}, \tag{68}$$

$$\tilde{U}_{\mathbf{k}_\parallel,\omega}(z) = \sqrt{\frac{\omega n_c^2}{k_z^{(r)} c^2}} \begin{cases} \tilde{T}(\mathbf{k}_\parallel,\omega) e^{-ik_z^{(l)} z} & z \in I \\ \tilde{I}(\mathbf{k}_\parallel,\omega) e^{-ik_z z} + \tilde{J}(\mathbf{k}_\parallel,\omega) e^{ik_z z} & z \in II \\ e^{-ik_z^{(r)} z} + \tilde{R}(\mathbf{k}_\parallel,\omega) e^{ik_z^{(r)} z} & z \in III \end{cases}. \tag{69}$$

Regions I, II and III are defined in Fig. 22, $(k_z^{(l)})^2 = (\omega^2/c^2) n_l^2 - k_\parallel^2$, $(k_z^{(r)})^2 = (\omega^2/c^2) n_r^2 - k_\parallel^2$ and $k_z^2 = (\omega^2/c^2) n_c^2 - k_\parallel^2$. These modes may be obtained by imposing Maxwell boundary conditions at the two mirror boundaries. The task is simplified by the use of transfer matrices (27) and (29). The problem is thus reduced to the solution of a set of four linear equations for the coefficients $R(\mathbf{k}_\parallel,\omega)$, $T(\mathbf{k}_\parallel,\omega)$, $I(\mathbf{k}_\parallel,\omega)$ and $J(\mathbf{k}_\parallel,\omega)$. In the following we will be interested in the explicit expression of the two coefficients I and J for the intracavity field, that we report here:

---

[10] This is a Kroeneker delta symbol. In fact, for the two in-plane directions, we use a quantization on a finite surface S, both for the exciton and for the photon modes. The k_z variable, on the contrary, is quantized in the continuum limit.

$$I(\mathbf{k}_\|,\omega) = \frac{k_z^{(l)}}{k_z} \frac{t_1(\mathbf{k}_\|,\omega)e^{i(k_z-k_z^{(l)})L_c/2}}{1 - r_1(\mathbf{k}_\|,\omega)r_2(\mathbf{k}_\|,\omega)e^{2ik_zL_c}} ,$$
$$J(\mathbf{k}_\|,\omega) = r_2(\mathbf{k}_\|,\omega)I(\mathbf{k}_\|,\omega)e^{ik_zL_c} , \qquad (70)$$

$$\tilde{I}(\mathbf{k}_\|,\omega) = \frac{k_z^{(r)}}{k_z} \frac{t_2(\mathbf{k}_\|,\omega)e^{i(k_z-k_z^{(r)})L_c/2}}{1 - r_1(\mathbf{k}_\|,\omega)r_2(\mathbf{k}_\|,\omega)e^{2ik_zL_c}} ,$$
$$\tilde{J}(\mathbf{k}_\|,\omega) = r_2(\mathbf{k}_\|,\omega)\tilde{I}(\mathbf{k}_\|,\omega)e^{ik_zL_c} . \qquad (71)$$

Here, $r_{1,2}$ and $t_{1,2}$ are the two DBR reflection and transmission coefficients. The modes defined in this way obey the orthonormality condition in the ω variable [Ley (1987), Savona (1996a)]. We can thus write the vector potential in second quantization form as[11]

$$\mathbf{A}(\mathbf{r},z) = \sum_{\mathbf{k}_\|} \boldsymbol{\epsilon}_{\mathbf{k}_\|} \int_0^\infty d\omega \sqrt{\frac{\hbar c^2}{S\omega n_c}} (\hat{a}_{\mathbf{k}_\|,\omega} U_{\mathbf{k}_\|,\omega}(z) + \hat{a}^\dagger_{\mathbf{k}_\|,\omega} U^*_{\mathbf{k}_\|,\omega}(z))e^{i\mathbf{k}_\| \cdot \mathbf{r}} ,$$
$$(72)$$

where $\boldsymbol{\epsilon}_{\mathbf{k}_\|}$ is the electric field polarization, S is the in–plane normalization surface and $\hat{a}_{\mathbf{k}_\|,\omega}$ is the photon Bose operator obeying the commutation rule $[\hat{a}_{\mathbf{k}_\|,\omega}, \hat{a}^\dagger_{\mathbf{k}'_\|,\omega'}] = 2\pi\delta(\omega-\omega')\delta_{\mathbf{k}_\|,\mathbf{k}'_\|}$. Obviously, the Hamiltonian of the photon field is

$$H_{em} = \sum_{\mathbf{k}_\|} \int_0^\infty d\omega\, \hbar\omega \hat{a}^\dagger_{\mathbf{k}_\|,\omega} \hat{a}_{\mathbf{k}_\|,\omega} , \qquad (73)$$

The linear exciton-photon interaction Hamiltonian is

$$H_I = \sum_{\mathbf{k}_\|} \int_0^\infty d\omega\, iC_{\mathbf{k}_\|,\omega}(\hat{a}_{\mathbf{k}_\|,\omega} + \hat{a}^\dagger_{-\mathbf{k}_\|,\omega})(\hat{A}_{-\mathbf{k}_\|} - \hat{A}^\dagger_{\mathbf{k}_\|}) , \qquad (74)$$

where the interaction coefficient is expressed in terms of the exciton envelope function and dipole matrix element, defined in the previous chapter, as

$$C_{\mathbf{k}_\|,\omega} = \frac{\omega(\mathbf{k}_\|)}{n_c}\sqrt{\frac{\hbar}{\omega}} F(0)\boldsymbol{\epsilon} \cdot \boldsymbol{\mu}_{cv} \int dz\, U_{\mathbf{k}_\|,\omega}(z)\rho(z) . \qquad (75)$$

We have neglected the so called A^2 interaction term [Savona (1996a)] which enters the interaction Hamiltonian and is known to give a negligible contribution to the cavity polariton dispersion. The diagonalization of this set of Hamiltonians gives exactly the polariton dispersion equation (62), as already observed [Savona (1996a)].

In order to introduce our quasimode approximation, we now focus on a simplified situation with equal mirrors. Moreover, we assume that the spacer

---

[11] From now on it is implicitly assumed that every sum over the photon modes spans both U and $\tilde{U}$ modes. Moreover, a polarization index is not explicitly indicated and the sum over different polarizations is always assumed.

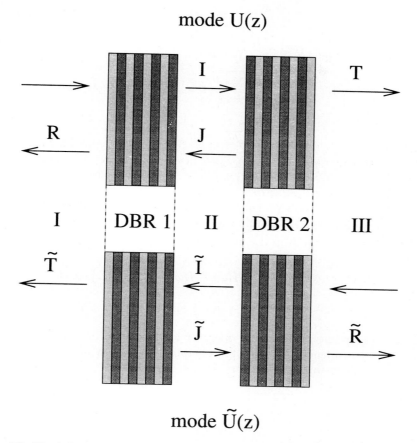

Fig. 22. The left and right propagating modes of a Fabry-Pérot, denoted by $U(z)$ and $\tilde{U}(z)$ respectively in the text. The materials in regions I, II, III, have refraction indices n_l, n_c, and n_r respectively.

refraction index is equal to the external refraction indices on both sides, namely $n_c = n_1 = n_2$. Under these assumptions the cavity is symmetric under space inversion along the z coordinate and the modes U and $\tilde{U}$ are obey $U(z) = \tilde{U}(-z)$. Correspondingly, the coefficients I and J are given by

$$I(\mathbf{k}_\|,\omega) = \frac{t(\mathbf{k}_\|,\omega)}{1 - r^2(\mathbf{k}_\|,\omega)e^{2ik_zL_c}} ,$$
$$J(\mathbf{k}_\|,\omega) = r(\mathbf{k}_\|,\omega)I(\mathbf{k}_\|,\omega)e^{ik_zL_c} , \qquad (76)$$

where r and t now denote the reflection and transmission coefficients of both mirrors. We now introduce a change of basis within each two dimensional subspace of modes with given $\mathbf{k}_\|$, by defining the symmetric and antisymmetric modes:

$$U^{(s)} = \frac{U + \tilde{U}}{\sqrt{2}}, \tag{77}$$

$$= \sqrt{\frac{2\omega n_c^2}{k_z c^2}}(I + J)\cos(k_z z)$$

$$U^{(a)} = \frac{U - \tilde{U}}{i\sqrt{2}}. \tag{78}$$

$$= \sqrt{\frac{2\omega n_c^2}{k_z c^2}}(I - J)\sin(k_z z)$$

We may write the vector potential field in terms of these modes by defining the two new Bose operators $\hat{a}^{(s)}_{\mathbf{k}_\parallel,\omega}$ and $\hat{a}^{(a)}_{\mathbf{k}_\parallel,\omega}$ in an analogous way. We now restrict to $k_\parallel = 0$ and drop the in-plane wave vector index for clarity.

$$\mathbf{A}(z) = \epsilon \int_0^\infty d\omega \sqrt{\frac{2\hbar c}{\omega}} \left[\left((I(\omega) + J(\omega))\, \hat{a}^{(s)}_\omega + \text{h.c.}\right) \cos\left(\frac{\omega n_c}{c} z\right) \right.$$
$$\left. + \left((I(\omega) - J(\omega))\, \hat{a}^{(a)}_\omega + \text{h.c.}\right) \sin\left(\frac{\omega n_c}{c} z\right) \right]. \tag{79}$$

The corresponding electric field operator is given by $\mathbf{E} = -(1/c)(\partial \mathbf{A}/\partial t)$. The time dependence of the photon operator is $\hat{a}(t) = \hat{a}(0)e^{i\omega t}$ and $\hat{a}^\dagger(t) = \hat{a}^\dagger(0)e^{-i\omega t}$. Then we easily obtain

$$\mathbf{E}(z) = i\epsilon \int_0^\infty d\omega \sqrt{2\hbar c\omega} \left[\left((I(\omega) + J(\omega))\, \hat{a}^{(s)}_\omega - \text{h.c.}\right) \cos\left(\frac{\omega n_c}{c} z\right) \right.$$
$$\left. + \left((I(\omega) - J(\omega))\, \hat{a}^{(a)}_\omega - \text{h.c.}\right) \sin\left(\frac{\omega n_c}{c} z\right) \right]. \tag{80}$$

These expressions contain no approximations and represent the z-dependent part of the vector potential and electric field operators for the symmetric structure we are considering. In the next section we will compare these expressions with the approximate ones obtained from the quasimode formalism. This comparison will allow to univocally define the quasimode coupling coefficient in the limit of high mirror reflectivity.

5.2 The quasimode approximation

We consider again the symmetric microcavity defined in the previous section. Because of our assumption of high reflectivity mirrors, as a first approximation we consider ideal totally reflecting mirrors, namely $r(\omega) = 1$. In this situation, the electromagnetic field inside the resonator and the one outside the resonator are totally decoupled. They thus can be described as two distinct fields. Taking into account the lowest cavity mode and neglecting all higher order modes, the electric field inside the spacer reads

$$\mathbf{E}(z) = i\epsilon \sqrt{\frac{2\pi \hbar c^2 \omega_c}{n_c L_c}} (\hat{a}_c - \hat{a}_c^\dagger) \cos\left(\frac{n_c \omega_c}{c} z\right) , \qquad (81)$$

where $\hat{a}_c$ is the Bose operator for the discrete cavity mode at $k_\parallel = 0$. Since we have assumed reflectivity equal to one, with zero phase change upon reflection, only symmetric modes exist inside the cavity. In the realistic situation of finite mirror transmission, the antisymmetric mode is strongly suppressed close to resonance, as can be seen by comparing the field strengths $|I(\omega) + J(\omega)|$ and $|I(\omega) - J(\omega)|$ appearing in expressions (79) and (80). The corresponding Hamiltonian is

$$H_{\text{cav}} = \hbar \omega_c \hat{a}_c^\dagger \hat{a}_c . \qquad (82)$$

The external electromagnetic field is represented by a continuum of photon modes that we denote with the Bose operators $\hat{b}_\omega$, obeying the commutation relation $[\hat{b}_\omega, \hat{b}_{\omega'}^\dagger] = 2\pi \delta(\omega - \omega')$. For the present treatment we do not need to specify the spatial dependence of these modes. The Hamiltonian for the external electromagnetic field reads

$$H_{\text{ext}} = \int d\omega \, \hbar\omega \, \hat{b}_\omega^\dagger \hat{b}_\omega , \qquad (83)$$

The quasimode approximation [Barnett (1988)] consists in introducing the finite mirror transmission through a phenomenological linear coupling Hamiltonian between the field inside and outside the resonator. This is a standard technique in the theory of laser operation and, as we will see later, largely simplifies the equations for the time evolution of the quantum operators (Heisenberg equations of motion). The form of this coupling is expressed through the Hamiltonian

$$H_{\text{qm}} = \int d\omega \, [v(\omega) \hat{a}_c \hat{b}_\omega^\dagger + v^*(\omega) \hat{a}_c^\dagger \hat{b}_\omega] , \qquad (84)$$

where $v(\omega)$ is the still unknown coupling coefficient which is assumed to be complex valued and frequency dependent. The problem defined by the three Hamiltonians (82), (83) and (84) is a typical problem of a discrete state coupled to a continuum. It can be diagonalized analytically using the standard Fano approach [Fano (1961)]. The operators corresponding to the eigenmodes of the total Hamiltonian $H = H_{\text{cav}} + H_{\text{ext}} + H_{\text{qm}}$ are mixed modes of the two fields written as

$$\hat{a}_\omega = \alpha(\omega) \hat{a}_c + \int \beta(\omega, \omega') \hat{b}_{\omega'} d\omega' , \qquad (85)$$

where the coefficient $\alpha(\omega)$ is given by [Fano (1961)]

$$\alpha(\omega) = \frac{v^*(\omega)}{\pi |v(\omega)|^2 + i(\omega - \omega_c - F(\omega))} \qquad (86)$$

and
$$F(\omega) = \mathcal{P} \int d\omega' \frac{|v(\omega')|^2}{\omega - \omega'} \; , \tag{87}$$

with $\mathcal{P}$ denoting the principal part. In this formalism, the operators $\hat{a}_\omega$ denote the eigenmodes of the total electromagnetic field, originating from the finite mirror transmission According to the Fano theory, the inverse transformation which gives the operator $\hat{a}_c$ in terms of the operators $\hat{a}_\omega$ is simply expressed in terms of the complex conjugate of the coefficient $\alpha(\omega)$ as

$$\hat{a}_c = \int d\omega \; \frac{v(\omega)}{\pi |v(\omega)|^2 - i(\omega - \omega_c - F(\omega))} \; \hat{a}_\omega \; . \tag{88}$$

Inserting this equality into the expression (81) for the intracavity field we obtain the expression of $\mathbf{E}(z)$ in terms of the field operators a_ω. This expression reads

$$\mathbf{E}(z) = i\epsilon \int d\omega \; \sqrt{\frac{2\pi\hbar c^2 \omega_c}{n_c L_c}} \left(\frac{v(\omega)}{\pi |v(\omega)|^2 - i(\omega - \omega_c - F(\omega))} \; \hat{a}_\omega - h.c. \right)$$
$$\times \cos\left(\frac{n_c \omega_c}{c} z\right) \; . \tag{89}$$

The approximation implied by the quasimode formalism appears clearly in this expression for the electric field operator, where all the field eigenmodes have the same spatial dependence inside the cavity. This is approximately valid only if the photon density of states is strongly peaked around the cavity resonance.

The coupling coefficient $v(\omega)$ in the quasimode approach is determined by comparison of Eqs. (89) and (80)[12]. Of course, we neglect the asymmetric field part in (80) because we are assuming that it vanishes with respect to the symmetric part, close to resonance. In order for the two expressions to be equal, we must have

$$\sqrt{\frac{2\pi\hbar c^2 \omega_c}{n_c L_c}} \; \frac{v(\omega)}{\pi |v(\omega)|^2 - i(\omega - \omega_c - F(\omega))} = \sqrt{\hbar c \omega} \left(I(\omega) + J(\omega) \right) \; ,$$
$$= \sqrt{\hbar c \omega} \; \frac{t(\omega)}{1 - r(\omega) e^{ik_z L_c}} \; . \tag{90}$$

We must not forget, however, that our approximation is valid only for high mirror reflectivities. We may thus develop the denominator on the right side of (90) to first order around the resonant cavity frequency. By introducing the complex representation $r(\omega) = |r(\omega)| e^{i\phi(\omega)}$, we get

---

[12] In Eq. (80) the integral runs over positive frequency values, while the Fano formalism used in the present section involves integrals over the whole ω axis. This difference can be overcame by extending the field (80) to negative frequencies and properly renormalizing the quantities under integral. It may be verified that this amounts to keeping the same expression (80) divided by $\sqrt{2}$.

$$\frac{v(\omega)}{\pi|v(\omega)|^2 - i(\omega - \omega_c - F(\omega))} \tag{91}$$

$$\simeq \sqrt{\frac{c}{2\pi n_c L_c}} \frac{t(\omega)}{(1 - |r(\omega)|)\frac{c}{n_c L_c} - i\left(\omega - \omega_c + \frac{c}{n_c L_c}\phi(\omega)\right)}.$$

It turns out that the equality (91) is satisfied by the following expressions for $v(\omega)$ and $F(\omega)$

$$v(\omega) = \sqrt{\frac{c}{2\pi n_c L_c}}\, t(\omega)\,, \tag{92}$$

$$F(\omega) = -\frac{c}{n_c L_c}\phi(\omega)\,. \tag{93}$$

The last step consists in checking the consistency between the pair of relations (92) and (93) and expression (87) which derives from the Fano treatment. We perform this check explicitly using the Kramers-Kronig relations for the frequency dependent reflection and transmission coefficients [Bassani (1983)]. The Kramers-Kronig relation that links the argument and the modulus of the reflection coefficient is given by

$$\phi(\omega) = -\frac{2\omega}{\pi}\mathcal{P}\int_0^\infty d\omega'\,\frac{\ln|r(\omega')|}{\omega'^2 - \omega^2}\,. \tag{94}$$

Introducing $R(\omega) = |r(\omega)|^2$ and $T(\omega) = |t(\omega)|^2$, and assuming that $R(\omega) \simeq 1$ in the frequency region of interest (close to the singularity in (94)), we may write

$$\phi(\omega) \simeq \frac{\omega}{\pi}\mathcal{P}\int_0^\infty d\omega'\,\frac{|T(\omega')|}{\omega'^2 - \omega^2}\,. \tag{95}$$

Turning the integral into an integral from $-\infty$ to $+\infty$ with the assumption $|T(-\omega)| = |T(\omega)|$, we finally get

$$\phi(\omega) \simeq \frac{1}{2\pi}\mathcal{P}\int_{-\infty}^{+\infty} d\omega'\,\frac{|t(\omega')|}{\omega' - \omega}\,. \tag{96}$$

which corresponds exactly to the consistency condition we were looking for. To resume, we have assumed that the intracavity field and the external field are distinct in the limit of vanishing mirror transmission. For a small but finite transmission, we have introduced a linear coupling between the two fields. We have diagonalized the Hamiltonian for the coupled system and compared the density of photon states with the exact one. This comparison, together with the requirement for causality introduced by means of Kramers-Kronig relations, has allowed to univocally define the quasimode coupling coefficient in terms of the mirror optical constants. A further remark concerns the specific case of DBRs, where the argument of the mirror reflectivity depends linearly on frequency according to (42). By inserting (42) into (87) and assuming constant reflectance R, some algebra shows that the whole quasimode formalism can be rewritten in terms of a frequency independent quasimode coupling coefficient

$$v^2 = \frac{c}{2\pi} \frac{1-R}{n_c L_{\text{eff}}} , \tag{97}$$

provided that we assume an effective Fabry-Pérot thickness

$$L_{\text{eff}} = L_c + L_{\text{DBR}} . \tag{98}$$

This replacement must be done in particular in the expression for the intracavity field (81) and in all those that follow. Of course, for this frequency independent quasimode coefficient, the frequency shift $F(\omega)$ is strictly zero.

The true advantage of this formalism stays in the comparison between the quasimode coupling coefficient v and the cavity mode broadening γ_c (44). It turns out that

$$v^2 = \frac{\gamma_c}{\pi} \tag{99}$$

namely that the coupling coefficient between the inner and outer regions is proportional to the escape rate of a photon from the cavity.

This result is intuitive at first view. It states that, in our artificial picture describing the electromagnetic field as two distinct fields, their linear coupling must be proportional to the cavity photon escape rate. The purpose of this section was to prove that this result is rigorous in the limit of high finesse for the DBR microcavity. As a matter of fact, this result is very important for practical application. In fact, now we are allowed to treat the external photon continuum as a reservoir and perform any kind of quantum calculation using the discrete cavity mode only. This implies a substantial simplification in the calculations. As a first example, we may write the master equations for the cavity mode, assuming a "cold" reservoir, namely zero photons in the external field. Physically, this is the correct assumption when we are interested in the spontaneous emission only. The density matrix of the reservoir is thus $\sigma_R = |0\rangle\langle 0|$, where $|0\rangle$ denotes the empty state of the external field. Applying the standard master equation formalism [Cohen-Tannoudji (1988)], we end up with the following equations for the density matrix σ_{ij} describing the intracavity field mode

$$\frac{d\sigma_{11}}{dt} = -\Gamma \sigma_{11} , \tag{100}$$

$$\frac{d\sigma_{00}}{dt} = +\Gamma \sigma_{11} , \tag{101}$$

$$\frac{d\sigma_{10}}{dt} = -i(\omega_c + \Delta)\sigma_{10} - \frac{\Gamma}{2}\sigma_{10} , \tag{102}$$

where the quantities Δ and Γ are expressed in terms of the interaction Hamiltonian as

$$\Gamma = \frac{2\pi}{\hbar} \int d\omega \, |\langle 1_\omega, 0_c | H_{\rm qm} | 0_\omega, 1_c \rangle|^2 \delta(\hbar\omega - \hbar\omega_c) \, , \qquad (103)$$

$$\Delta = \frac{1}{\hbar} \mathcal{P} \int d\omega \, \frac{|\langle 1_\omega, 0_c | H_{\rm qm} | 0_\omega, 1_c \rangle|^2}{\hbar\omega_c - \hbar\omega} \, . \qquad (104)$$

These quantities are the energy damping and shift of the discrete level due to the coupling with the reservoir continuum. Simple calculations show that $\Delta = 0$, because of the principal part integration over the quantity v^2 which does not depend on ω, and $\Gamma = 2\gamma_c$. Thus, the decay times for the population of the cavity mode and for the coherence between the cavity mode and the vacuum state are correctly accounted for by our quasimode approach.

When writing a quantum model which includes the coupling to the exciton level, the coupling can be expressed in terms of the intracavity discrete photon mode as

$$H_{\rm pol} = V(\hat{A}\hat{a}_c^\dagger + \hat{A}^\dagger \hat{a}_c) \, , \qquad (105)$$

where $\hat{A}$ and $\hat{A}^\dagger$ are the Bose operators for the exciton level[13] and V is the coupling coefficient derived in the previous chapter.

The next section is devoted to a model of microcavity polariton photoluminescence that we develop as an example of application of the concepts and tools learned in this chapter.

5.3 Simple model of microcavity photoluminescence

One of the most direct optical characterizations of exciton and polariton states is the measurement of the photoluminescence spectrum under non resonant excitation. In this kind of experiment, the system is excited at energies much higher than the exciton level. The system then relaxes down to the radiative levels which emit light. The relaxation process is in general very complex and involves several different mechanisms. At low excitation densities, the most important mechanism is the exciton scattering through absorption or emission of optical and acoustic phonons [Piermarocchi (1996)]. In particular, for low enough temperatures only the acoustic phonon modes are thermally populated and consequently only acoustic phonon scattering processes can take place. In typical GaAs quantum wells, this situation corresponds to temperatures lower than about 50 K [Piermarocchi (1997)]. We want to describe the photoluminescence spectra in this regime of physical parameters. The main steps of relaxation and radiative recombination are illustrated in Fig. 23.

The first step is the exciton formation process. By means of carrier-carrier and carrier-phonon scattering, the free carriers created by the initial excitation form exciton bound states. This process is quasielastic at low temperature and excitation density [Piermarocchi (1997)] as shown in the picture.

---

[13] for a given $\mathbf{k}_\parallel$. We remark that until now we have never considered a situation in which different in-plane wave vectors are coupled to each other. Then, the microcavity polariton problem is separate in the $\mathbf{k}_\parallel$ variable.

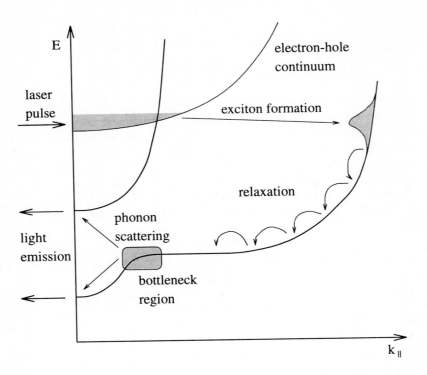

Fig. 23. A sketch of the excitation, exciton formation and relaxation processes in a photoluminescence experiment under nonresonant excitation.

The formation process is also very rapid compared to the characteristic time of the acoustic phonon relaxation that follows. This relaxation process brings the exciton population to the radiative region where a competition between radiative recombination and further relaxation takes place. Several effect have to be considered at this point. First, the leaky modes described in the previous chapter act as a sink for the exciton population, because a large part of the excitons recombine by emitting photons through the leaky mode channels [Tassone (1996), Tassone (1997)]. The remaining excitons further relax and eventually reach the strong coupling region where the slope of the lower polariton branch suddenly increases. This increased slope dramatically slows down the relaxation to lower $k_\parallel$, giving rise to the so called "bottleneck effect" [Tassone (1997)]. In addition, the radiative recombination rate within the strong coupling region is very fast in typical samples, being of the order of 1 ps as inferred from the radiative rates calculated in the previous chapter. The radiative recombination turns out to be much faster compared to the relaxation timescale which is of the order of 100 ps [Piermarocchi (1996)]. This implies that, in a stationary regime like under CW excitation, the polariton population builds up at the bottleneck level while the polariton levels in the strong coupling have, on the average, a vanishing population. This very qualitative picture is supported by several theoretical and experimental investigations [Tassone (1996), Tassone (1997), Sermage (1996)]. For the purpose of the present treatment, the important feature is the buildup of the population at the bottleneck. Since we want to describe a situation under stationary CW excitation, we will neglect all the details of the relaxation and formation processes and assume that the population is peaked right above the bottleneck region.

Another important feature of a realistic quantum well is the inhomogeneous broadening of the exciton level. We have already discussed the implications of inhomogeneous broadening and the problem of its inclusion in a microcavity polariton model. Here, we will make one step further and introduce the linear dispersion theory, previously mentioned, at the level of our quantum treatment. In practice, we will assume a set of exciton levels with different energies at $k_\parallel = 0$, whose optical density (oscillator strength) is weighed by a Gaussian lineshape[14]. We point out that in reality the presence of disorder-induced inhomogeneous broadening implies a partial lifting of the $k_\parallel$ selection rule. Then, our picture of the formation-relaxation mechanisms is in principle oversimplified and a much more complex scenery must be employed to correctly describe the photoluminescence process [Zimmermann (1997)]. In particular, relaxation can occur between the closely spaced energy levels within the inhomogeneous distribution (the so called spectral diffusion) and the dynamics of the photoluminescence process can be consequently modified. The model that we are going to present is based on a much simpler assumption concerning inhomogeneous broadening. An a priori justification might come from the assumption of small inhomogeneous broadening with respect to the Rabi splitting. In the end, however, the only valid justification comes from the comparison with existing experimental results which will turn out to be quite satisfying within the range of parameters considered.

We illustrate in Fig. 24 the different sets of bosonic levels considered in the present model. The model consists in a set of bosonic levels, distributed in energy, representing the exciton levels at $k_\parallel = 0$. In order to simplify calculations, this energy distribution as well as those related to the phonon bath and to the external photon continuum will be assumed as continuous. Then, the exciton levels are described by the Bose operators $\hat{A}_\omega$ and $\hat{A}_\omega^\dagger$ and the bare exciton Hamiltonian is

$$H_{\text{exc}} = \int d\omega \, \hbar\omega \, \hat{A}_\omega^\dagger \hat{A}_\omega \ . \tag{106}$$

All the integrals in the variable ω are performed in the interval $[-\infty, +\infty]$. The cavity mode has frequency ω_c and is described by the Hamiltonian

$$H_c = \hbar\omega_c \, \hat{a}_c^\dagger \hat{a}_c \ , \tag{107}$$

---

[14] It will appear from the equations that the same treatment can be developed without additional difficulties using an arbitrary exciton lineshape.

where a_c and $a_c^\dagger$ are the corresponding Bose operators. We further consider the external photon continuum, whose interpretation has been given in the section on the quasimode formalism, a continuum of levels representing the acoustic phonon bath and one discrete level at energy $\hbar\omega_b$ that represents the population buildup level in the bottleneck region. The corresponding Hamiltonians are analogously defined as

$$H_{\text{ext}} = \int d\omega\, \hbar\omega\, \hat{b}_\omega^\dagger \hat{b}_\omega , \quad (108)$$

$$H_{\text{ph}} = \int d\omega\, \hbar\omega\, \hat{c}_\omega^\dagger \hat{c}_\omega , \quad (109)$$

$$H_b = \hbar\omega_b\, \hat{D}^\dagger \hat{D} , \quad (110)$$

where the operators $\hat{b}_\omega$, $\hat{c}_\omega$ and $\hat{D}$ are the corresponding annihilation operators.

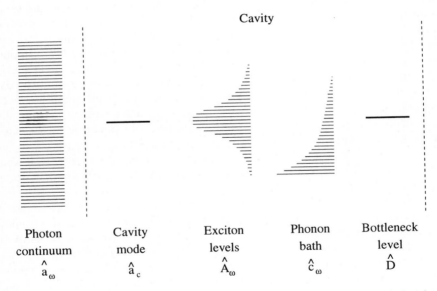

Fig. 24. The different levels considered in the calculation of this section are sketched here. The picture suggests the assumption of a thermalized phonon bath, employed in our derivation.

Three distinct interactions are assumed. First, the quasimode interaction between the intracavity and external photon modes, given by the Hamiltonian (84). Second, the polariton interaction between exciton and cavity modes. This interaction is described by a generalization of (105) that takes into account the inhomogeneous broadening in the optical coupling as

$$H_{\text{pol}} = \hbar V \int d\omega\, \alpha(\omega)[\hat{a}_c \hat{A}_\omega^\dagger + \hat{a}_c^\dagger \hat{A}_\omega]\,, \tag{111}$$

where the function $\alpha(\omega)$ describes the weight of the different exciton levels in the optical coupling and is defined as

$$\alpha(\omega) = \left[\frac{1}{\sqrt{2\pi\gamma_{\text{inh}}^2}} e^{-\frac{(\omega-\omega_0)^2}{2\gamma_{\text{inh}}^2}}\right]^{1/2}. \tag{112}$$

Here, $\hbar\omega_0$ is the central exciton energy and γ_{inh} is the inhomogeneous broadening parameter. The coefficient $\alpha(\omega)$ redistributes the exciton oscillator strength over the broadened exciton line and satisfies the normalization condition $\int d\omega\, \alpha^2(\omega) = 1$. The third interaction is given by the scattering process that brings the excitation from the bottleneck level to the radiative polariton levels through phonon emission and absorption. The corresponding Hamiltonian is

$$H_{\text{scat}} = \hbar \int d\omega \int d\omega'\, \beta(\omega)\alpha(\omega')[\hat{D}\hat{c}_\omega^\dagger \hat{A}_\omega^\dagger + \hat{D}\hat{c}_\omega \hat{A}_\omega^\dagger + \hat{D}^\dagger \hat{c}_\omega \hat{A}_\omega + \hat{D}^\dagger \hat{c}_\omega^\dagger \hat{A}_\omega]\,, \tag{113}$$

where $\alpha(\omega)$ is the inhomogeneous exciton distribution (112) which accounts for the density of exciton states. The function $\beta(\omega)$ is a scattering matrix element whose form will be specified later. The first two terms in square brackets describe the transfer of excitation from the bottleneck level to the exciton levels by means of emission and absorption of a photon respectively. The two other terms are the reverse processes that bring the excitation from the exciton levels to the bottleneck level.

We work in the Heisenberg representation, thus all the creation and annihilation operators are time dependent. The total Hamiltonian of the system is then used to obtain the Heisenberg equations for these operators, according to the formula for the time evolution of an arbitrary operator $\hat{O}$

$$\frac{d\hat{O}}{dt} = \frac{1}{i\hbar}[\hat{O}, H]\,. \tag{114}$$

The Heisenberg equations for the time evolution of the cavity operator $\hat{a}$ the exciton operators $\hat{A}_\omega$ and the external photon continuum $\hat{a}_\omega$ are given by

$$\dot{\hat{a}}_c(t) = -i\omega_c \hat{a}_c(t) - iV\int d\omega\, \alpha(\omega)\hat{A}_\omega(t) - iv\int d\omega\, \hat{b}_\omega(t)\,, \tag{115}$$

$$\dot{\hat{A}}_\omega(t) = -i\omega \hat{A}_\omega(t) - iV\alpha(\omega)\hat{a}_c(t) \tag{116}$$

$$- i\alpha(\omega)\int d\omega'\, \beta(\omega')\hat{D}(t)[\hat{c}_{\omega'}^\dagger(t) + \hat{c}_{\omega'}(t)]\,, \tag{117}$$

$$\dot{\hat{b}}_\omega(t) = -i\omega \hat{b}_\omega(t) - iv\, \hat{a}_c(t)\,. \tag{118}$$

Two further equations could be derived to describe the time evolution of operators $\hat{D}$ and $\hat{c}_\omega$. Our next assumption consists in neglecting the memory of the interactions on these levels and considering their evolution as free. Thus we have

$$\hat{D}(t) = \hat{D}_0 e^{-i\omega_b t} , \quad (119)$$

$$\hat{c}_\omega(t) = \hat{c}_{0,\omega} e^{-i\omega t} . \quad (120)$$

It could be shown that these assumptions are equivalent to the usual approximation, made within the master equations formalism, that the time evolution of the density matrix for the reservoir is not affected by the interaction with the system. This is a standard assumption for the phonon reservoir [Cohen-Tannoudji (1988)]. For the bottleneck level, instead, it corresponds to the physical assumption of stationarity of the system: the population of the bottleneck level is kept constant by the CW excitation.

Integration of Eqs. (117) and (118) gives

$$\hat{b}_\omega(t) = -iv \int_0^t dt' \hat{a}_c(t') e^{i\omega(t'-t)} + \hat{b}_{0,\omega} e^{-i\omega t} , \quad (121)$$

$$\hat{A}_\omega(t) = -iV\alpha(\omega) \int_0^t dt' \hat{a}_c(t') e^{i\omega(t'-t)} + \hat{A}_{0,\omega} e^{-i\omega t}$$
$$- i\alpha(\omega) \int_0^t dt' \int d\omega' \beta(\omega') \hat{D}(t') [\hat{c}^\dagger_{\omega'}(t') + \hat{c}_{\omega'}(t')] e^{i\omega(t'-t)} , \quad (122)$$

where $\hat{b}_{0,\omega}$ and $\hat{A}_{0,\omega}$ are the corresponding freely evolving operators at $t = 0$ and represent the "initial condition" for the solution of the operatorial equations. They are necessary because they preserve the Bose commutation rule, but they do not contribute to the expectation values calculated in what follows.

By placing these two expressions in Eq. (115), some straightforward algebra leads to the following Langevin equation for the cavity mode operator

$$\dot{\hat{a}}_c(t) = -i(\omega_c - i\gamma_c)\hat{a}_c(t) - V^2 \int_0^t dt' e^{-\frac{\gamma_{inh}^2 (t-t')^2}{2}} e^{-i(\omega_e - i\gamma)(t-t')} \hat{a}_c(t') \quad (123)$$
$$- iv \int d\omega \hat{b}_{0,\omega} e^{-i\omega t} - iV \int d\omega \hat{A}_{0,\omega} \alpha(\omega) e^{-i\omega t}$$
$$- iV \int d\omega \int d\omega' \alpha^2(\omega') \beta(\omega)$$
$$\times \left[\frac{e^{-i(\omega_b - \omega)t} - e^{-i(\omega' - i\gamma)t}}{\omega_b - \omega - \omega' + i\gamma} \hat{D}_0 \hat{c}^\dagger_{0,\omega} + \frac{e^{-i(\omega_b + \omega)t} - e^{-i(\omega' - i\gamma)t}}{\omega_b + \omega - \omega' + i\gamma} \hat{D}_0 \hat{c}_{0,\omega} \right] ,$$

where the relation (99) for the coefficient v was used. We remark that, in addition to the inhomogeneous exciton distribution, we have included an additonal small exciton homogeneous broadening through the quantity γ, which was added to the energy of each exciton level. This completes our description of the bare exciton lineshape which is now given by a convolution of the Gaussian lineshape $\alpha^2(\omega)$ and a Lorentzian of width γ. Since we want to obtain a result for the luminescence under stationary conditions, we look for a stationary solution of Eq. (123). Thus, we let to zero the terms with the broadening γ appearing in the exponentials and we make the ansatz (we neglect the terms proportional to $\hat{b}_{0,\omega}$ and $\hat{A}_{0,\omega}$ because, as already mentioned, they do not contribute to the final result)

$$\lim_{t\to\infty} \hat{a}_c(t) = \xi(\omega)\hat{D}_0\hat{c}^\dagger_{0,\omega}e^{-i(\omega_b-\omega)t} + \zeta(\omega)\hat{D}_0\hat{c}_{0,\omega}e^{-i(\omega_b+\omega)t} \ . \qquad (124)$$

By replacing this expression into (123), one finally gets the following result for the two unknown functions $\xi(\omega)$ and $\zeta(\omega)$

$$\xi(\omega) = \frac{V\sigma_-(\omega)\beta(\omega)}{\omega_b - \omega - \omega_c + i\gamma_c - V^2\sigma_-(\omega)} \ , \qquad (125)$$

$$\zeta(\omega) = \frac{V\sigma_+(\omega)\beta(\omega)}{\omega_b + \omega - \omega_c + i\gamma_c - V^2\sigma_+(\omega)} \ , \qquad (126)$$

where

$$\sigma_\pm(\omega) = \int d\omega' \frac{\alpha^2(\omega')}{\omega_b \pm \omega - \omega' + i\gamma} \qquad (127)$$

The frequency spectrum of a light emitting system is calculated, according to the theory of the physical spectrum of light [Eberly (1977)], from the two-times correlation function of the electric field

$$\mathcal{D}(\tau) = \langle \mathbf{E}^{(-)}(t) \cdot \mathbf{E}^{(+)}(t-\tau) \rangle \ , \qquad (128)$$

where $\langle \cdots \rangle$ indicates the trace over the density matrices of the phonon bath and the bottleneck level, and the superscripts "+" and "-" denote the positive (annihilation operator) and negative (creation operator) energy terms in the quantum field. In particular, we can calculate the intracavity electric field (81) using the solution (124) because the external field is simply proportional to the intracavity field through the quasimode coefficient. We assume a thermalized phonon bath and a bottleneck level in a pure state with occupation number N. The first assumption implies a density matrix $\rho_{ph} = \int d\omega \bar{n}_\omega \hat{c}^\dagger_\omega |0\rangle\langle 0|\hat{c}_\omega$, where $\bar{n}_\omega$ is the Bose distribution. The second one has the physical meaning of a stationary situation in which the population of the levels in the bottleneck region is kept constant by the relaxation from higher wave vectors. The spectrum of the emitted light is given [Eberly (1977)] by the following expression

$$I(\omega) = 2\mathcal{R}e \int_0^\infty d\tau \mathcal{D}(\tau)e^{-i\omega\tau} \ . \qquad (129)$$

An explicit expression for this latter quantity can be obtained after some straightforward algebra using the equations derived in this section. We give only the final expression for the luminescence spectrum

$$I(\omega) \propto N[(\bar{n}(\omega_b-\omega)+1)|\xi(\omega_b-\omega)|^2\theta(\omega_b-\omega)+\bar{n}(\omega-\omega_b)|\zeta(\omega-\omega_b)|^2\theta(\omega-\omega_b)] ,\tag{130}$$

where $\theta(\omega)$ is the Heavyside step function.

A few remarks can be drawn from this result. First, the overall emitted intensity is proportional to the population of the bottleneck level, which is an obvious consequence of the assumption of stationarity. Then, we see that the spectrum is given by the sum of two contributions, one for frequencies lower than ω_b and the other for frequencies higher than ω_b. These two contributions correspond to the two processes of scattering through phonon emission and absorption respectively. In fact, they are proportional to the factors $\bar{n}+1$ and $\bar{n}$. In the present model, thus, the excitation is transferred from the bottleneck level to the radiative polariton levels through the absorption or the emission of a phonon of the thermal bath. The approximation of free evolution of the phonon and bottleneck level operators implies also that the scattering in the opposite direction, from the radiative levels to the bottleneck level, is neglected. This approximation is valid for sufficiently low temperatures and for fast polariton recombination rate compared to the scattering rate. Before deriving the limits of validity of the model, however, we discuss the form of the coefficient $\beta(\omega)$ which appears in (113). First, from the scattering Hamiltonian it turns out that $\beta(\omega)$ has the dimensions of the square root of an energy. The correct form for this coefficient can be deduced from a microscopic model for the exciton-phonon scattering described in Ref. [Piermarocchi (1996)]. In turns out that the scattering coefficient is proportional to the square root of the exchanged momentum $|\mathbf{q}|$ in the true scattering process and, consequently, to the square root of the exchanged energy. We can thus assume that $\beta(\omega) = \beta_0 \omega^{\frac{1}{2}}$, where β_0 is a c-number. We see from (130) that, according to our approximations, the spectrum does not depend on β_0[15]. At this point, an estimation of the validity range can be made by remarking that the first correction to expressions (119) and (120) is proportional to $\beta(\omega)$. Consequently, a term of the order of $V\beta_0^2$ would appear in (123). This term has to be much smaller than the last term of (123), which is proportional to $V\beta_0$. The estimate however can be obtained only after the thermal average of the operators is computed. It is then easy to show that the condition of validity is given by

$$\beta_0\sqrt{\bar{n}} \ll 1 , \tag{131}$$

where $\bar{n}$ is the average number of phonons in the bath at a given temperature. The quantity β_0 can be estimated as follows. The polariton scattering coefficient is approximately given by $\beta_0^2 \Delta E \bar{n}$, where we indicate with ΔE the

---

[15] Actually, it can be shown that the overall expression for the spectrum is proportional to β_0. This factor is important only when we compare the emitted intensity for different values of the other parameters. This model, however, is not able to predict absolute emission intensities, as will be discussed at the end of the section.

exchanged energy. This coefficient is known for most semiconductor materials. In the case of GaAs quantum well excitons, it has been measured and amounts to $5\mu eV/K \cdot T$ [Schulteis (1986)]. The Bose occupation number at low temperature is $\bar{n} \simeq k_B T/\Delta E$. By comparing the approximated expression $\beta_0^2 \Delta E \bar{n}$ with this measured value, one obtains

$$\beta_0^2 \simeq k_B^{-1} \cdot 5 \left[\frac{\mu eV}{K}\right] \simeq 0.25 \ . \tag{132}$$

Finally, from (131) and assuming that the maximum exchanged energy is of the order of the Rabi splitting Ω_R, an upper limit for the temperature is provided by the relation $\beta_0 k_B T/\Omega_R \ll 1$, which corresponds to a maximum temperature of the order of $60\ K$ in GaAs quantum well excitons and for $\Omega_R = 5$ meV.

We compare the results of (130) with luminescence measurements performed by Stanley [Stanley]. The measured and calculated spectra for different temperatures are plotted in Fig. 25. The calculation has been made using the nominal parameters of the sample, except for the inhomogeneous broadening parameter γ_{inh} and the two energies $\hbar\omega_c$ and $\hbar\omega_0$. In particular, the measured value of the temperature was used. Since the overall intensity of the luminescence is not predicted by (130), the proportionality factor appearing in that expression has been adjusted for each spectrum in order to match the experimental data. The different spectra were taken from the same position of the sample, which implies that the cavity mode frequency is unchanged. We see that the agreement between Eq. (130) and the experimental data is very satisfactory, especially for low temperature. The larger discrepancy occurs on the high energy tail of the higher polariton peak. This is explained if we consider that the luminescence from the lower tail of the electron-hole continuum also contributes to the spectrum. This contribution is not considered in the model. The value of the parameter γ_{inh} obtained by the fit is $\gamma_{inh} = 4$ meV, which corresponds to a full width at half maximum of the bare exciton resonance equal to 8.5 meV. This value compares well with the same quantity obtained from the emission spectrum along the quantum well plane.

The most striking feature of the data in Fig. 25 is the thermal shape of the luminescence spectra. In fact it has been checked that each luminescence spectrum matches the corresponding absorption spectrum (which, in turn is proportional to the density of states) weighted by a Boltzmann factor. This fact, however, is in contrast with what expected in microcavities, where the radiative rates are extremely fast and should not leave enough time for the system to reach a thermal population in the radiative region. Deviations from the thermal distribution have already been predicted as a result of the competition between relaxation and radiative processes for quantum well excitons [Piermarocchi (1996)]. Furthermore, phenomena such as the bottleneck effects have been known since long in the case of bulk polariton photoluminescence [Sumi (1976), Weisbuch (1977), Weisbuch (1979)]. Within the present

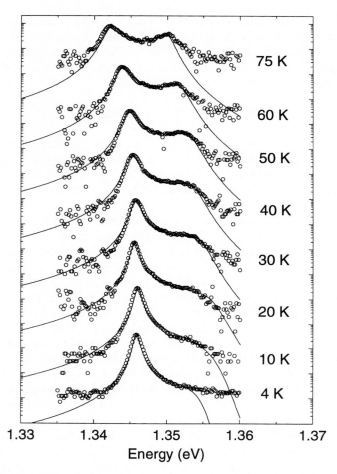

Fig. 25. Comparison between the measured (circles) and calculated (full lines) luminescence spectra for different temperatures. The spectra are vertically shifted with respect to each other for clarity.

model, the photoluminescence lineshape is naturally explained. The thermal factor in the spectra originates from the function $\bar{n}(\omega)$ appearing in (130), while the functions $\xi(\omega)$ and $\zeta(\omega)$ provide the polariton density of states. The factors depending on $\bar{n}(\omega)$ reflect the thermal population of the phonon bath. The scattering coefficients are thus weighted by these factors and, as a consequence of the stationary regime, the same terms occur in the expression for the emitted spectrum. However, because of the very fast radiative rates of the polaritons in the radiative region, there is no time for population buildup and these levels are practically empty. Thus, it is not correct to speak of thermalized polariton population because the thermal distribution enters the polariton dynamics through the phonon population only.

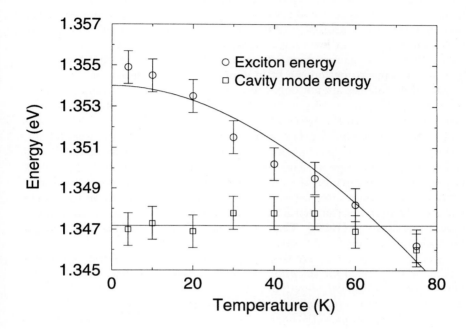

Fig. 26. The fitted values of the uncoupled energy and cavity modes are reported. The curves represent the constant cavity energy and the variation of the exciton energy predicted by the bandgap renormalization law.

A confirmation of these results is obtained by plotting the fitted values of the exciton and cavity mode energy as a function of the temperature. This is made in Fig. 26. The cavity mode energy should be temperature-independent, while the exciton energy is expected to vary according to the empirical law for the bandgap renormalization [Hellwege (1982)], which reads

$$E_{\text{g}}(T) = E_{\text{g},0} - 5.8 \times 10^{-4} \frac{T^2}{T+300} , \qquad (133)$$

with the temperature T expressed in Kelvin. This curve is plotted in Fig. 26 and matches well the fitted values.

In addition to the upper limit on the temperature, this model has another limitation. The shape of each spectrum is well reproduced, but the relative intensity of two spectra under different quantum well and cavity parameters (in particular the exciton-cavity detuning) can not be predicted. The reason of this restriction comes both from the approximations (119) and (120), and from the assumption of a single in–plane wave vector for the radiative levels. The free evolution of the phonon and bottleneck level operators implies, as was already remarked, that the scattering processes from the radiative to the non radiative levels are not included in the calculated dynamics. This processes are negligible at low temperatures but, being temperature dependent, they influence the temperature dependence of the absolute emitted intensity. The second approximation is a little more restrictive. The strongly radiative polariton resonances are not only those at $\mathbf{k}_\parallel = 0$. A whole cone in the range of about $k_0/10$ consists of polariton levels with enhanced spontaneous emission with respect to the bare exciton. When the exciton-cavity detuning is varied, the relative importance of these radiative channels changes and this change is reflected in the balance of the polariton populations under a stationary excitation regime. This implies that the dependence of the overall luminescence intensity on the detuning is not reproduced either. The integrated luminescence intensity as a function of the detuning is an important feature which allows the characterization of the polariton interaction in microcavities [Stanley (1996)]. The restriction discussed above is thus a major limitation for our model of polariton luminescence. Nevertheless, the assumptions of the present model are very simple and they allow to obtain a compact analytical expression for the luminescence spectrum. Furthermore, the comparison between the calculated and the measured data shows that the shape of each spectrum is very well reproduced, and the thermalization of the spectrum, which was apparently unexpected, can be explained on a very simple basis.

6 Outlook

The optical response of semiconductor microcavities is characterized by a wide phenomenology. In fact, in addition to the variety of linear and non-linear features related to the pure excitonic response, the strong coupling regime and the peculiar polariton dispersion are responsible of a number of new interesting fetures. Among these, we have described the bottleneck effect and the persistence of the thermal shape in photoluminescence spectra of highly inhomogeneous polaritons at low temperatures (Section 5.3). We have also mentioned in Section 4.5 the problem of polariton linewidths, pointing out that, at least in the two cases of phonon and excitation broadening, the broadening mechanism might be strongly influenced by the polariton dispersion and the lower polariton linewidth consequently suppressed. Other very recent results include, as an example, the first observation of the Purcell effect [Purcell (1946)] in microcavity embedded semiconductor quantum boxes [Gérard (1998)]. Among the first experimental findings in the nonlinear regime, which reveal unexpected new behaviours and should stimulate further investigation, we underline the first evidence of the suppression of the lower polariton excitation induced linewidth [Baumberg (1998)] and the first experimental observation of the resonant optical Stark effect [Quochi (1998)] which has been made possible by the pulse shaping induced by the Fabry–Pérot filter. Apart from the fundamental aspects of the physics of microcavities, we should recall how the technology of optical devices has progressed in the last ten years with the development of vertical cavity lasers (VCSEL) [Ebeling (1993)] and light emitting diodes [Benisty (1998)].

In conclusion, after the early enthusiasm brought by the measurement of modified exciton spontaneous emission [Yokoyama (1990)] and vacuum field Rabi splitting [Weisbuch (1992)], the scientific investigation in the field of semiconductor microcavities is experiencing a second birth and will probably represent a mainstream domain of the physics of semiconductors in the years to come. In the present lecture notes we have only addressed the basic problem of the linear optical response of semiconductor microcavities. Our aim was, apart from providing the basic theoretical tools, to stimulate further reading and discussion, and to encourage young scientists who are willing to address these problems. We hope to have succeded in this purpose.

Acknowledgements - I am very grateful to my collaborators, Antonio Quattropani, Paolo Schwendimann, Carlo Piermarocchi and Cristiano Ciuti, for the strong support, both scientific and moral, while preparing these lectures. I am particularly indebted to Claudio Andreani with whom I have done a large part of the work on the semiclassical response of microcavity polaritons. He also showed me for the first time the derivation of the general transfer matrix in terms of the reflection and transmission coefficients using the time reversal invariance. Particular thanks go to Romuald Houdré and Ross Stanley for the numerous discussions concerning these topics, and to Claude Weisbuch and Henri Benisty for all the suggestions and, mostly, for giving me the opportunity to write these lecture notes.

References

Abram (1996). Abram I., Sermage B., Long S., Bloch J., Planel R., Thierry-Mieg V. (1996): Spontaneous emission dynamics in planar semiconductor microcavities, in Microcavities and Photonic Bandgaps, J. Rarity and C. Weisbuch eds., 69 (Kluwer, Dordrecht)

Agarwal (1984). Agarwal G. S., Phys. Rev. Lett. **51**, 550 (1984).

Agranovich (1966). Agranovich V. M., Dubowskii A. O. (1966): Effect of retarded interaction on the exciton spectrum in one-dimensional and two-dimensional crystals. Pis'ma Zh. Eksp. Teor. Fiz. **3**, 345 (JETP Lett. **3**, 233)

Andreani (1991). Andreani L. C., Tassone F., Bassani F. (1991): Radiative lifetime of free excitons in quantum wells. Solid State Commun. **77**, 641

Andreani (1994). Andreani L. C. (1994): Exciton–polaritons in superlattices. Physics Letters A **192**, 99

Andreani (1995). Andreani L. C. (1995): Optical transitions, excitons and polaritons in bulk and low–dimensional semiconductor structures: in E. Burstein and C. Weisbuch (Eds.), Confined Electrons and Photons: new Physics and Devices, Plenum Press, New York

Andreani (1998). Andreani L. C., Panzarini G., Kavokin A. V., Vladimirova M. V. (1998): Effect of inhomogeneous broadening on optical properties of excitons in quantum wells. Phys. Rev. B **57**, 4670

Barnett (1988). Barnett S. M., Radmore P. M. (1988): Quantum theory of cavity quasimodes. Optics Commun. **68**, 364

Bassani (1983). Bassani F., Altarelli M. (1983): Interaction of radiation with condensed matter. In E. E. Koch (Ed.), Handbook on Syncrotron Radiation, North Holland, Amsterdam, 463–606

Baumberg (1998). Baumberg J. J., Armitage A., Skolnick M. S., Roberts J. S. (1998): Suppressed polariton scattering in semiconductor microcavities. Phys. Rev. Lett. **81**, 661

Bastard (1989). Bastard G. (1989). Wave Mechanics Applied to Semiconductor Heterostructures. Les Editions de Physique, Les Ulis, Paris

Benisty (1998). Benisty H., De Neve H., Weisbuch C. (1998): Impact of planar microcavity effects on light extraction. IEEE J. of Quantum Elec. **34**, 146

Born and Wolf (1993). Born M., Wolf E. (1993), Principles of Optics, sixth ed., (Pergamon Press, Oxford)

Brinkmann (1996). Brinkmann D., Rossi F., Kock S. W., Thomas P. (1996): Phonon–induced dephasing of localized optical excitations. Phys. Rev. B **54**, 2561

Brinkmann (1998). Brinkmann D., Bott K., Koch S. W., Thomas P. (1998): Disorder–induced dephasing of excitons in semiconductor heterostructures. Phys. Stat. Sol. (b) **206**, 493

Burstein (1995). E. Burstein and C. Weisbuch (Eds.), *Confined Electrons and Photons: new Physics and Devices*, (Plenum Press, New York, 1995).

Carmichael (1989). Carmichael H. J., Brecha R. J., Raizen M. G., Kimble H. J., Rice P. R. (1989): Subnatural linewidth averaging for coupled atomic and cavity-mode oscillators. Phys. Rev. A **40**, 5516

Citrin (1993). Citrin D. S. (1993): Radiative lifetimes of excitons in quantum wells: Localization and phase-coherenece effects. Phys. Rev. B **47**, 3832

Citrin (1994). Citrin D. S. (1994): Exciton radiative decay and polaritons in multiquantum wells: quantum well to superlattice crossover. Solid State Commun. **89**, 139 (1994).

Citrin (1996a). Citrin D. S. (1996): Time-domain theory of resonant Rayleigh scattering by quantum wells: Early-time evolution. Phys. Rev. B **54**, 14572

Citrin (1996b). Citrin D. S. (1996): Coherence transfer via resonance Rayleigh scattering of exciton polaritons in a semiconductor microcavity. Phys. Rev. B **54**, 16425

Ciuti (1998). Ciuti C., Savona V., Piermarocchi C., Quattropani A., Schwendimann P. (1998): Threshold behavior in the collision broadening of microcavity polaritons. Phys. Rev. B. **58**, R10123

Cohen-Tannoudji (1988). Cohen-Tannoudji C., Dupont-Roc J., Grynberg G. (1988): Processus d'interaction entre photons et atomes, Editions du CNRS, Paris

Ebeling (1993). Ebeling K. J. (1993): Integrated optoelectronics: waveguide optics, photonics, semiconductors. Springer-Verlag, Berlin

Eberly (1977). Eberly J. H., Wódkiewicz K. (1977): The time-dependent physical spectrum of light. J. Opt. Soc. Am. **67**, 1252

Ell (1998). C. Ell, J. Prineas, T. R. Nelson, Jr., S. Park, H. M. Gibbs, G. Khitrova, S. W. Koch, and R. Houdré. Influence of structural disorder and light coupling on the excitonic response of semiconductor microcavities. Phys. Rev. Lett. **80**, 4795 (1998).

Fano (1961). Fano U. (1961): Effects of configuration interaction on intensities and phase shifts. Phys. Rev. **124**, 1866

Fisher (1996). Fisher T. A., Afshar A. M., Whittaker D. M., Skolnick M. S., Kinsler P., Roberts J. S., Hill G., Pate M. A. (1996): Magnetic and electric field effects in semiconductor quantum microcavity structures. In J. Rarity and C. Weisbuch (eds.), Microcavities and Photonic Bandgaps, Kluwer, The Netherlands

Gérard (1998). Gérard J. M., Sermage B., Gayral B., Legrand B., Costard E., Thierry-Mieg V. (1998): Enhanced spontaneous emission by quantum boxes in a monolithic optical microcavity. Phys. Rev. Lett. **81**, 1110

Glutsch (1994). Glutsch S., Bechstedt F. (1994): Theory of asymmetric broadening and shift of excitons in quantum structures with rough interfaces. Phys. Rev. B **50**, 7733

Glutsch (1996). Glutsch S., Chemla D. S., Bechstedt F. (1996): Numerical calculation of the optical absorption in semiconductor quantum structures. Phys. Rev. B **54**, 11592

Haroche (1992). Haroche S. (1992): Cavity quantum electrodynamics. In Dalibard J., Raymond J. M., Zinn-Justin J. (eds.), Fundamental systems in quantum optics, London Science Publishers, Amsterdam

Haug (1994). Haug H., Koch S. W. (1994): Quantum theory of the optical and electronic properties of semiconductors. 3rd edn., World Scientific, Singapore

Hellwege (1982). Hellwege A. M., Madelung O. (eds.) (1982). Physics of Group IV Elements and III-V Compounds, Landolt-Börnstein, New Series, Group III, **17**, Springer-Verlag, Berlin

Hood (1998). Hood C. J., Chapman M. S., Lynn T. W., Kimble H. J. (1998): Real–time cavity QED with single atoms. Phys. Rev. Lett. **80**, 4157

Hopfield (1958). Hopfield J. J. (1958): Theory of the contribution of excitons to the complex dielectric constant of crystals. Phys. Rev. **112**, 1555

Houdré (1995). Houdré R., Gibernon J. L., Pellandini P., Stanley R. P., Oesterle U., Weisbuch C., O'Gorman J., Roycroft B., Ilegems M. (1995): Saturation of the strong-coupling regime in a semiconductor microcavity: Free-carrier bleaching of cavity polaritons. Phys. Rev. B **52**, 7810

Jackson (1975). Jackson J. D., Classical Electrodynamics. Wiley, New York

Jahnke (1996). Jahnke F., Kira M. Koch S. W., Khitrova G., Lindmark E. K., Nelson T. R., Jr., Wick D. V., Berger J. D., Lyngnes O., Gibbs H. M., Tai K. (1996): Excitonic nonlinearities of semiconductor microcavities in the nonperturbative regime. Phys. Rev. Lett. **77**, 5257

Jahnke (1997). Jahnke F., Kira M., Koch S. W. (1997): Linear and nonlinear optical properties of excitons in semiconductor quantum wells and microcavities. Z. Phys. B **104**, 559

Jorda (1993). Jorda S., Rössler U., Broido D. (1993): Fine structure of excitons and polariton dispersion in quantum wells. Phys. Rev. B **48**, 1669

Kaluzny, 1983. Kaluzny, Y., P. Goy, M. Gross, J. M. Raymond and S. Haroche (1983). Observation of self-induced Rabi oscillations in two-level atoms excited inside a resonant cavity: The ringing regime of superradiance. *Phys. Rev. Lett.*, **51**, 1175

Kavokin (1998). A. V. Kavokin. Motional narrowing of inhomogeneously broadened excitons in a semiconductor microcavity: semiclassical treatment. Phys. Rev. B **57**, 3757 (1998)

Knox (1963). Knox, R. S. (1963). Theory of Excitons. In F. Seitz and D. Turnbull (eds.), Solid State Physics, Academic Press, New York

Ley (1987). Ley M., Loudon R. (1987): Quantum theory of high–resolution length measurement with a Fabry–Pérot interferometer. J. of Mod. Optics **34**, 227

Lindberg (1988). Lindberg M., Koch S. W. (1988): Effective Bloch equations for semiconductors. Phys. Rev. B **38**, 3342

Lyngnes (1997). Lyngnes O., Berger J. D., Prineas J. P., Park S., Khitrova G., Gibbs H. M., Jahnke F., Kira M., Koch S. W. (1997): Nonlinear emission dynamics from semiconductor microcavities in the nonperturbative regime. Solid State Commun. **104**, 297

McLeod (1986). McLeod H. A. (1986): Thin-Film Optical Filters. second ed., Hilger

Meystre (1992). Meystre P. (1992): Cavity quantum optics and the quantum measurement process. Progress in Optics XXX 261

Piermarocchi (1996). Piermarocchi C, Tassone F., Savona V., Quattropani A., Schwendimann P. (1996): Nonequilibrium dynamics of free quantum-well excitons in time-resolved photoluminescence. Phys. Rev. B **53**, 15834

Piermarocchi (1997). Piermarocchi C, Tassone F., Savona V., Quattropani A., Schwendimann P. (1997): Exciton formation rates in $GaAs/Al_xGa_{1-x}As$ quantum wells. Phys. Rev. B **55**, 1333

Purcell (1946). Purcell E. M. (1946): Spontaneous emission probabilities at radiofrequencies. Phys. Rev. **69**, 681

Quochi (1998). Quochi F., Bongiovanni G., Mura A., Staehli J. L., Deveaud B., Stanley R. P., Oesterle U., Houdré R. (1998): Strongly driven semiconductor microcavities: From the polariton doublet to an ac Stark triplet. Phys. Rev. Lett. **80**, 4733

Rarity (1996). Microcavities and Photonic Bandgaps, J. Rarity and C. Weisbuch eds., (Kluwer, Dordrecht, 1996)

Rhee (1996). Rhee J.-K., Citrin D. S., Norris T. B., Arakawa Y., Nishioka M. (1996): Femtosecond dynamics of semiconductor-microcavity polaritons in the nonlinear regime. Solid. State Commun. **97**, 941

Sanchez-Mondragon (1983). Sanchez-Mondragon J. J., Narozhny N. B., Eberly J. H. (1983): Theory of spontaneous emission line shape in an ideal cavity. Phys. Rev. Lett. **51**, 550

Savona (1995). Savona V., Andreani L. C., Schwendimann P., Quattropani A. (1995): Quantum well excitons in semiconductor microcavities: Unified treatment of weak and strong coupling regimes. Solid State Commun. **93** (1995), 733

Savona (1996a). Savona V., Tassone F., Piermarocchi C., Quattropani A., Schwendimann P. (1996): Theory of polariton photoluminescence in arbitrary semiconductor microcavity structures. Phys. Rev. B **53**, 13051

Savona (1996b). Savona V., Weisbuch C. (1996): Theory of time-resolved light emission from polaritons in a semiconductor microcavity under resonant excitation. Phys. Rev. B **54**, 10835

Savona (1997a). Savona V., Piermarocchi C., Quattropani A., Tassone F., Schwendimann P. (1997): Microscopic theory of motional narrowing of microcavity polaritons in a disordered potential. Phys. Rev. Lett. **78**, 4470

Savona (1997b). Savona V., Piermarocchi C. (1997): Microcavity polaritons: homogeneous and inhomogeneous broadening in the strong coupling regime. Phys. Stat. Sol. (a) **164**, 45

Schnabel (1992). Schnabel R. F., Zimmermann R., Bimberg D., Nickel H., Lösch R., Schlapp W. (1992): Influence of exciton localization on recombination line shapes: $In_xGa_{1-x}As/GaAs$ quantum wells as a model. Phys. Rev. B **46**, 9873

Schulteis (1986). Schulteis L., Honold A., Kuhl J., Köler K., Tu C. W. (1986): Optical dephasing of homogeneously broadened two-dimensional exciton transitions in GaAs quantum wells. Phys. Rev. B **34**, 9027

Sermage (1996). Sermage B., Long S., Abram I., Marzin J. Y., Bloch J., Planel R., Thierry-Mieg V. (1996): Time resolved spontaneous emission of excitons in a microcavity: Behavior of the individual exciton–photon mixed states. Phys. Rev. B **53**, 16516

Stanley. Stanley R. P.: Unpublished

Stanley (1993). Stanley R. P., Houdré R., Oesterle U., Ilegems M., Weisbuch C. (1993): Impurity modes in one–dimensional periodic systems: The transition from photonic band gaps to microcavities. Phys. Rev. A **48**, 2246

Stanley (1996). Stanley R. P., Houdré R., Weisbuch C., Oesterle U., Ilegems M. (1996): Cavity–polariton photoluminescence in semiconductor microcavities: Experimental evidence, Phys. Rev. B **53**, 10995

Stanley (1997). Stanley R. P., Houdré R., Oesterle U., Ilegems M. (1997): Semiconductor microcavity polaritons: photoquenching of inhomogeneous broadening and acoustic phonon broadening. Presented to the OSA-ILX Interdisciplinary Laser Conference, Long Beach, October 1997

Sumi (1976). Sumi H. (1976): On the exciton luminescence at low temperatures: Importance of the polariton viewpoint. J. Phys. Soc. Jpn **21**, 1936

Tassone (1990). Tassone F., Bassani F., Andreani, L. C. (1990): Resonant and surface polaritons in quantum wells. Il Nuovo Cimento **12D**, 1673

Tassone (1992). Tassone F., Bassani F., Andreani L. C. (1992): Quantum-well reflectivity and exciton-polariton dispersion. Phys. Rev. B **45**, 6023

Tassone (1996). Tassone F., Piermarocchi C., Savona V., Schwendimann P., Quattropani A. (1996): Photoluminescence decay times in strong-coupling semiconductor microcavities. Phys. Rev. B **53**, R7642

Tassone (1997). Tassone F., Piermarocchi C., Savona V., Quattropani A., Schwendimann P. (1997): Bottleneck effects in the relaxation and photoluminescence of microcavity polaritons. Phys. Rev. B **56**, 7554

Thompson (1992). Thompson R. J., Rempe G., Kimble H. J. (1992): Observation of normal-mode splitting for an atom in an optical cavity. Phys. Rev. Lett. **68**, 1132

Ulbrich (1979). Ulbrich R. G., Fehrenbach G. W. (1979): Polariton wave packet propagation in the exciton resonance of a semiconductor. Phys. Rev. Lett. **43**, 963

Weisbuch (1977). Weisbuch C., Ulbrich R. G. (1977): Resonant polariton fluorescence in gallium arsenide. Phys. Rev. Lett. **39**, 654

Weisbuch (1979). Weisbuch C., Ulbrich R. G. (1979): Spatial and spectral features of polariton fluorescence. J. Lumin. **18/19**, 27

Weisbuch (1981). Weisbuch C., Dingle R., Gossard A. C., Wiegmann W. (1981): Optical characterization of interface disorder in GaAs-$Ga_{1-x}Al_x$As multi-quantum well structures. Solid State Commun. **38**, 709

Weisbuch (1982). Weisbuch C., Ulbrich R. G. (1982): Resonant light scattering mediated by excitonic polaritons in semiconductors. In M. Cardona and G. Güntherodt (eds.), *Light scattering in solids III*, Springer-Verlag, Berlin

Weisbuch (1992). Weisbuch C., Nishioka M., Ishikawa A., Arakawa Y. (1992): Observation of the coupled exciton-photon mode splitting in a semiconductor quantum microcavity. *Phys. Rev. Lett.*, **69**, 3314

Whittaker (1997). Whittaker D. M., Skolnick M. S., Fisher T. A., Armitage A., Baxter D., Astratov V. N. (1997): Excitons and polaritons in semiconductor microcavities. Phys. Stat. Sol. (a) **164**, 13

Whittaker (1998). Whittaker D. M. (1998): What determines inhomogeneous linewidth in semiconductor microcavities? Phys. Rev. Lett. **80**, 4791

Yokoyama (1990). Yokoyama H., Nishi K., Anan T., Yamada H., Bronson S. D., Ippen E. P. (1990): Enhanced spontaneous emission from GaAs quantum wells in monolithic microcavities. Appl. Phys. Lett. **57**, 2814

Zhu (1990). Zhu Y., Gauthier D. J., Morin S. E., Wu Q., Carmichael H. J., Mossberg T. W. (1990): Vacuum Rabi splitting as a feature of linear-dispersion theory: Analysis and experimental observations. Phys. Rev. Lett. **64**, 2499

Zimmermann (1992). Zimmermann R. (1992): Theory of dephasing in semiconductor optics. Phys. Stat. Sol. (b) **173**, 129

Zimmermann (1995). Zimmermann R. (1995): Theory of resonant Rayleigh scattering of excitons in semiconductor quantum wells. Il Nuovo Cimento D **17**, 1801

Zimmermann (1997). Zimmermann R., Runge E. (1997): Excitons in narrow quantum wells: Disorder localization and luminescence kinetics. Phys. Status Sol. (a) **164**, 511

Spontaneous Emission Control and Microcavity Light Emitters

Seng Tiong Ho, Liwei Wang, and Seoijin Park

Electrical and Computer Engineering Department, Northwestern University, 2145 Sheridan Road, Evanston, Illinois 60208-3118, USA

Introduction

In recent years, there has been much interest in showing that spontaneous emission is not an immutable property of atoms, but can be modified depending on the environment around the atoms.

Physically, the emission of a photon from an excited atom can come about only through interaction between the excited atom and quantized vacuum field. Hence, spontaneous emission is affected by the nature of quantized vacuum field around the atom. The spatial propagation of quantized electromagnetic field is governed by the same set of equations as classical electromagnetic field, that is, the Maxwell's equations. Just as classical electromagnetic field can be modified by optical cavities and various types of optical structures such as waveguides and gratings, so can be quantized vacuum field.

Modification of the nature of spontaneous emission can happen in two essential ways, namely the spatial pattern of emission and the total rate of emission. In certain applications, it is the modification of the pattern of emission that is useful. Whereas, in some other applications, one may desire the modification of the total emission rate.

A few types of optical structure have been identified as useful for causing large modification in spontaneous emission.[1] They include microcavity structures, low dimensional photonic structures (e.g. small planar and channel optical waveguides, which are called photonic wells and wires), and optical grating structures (e.g. photonic bandgap structures). Currently, a few types of devices have taken advantages of these spontaneous-emission modifying optical structures, with the purpose of improving device performances. These devices include vertical-cavity surface emission lasers, microdisk lasers,[2] photonic-wire lasers,[3] and lasers based on photonic-bandgap structures.[4]

The purpose of this chapter is to provide a tutorial to address the issues of how one can control spontaneous emission using microcavities and low-dimensional photonic structures, and to point out some useful applications to devices.

In section one, we will review the theory of spontaneous emission and show how spontaneous emission rate can be calculated using Fermi's golden rule. This section also establishes various symbols and emission rate formulas for applications to later sections.

In sections two and three, we will study modification of spontaneous emission in metallic and dielectric planar and channel waveguides and microcavities. The modification of spontaneous emission becomes significant only when the waveguide dimensions are small compared to optical wavelength. Planar and channel waveguides can be seen as low-dimensional photonic structures that confine the motion of photons to a 2-D plane or a 1-D channel. The modification of spontaneous emission is due partially to the modification in the photonic density of states in these low-dimensional photonic structures. This is analogous to the modification of spontaneous emission in low-dimensional electronic structures such as semiconductor quantum wells, wires, and dots for which the electronic density of states are modified. In this sense, low-dimensional photonic structures are photonic analogues of electronic quantum wells, wires, and dots, and may be referred to as photonic wells, wires, and dots. Physically, they are in the form of strongly guiding planar waveguides (photonic wells), strongly-guiding channel waveguides (photonic wires), and 3-D enclosed cavities (photonic dots). It is interesting to point out that a 3-D enclosed cavity can be made by enclosing a strongly-guiding waveguide with two highly-reflecting end mirrors. As a result, its spontaneous emission modifying behavior can be related to that of a channel waveguide. We note that the strongly-guiding property mentioned above is important as, for example, a planar waveguide will not have the effect of a photonic well (i.e. will not modify spontaneous emission significantly), unless it confines photons tightly.

A main difference between the basic properties of electrons and photons is that photons are Bosons and are intrinsically more wave like while electrons are Fermions and are intrinsically more particle like. As a result, the wave behavior of photons is usually regarded as classical behavior while the wave behavior of electrons is usually regarded as quantum behavior. Hence, we chose not to use the word "quantum" when we called these low-dimensional photonic structures as photonic wells, wires, and dots.

It is useful to make a few comments here on the different "pictures" that can be used in spontaneous emission calculations. It is true that a classical dipole in these low-dimensional structures will also experience the same modification in its emission behavior. Hence, the effect can be predicted based on classical electromagnetic field theory. However, a quantized emission source such as an atom will not radiate spontaneously unless the electromagnetic field is also quantized. This is simply due to the fact that without field quantization, the eigen-energy states of the quantized source are also the eigen-energy states of the entire source and field system, and hence are stationary states that do not evolve with time (i.e. no transitions) at zero classical field. It is interesting to note that a quantized source will have stimulated emission behavior under classical electromagnetic field. Quantum mechanics predicted that spontaneous emission rate into an electromagnetic field mode is equal to stimulated emission rate with one photon in that mode. Using this relation, one can "calculate" spontaneous emission by placing one real photon in very mode and compute the total stimulated emission rate. In this sense, spontaneous emission can be calculated "semi-classically" (i.e. with quantized source and classical E&M, but with the spontaneous-stimulated emission connection, which is arguably a quantum result).

In the quantum fluctuation picture, spontaneous emission can be viewed as "stimulated emission" caused by background vacuum fluctuations. Operationally, this is identical to the above "placing-one-real-photon picture". In the fully quantized treatment, the quantized vacuum field modes are modified when they propagate into a cavity (or a waveguide) structure and that causes modification in spontaneous emission. As will be pointed out later, this spatial propagation of field mode operator can be made rather rigorous using localized-photon operators.[5] Thus, operationally, in terms of calculation algorithm, the last three pictures ("fully quantum picture", "vacuum fluctuation picture", and "placing-one-real-photon picture") are identical.

In section two, we show that drastic modification in spontaneous emission rates and patterns can be achieved with metallic photonic wires, while more moderate modification can be achieved with metallic photonic wells. We also consider cases of metallic planar microcavities and metallic channel-waveguide microcavities, which are identical in structures to metallic photonic wells and metallic photonic dots, respectively. The possibility of using metallic structures to realize very small cavities that allow one to capture all the spontaneous emission into a single cavity mode is discussed. In order to model the medium geometry more realistically, we have explored cases for which the active medium is in the form of a thin layer such as a quantum well.

In section three, we show that strong modification in spontaneous emission rates and patterns can be achieved with dielectric photonic wires, while more moderate modification can be achieved with dielectric photonic wells. The main difference between metallic and dielectric cases is that the metallic structures tend to achieve enhancement (i.e. increase) in spontaneous emission rates while the dielectric structures tend to achieve suppression in spontaneous emission rates and cannot provide much enhancement. This difference can be traced to the difference in boundary conditions: the field changes sign when bounced from a metallic surface while it does not change sign when bounced from a dielectric interface if the field is incident from the high-refractive-index medium side. We show that for the "reverse situation" of a leaky dielectric waveguide formed with a low-refractive-index waveguide core surrounded by a high-refractive-index medium surrounding, the behavior for a dipole in the waveguide is again similar to the metallic case. An application example is given at the end of this section to show how one can realize a dielectric photonic-wire microcavity laser with a high spontaneous emission coupling factor.

1 Theory of Spontaneous Emission

In this section, we review the typical approach to calculate spontaneous emission from an excited atom based on quantum electrodynamics.

1.1 Atom-Field Interaction

The problem of interest is to calculate the transition rate of an excited atom in space. Let us take the Hamiltonian governing the system of interest to be that of a single-electron atom interacting with quantized fields. The Hamiltonian operator for this sys-

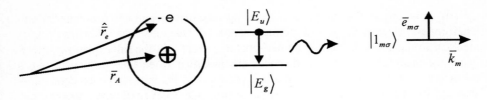

Fig. 1.1. An excited atom can emit a photon into a quantized field mode through atom-field interaction.

tem in Coulomb gauge is given as follows:

$$\hat{H} = \frac{1}{2m_e}\left[\hat{\bar{P}}_e + e\hat{\bar{A}}(\hat{\bar{r}}_e,t)\right]^2 - \frac{e^2}{4\pi\varepsilon_0|\hat{\bar{r}}_e - \bar{r}_A|} \\ + \frac{1}{2}\int d^3\bar{r}\left[\varepsilon_0\hat{\bar{E}}^2(\bar{r},t) + \mu_0\hat{\bar{H}}^2(\bar{r},t)\right], \quad (1\text{-}1)$$

where $\hat{\bar{r}}_e$ is the position operator for the electron, $\hat{\bar{P}}_e$ is the canonical momentum operator for the electron, $\bar{r}_A$ is the position of the nucleus treated as a classical variable, $-e$ is the charge of electron (i.e. $e > 0$ as defined), and m_e is the mass of electron. Note that in this Hamiltonian, we have neglected the dynamics of the nucleus, as it is usually much heavier than the electron. Hence the position operator of the nucleus has been replaced by a fixed classical variable $\bar{r}_A$.

1.2 Quantized Field in Plane-Wave Modes

In the usual quantization procedure for electromagnetic field,[6] one first considers a box of finite spatial volume as shown in Fig. 1.2. The vector potential in the volume is expanded in terms of a discrete set of orthogonal modes that form a complete set of spatial functions. After the quantization procedure, the expansion coefficients of the vector potential in terms of these modes then become the creation and annihilation operators for photons in these modes. The volume is taken to be arbitrarily large later.

As shown in Fig. 1.2, the volume of the box of interest is given by $V_Q = L_x L_y L_z$. The volume V_Q will be referred to as the volume of quantization.

1.3 The Vector Potential Operator

One convenient set of orthogonal modes is the travelling plane-wave modes $\{e^{i\bar{k}_m \cdot \bar{r}}\}$ with periodic bound-

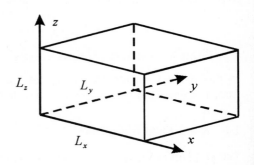

Fig. 1.2. Box of quantization for the field.

ary conditions (e.g. $u(x) = u(x+L_x)$ etc), in terms of which the vector potential operator can be shown to be given by:

$$\hat{\vec{A}}(\vec{r},t) = \sum_m \sum_\sigma \vec{e}_{m\sigma} \left(\frac{\xi_m}{\omega_m}\right) \left[\hat{a}_{m\sigma}(t)e^{i\vec{k}_m\cdot\vec{r}} + \hat{a}^\dagger_{m\sigma}(t)e^{-i\vec{k}_m\cdot\vec{r}}\right], \quad (1\text{-}2)$$

$$\xi_m = \sqrt{\frac{\hbar\omega_m}{2\varepsilon_0 V_Q}}, \quad \omega_m = |\vec{k}_m|c, \quad (1\text{-}3)$$

where the modes are labeled by m, which is a short-form for a set of three mode indices $m = \{m_x, m_y, m_z\}$. Here $m_x, m_y, m_z \in \{0, \pm 1, \pm 2, ..., \pm \infty\}$. The $\vec{k}_m$ vectors are quantized and given by:

$$\vec{k}_m = k_{mx}\vec{e}_x + k_{my}\vec{e}_y + k_{mz}\vec{e}_z, \quad (1\text{-}4)$$

$$k_{mx} = \frac{2\pi m_x}{L_x}, \quad k_{my} = \frac{2\pi m_y}{L_y}, \quad k_{mz} = \frac{2\pi m_z}{L_z}. \quad (1\text{-}5)$$

The $\vec{e}_{m\sigma}$ vectors, with $\sigma = 1$ or 2, are two mutually orthogonal polarization vectors for plane-wave mode m. The vectors $\vec{e}_{m1}$ and $\vec{e}_{m2}$ are orthogonal to $\vec{k}_m$ as shown in Fig. 1.3.

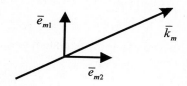

Fig. 1.3. A plane-wave mode has two orthogonal polarization vectors.

1.4 The Electric Field and Magnetic Field Operators

As Coulomb gauge is used, the electric field and magnetic field operators can be obtained from the vector potential operator via:

$$\hat{\vec{E}}(\vec{r},t) = -\frac{\partial \hat{\vec{A}}(\vec{r},t)}{\partial t} = \sum_{m\sigma} i\vec{e}_{m\sigma}\xi_m\left[\hat{a}_{m\sigma}(t)e^{i\vec{k}_m\cdot\vec{r}} - \hat{a}^\dagger_{m\sigma}(t)e^{-i\vec{k}_m\cdot\vec{r}}\right], \quad (1\text{-}6)$$

$$\mu_0\hat{\vec{H}}(r,t) = \vec{\nabla}\times\hat{\vec{A}}(\vec{r},t) = \sum_{m\sigma} i(\vec{k}_m\times\vec{e}_{m\sigma})\frac{\xi_m}{\omega_m}\left[\hat{a}_{m\sigma}(t)e^{i\vec{k}_m\cdot\vec{r}} - \hat{a}^\dagger_{m\sigma}(t)e^{-i\vec{k}_m\cdot\vec{r}}\right]. \quad (1\text{-}7)$$

1.5 Atom-Field Interaction Energy

Consider the $\frac{1}{2m_e}[\hat{\vec{P}}_e + e\hat{\vec{A}}]^2$ term in the Hamiltonian and expanding it:[7]

$$\frac{1}{2m_e}\left[\hat{\vec{P}}_e + e\hat{\vec{A}}\right]^2 = \frac{1}{2m_e}\left[\hat{\vec{P}}_e^2 + e\hat{\vec{P}}_e\cdot\hat{\vec{A}} + e\hat{\vec{A}}\cdot\hat{\vec{P}}_e + e^2\hat{\vec{A}}^2\right]. \quad (1\text{-}8)$$

We can ignore the $\frac{e^2}{2m_e}\hat{\vec{A}}^2$ term in the expansion in general for weak field. Further, using $[\hat{\vec{P}}_e, \hat{\vec{A}}] = \vec{\nabla}\cdot\hat{\vec{A}}$ (derived from the general commutator $[\hat{P}, \hat{F}(x)] = \frac{d\hat{F}(x)}{dx}$) and the fact that $\vec{\nabla} \cdot \hat{\vec{A}} = 0$ in Coulomb gauge, we conclude that $\hat{\vec{P}}_e$ and $\hat{\vec{A}}$ commute with each other as $[\hat{\vec{P}}_e, \hat{\vec{A}}] = 0$. This gives:

$$\frac{1}{2m_e}\left[\hat{\vec{P}}_e + e\hat{\vec{A}}\right]^2 \approx \frac{1}{2m_e}\hat{\vec{P}}_e^{\,2} + \frac{e}{m_e}\hat{\vec{P}}_e \cdot \hat{\vec{A}}. \qquad (1\text{-}9)$$

The first term above corresponds to the kinetic energy of the electron in the absence of field, and the second term involves both the atomic operator and the field operator and would be responsible for the interaction energy between the atom and the field. Hence we can write:

$$\hat{H}_{\text{AFint}} = \frac{e}{m_e}\hat{\vec{P}}_e \cdot \hat{\vec{A}}(\hat{\vec{r}}_e, t), \qquad (1\text{-}10)$$

where $\hat{H}_{\text{AFint}}$ is called the atom-field interaction energy.

1.6 Free-Atom and Free-Field Hamiltonians and Their Eigenstates

We can now write the total Hamiltonian as

$$\hat{H} = \hat{H}_0 + \hat{H}_{\text{AFint}}, \qquad (1\text{-}11)$$

where $\hat{H}_0$ is the free-atom plus free-field Hamiltonian:

$$\begin{aligned}
\hat{H}_0 &= \hat{H}_{\text{Free-Atom}} + \hat{H}_{\text{Free-Field}}, \\
\hat{H}_{\text{Free-Atom}} &= \frac{1}{2m_e}\hat{\vec{P}}_e^{\,2} - \frac{e^2}{4\pi\varepsilon_0|\hat{\vec{r}}_e - \vec{r}_A|}, \\
\hat{H}_{\text{Free-Field}} &= \frac{1}{2}\int d^3\vec{r}\left[\varepsilon_0 \hat{\vec{E}}^{\,2}(\vec{r},t) + \mu_0 \hat{\vec{H}}^{\,2}(\vec{r},t)\right] \\
&= \sum_{m\sigma}\hbar\omega_m(\hat{a}_{m\sigma}^\dagger \hat{a}_{m\sigma} + \tfrac{1}{2}).
\end{aligned} \qquad (1\text{-}12)$$

The eigenstates of $\hat{H}_{\text{Free-Atom}}$ are the atomic wave functions $\{|E_l\rangle\}$:

$$\hat{H}_{\text{Free-Atom}}|E_l\rangle = E_l|E_l\rangle. \qquad (1\text{-}13)$$

The eigenstates of $\hat{H}_{\text{Free-Field}}$ are the photon number states

$$\hat{H}_{\text{Free-Field}}|n_{m\sigma}\rangle = \hbar\omega_m(n_{m\sigma} + \tfrac{1}{2})|n_{m\sigma}\rangle, \qquad (1\text{-}14)$$

where $|n_{m\sigma}\rangle$ is the state of n photons in plane-wave mode of $\vec{k}$-vector $\vec{k}_m$ and polarization $\vec{e}_{m\sigma}$.

1.7 Electric-Dipole Approximation

Consider the atom-field interaction energy term:

$$\hat{H}_{\text{AFint}} = \frac{e}{m_e} \hat{\vec{P}}_e \cdot \hat{\vec{A}}(\hat{\vec{r}}_e, t), \tag{1-15}$$

where

$$\hat{\vec{A}}(\hat{\vec{r}}_e, t) = \sum_{m\sigma} \vec{e}_{m\sigma} \left(\frac{\xi_m}{\omega_m} \right) \left(\hat{a}_{m\sigma}(t) e^{i\vec{k}_m \cdot \hat{\vec{r}}_e} + \hat{a}^\dagger_{m\sigma}(t) e^{-i\vec{k}_m \cdot \hat{\vec{r}}_e} \right). \tag{1-16}$$

We can add $\vec{r}_A$ to the exponential term and subtract it away, which will allow us to expand the exponential term in terms of the small quantity $i\vec{k}_m \cdot (\hat{\vec{r}}_e - \vec{r}_A)$ as follows:

$$e^{i\vec{k}_m \cdot \hat{\vec{r}}_e} = e^{i\vec{k}_m \cdot (\hat{\vec{r}}_e - \vec{r}_A)} e^{i\vec{k}_m \cdot \vec{r}_A} \approx \left[1 + i\vec{k}_m \cdot (\hat{\vec{r}}_e - \vec{r}_A) + \cdots \right] e^{i\vec{k}_m \cdot \vec{r}_A}. \tag{1-17}$$

The variable $i\vec{k}_m \cdot (\hat{\vec{r}}_e - \vec{r}_A)$ is typically much smaller than unity based on the facts that at optical frequencies we have $\lambda_m \sim 0.5$ µm = 5000Å and $|\hat{\vec{r}}_e - \vec{r}_A| \sim$ Bohr radius $\approx$ 0.5Å, giving $\vec{k}_m \cdot (\hat{\vec{r}}_e - \vec{r}_A) \approx \frac{2\pi}{\lambda_m} |\hat{\vec{r}}_e - \vec{r}_A| \ll 1$. To the lowest order we may approximate $e^{i\vec{k}_m \cdot (\hat{\vec{r}}_e - \vec{r}_A)} \approx 1$, so that $e^{i\vec{k}_m \cdot \hat{\vec{r}}_e} \approx e^{i\vec{k}_m \cdot \vec{r}_A}$. This is called the Electric Dipole Approximation.[7] The inclusion of higher order terms will give rise to other multipole interactions (electric and magnetic quadrupoles etc). We shall denote this approximated Hamiltonian as $\hat{H}_{\text{EDint}}$:

$$\hat{H}_{\text{AFint}} \approx \frac{e}{m_e} \hat{\vec{P}}_e \cdot \hat{\vec{A}}(\vec{r}_A, t) \equiv \hat{H}_{\text{EDint}}. \tag{1-18}$$

1.8 The $\hat{\vec{E}} \cdot \hat{\vec{r}}_e$ Form for Electric Dipole Energy

Near atomic resonance, this electric-dipole atom-field interaction energy in the form of $\hat{\vec{A}} \cdot \hat{\vec{P}}_e$ can be approximated in a $\hat{\vec{E}} \cdot \hat{\vec{r}}_e$ form. Near resonance, $\hat{\vec{r}}_e$ and $\hat{\vec{A}}$ will be of the form given by $\hat{\vec{r}}_e \approx (\hat{\vec{\gamma}} e^{-i\omega_A t} + \hat{\vec{\gamma}}^\dagger e^{i\omega_A t})$ and $\hat{\vec{A}} \approx \hat{\vec{\alpha}} e^{-i\omega_m t} + \hat{\vec{\alpha}}^\dagger e^{i\omega_m t}$, where ω_A is the atomic resonance frequency. Using Heisenberg equation of motion, it can be shown that $\hat{\vec{P}}_e + e\hat{\vec{A}} = m_e \frac{\partial \hat{\vec{r}}_e}{\partial t}$. If we neglect the $\hat{\vec{A}}^2$ term in the $\hat{\vec{A}} \cdot \hat{\vec{P}}_e$ interaction, we have:

$$\hat{\vec{P}}_e \cdot \hat{\vec{A}} \approx m_e \frac{\partial \hat{\vec{r}}_e}{\partial t} \cdot \hat{\vec{A}} \approx m_e (-i\omega_A \hat{\vec{\gamma}} e^{-i\omega_A t} + i\omega_A \hat{\vec{\gamma}}^\dagger e^{i\omega_A t}) \cdot \hat{\vec{A}}, \tag{1-19}$$

$$\hat{\vec{E}} \cdot \hat{\vec{r}}_e = -\frac{\partial \hat{\vec{A}}}{\partial t} \cdot \hat{\vec{r}}_e \approx -(-i\omega_m \hat{\vec{\alpha}} e^{-i\omega_m t} + i\omega_m \hat{\vec{\alpha}}^\dagger e^{i\omega_m t}) \cdot \hat{\vec{r}}_e. \tag{1-20}$$

Hence with $\omega_A \approx \omega_m$ (near atomic resonance), we have: $\frac{e}{m_e} \hat{\vec{P}}_e \cdot \hat{\vec{A}} \approx e \frac{\partial \hat{\vec{r}}_e}{\partial t} \cdot \hat{\vec{A}} \approx e\hat{\vec{E}} \cdot \hat{\vec{r}}_e$, where we have made the rotating-wave approximation and neglected terms rotating at $2\omega_A$ (valid at near-resonance). This then gives:

$$\hat{H}_{\text{EDint}} \approx e\hat{\vec{E}}(\vec{r}_A, t) \cdot \hat{\vec{r}}_e. \tag{1-21}$$

This form of interaction Hamiltonian is called the $\hat{\vec{E}} \cdot \hat{\vec{r}}_e$ form.[7]

1.9 Second Quantization

Let $\{|E_l\rangle\}$ be the complete set of atomic eigenkets for $\hat{H}_{\text{Free-Atom}}$, so that we have $\hat{H}_{\text{Free-Atom}}|E_l\rangle = E_l|E_l\rangle$. As $\{|E_l\rangle\}$ is a complete set of states, we have

$$\sum_l |E_l\rangle\langle E_l| = \hat{1}, \tag{1-22}$$

where $\hat{1}$ is the unity operator for atomic operators such that $\hat{1}\hat{\Theta} = \hat{\Theta} = \hat{\Theta}\hat{1}$, with $\hat{\Theta}$ being any atomic operator. Using this completeness relation allows us to write atomic operators based on these atomic eigenstates, which is a process called "second quantization". We can then write:

$$\begin{aligned}\hat{H}_0 &= \hat{1}\hat{H}_0\hat{1} = \sum_k |E_k\rangle\langle E_k|\hat{H}_0 \sum_l |E_l\rangle\langle E_l| \\ &= \sum_{k,l} |E_k\rangle\langle E_k|\hat{H}_0|E_l\rangle\langle E_l| = \sum_k E_k |E_k\rangle\langle E_k|,\end{aligned} \tag{1-23}$$

where we have used $\hat{H}_0|E_l\rangle = E_l|E_l\rangle$, and $\langle E_k|E_l\rangle = \delta_{kl}$. Similarly

$$e\hat{\vec{r}}_e = \sum_{k,l} |E_k\rangle\langle E_k|e\hat{\vec{r}}_e|E_l\rangle\langle E_l| = \sum_{k,l} \overline{\mu}_{kl} |E_k\rangle\langle E_l|, \tag{1-24}$$

where $\overline{\mu}_{kl} = \langle E_k|e\hat{\vec{r}}_e|E_l\rangle$ is the matrix element for the dipole operator $e\hat{\vec{r}}_e$.

1.10 Two-Level System

Let us assume that there are two dominant atomic energy levels of interest. Let the upper level with energy E_u be $|E_u\rangle$ and the ground level with energy E_g be $|E_g\rangle$. In this case, the second quantized operator for the total Hamiltonian will be reduced to:

$$\hat{H} = \hbar\omega_A \hat{N}_u + (\hat{V}^\dagger \overline{\mu} + \hat{V}\overline{\mu}^*) \cdot \hat{\vec{E}}(\vec{r}_A, t) + \sum_{m\sigma} \hbar\omega_m (\hat{a}_{m\sigma}^\dagger \hat{a}_{m\sigma} + \tfrac{1}{2}), \tag{1-25}$$

where $\hbar\omega_A = (E_u - E_g)$. $\hat{N}_u = |E_u\rangle\langle E_u|$ is the upper level population operator, $\hat{V} = |E_g\rangle\langle E_u|$ is the atomic down-transition operator, $\hat{V}^\dagger = |E_u\rangle\langle E_g|$ is the atomic up-transition operator, and $\overline{\mu} = \overline{\mu}_{ug} = \langle E_u|e\hat{\vec{r}}_e|E_g\rangle$ is the dipole matrix element. We have used $\overline{\mu}_{gg} = \overline{\mu}_{uu} = 0$ as $\langle E_u|\hat{\vec{r}}_e|E_u\rangle = \langle E_g|\hat{\vec{r}}_e|E_g\rangle = 0$, and used $\hat{N}_u + \hat{N}_g = \hat{1}$.

1.11 Transition Rate Calculation via Fermi's Golden Rule

Let $\hat{H} = \hat{H}_0 + \hat{H}_{\text{int}}(t)$ be the Hamiltonian of the system in the Schrödinger picture, where $\hat{H}_0$ is the unperturbed system Hamiltonian with a known solution and $\hat{H}_{int}(t)$ is time-dependant interaction energy. Suppose $\hat{H}_{int}(t)$ has the form of a sinusoidal excitation:

$$\hat{H}_{int}(t) = \hat{M}_{int} e^{-i\omega t} + \hat{M}_{int}^\dagger e^{i\omega t}, \tag{1-26}$$

then one can compute the transition rate between an initial state $|I\rangle$ and a final state $|F\rangle$ using perturbation theory.[6,8] The transition-rate result is called Fermi's golden rule. In the derivation of Fermi's golden rule, we start with the Schrödinger equation

$$i\hbar \frac{\partial}{\partial t}|\psi(t)\rangle = (\hat{H}_0 + \hat{H}_{int})|\psi(t)\rangle, \tag{1-27}$$

and let $|\psi(t)\rangle = C(t)e^{-i\frac{E_I}{\hbar}t}|I\rangle + D(t)e^{-i\frac{E_F}{\hbar}t}|F\rangle$, from which we obtain:

$$\frac{d}{dt}D(t) \approx \frac{1}{i\hbar}\langle F|\hat{M}_{int}^\dagger|I\rangle C(t)e^{i(\omega-\Omega)t}, \tag{1-28}$$

where $\Omega = (E_I - E_F)/\hbar$, and we have used the rotating wave approximation. The transition rate is then given by $\gamma = \frac{d}{dt}|D(t)|^2$. For times shorter than the decay time $t < (1/\gamma)$, $C(t)$ can be taken as 1. The direct integration of $D(t)$ gives the atomic decay rate at time t:

$$\gamma \approx \frac{2}{\hbar^2}\left|\langle F|\hat{M}_{int}^\dagger|I\rangle\right|^2 \frac{\sin(\Omega-\omega)t}{(\Omega-\omega)}. \tag{1-29}$$

In the case where there are many final states with the same value for the matrix element $\langle F|\hat{M}_{int}^\dagger|I\rangle$, we have:

$$\gamma = \int \frac{2}{\hbar^2}\left|\langle F|\hat{M}_{int}^\dagger|I\rangle\right|^2 \frac{\sin(\Omega-\omega)t}{(\Omega-\omega)}\rho_E(E_F)dE_F, \tag{1-30}$$

where $\rho_E(E_F)$ is the number of states per unit energy. At long time t, the integrand can be approximated by a delta function,

$$\frac{\sin(\Omega-\omega)t}{(\Omega-\omega)} \approx \pi\delta(\Omega-\omega) = \pi\hbar\delta(\hbar\Omega-\hbar\omega). \tag{1-31}$$

This approximation is justified only if other E_F-dependent functions in the integrand

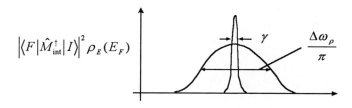

Fig. 1.4. Relative widths of the decay rate and E_F-dependent function.

are slowly varying in Ω within the width of $\frac{\sin(\Omega-\omega)t}{(\Omega-\omega)}$. The width of $\frac{\sin(\Omega-\omega)t}{(\Omega-\omega)}$ is given by $\Delta\Omega = \Delta E_F/\hbar \approx \pi/t$. The other terms in the integrand, $\rho_E(E_F)$ and $\left|\langle F|\hat{M}_{int}^\dagger|I\rangle\right|^2$, are generally functions of E_F. Let the characteristic width of $\rho_E(E_F)\left|\langle F|\hat{M}_{int}^\dagger|I\rangle\right|^2$ be $\Delta E_\rho = \hbar\Delta\omega_\rho$. The validity of the approximation then requires $\Delta\Omega < \Delta\omega_\rho$ or $t > (\pi/\Delta\omega_\rho)$. Combining with the requirement that $(1/\gamma) > t$, we have $(1/\gamma) > t > (\pi/\Delta\omega_\rho)$ or $\Delta\omega_\rho/\pi > \gamma$ as shown in Fig. 1.4.

Thus the validity of the application of Fermi's golden rule requires that

$$\frac{\Delta\omega_\rho}{\pi} > \gamma. \tag{1-32}$$

A more rigorous treatment in quantum optics shows that the decay rate formula is valid under Markov approximation, or when there is no long-time memory for the atom emitting the field, which would be the case if the atom is coupled to many field modes.[9]

1.12 Summary of Decay Rate Formulas

In summary, the specific case discussed above assumes that we have transitions between one initial state $|I\rangle$ and one group of final states $|F_1\rangle, |F_2\rangle, \ldots, |F_n\rangle$ with energies around the value E_F, and that these states have similar values for their matrix elements so that $\langle F_1|\hat{M}_{int}^\dagger|I\rangle \approx \langle F_k|\hat{M}_{int}^\dagger|I\rangle \equiv \langle F|\hat{M}_{int}^\dagger|I\rangle$, where $\langle F|\hat{M}_{int}^\dagger|I\rangle$ represents the matrix element for this group. For this case we have

$$\gamma = \int \frac{2\pi}{\hbar}\left|\langle F|\hat{M}_{int}^\dagger|I\rangle\right|^2 \delta(E_I - E_F - \hbar\omega)\frac{\delta N}{\delta E_F}dE_F, \tag{1-33}$$

where δN is the number of states within the energy width δE_F so that

$$\frac{\delta N}{\delta E_F} = \text{number of states per unit energy} \equiv \rho_E(E_F). \tag{1-34}$$

In a more general case, we can have transitions between one initial state $|I\rangle$ and many groups of final states with different matrix elements. Let us label the representative matrix element for group j as $\langle F^{(j)}|\hat{M}_{int}^\dagger|I\rangle$, then

$$\gamma = \sum_j \delta\gamma_j, \tag{1-35}$$

$$\delta\gamma_j = \int \frac{2\pi}{\hbar}\left|\langle F^{(j)}|\hat{M}_{int}^\dagger|I\rangle\right|^2 \delta(E_I - E_F^{(j)} - \hbar\omega)\frac{\delta N^{(j)}}{\delta E_F^{(j)}}dE_F^{(j)}, \tag{1-36}$$

where $\delta\gamma_j$ is the (incremental) contribution to γ from group j and $\delta N^{(j)}/\delta E_F^{(j)} \equiv \rho_E^{(j)}(E_F^{(j)})$ is the density of states for group j at energy $E_F^{(j)}$. In the situation where the group j can be parameterized by a continuous parameter, we can replace Σ_j by an integral giving

$$\gamma = \int d\gamma. \tag{1-37}$$

1.13 Spontaneous Emission Rate Calculation

To calculate the spontaneous emission rate of an excited two-level atom we start with the two-level atom Hamiltonian:

$$\begin{aligned}
\hat{H} &= \hat{H}_0 + \hat{H}_{int}(t), \\
\hat{H}_0 &= \hbar\omega_A \hat{N}_u + \sum_{m\sigma} \hbar\omega_m (\hat{a}^\dagger_{m\sigma} \hat{a}_{m\sigma} + \tfrac{1}{2}), \\
\hat{H}_{int} &= (\hat{V}^\dagger \overline{\mu} + \hat{V}\overline{\mu}^*) \cdot \hat{\vec{E}}(\vec{r}_A, t).
\end{aligned} \tag{1-38}$$

Let us consider one initial state $|I\rangle = |0\rangle|E_u\rangle$ where there is no photon and the atom is in the upper level, and many final states $|F\rangle = |1_{m\sigma}\rangle|E_g\rangle$ where 1 photon is emitted into one of the plane-wave modes and the atom makes a transition to the ground level. Specifically, let us consider one group of final photon states with $\bar{k}$-vectors around a particular value $\bar{k}_m$ and within a solid angle $d\Omega$. This group of states spans a range of photon energy $\hbar\omega_m$, and Fermi's golden rule has to be rewritten to integrate over $\hbar\omega$. The (incremental) contribution to the emission rate due to this group of photon final states and the associated atomic states is then given by

$$d\gamma(\bar{k}_m) = \int \frac{2\pi}{\hbar} |M_{FI}|^2 \delta(E_I - E_F - \hbar\omega) \rho_{ph}(\hbar\omega,\Omega) d\Omega\, d\hbar\omega, \tag{1-39}$$

where $\rho_{ph}(\hbar\omega,\Omega)d\Omega = \delta N_{ph}/\delta\hbar\omega$ is the number of photon states per unit photon energy $\hbar\omega$ within the solid angle $d\Omega$. Let $E_I - E_F = \hbar\omega_A$, where ω_A is the atomic resonance frequency. After performing the integration, we get

$$d\gamma(\bar{k}_m) = \frac{2\pi}{\hbar} |M_{FI}|^2 \rho_{ph}(\hbar\omega_A,\Omega) d\Omega. \tag{1-40}$$

The matrix element $M_{FI} = \langle F|\hat{M}^\dagger_{int}|I\rangle$ is also evaluated at $\omega = \omega_A$, where $\langle F|\hat{M}^\dagger_{int}|I\rangle$ can be calculated from:

$$\begin{aligned}
\langle F|\hat{H}_{EDint}|I\rangle &= \langle 1_{m\sigma}|\langle E_g|\hat{H}_{EDint}|E_u\rangle|0\rangle \\
&= \langle 1_{m\sigma}|\langle E_g|\hat{V}\overline{\mu}^* \cdot \vec{e}_{m\sigma}(-i\xi_m) e^{-i\bar{k}_m\cdot\vec{r}_A} \hat{a}^\dagger_{m\sigma}(0) e^{i\omega_m t}|E_u\rangle|0\rangle \\
&= (-i\xi_m)\overline{\mu}^* \cdot \vec{e}_{m\sigma} e^{i\omega_m t} e^{-i\bar{k}_m\cdot\vec{r}_A} \equiv M_{FI} e^{i\omega_A t}.
\end{aligned} \tag{1-41}$$

Thus $|M_{FI}|^2 = |\xi_m|^2 |\overline{\mu}^* \cdot \vec{e}_{m\sigma}|^2$ with $\overline{\mu}^* = \langle E_g|e\vec{r}_e|E_u\rangle = \mu^*_z \vec{e}_z$ if the upper-level is $2P_z$ state of the hydrogen (i.e. we have z-oriented dipole). The emission rate into a particular polarization mode is then given by

$$d\gamma_\sigma(\bar{k}_m) = \frac{2\pi}{\hbar}|\xi_m|^2|\mu_z|^2|\bar{e}_z \cdot \bar{e}_{m\sigma}|^2 \rho_{ph}(\hbar\omega_A, \Omega)d\Omega. \qquad (1\text{-}42)$$

1.14 Spontaneous Emission Rate in Free Space

Let us apply the formula above to calculate the emission rate in free space. First we need to find $\rho_{ph}(\hbar\omega_A, \Omega)$. Consider a group of photon states with $\bar{k}$-vectors around $\bar{k}_m$. Let θ, ϕ be the direction of $\bar{k}_m$. Let us consider all modes with $\bar{k}$-vectors lying around $\bar{k}_m$ within a small $\bar{k}$-space volume $d^3\bar{k}$ defined by a small solid angle around θ, ϕ in $\bar{k}$-space as shown.

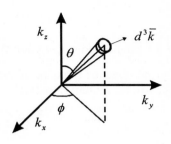

Fig. 1.5. The volume element $d^3\bar{k}$ in $\bar{k}$ space.

Since

$$k_{mx} = \frac{2\pi m_x}{L_x}, \; k_{my} = \frac{2\pi m_y}{L_y}, \; k_{mz} = \frac{2\pi m_z}{L_z}, \qquad (1\text{-}43)$$

the incremental $\bar{k}$-space volume $d^3\bar{k}$ is given by:

$$d^3\bar{k} = dk_{mx}dk_{my}dk_{mz} = \frac{(2\pi)^3}{L_xL_yL_z}\delta m_x \delta m_y \delta m_z. \qquad (1\text{-}44)$$

But $\delta m_x \delta m_y \delta m_z = \delta N_{ph}$ is the total number of photon states in that volume, hence

$$d^3\bar{k} = \frac{(2\pi)^3}{V_Q}\delta N_{ph}, \qquad (1\text{-}45)$$

where $V_Q = L_xL_yL_z$ is the volume of quantization. Now $d^3\bar{k} = k^2 d\Omega dk = k^2 \sin\theta d\theta d\phi dk$ with $k = |\bar{k}| = \omega/c$. Hence, the density of states for these free-space photon modes is:

$$\rho_{ph}(\hbar\omega, \Omega)d\Omega = \frac{\delta N_{ph}}{\delta \hbar\omega} = \frac{V_Q}{(2\pi)^3}\frac{d^3\bar{k}}{\delta\hbar\omega} = \frac{V_Q}{(2\pi)^3}\frac{1}{\hbar}k^2\frac{dk}{d\omega}\sin\theta \, d\theta d\phi. \qquad (1\text{-}46)$$

We see that in this case $\rho_{ph}(\hbar\omega, \Omega)$ is actually independent of the solid angle Ω.

This gives for the polarization σ:

$$d\gamma_\sigma(\bar{k}_m) = d\gamma_\sigma(\theta,\phi) = \frac{2\pi}{\hbar}\frac{|\mu_z|^2 \hbar\omega}{2\varepsilon_0 V_Q}|\bar{e}_{m\sigma} \cdot \bar{e}_z|^2 \frac{V_Q}{(2\pi)^3}\frac{1}{\hbar}k^2 \frac{dk}{d\omega}\sin\theta \, d\phi \, d\theta \qquad (1\text{-}47)$$

where ω is to be evaluated at $\omega_A = (E_u - E_g)/\hbar$. To find the dipole polarization factor $|\bar{e}_{m\sigma} \cdot \bar{e}_z|$, we define the two orthogonal polarization vectors for each $\bar{k}_m$ mode, $\bar{e}_{m1}$ and $\bar{e}_{m2}$ such that $\bar{e}_{m1}$ is in the direction of θ and $\bar{e}_{m2}$ is in the direction of ϕ as shown in Fig. 1.6.

Clearly $\bar{e}_{m1} \cdot \bar{e}_z = \sin\theta$ and $\bar{e}_{m2} \cdot \bar{e}_z = 0$. The dipole spontaneous emission pattern as a function of angle is then given by:

$$d\gamma_{\sigma=1}(\theta,\phi) = \frac{|\mu_z|^2 \omega_A^3}{2(2\pi)^2 \hbar \varepsilon_0 c^3} \sin^3\theta \, d\phi \, d\theta, \text{ and } d\gamma_{\sigma=2}(\theta,\phi) = 0. \quad (1\text{-}48)$$

Thus the emission into the ϕ polarized $\bar{e}_{m2}$ is zero. The total emission rate is:

$$\gamma = \sum_\sigma \int d\gamma_\sigma(\theta,\phi). \quad (1\text{-}49)$$

After integrating over θ and ϕ, we can show that the spontaneous emission rate in free space is:[6]

$$\gamma_{FS} = \frac{|\mu_z|^2 \omega_A^3}{3\pi \hbar \varepsilon_0 c^3}, \quad (1\text{-}50)$$

where we have used $\int_0^\pi \sin^3\theta \, d\theta = \frac{4}{3}$ and $\omega = kc$.

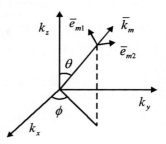

Fig. 1.6 The two orthogonal polarization vectors.

1.15 Density of States per Unit Volume, Unit Area, and Unit Length

From above, we have

$$d\gamma_\sigma(\bar{k}_m) = \frac{2\pi}{\hbar} |\xi_m|^2 |\mu_z|^2 |\bar{e}_z \cdot \bar{e}_{m\sigma}|^2 \rho_{ph}(\hbar\omega_A, \Omega) d\Omega. \quad (1\text{-}51)$$

Let us divide the density of states by the volume of quantization and define the density of states per unit volume as $\rho_{ph/V}(\hbar\omega_A) \equiv \rho_{ph}(\hbar\omega_A)/V_Q$. In terms of which

$$d\gamma_\sigma(\bar{k}_m) = \frac{2\pi}{\hbar} |\xi_m|^2 |\mu_z|^2 |\bar{e}_z \cdot \bar{e}_{m\sigma}|^2 V_Q \rho_{ph/V}(\hbar\omega_A, \Omega) d\Omega. \quad (1\text{-}52)$$

This allows us to express the emission rate in terms of quantities that are independent of the volume of quantization. For example, $|\xi_m|^2 V_Q = \hbar\omega_m/2\varepsilon_0$ is independent of V_Q and has a value determined by the quantized photon energy $\hbar\omega_m$. Likewise $\rho_{ph/V}$ is also independent of V_Q. Thus, the physically meaningful quantities that affect the spontaneous emission rate in 3-D space are the magnitude of the quantized photon energy and the density of states per unit volume. As will be seen later, it is also of interest to describe spontaneous emission rate into a 2-D or 1-D waveguide mode. For these low-dimensional photonic structures, the physical variables of interest will be the density of states per unit area and the density of states per unit length. As will be seen in the next two sections, the density of states factor takes care of those mode parameters with continuous degrees of freedom. The discrete guided mode parameters are dealt with individually.

1.16 Quantized Mode Amplitudes for General Modes

For modes that are not plane waves, we can write in general

$$\hat{\bar{E}}_{m\sigma}(\bar{r},t) = \left[iz_m \bar{F}_{m\sigma}(\bar{r}) \hat{a}_{m\sigma}(t) + \text{H.c.} \right],$$
$$\hat{\bar{E}}(\bar{r},t) = \sum_{m\sigma} \hat{\bar{E}}_{m\sigma}(\bar{r},t), \qquad (1\text{-}53)$$

where H.c. denotes Hermitian conjugation, $\bar{F}_{m\sigma}(\bar{r})$ is the mode spatial function, and z_m is a normalization constant whose value is to be determined by field quantization. The value for $|z_m|^2$ can be determined by imposing the total vacuum field energy (expectation value of the total field energy in vacuum) for each mode to be $\tfrac{1}{2}\hbar\omega_m$:

$$\begin{aligned}
\frac{\hbar\omega_m}{2} &= \langle 0 | \hat{H}_{\text{Free-Field}} | 0 \rangle \\
&= \int_{V_Q} d^3\bar{r} \left[\tfrac{1}{2}\varepsilon(\bar{r}) \langle 0 | \hat{\bar{E}}_{m\sigma}^2(\bar{r},0) | 0 \rangle + \tfrac{1}{2}\mu(\bar{r}) \langle 0 | \hat{\bar{H}}_{m\sigma}^2(\bar{r},0) | 0 \rangle \right] \\
&= \int_{V_Q} d^3\bar{r}\, \varepsilon(\bar{r}) \langle 0 | \hat{\bar{E}}_{m\sigma}^2(\bar{r},0) | 0 \rangle,
\end{aligned} \qquad (1\text{-}54)$$

where we have taken the electric and magnetic field energies to be equal, and $\varepsilon(\bar{r})$ is the dielectric constant at $\bar{r}$. Now $\langle 0 | \hat{\bar{E}}_{m\sigma}^2(\bar{r},0) | 0 \rangle = |z_m|^2 |\bar{F}_m(\bar{r})|^2$ so that $\tfrac{1}{2}\hbar\omega_m = \int_{V_Q} d^3\bar{r}\, \varepsilon(\bar{r}) |z_m|^2 |\bar{F}_m(\bar{r})|^2$, giving

$$|z_m|^2 = \frac{\hbar\omega_m}{2 \int_{V_Q} d^3\bar{r}\, \varepsilon(\bar{r}) |\bar{F}_m(\bar{r})|^2}. \qquad (1\text{-}55)$$

1.17 The Mode Volume Factor

For the general case, $|\xi_m|^2$ in the decay rate formula will be replaced by $|z_m \bar{F}_{m\sigma}(\bar{r})|^2$. (Note that for plane-wave modes $|\bar{F}_{m\sigma}(\bar{r})|^2 = |e^{i\bar{k}_{m\sigma}\cdot\bar{r}}|^2 = 1$.) It is useful to cast $|z_m|^2$ in terms of the effective mode volume, which can be defined as follows. In general $|\bar{F}_m(\bar{r})|$ has a peak value at $\bar{r}_{\text{peak}}$. We can write

$$\int_{V_Q} d^3\bar{r}\, \varepsilon(\bar{r}) |\bar{F}_m(\bar{r})|^2 = \varepsilon(\bar{r}_{\text{peak}}) |\bar{F}_m(\bar{r}_{\text{peak}})|^2 V_{\text{mode}}, \qquad (1\text{-}56)$$

where V_{mode} is called the effective mode volume. In terms of V_{mode} we have from Eqs. (1-55) and (1-56):

$$|z_m|^2 |\bar{F}_m(\bar{r})|^2 = \frac{\hbar\omega_m}{2\varepsilon(\bar{r}_{\text{peak}}) V_{\text{mode}}} \frac{|\bar{F}_m(\bar{r})|^2}{|\bar{F}_m(\bar{r}_{\text{peak}})|^2} \equiv |\xi_m|^2. \qquad (1\text{-}57)$$

This is a direct analogue of the free-field case for which we have

$$|\xi_m|^2 = \frac{\hbar\omega_m}{2\varepsilon_0 V_Q}. \qquad (1\text{-}58)$$

1.18 Emission Rate for the General Case

In the general case V_{mode} plays the role of V_Q. The emission rate for the general case can be written as:

$$d\gamma_{m\sigma} = \frac{2\pi}{\hbar}|\xi_m|^2|\mu_z|^2|\bar{e}_z \cdot \bar{e}_{m\sigma}(\bar{r}_{\text{dipole}})|^2 \, \rho_{\text{ph}}(\hbar\omega_A, \Omega)d\Omega, \qquad (1\text{-}59)$$

where $\bar{r}_{\text{dipole}}$ is the location of the z-dipole, the polarization vector $\bar{e}_{m\sigma}(\bar{r}_{\text{dipole}})$ is the unit vector of $\bar{F}_m(\bar{r})$ at $\bar{r} = \bar{r}_{\text{dipole}}$, and $|\xi_m|^2$ is given by

$$|\xi_m|^2 = |z_m|^2|\bar{F}_m(\bar{r}_{\text{dipole}})|^2 = \frac{\hbar\omega_m}{2\varepsilon(\bar{r}_{\text{peak}})V_{\text{mode}}} \frac{|\bar{F}_m(\bar{r}_{\text{dipole}})|^2}{|\bar{F}_m(\bar{r}_{\text{peak}})|^2}$$
$$= \frac{\hbar\omega_m}{2} \frac{|\bar{F}_m(\bar{r}_{\text{dipole}})|^2}{\int_{V_Q} d^3\bar{r}\,\varepsilon(\bar{r})|\bar{F}_m(\bar{r})|^2}. \qquad (1\text{-}60)$$

Thus the emission rate is related to the vacuum field intensity of the mode at the location of the dipole. In a typical calculation, once we know the mode function $\bar{F}_m(\bar{r})$, we can compute the mode volume factor by computing $\int_{V_Q} d^3\bar{r}\,\varepsilon(\bar{r})|\bar{F}_m(\bar{r})|^2$ and then obtain $|\xi_m|^2$.

2 Spontaneous Emission in Metallic Waveguides (Metallic Photonic Wells, Wires, and Dots) and Microcavities

2.1 Introduction

Both the rate and pattern of spontaneous emission can be modified in metallic waveguides and microcavities. In this section, we will consider the modification of spontaneous emission in different types of metallic waveguide and microcavity.[1,10] We will first study the modification of spontaneous emission in planar metallic waveguides (metallic photonic wells as shown in Fig. 2.1a). It will include both the ideal lossless waveguides and lossy waveguides. We will then study the case of channel metallic waveguides with a rectangular cross section (metallic photonic wires as

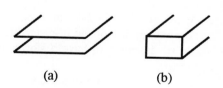

Fig. 2.1. Two types of metallic waveguides: (a) planar, (b) rectangular.

shown in Fig. 2.1b). The active medium is typically assumed to be in the form of a thin layer such as a quantum well situated at the center of the structure.

The results for microcavities can be obtained from waveguide results. For example, A "planar-mirror microcavity" (see Fig. 2.2a) can be formed with two partially-transmitting metal plates spaced by a distance from each other, which of course is the same configuration as a lossy planar waveguide. A "channel-waveguide microcavity" can be formed by enclosing the two ends of a short rectangular waveguide with two partially-transmitting parallel metal plates (see Fig. 2.2b), or by forming a ring cavity with waveguide. We may refer to such type of microcavity as photonic-wire microcavity. A metallic-disk microcavity (see Fig. 2.2c) for which the cavity modes are the whispering gallery guided modes can be regarded as a "planar-waveguide microcavity" when the disk diameter is large (a large disk diameter gives a wide mode size and would behave like a planar waveguide). We may refer to such type of microcavity as photonic-well microcavity. We note that a "channel-waveguide microcavity" is also a photonic dot (3-D enclosed cavity).

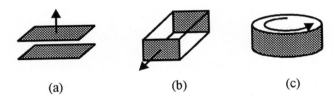

Fig. 2.2. Different types of microcavities: (a) Planar, (b) Photonic-Wire, (c) Photonic Well.

In terms of applications to microcavity lasers, one quantity of interest is the fraction of spontaneous emission that can be channeled into a single cavity mode, which is often referred to as the spontaneous-emission coupling factor or the β value. A large spontaneous-emission coupling factor can lead to lower lasing threshold for lasers and is also desirable in terms of realizing high-efficiency light-emitting devices (LEDs etc). With a unity β value, every photon emitted spontaneously will be captured into a single cavity mode. Such will be the ideal microcavity. Another quantity of interest is the modification in the total decay rate, which can be used to change the transparency pumping rate of the laser medium. To lower the transparency pumping rate, it would be desirable to have a slower decay rate (i.e. to have inhibited spontaneous emission).

The results in this section show that a large β value approaching the ideal limit of unity can in principle be achieved with a metallic channel-waveguide microcavity (made of ideal metal). In addition, it is also possible to achieve a large modification in the spontaneous decay rate, which can be either inhibition or enhancement. In the case of real metal, the inhibited spontaneous emission effect may become less effective.

A metallic photonic well (or a photonic-well microcavity) does not have as strong an effect as a metallic photonic wire (or a photonic-wire microcavity). Nevertheless, it may still be utilized in some situations of interest.

2.2 Emission Rate in Lossless Metallic Photonic Well

Let us consider the case of a perfectly conducting planar waveguide (metallic photonic well) formed by two infinite metallic plates in air. The plates are separated by a distance w, which defines the waveguide width. We use a coordinate system as shown in Fig. 2.3, and label the three different spatial regions as region 1, 2, and 3, respectively. Region 2 is the guiding region.

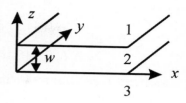

Fig. 2.3. The coordinate system we use for metallic planar waveguide.

There are many discrete planar waveguide modes, corresponding to a discrete set of k_{mz}. The allowed values of k_{mz} are determined by the waveguide structure, $k_{mz} = m_z \pi / w$. Here m_z is an integer ranging from 1, 2, ... up to a maximum integer value not over $2w/\lambda$. These modes have finite extend in the z direction. In the x-y plane, we can impose a box of quantization with an area $L_x L_y$. The photon density of states $\rho_{\text{ph}}(\hbar\omega)$ can be derived by using:

$$k_{mx} = \frac{2\pi n_x}{L_x}, \qquad k_{my} = \frac{2\pi n_y}{L_y},$$

$$\delta k_{mx} \delta k_{my} = \frac{(2\pi)^2}{L_x L_y} \delta m_x \delta m_y = \frac{(2\pi)^2}{L_x L_y} \delta N_{\text{ph}}, \qquad (2\text{-}1)$$

$$\delta k_{mx} \delta k_{my} = k_{mxy} dk_{mxy} d\phi$$

where k_{mxy} is the projection of the $\bar{k}$-vector in the x-y plane given by $k_{mxy} = k \sin\theta$, and θ is the angle between $\bar{k}_m$ and the z axis, and δN_{ph} is the number of states within the $\bar{k}$-space area of $\delta k_{mx} \delta k_{my}$. This gives:

$$\rho_{\text{ph}}(\hbar\omega, \phi) d\phi = \frac{\delta N_{\text{ph}}}{\delta \hbar\omega} = \frac{1}{\hbar} \cdot \frac{L_x L_y}{(2\pi)^2} k_{mxy} \frac{dk_{mxy}}{d\omega} d\phi. \qquad (2\text{-}2)$$

The incremental contribution to the decay rate for a z-dipole due to guided mode m_z is then:

$$\gamma_{gm_z} = \int_0^{2\pi} \frac{2\pi}{\hbar^2} |\xi_m|^2 |\mu_z|^2 |\bar{e}_{m\sigma}(z_{\text{dipole}}) \cdot \bar{e}_z|^2 \frac{L_x L_y}{(2\pi)^2} k_{mxy} \frac{dk_{mxy}}{d\omega} d\phi. \qquad (2\text{-}3)$$

For x-dipole or y-dipole, we replace the polarization factor by $\bar{e}_{m\sigma}(z_{\text{dipole}}) \cdot \bar{e}_x$ and $\bar{e}_{m\sigma}(z_{\text{dipole}}) \cdot \bar{e}_y$ respectively. The total decay rate is given by

$$\gamma = \sum_{m_z} \gamma_{gm_z}. \qquad (2\text{-}4)$$

To calculate γ_{gm_z}, we need $|\xi_m|^2$ which can be obtained by first computing:

$$\int_{-\infty}^{\infty} d^3\vec{r}\,\varepsilon(\vec{r})|\overline{F}_m(\vec{r})|^2 = \varepsilon_0 L_x L_y \int_{-\infty}^{\infty} dz |\overline{F}_m(z)|^2, \qquad (2\text{-}5)$$

giving:

$$|\xi_m|^2 = \frac{\hbar\omega_m |\overline{F}_m(z_{\text{dipole}})|^2}{2\varepsilon_0 L_x L_y L_{\text{mode}} |\overline{F}_m(z_{\text{peak}})|^2} \qquad (2\text{-}6)$$

where L_{mode} is the effective width given by:

$$L_{\text{mode}} = \frac{\int_{-\infty}^{\infty} dz |\overline{F}_m(z)|^2}{|\overline{F}_m(z_{\text{peak}})|^2}. \qquad (2\text{-}7)$$

The expression for $\overline{F}_m(z)$ in region 2 is

$$\overline{F}_m = \begin{cases} E_0 \overline{e}_z & \text{(TEM mode)} \\ \sin(k_{mz}z)\overline{e}_{mxy} \times \overline{e}_z & \text{(TE modes)} \\ k_{mxy}\cos(k_{mz}z)\overline{e}_z - ik_{mz}\sin(k_{mz}z)\overline{e}_{mxy} & \text{(TM modes)} \end{cases} \qquad (2\text{-}8)$$

where $\overline{e}_{mxy}$ is a unit vector in the direction of $\overline{k}_{mxy}$.

Let γ_{FS} be the decay rate for the dipole in free-space, where $\gamma_{\text{FS}} = |\mu|^2 k_A^3 / 3\pi\varepsilon_0 \hbar$, $\omega_A = k_A c$. We shall present the results by normalizing it to the free-space rate and define the normalized rate as $R_{gm_z} = \gamma_{gm_z} / \gamma_{\text{FS}}$.

For a vertical dipole (z-dipole), we get with $\lambda = 2\pi/k$:

$$R_{gm_z}^{\text{vert}} = \begin{cases} \dfrac{3}{4(w/\lambda)} & \text{TEM mode} \\ 0 & \text{TE mode } (m_z = 1,2,3,\ldots) \\ \dfrac{3\lambda^2(k^2 - k_{mz}^2)}{8\pi^2(w/\lambda)}\cos^2 k_{mz}z & \text{TM mode } (m_z = 1,2,3,\ldots) \end{cases} \qquad (2\text{-}9)$$

For a horizontal dipole (x, y-dipole), we get:

$$R_{gm_z}^{\text{hor}} = \begin{cases} 0 & \text{TEM mode} \\ \dfrac{3}{4(w/\lambda)}\sin^2 k_{mz}z & \text{TE mode } (m_z = 1,2,3,\ldots) \\ \dfrac{3\lambda^2(k^2 - k_{mz}^2)}{16\pi^2(w/\lambda)}\sin^2 k_{mz}z & \text{TM mode } (m_z = 1,2,3,\ldots) \end{cases} \qquad (2\text{-}10)$$

We see that the emission rate is inversely proportional to the normalized width of the waveguide (w/λ).

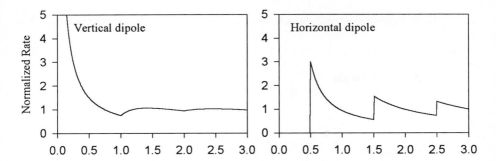

Fig. 2.4. Normalized spontaneous emission rate vs. w/λ for lossless planar waveguide with dipole at the center.

2.3 Results for Lossless Metallic Photonic Well

The results of calculation are shown in Fig. 2.4. Note that
1. At $w = \frac{1}{2}\lambda = 0.5\lambda$ cavity size, the horizontal dipole emission rate is *3 times* higher than in free space.
2. The horizontal dipole emission is *completely cut off* at $w < \frac{1}{2}\lambda$.
3. The vertical dipole emission is close to the free space rate at $w > \frac{1}{2}\lambda$. It grows rapidly to 5 times free space rate at $w = 0.2\lambda$ and to infinity at $w \to 0$. This is due to strong emission into the TEM mode, which is never cutoff and has a high vacuum field intensity when the plate separation approaches zero.
4. The total emission rates of all the vertical and horizontal dipoles approach the free space value of 1 when $w \gg 5\lambda$.

2.4 Emission Rate in Lossy Metallic Photonic Well and Planar Microcavity

Suppose the metallic walls of the planar waveguide are lossless but are so thin that they are partially transmitting. Then the metallic waveguide will be a lossy waveguide. As all the fields in such a lossy metallic planar waveguide will eventually leak to regions 1 and 3, we can make a reverse process and propagate vacuum modes in regions 1 and 3 from infinity into the waveguide and excite the waveguide modes. This provides us with a way to obtain the vacuum field amplitude in lossy metallic planar waveguide. As mentioned in the introduction to Section 1, in the fully quantized treatment, this approach would involve

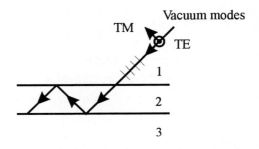

Fig. 2.5. Propagation of vacuum modes into a lossy metallic planar waveguide.

the spatial propagation of field mode operators, which can be made rigorous by using the localized photon operators.[5]

The density of states and intensity of the incident field are those for the vacuum field in free-space, namely:

$$\left|\xi_m^{FS}\right|^2 = \frac{\hbar\omega}{2\varepsilon_0 V_Q} \qquad (2\text{-}11)$$

and

$$\rho_{ph}(\hbar\omega,\Omega)d\Omega = \frac{V_Q}{(2\pi)^3}\frac{1}{\hbar}k^2\frac{dk}{d\omega}\sin\theta d\theta d\phi \equiv \rho_s \sin\theta d\theta d\phi. \qquad (2\text{-}12)$$

The incremental contribution to γ is then given by

$$d\gamma_R(\Omega) = \frac{2\pi}{\hbar^2}\left|\xi_m\right|^2\left|\overline{e}_{m\sigma}\cdot\overline{e}_{\text{dipole}}(z_{\text{dipole}})\right|^2\left|\overline{\mu}\right|^2 \rho_s \sin\theta d\theta d\phi, \qquad (2\text{-}13)$$

where ξ_m is the mode's vacuum amplitude at the location of the dipole in the waveguide. However, as every mode ξ_m actually comes from a quantized mode ξ_m^{FS} in the free-space regions 1 and 3, we can obtain ξ_m by propagating ξ_m^{FS} into the structure. Let r and t be the reflection and transmission coefficients for the field amplitude, and let us define G_m as the ratio $G_m = \xi_m / \xi_m^{FS}$. The total decay rate is then given by:

$$\gamma_R = 2\int_0^{2\pi}d\phi\int_0^{\pi/2}d\theta\frac{2\pi}{\hbar^2}\left|\xi_m^{FS}\right|^2\left|G_m\right|^2\left|\overline{e}_{m\sigma}\cdot\overline{e}_{\text{dipole}}\right|^2\left|\overline{\mu}\right|^2\rho_s \sin\theta. \qquad (2\text{-}14)$$

We note that for vertical dipole pointing along z:

$$\left|\overline{e}_{m\sigma}\cdot\overline{e}_{\text{dipole}}\right|^2 = \begin{cases}\sin^2\theta & \text{TM mode} \\ 0 & \text{TE mode}\end{cases} \qquad (2\text{-}15)$$

For horizontal dipole pointing along x:

$$\left|\overline{e}_{m\sigma}\cdot\overline{e}_{\text{dipole}}\right|^2 = \begin{cases}\cos^2\theta\sin^2\phi & \text{TM mode} \\ \sin^2\phi & \text{TE mode}\end{cases} \qquad (2\text{-}16)$$

For a vertical dipole the decay rate is

$$R_{TM}^{\text{vert}} = \frac{3}{2}\int_0^1 \frac{\frac{1-r}{1+r}+\frac{2r}{1-r^2}[\cos^2 kuz + \cos^2 ku(w-z)]}{1+\left(\frac{2r}{1-r^2}\right)^2\sin^2 kuw}(1-u^2)du, \qquad (2\text{-}17)$$

where the integration variable $u = \cos\theta$. Note that $ku = k\cos\theta = k_z$. For a horizontal dipole, from the TE-polarized field we get

$$R_{TE}^{\text{hor}} = \frac{3}{4}\int_0^1 \frac{\frac{1-r}{1+r}+\frac{2r}{1-r^2}[\sin^2 kuz + \sin^2 ku(w-z)]}{1+\left(\frac{2r}{1-r^2}\right)^2\sin^2 kuw}du; \qquad (2\text{-}18)$$

and from the TM-polarized field we get

$$R_{TM}^{\text{hor}} = \frac{3}{4}\int_0^1 \frac{\frac{1-r}{1+r}+\frac{2r}{1-r^2}[\sin^2 kuz + \sin^2 ku(w-z)]}{1+\left(\frac{2r}{1-r^2}\right)^2\sin^2 kuw}u^2 du. \qquad (2\text{-}19)$$

Compared with the lossless case, we are now integrating over a continuous range of k_z instead of summing over discrete values of k_{mz}.

For r close to 1, the integrands are sharply peaked at the original values of $k_z \approx k_{mz}$ due to large variation in their denominators at such values of k_z, and we can approximate the common denominator by:

$$\left. \frac{1}{1+\left(\frac{2r}{1-r^2}\right)^2 \sin^2 kuw} \right|_{ku \sim k_{mz}} \sim \frac{1}{1+w^2\left(\frac{2r}{1-r^2}\right)^2 (k_z - k_{mz})^2} \qquad (2\text{-}20)$$

i.e. it can be approximated by a Lorentzian lineshape. In fact, if we take the limit $r \to 1$, we will be able to recover the results for the lossless waveguide. In the neighborhood of a peak, using the above expansion, we get

$$\int_{-\Delta+u_m}^{\Delta+u_m} \frac{\frac{2r}{1-r^2}}{1+\left(\frac{2r}{1-r^2}\right)^2 \sin^2 kuw} du \to \int_{-\infty}^{+\infty} \frac{\frac{2r}{k(1-r^2)}}{1+w^2\left(\frac{2r}{1-r^2}\right)^2 (k_z - k_{mz})^2} dk_z = \frac{\pi}{kw}, \qquad (2\text{-}21)$$

where $u_m = k_{mz}/k$. In the numerators of the integrands, we can replace the variable u by the constant value $u_m = k_{mz}/k$, because they do not change appreciably in the peak regions when $r \to 1$. Hence, for the vertical dipole the expression is reduced to

$$R_{TM}^{vert} \approx \frac{3}{2} \int_0^1 \frac{\frac{2r}{1-r^2}[\cos^2 kuz + \cos^2 ku(w-z)]}{1+\left(\frac{2r}{1-r^2}\right)^2 \sin^2 kuw}(1-u^2) du$$

$$\approx \sum_m \frac{3}{2} \int_0^1 \frac{\frac{2r}{1-r^2}}{1+\left(\frac{2r}{1-r^2}\right)^2 (kw)^2 (u-u_m)^2} du\,[\cos^2 ku_m z + \cos^2 ku_m(w-z)](1-u_m^2)$$

$$= \sum_m \frac{3}{2} \int_0^1 \frac{\frac{2r}{1-r^2}}{1+\left(\frac{2r}{1-r^2}\right)^2 w^2 (k_z - k_{mz})^2} d\left(\frac{k_z}{k}\right)[\cos^2 k_{mz} z + \cos^2 k_{mz}(w-z)]\left(1-\frac{k_{mz}^2}{k^2}\right)$$

$$\approx \frac{3\pi}{2kw} \sum_m [\cos^2 k_{mz} z + \cos^2 k_{mz}(w-z)]\left(1-\frac{(k_{mz})^2}{k^2}\right)$$

$$= \sum_m \frac{3\lambda^2 k_{mxy}^2}{8\pi^2 (w/\lambda)} \cos^2 k_{mz} z = \sum_{m_z} R_{gm_z}^{vert}$$

$$(2\text{-}22)$$

which is the same as the TM mode contributions.

Note that the sum is over $m = 0, 1, 2, \ldots$. For the $m = 0$ case, R_{g0}^{vert} is the decay rate into the TEM mode. It corresponds to the peak at $u \approx 0$ (i.e. when $\cos\theta = 0$ or when the guiding angle for the TM wave is 90 degrees, for which the $\overline{E}$ field is perpendicular to the metallic planes). In this case we integrate over only half of the lineshape function, which results in

$$R_{g0}^{vert} = 3\pi/2kw \quad \text{(TEM mode)}. \qquad (2\text{-}23)$$

This agreement with the lossless case shows that the method of propagating vacuum fields into the waveguide structure is consistent with the method based on guided

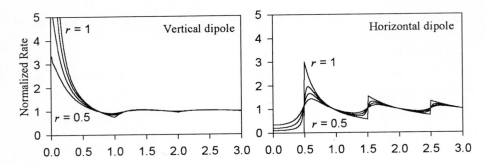

Fig. 2.6. Normalized spontaneous emission rate vs. w/λ. The curves are for r-values of 1, 0.82, 0.67, and 0.50.

modes. The method of propagating vacuum fields is more general as it can treat the lossy case.

2.5 Results for Lossy Metallic Photonic Well

The results of the calculation with $r = 1$, 0.82, 0.67, and 0.50 are shown in Fig. 2.6. Note that
1. Even for the $r = 0.5$ or $r^2 = 0.25$ case, the horizontal dipole emission rate at $w \approx 0.6\lambda$ is enhanced by 1.6 times and at $w = 0.25\lambda$ is reduced by 4 times.
2. Even for the $r = 0.5$ or $r^2 = 0.25$ case, the vertical dipole emission rate at $w = 0.25\lambda$ is enhanced by a factor of 2 times.
3. The total emission rates of all the vertical and horizontal dipoles approach the free-space value of 1 when $w \gg 5\lambda$.

2.6 Results for Metallic Planar Microcavity

The above results can be applied to predict spontaneous emission in a planar microcavity, say with a 0.5λ long cavity. In that case the total decay rate modification is as described above at $w = 0.5\lambda$. The spatial emission pattern can be studied by plotting the emission rates (per radian of θ) as a function of the angle θ. This is shown in Fig. 2.8 for the cases with $w = 0.5\lambda$ and $w = 0.55\lambda$, respectively. They describe the spatial emission into the cavity mode for a parallel-plate metallic microcavity. As pointed out earlier, the emission energy is sharply peaked at $k_z \approx k_{mz}$ as $r \to 1$.

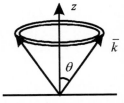

Fig. 2.7. Illustration of the angle θ between the $\bar{k}$-vector and the z-axis.

2.7 Emission Rate in Lossless Metallic Photonic Wire

For a lossless rectangular waveguide (metallic photonic wire), we again use the mode

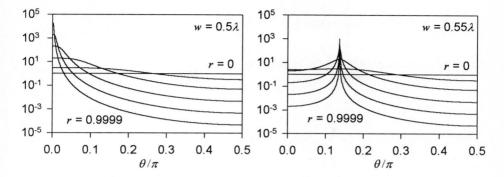

Fig. 2.8. Normalized spontaneous emission rate per unit angle for TE mode at two different plate separations. The curves corresponds to r = 0, 0.5, 0.9, 0.99, 0.999, and 0.9999 respectively.

approach to find its modification of spontaneous emission for a dipole placed inside the waveguide. Let a and b be the dimensions of the rectangular cross-section.

The electric fields for a TE mode are

$$\begin{cases} E_x = k_{my} \cos k_{mx} x \sin k_{my} y, \\ E_y = k_{mx} \sin k_{mx} x \cos k_{my} y, \\ E_z = 0; \end{cases} \quad (2\text{-}24)$$

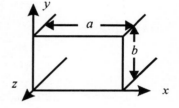

Fig. 2.9 The rectangular waveguide.

where $k_{mx} = m_x \pi / a$ and $k_{my} = m_y \pi / b$. The mode indices m_x and m_y are non-negative integers, but they cannot be both zero at the same time. They are also bounded by the condition $k_{mx}, k_{my} \leq k = n\omega/c$. For a TM mode we have

$$\begin{cases} E_x = \dfrac{ik_{mx} k_z}{k^2 - k_z^2} \cos k_{mx} x \sin k_{my} y, \\ E_y = \dfrac{ik_{my} k_z}{k^2 - k_z^2} \sin k_{mx} x \cos k_{my} y, \\ E_z = \sin k_{mx} x \sin k_{my} y. \end{cases} \quad (2\text{-}25)$$

To find the density of states, we quantize the field in the z direction along a line of length L_z. In this case

$$\rho_{ph}(\hbar\omega) = \frac{L_z}{2\pi} \frac{dk_z}{d\omega}, \quad (2\text{-}26)$$

and

$$|\xi_m|^2 = \frac{\hbar\omega |\overline{F}_m(x_{\text{dipole}}, y_{\text{dipole}})|^2}{2\varepsilon_0 L_z A_{\text{mode}}}, \tag{2-27}$$

where the mode area factor A_{mode} is given by:

$$A_{\text{mode}} = \frac{\int dx \int dy |\overline{F}_m(x,y)|^2}{|\overline{F}_m(\vec{r}_{\text{peak}})|^2}. \tag{2-28}$$

When normalized by the free-space value, the decay rate for an x-dipole is

$$R_{m_x,m_y}^{x\text{ dipole,TE}} = \frac{12\pi k_{my}^2}{abk|k_z|(k^2 - k_z^2)} \cos^2 k_{mx}x \sin^2 k_{my}y \quad (\text{TE mode, } m_x \neq 0)$$

$$R_{0,m_y}^{x\text{ dipole,TE}} = \frac{6\pi}{abk|k_z|} \sin^2 k_{my}y \quad (\text{TE mode, } m_x = 0) \quad (2\text{-}29)$$

$$R_{m_x,m_y}^{x\text{ dipole,TM}} = \frac{12\pi |k_z| k_{mx}}{abk^3(k^2 - k_z^2)} \cos^2 k_{mx}x \sin^2 k_{my}y \quad (\text{TM mode})$$

For a y-dipole,

$$R_{m_x,m_y}^{y\text{ dipole,TE}} = \frac{12\pi k_{mx}^2}{abk|k_z|(k^2 - k_z^2)} \sin^2 k_{mx}x \cos^2 k_{my}y \quad m_y \neq 0$$

$$R_{m_x,0}^{y\text{ dipole,TE}} = \frac{6\pi}{abk|k_z|} \sin^2 k_{mx}x \quad m_y = 0 \quad (2\text{-}30)$$

$$R_{m_x,m_y}^{y\text{ dipole,TM}} = \frac{12\pi |k_z| k_{my}}{abk^3(k^2 - k_z^2)} \sin^2 k_{mx}x \cos^2 k_{my}y.$$

For a z-dipole,

$$R_{m_x,m_y}^{z\text{ dipole,TM}} = \frac{12\pi(k^2 - k_z^2)}{abk^3|k_z|} \sin^2 k_{mx}x \sin^2 k_{my}y. \tag{2-31}$$

The z-dipole does not couple to the TE mode.

2.8 Results for Lossless Metallic Photonic Wire

The results of calculation are shown in several figures. They are prepared for a few cases of interest.

Case I: Dipole at the Center

In this case the dipoles are at the center of the waveguide. As an example, we present the case when $a = 2b$. The emission rate is plotted as a function of a in the normalized unit of a/λ. Note that:
1. At $a < \lambda/2$, the emission rates for all the x, y,

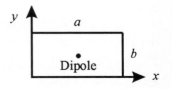

Fig. 2.10. Dipole at the center of a rectangular waveguide.

2. The emission from the x-dipole into TE mode at $a \approx 1.01\lambda$, and the emission from the z-dipole into TM mode at $a \approx 1.12\lambda$ are reduced by about 30%.

2.11 The Metallic Planar Waveguide (Metallic Photonic Well) Limit

If we take the limit $a \to \infty$ for the lossless rectangular channel waveguide discussed above while keeping b fixed, we shall expect to get the planar waveguide results. Using the results for the lossless rectangular waveguide, let us take $x = \frac{1}{2}a$, and sum over the index m_x for k_{mx}. Since k_{mx} ranges from 0 to a maximum value of $\sqrt{k^2 - k_{my}^2}$, let us define a variable

$$s = \frac{m_x \pi}{a\sqrt{k^2 - k_{my}^2}}, \text{ with } \frac{\Delta s}{\Delta m_x} = \frac{\pi}{a\sqrt{k^2 - k_{my}^2}}. \quad (2\text{-}34)$$

The summation over m_x becomes an integral over s. The symbols that depend on m_x become

$$k_{mx} = \frac{m_x \pi}{a} = s\sqrt{k^2 - k_{my}^2}, \text{ and } |k_z| = \sqrt{k^2 - k_{mx}^2 - k_{my}^2} = \sqrt{(k^2 - k_{my}^2)(1-s^2)}.$$

$$(2\text{-}35)$$

Putting them in the decay rate formulas, for the total decay rate of an x-dipole we get

$$R_{m_y}^{x\,\text{dipole}} = \sum_{m_x} \left(R_{m_x,m_y}^{x\,\text{dipole,TE}} + R_{m_x,m_y}^{x\,\text{dipole,TM}} \right)$$

$$\to \frac{6\sin^2 k_{my} y}{bk^3} \int_0^1 \frac{ds}{\sqrt{1-s^2}} \left[k^2(1-s^2) + k_{my}^2 s^2 \right]$$

$$= \frac{3\pi}{2bk^3}\left(k^2 + k_{my}^2\right)\sin^2 k_{my} y = \text{Results for planar waveguide.} \quad (2\text{-}36)$$

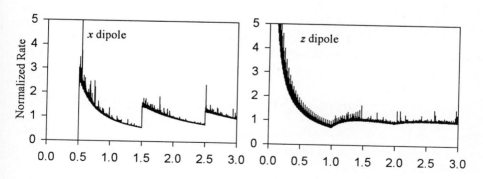

Fig. 2.16 Normalized total spontaneous emission rate vs. b/λ. The dipole is at the center, and we have set $a = 100b$. For a lossless rectangular waveguide, the rate approaches infinity at the cut-on of a mode, however, they show up as finite spikes in this figure due to numerical round off.

We have calculated the total emission rate for such a rectangular waveguide with $a = 100b$. The results are shown Fig. 2.16, which can be seen to agree with the planar waveguide results shown earlier in Fig. 2.4.

2.12 The Metallic Photonic-Wire Microcavity, Metallic Photonic Dot, and the Purcell Factor

Let us consider a cavity formed by placing two highly reflecting mirrors at $z = 0$ and $z = d$ of a lossless rectangular waveguide. This is a metallic photonic-wire microcavity, which can also be referred to as a metallic photonic dot. Let r and t be the reflection and transmission coefficient for the field amplitude. It can be shown that if the original field amplitude in the waveguide is $\overline{F}_m$, then the field amplitude in the cavity is given by

$$\left|\overline{F}_{mc}\right|^2 = t^2 \frac{(1-r)^2 + 4r\sin^2 k_{mz}(d-z)}{(1-r^2)^2 + 4r^2 \sin^2 k_{mz}d} \left|\overline{F}_m\right|^2. \tag{2-37}$$

This shows that for our cavity, the spontaneous emission rate is just that for the rectangular waveguide multiplied by the factor in front of $\left|\overline{F}_m\right|^2$.

For a high Q cavity, $r \to 1$, and we can neglect the first term $(1-r)^2$ in the numerator. If the dipole is on resonance with the cavity, then $\sin k_{mz}d = 0$, and we are left with $\frac{4r}{1-r^2}\sin^2 k_{mz}z = \frac{4F}{\pi}\sin^2 k_{mz}z$, where we have used $r^2 + t^2 = 1$, and the cavity finesse $F = \frac{\pi r}{1-r^2}$. On the other hand, the finesse is related to the cavity Q by $F = Q\Delta\omega/\omega_0$, where ω_0 is the resonance frequency, and $\Delta\omega$ is the free spectrum range. For simplicity we consider the case of $k_{mz} \approx k$, for which we can get $\Delta\omega = \pi c/d$. Hence the enhancement factor from the cavity is $\frac{4Q}{kd}\sin^2 k_{mz}z$.

Combine this with results for the waveguide, we can get the cavity enhancement factor. For example, for an x-dipole, TE mode of the field with $m_x = 0$,

$$R_c \approx \frac{24\pi Q}{V_c k^2 |k_z|} \sin^2 k_{my} y \sin^2 k_z z, \tag{2-38}$$

where $V_c = abd$ is the cavity volume. When averaged over the spatial coordinates, we have

$$R_c \approx \frac{6\pi Q}{k^3 V_c} = \frac{3Q\lambda^3}{4\pi^2 V_c}. \tag{2-39}$$

The Purcell factor describes the enhancement of spontaneous emission for a dipole placed inside a cavity.[1,10,11] It turns out that the above formula is just the result predicted by the Purcell factor.

2.13 Achieving Near-Unity Spontaneous-Emission Coupling Factor and Strong Decay Rate Modifications with Metallic Photonic-Wire Microcavity

From Fig. 2.13, it is clear that if we choose a rectangular waveguide dimension with a normalized value of about $0.5\lambda < a < \lambda$, the x-dipole and z-dipole emission will be almost completely suppressed and all the emission will come from the y-dipole emitting into the TE mode with y polarization. This, in theory, will provide near-unity value for the spontaneous-emission coupling factor. At the same time, we also see that the total decay rate for the y-dipole will be strongly inhibited at $a < 0.5\lambda$ and strongly enhanced at $a > 0.5\lambda$. It is important to note that the use of a rectangular waveguide with a 2 to 1 aspect ratio for the waveguide width vs. height is important for breaking the polarization mode degeneracy. This makes it possible to achieve the near unity β value.

2.14 Smallest Possible Metallic Microcavity?

The smallest microcavity may be realized with above metallic channel waveguide having a length forming a half-wavelength microcavity. This length, denoted as ℓ_c, would be dependent on the velocity of propagation for the guided mode. It turns out that at $a \approx 0.5\lambda$ the mode in the metallic channel waveguide will propagate with zero velocity and hence ℓ_c will be infinitely long. Thus, there is an optimal value for ℓ_c. It is not hard to show that to form a half-wavelength microcavity, we must have $\ell_c = a/\sqrt{4(\frac{a}{\lambda})^2 - 1}$. As $b = a/2$, the physical volume of the cavity would be given by $V_{cav}^{physical} = (\frac{1}{2}a^3)/\sqrt{4(\frac{a}{\lambda})^2 - 1}$. The sinusoidal variation of the mode will give a mode width approximately equal to half the physical width (i.e. half its zero point width). It turns out that the only dipole that is coupled significantly to the guided mode for $a < \lambda$ is the y dipole (see Fig. 2.13). Furthermore, the mode that the y dipole couples to is the TE mode polarized in the y-direction and has sinusoidal variation only along the x direction (i.e., along side "a"). There will also be sinusoidal field amplitude variation along the cavity length. Hence the mode volume is given approximately by the cavity physical volume multiplied by $(\frac{1}{2})^2$, giving $V_{cav}^{mode} = (\frac{1}{8}a^3)/\sqrt{4(\frac{a}{\lambda})^2 - 1}$. The smallest possible microcavity for the ideal metal case, it turns out, is achieved with $a \approx 0.61\lambda$. For $\lambda = 1.5\mu m$, we get $a = 0.92\mu m$, $b = 0.46\mu m$, $\ell_c = 1.316\mu m$, and $V_{cav}^{mode} = 0.139\mu m^3$ (This value is close to that of a cavity formed with a cubic half-wavelength box for which: $V_{cubic\ cav}^{mode} = (\frac{\lambda}{2n})^3 \times (\frac{1}{2})^2 \approx 0.105\mu m^3$). If we fill the cavity with a semiconductor active medium with a refractive index of 3.4, then we will have $V_{cav}^{mode} = 0.0035\mu m^3$ (with $a = 0.271\mu m$, $b = 0.135\mu m$, $\ell_c = 0.387\mu m$), which is to be compared with the dielectric case in the next section.

2.15 Real vs. Ideal Metals

As the above results assumed ideal metal, the use of real metal will modify the results somewhat. Part of the modification will come from the fact that real metal has a finite skin depth. This will modify the decay rate result somewhat when the two metallic plates forming the cavity are very closely spaced. For example, the inhibited spontaneous emission that occurred at small plate separation for the horizontal dipole may be somewhat reduced. This behavior is similar to the case of "leaky dielectric waveguides" discussed in section 3.3. Another part of the modification will arise from deviation from the π phase shift for reflection from the surface of a real metal. This deviation could lead to slight changes in the peak locations and their magnitudes for the various curves describing the modification of spontaneous emissions. The general characters of these curves, however, remain similar.[10]

3 Spontaneous Emission in Dielectric Photonic Wells, Wires, and Microcavities

In this chapter, we will discuss the spontaneous emission characteristics in dielectric photonic well structures and photonic wire structures. We first introduce the calculation method of the spontaneous emission rate in the dielectric photonic well structures. The photonic well structure is in the form of a two-dimensional dielectric planar waveguide, and the spontaneous emission rates are calculated for the excited dipoles inside the dielectric waveguide. We then present the modification of the spontaneous emission rates in a dielectric photonic wire structure in the form of a rectangular dielectric channel waveguide. The theoretical calculations were carried out for the rectangular dielectric waveguides with excitons along the center plane such as a single quantum well structure. We present some numerical examples and discuss about the optimum structures. Lastly, we discussed the spontaneous coupling factor that could be achieved by a dielectric photonic-wire microcavity ring laser.

3.1 Spontaneous Emission Rate in a Dispersive Lossless Bulk Dielectric Medium

Before we describe the modification of spontaneous emission rate in a dielectric photonic well or wire, it would be of interest to review the modification of spontaneous emission in a bulk dielectric medium.[12] The refractive index of the medium can be frequency dependent but the medium will be assumed to be lossless. As discussed in Landau and Liftshiz[13], such a medium can be described by the following Hamiltonian:

$$H = \int d^3x \frac{1}{2}\left[\frac{d\omega\varepsilon}{d\omega}E_m^2 + \mu_0 H_m^2\right]$$

$$H = \int d^3x \frac{cn_m}{v_m}\varepsilon_0 E_m^2, \tag{3-1}$$

where E_m and H_m are the electric and magnetic fields for that mode, v_m is the group velocity, and n_m is the refractive index at frequency Ω_{pm}. Ω_{pm} is the physical frequency for mode k_m in the medium. The normalized constant in the electric field operator is equal to the electric-field strength carried by half a photon energy. It can be found by setting $H = \hbar\Omega_{pm}/2$, giving

$$E_m = \left(\frac{\hbar\Omega_{pm} v_m}{2\varepsilon_0 V_\varrho n_m c}\right)^{1/2}, \qquad (3\text{-}2)$$

which is to be compared with $E_m = \left(\frac{\hbar\Omega_{pm}}{2\varepsilon V_\varrho}\right)^{1/2}$ for a dispersionless dielectric medium

and $E_m = \left(\frac{\hbar\omega_m}{2\varepsilon_0 V_\varrho}\right)^{1/2}$ for free space.

On the basis of the above argument, the macroscopic field operators for a dispersive dielectric medium is given by

$$\hat{\vec{E}}(\vec{r},t) = \sum_{m\sigma} i \left(\frac{\hbar\Omega_{pm} v_m}{2\varepsilon_0 V_\varrho n_m c}\right)^{1/2} \vec{e}_{m\sigma}\left[\hat{a}_{m\sigma}(t)e^{i\vec{k}_m\cdot\vec{r}} - \hat{a}^\dagger_{m\sigma}(t)e^{-i\vec{k}_m\cdot\vec{r}}\right], \qquad (3\text{-}3)$$

$$\mu_0 \hat{\vec{H}}(r,t) = \sum_{m\sigma} i \left(\frac{\hbar v_m}{2\varepsilon_0 V_\varrho n_m \Omega_{pm} c}\right)^{1/2} (\vec{k}_m \times \vec{e}_{m\sigma})\left[\hat{a}_{m\sigma}(t)e^{i\vec{k}_m\cdot\vec{r}} - \hat{a}^\dagger_{m\sigma}(t)e^{-i\vec{k}_m\cdot\vec{r}}\right]. \qquad (3\text{-}4)$$

Using these field operators, the decay rate for a dispersive dielectric bulk medium is given by

$$\gamma = \sqrt{(\varepsilon/\varepsilon_0)}\gamma_{FS}, \qquad (3\text{-}5)$$

where γ_{FS} is the decay rate of the same dipole in free space. Note that the group velocity term from the field amplitude factor is cancelled by another group velocity term from the $dk/d\omega$ factor in the density of states formula. This decay rate formula can also be obtained by propagating vacuum field into the bulk medium from all dielectrics.[12]

The above decay rate formula does not include the effect of local field. The inclusion of local field effect has been discussed in the literature.[14] If the emitting atom is small and occupy a lattice site in a crystalline dielectric medium with cubic lattice, then the inclusion of the local-field effect will lead to a decay rate of:

$$\gamma = \left[\frac{2+(\varepsilon/\varepsilon_0)}{3}\right]^2 \sqrt{(\varepsilon/\varepsilon_0)}\gamma_{FS} \qquad (3\text{-}6)$$

The factor in the right side of Eq. (3-6) is often referred to as the local-field correction factor. This factor can be derived if one assumed a "virtual cavity" model. In the case where the atom is relatively large or has a low polarizability, the local field effect tends to be given by a "real cavity" model, giving:

$$\gamma = \left[\frac{3(\varepsilon/\varepsilon_0)}{1+2(\varepsilon/\varepsilon_0)}\right]^2 \sqrt{(\varepsilon/\varepsilon_0)}\gamma_{FS} \qquad (3\text{-}7)$$

3.2 Spontaneous Emission Rate in Dielectric Photonic Well

The dielectric photonic well structure consists of a high-refractive-index guiding layer with refractive index n_D and thickness l (labeled as region 2) and low-refractive-index claddings with refractive index n (labeled as region 1 and 3) as shown in Fig 3.1. With similar derivation as Eq. (2-4) in the case of a metallic planar waveguide, one can show that the spontaneous emission rate contribution to γ_g from mode m (i.e. due to emission into mode m) is given by:

$$\gamma_{gm} = \int_0^{2\pi} \frac{2\pi}{\hbar^2} |\xi_m|^2 |\mu_z|^2 |\overline{e}_{m\sigma}(z_{\text{dipole}}) \cdot \overline{e}_z|^2 \frac{L_x L_y}{(2\pi)^2} k_{mxy} \frac{dk_{mxy}}{d\omega} d\phi, \qquad (3\text{-}8)$$

where $|\xi_m|^2 = \dfrac{\hbar \omega_A |\overline{F}_m(z_{\text{dipole}})|^2}{2\varepsilon(z_{peak}) L_x L_y L_{\text{mode}}}$, $L_{\text{mode}} = \dfrac{\int_{-\infty}^{\infty} dz \varepsilon(z) |\overline{F}_m(z)|^2}{\varepsilon(z_{peak}) |\overline{F}_m(z_{peak})|^2}$.

The total spontaneous emission rate into guided modes γ_g is given by

$$\gamma_g = \sum_m \gamma_{gm}. \qquad (3\text{-}9)$$

The expressions for $\overline{F}_m$ in region 1 and 2 are:
For even TE modes

$$\overline{F}_m = \begin{cases} \overline{e}_x \exp[-q(z - l_D/2)] \to (\text{region1}) \\ \overline{e}_x \cos(k_{mz}) \Big/ \cos(k_{mz} l_D / 2) \to (\text{region2}) \end{cases}, \qquad (3\text{-}10)$$

For even TM modes

$$\overline{F}_m = \begin{cases} (\overline{e}_x i \cos\theta + \overline{e}_z \sin\theta) \exp[-q(z - l_D/2)] \to (\text{region1}) \\ \overline{e}_x i \cos\theta \sin(k_{mz} z)/S + \overline{e}_z (n/n_D)^2 \sin\theta \cos(k_{mz} z)/C \to (\text{region2}) \end{cases},$$

(3-11)

where $C = \cos(k_{mz} l_D / 2)$, $S = \sin(k_{mz} l_D / 2)$, and q is the decay constant in region 1. For odd modes, the $\cos(k_{mz} z)$ and $\sin(k_{mz} z)$ (also C and S) in region 2 are interchanged.

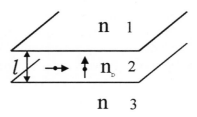

Fig. 3.1. The schematic diagram of a photonic well.

Using the relation $k = \sqrt{k_{mxy}^2 + k_{mz}^2}$ and $k = \omega_A n_D / c$, where k_{mz} is also a function of k_{mxy}, we can solve for $\dfrac{kdk}{k_{mxy}dk_{mxy}}$, which then enables us to replace $k_{mxy}dk_{mxy}$ by kdk. To express k_{mz} in terms of k_{mxy}, we use the pair of equations normally used in solving for the wave-guide dispersion relation. They are given for example by:[15] $q^2 = k^2(1-(n/n_D)^2)$, where $q = k_{mz}\tan(k_{mz}l_D/2)$ for odd TE modes. The result is

$$\frac{k_{mxy}dk_{mxy}}{kdk} = \frac{1-n_{12}^2}{C_1 C_2^2[1+C_3\gamma/(\cos\gamma\sin\gamma)]+n_{12}^2}, \tag{3-12}$$

where $n_{12} = n/n_D$, $C_1 = 1$ for TE modes, $C_1 = (n_{12})^4$ for TM modes. The values of C_2 and C_3 are dependant on whether the mode is odd or even. For even modes, $C_2 = \tan\gamma$, $C_3 = 1$. For odd modes, $C_2 = \cot\gamma$, $C_3 = -1$. By taking $k = \omega_A n_D/c$, and $dk/d\omega = n_D/c$, we can compute the value for γ_{gm}.

The spontaneous emission rate into the radiation modes is given by

$$\gamma_R = 2\int_0^{2\pi}d\phi_n \int_0^{\pi/2}d\theta_n \frac{2\pi}{\hbar^2}|\varepsilon_m^{FS}|^2|G_m|^2|\vec{e}_{m\sigma}\cdot\vec{e}_{dipole}|^2|\vec{\mu}|^2 \rho_n \sin\theta_n, \tag{3-13}$$

where

$$G_m = \frac{(R_{21}/T_{21})\exp(ik_{mz}l_D/2)}{R_{21}(T_{21}T_{12} - R_{21}R_{12}/T_{12})\exp(ik_{mz}l_D)+(R_{12}/T_{12}T_{21})\exp(-ik_{mz}l_D)}, \tag{3-14}$$

where $k_{mz} = 2\pi\cos(\theta_s)/(\lambda_{fs}/n_D)$, $\lambda_{fs} = 2\pi c/\omega_A$, $R_{21} = (h-g)/(g+h) = -R_{12}$, $T_{21} = 2h/(g+h) = T_{12}g/h$, and θ_n is the incident angle in the low-index region. Depending on mode polarization, g and h are given by $g = 2\pi n\cos\theta_n[TE] = 2\pi\cos\theta_n/n[TM]$ and $h = 2\pi n_D\cos\theta_s[TE] = 2\pi\cos\theta_s/n_D[TM]$. The angles θ_n and θ_s are related simply via the Snell's law of refraction, i.e. $\theta_s = \sin^{-1}(\sin(\theta_n)n/n_s)$ and $\phi_n = \phi_s$. We have labeled the low-refractive-index medium as region 1 and the high-refractive-index medium as region 2. R_{12} is the reflection coefficient for the wave in region 1 going into region 2.

Thin Medium Limit:

In the limit when $l_D \ll \lambda/n_D$, we can show that

$$\gamma_{hori} = \left(\frac{n}{n_D}\right)\times\gamma_\infty; \quad \gamma_{vert} = \left(\frac{n}{n_D}\right)^5\times\gamma_\infty. \tag{3-15}$$

This gives for n=1 and n_D=3.4: $\gamma_{hori} = 0.294\gamma_\infty$ and $\gamma_{vert} = 0.0022\gamma_\infty$, where γ_∞ is the decay rate in the bulk medium.

Numerical examples:

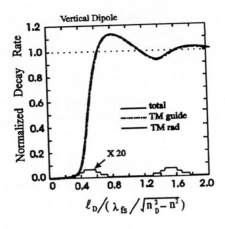

Fig. 3.2. Spontaneous emission rates for the case $n_D=3.4$ and $n=1.0$: The normalized spontaneous emission rate R_g(TM) and R_R(TM) (x20) of the vertical dipole vs. the normalized photonic well thickness from 0 to 2. The total normalized spontaneous emission rate R_{sp} of the vertical dipole is also given.

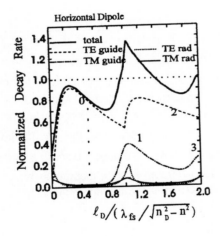

Fig. 3.3. Spontaneous emission rates for the case $n_D=3.4$ and $n=1.0$: The normalized spontaneous emission rate R_g(TE), R_g(TM), R_R(TE), and R_R(TM) of the horizontal dipole vs. the normalized photonic well thickness from 0 to 2. The total normalized spontaneous emission rate R_{sp} of the horizontal dipole is also given. At 0, the contribution is from TE0 mode; at 2 is from TE0 and TE2 modes; at 1 is from TM1 mode; at 3 from TM1 and TM3 modes.

The spontaneous emission rates of the horizontal and vertical dipoles in various cases are investigated as a function of the waveguide thickness. Fig. 3.2 shows the spontaneous emission rates of vertical dipole for the case of $n_D=3.4$ and $n=1.0$. We note that the vertical dipole does not emit into the TE modes or TM odd-order modes. The spontaneous emission rates into the TM even-order guided modes and TM radiation modes and the total spontaneous emission rate are shown by the dotted long-dashed line, the thin solid line and the thick solid line, respectively. The total spontaneous emission rate for the vertical dipole begins to reduce substantially at a waveguide thickness below $l_D \cong 0.5(\lambda_{fs}/\sqrt{n_D^2-n^2})$ (half-wave thick). It reaches the value of $(n/n_D)^5 \cong 0.002$ at $l_D = 0$. Note that in the region between $0 < l_D < 0.5(\lambda_{fs}/\sqrt{n_D^2-n^2})$, the vertical dipole emission is highly suppressed.

Fig. 3.3 shows the spontaneous emission rate of the horizontal dipole for the case of $n_D=3.4$ and $n=1.0$. We show the spontaneous emission rates into the TM guided modes (thin solid line), the TE guided modes (dotted line), the TM radiation modes (dotted long-dashed line) and TE radiation modes (short-dashed line) and the total spontaneous emission rate. In this figure, we see that the total spontaneous emission rate for the horizontal dipole begins to reduce substantially at a waveguide thickness below $l_D \cong 0.25(\lambda_{fs}/\sqrt{n_D^2-n^2})$ (quarter-wave thick). It reaches the value of $(n/n_D) \cong 0.3$ at $l_D = 0$. Note that in the region between

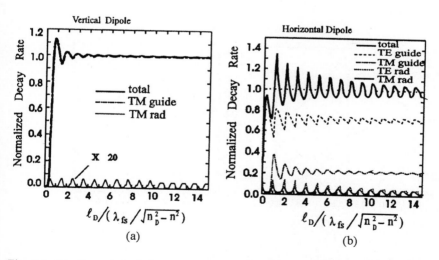

Fig. 3.4. Spontaneous emission rates for the case $n_D=3.4$ and $n=1.0$: (a) The normalized spontaneous emission rate R_g(TM) and R_R(TM) (x20) of the vertical dipole vs. the normalized photonic well thickness from 0 to 15. The total normalized spontaneous emission rate R_{sp} of the vertical dipole is also given. (b) The normalized spontaneous emission rate R_g(TE), R_g(TM), R_R(TE), and R_R(TM) of the horizontal dipole vs. the normalized photonic well thickness from 0 to 15. The total normalized spontaneous emission rate R_{sp} of the horizontal dipole is also given.

$0.15(\lambda_{fs}/\sqrt{n_D^2-n^2}) < l_D < 0.5(\lambda_{fs}/\sqrt{n_D^2-n^2})$, a high percentage (>95%) of the emission goes into the lowest-order TE guided mode. Combined with the result for the vertical dipole, we see that even for a randomly oriented dipole, a high percentage of the emission can be made to go into the lowest-order TE mode with $0.15(\lambda_{fs}/\sqrt{n_D^2-n^2}) < l_D < 0.5(\lambda_{fs}/\sqrt{n_D^2-n^2})$. Fig. 3.4(a) and 3.4(b) respectively show the spontaneous emission rates into the vertical dipole and the horizontal dipole, respectively, in the range of the normalized thickness from 0 to 15.

Fig. 3.5 shows the spontaneous emission rate of the vertical and horizontal dipoles for the case where $n_D=100$. In Fig. 3.5(a), the spontaneous emission rate from the horizontal dipole into the TE guided modes is given as the short-dashed line and the TM guided mode as the dotted long-dashed line. The spontaneous emission rate into the radiation modes is too small to be shown in the figure. The total spontaneous emission rate is shown as the solid line. In Fig 3.5(b), the spontaneous emission rate of the vertical dipole into the TM even-order guided modes is given. It overlaps with the curve for the total spontaneous emission rate, as the spontaneous emission rate into the radiation mode is too small to be seen. From Fig. 3.5, we see that in the limit of high refractive index, the emission into the radiation modes is negligible. Otherwise its behavior is similar to the $n_D=3.4$ case, in that we can make the spontaneous emission go into mainly the lowest-order TE guided mode with $0.15(\lambda_{fs}/\sqrt{n_D^2-n^2}) < l_D < 0.5(\lambda_{fs}/\sqrt{n_D^2-n^2})$, and the emission of the vertical dipole is nearly completely suppressed in that region.

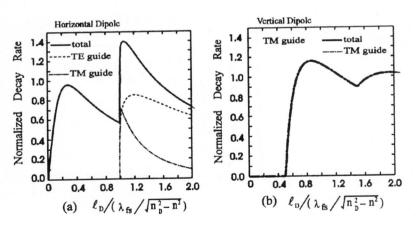

Fig. 3.5. Spontaneous emission rates for the case $n_D=100$ and $n=1.0$: (a) The normalized spontaneous emission rate $R_g(TE)$ and $R_g(TM)$ of the horizontal dipole vs. the normalized photonic well thickness from 0 to 2. The total normalized spontaneous emission rate R_{sp} of the horizontal dipole is also given. The value $R_R(TE)$ and $R_R(TM)$ is too small to be observed; (b) The normalized spontaneous emission rate $R_g(TM)$ of the vertical dipole vs. the normalized photonic well thickness from 0 to 2. The value $R_R(TM)$ is too small to be observed.

3.3 Comparison between dielectric and metallic planar waveguides

It is interesting to compare modification of spontaneous emission in dielectric and metallic waveguides. The main difference between the metallic and dielectric cases is that the metallic structures tend to achieve enhancement (i.e. increase) in the spontaneous emission rates while the dielectric structures tend to achieve suppression in the spontaneous emission rates and cannot provide much enhancement. This difference can be traced to the difference in boundary conditions: the field changes sign when bounced from a metallic surface while it does not change sign when bounced from a dielectric interface if the field is incident from the high-refractive-index medium side. For this reason, a lossy planar metallic waveguide should really be compared with a leaky dielectric waveguide with the middle layer having a lower refractive index.

From another point of view, we see that when the middle layer of a dielectric structure has lower refractive index, the field in it will bounce between the two interface boundaries and gradually leak out. This is just what happens in a lossy metallic waveguide. On the other hand, with a high refractive index guiding layer, radiation modes can leak out as in a lossy metallic waveguide, but guided modes propagate without any attenuation. There is no counterpart for such a mode behavior in the case of a metallic waveguide.

We have calculated spontaneous emission rate for both a lossy metallic waveguide and a leaky dielectric waveguide under comparable conditions. The results are shown in Fig. 3.6. Evidently they agree with each other quite well when the width $w > 0.5\lambda$. When the width gets smaller than half a wavelength, we see obvious differences between the two cases. This can be explained if we notice that when the dipole is close to the interfaces, it can sense evanescent fields in addition to what can propagate into the middle layer in the first place (i.e. the fields that are allowed by Snell's law). Thus the total vacuum fields become stronger, resulting in stronger spontaneous emission.

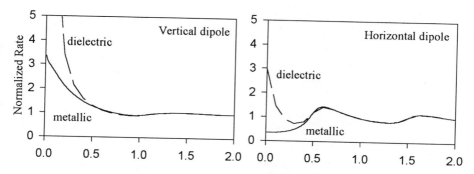

Fig. 3.6. Normalized spontaneous emission rate vs. w/λ. The solid lines are for planar metallic waveguide with $r = 0.5$, and the broken lines are for a dielectric structure with an $n = 1$ layer sandwiched between $n = 3$ layers. The dipole is at the center in both cases.

3.4 Spontaneous Emission Rate in Dielectric Photonic Wire

The modification of the spontaneous emission rate in the dielectric photonic wire structure is investigated by using the rectangular waveguide with width a and height b as shown in Fig. 3.7. The high-refractive-index waveguiding region has a refractive index n_g, and the surrounding cladding region has a refractive index n_s. The excitons are located at the center plane inside the waveguide, respectively. The excitons are modeled as three independent dipoles oscillating along the x, y, and z-axis. The spontaneous emission rate into mode m is given by

$$\gamma_{gm} = \frac{2\pi}{\hbar^2} \xi_m^2 \left| \vec{e}_{m\sigma} \cdot \vec{e}_{dipole}(x,y,z) \right|^2 |\vec{\mu}|^2 \frac{L_z}{2\pi} \frac{dk_{mz}}{d\omega}, \quad (3\text{-}16)$$

where
$$\xi_m^2 = \frac{\hbar \omega_A \left| \vec{F}_m(\vec{r}_{dipole}) \right|^2}{2\varepsilon(\vec{r}_{peak}) L_z A_{mode}}$$

with A_{mode} being the effective mode area.

In order to calculate accurate solutions in strongly guiding waveguides, we cannot use the usual effective index methods for weakly guiding waveguides and have to develop a numerically more rigorous method. For that, we used the FDM (finite difference method) to calculate the full-vectorial mode fields and the propagation constants.[16,17,18] The wave equations for the full-vectorial mode can be derived by taking the curl of Maxwell's equations and using the vector identity.

$$\nabla^2 E + \omega^2 \varepsilon \mu_0 E = \nabla(\nabla \cdot E) \quad (3\text{-}17)$$

The longitudinal field component is decoupled from the transverse components. Therefore, the modal fields can be expressed in the form.

$$E = (E_t + E_z)\exp(-i\beta z) \quad (3\text{-}18)$$

Substituting (3-18) into (3-17), we can obtain the vector wave equations for the transverse fields.

$$\frac{\partial^2 E_x}{\partial y^2} + (\omega^2 \varepsilon \mu_0 - \beta^2) E_x = \frac{\partial^2 E_y}{\partial x \partial y} - i\beta \frac{\partial E_z}{\partial x}$$

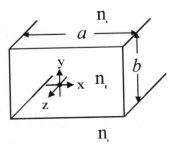

Fig. 3.7. The schematic diagram of a photonic wire.

$$\frac{\partial^2 E_y}{\partial x^2} + (\omega^2 \varepsilon \mu_0 - \beta^2) E_y = \frac{\partial^2 E_x}{\partial y \partial x} - i\beta \frac{\partial E_z}{\partial y} \qquad (3\text{-}19)$$

From the divergence relation $\nabla \cdot (\varepsilon E) = 0$, E_z can be expressed as

$$i\beta E_z \varepsilon = \frac{\partial(\varepsilon E_x)}{\partial x} + \frac{\partial(\varepsilon E_y)}{\partial y} \qquad (3\text{-}20)$$

Substituting (3-20) into (3-19), we obtain the full-vectorial wave equations that can be easily discretized.

$$\frac{\partial}{\partial x}\left[\frac{1}{\varepsilon}\frac{\partial(\varepsilon E_x)}{\partial x}\right] + \frac{\partial^2 E_x}{\partial y^2} + \omega^2 \varepsilon \mu_0 E_x + \frac{\partial}{\partial x}\left[\frac{1}{\varepsilon}\frac{\partial(\varepsilon E_y)}{\partial y}\right] - \frac{\partial^2 E_y}{\partial x \partial y} = \beta^2 E_x \qquad (3\text{-}21a)$$

$$\frac{\partial}{\partial y}\left[\frac{1}{\varepsilon}\frac{\partial(\varepsilon E_y)}{\partial y}\right] + \frac{\partial^2 E_y}{\partial x^2} + \omega^2 \varepsilon \mu_0 E_y + \frac{\partial}{\partial y}\left[\frac{1}{\varepsilon}\frac{\partial(\varepsilon E_x)}{\partial x}\right] - \frac{\partial^2 E_x}{\partial y \partial x} = \beta^2 E_y \qquad (3\text{-}21b)$$

If we know E_x and E_y, then E_z can be obtained using the relation in Eq. (3-20).

Numerical method for discretization:

We can discretize the components of Eq. (3-21) based on the finite difference scheme[19] either using an uniform or a nonuniform mesh. The nonuniform mesh with a dense grid around the dielectric boundary reduces computing time and achieves desirable accuracy. However, the uniform mesh gives rise to severe penalty in the computing time and memory for the same accuracy. After direct discretization based on the following figure (Fig. 3.8), the finite difference equations are given below. Fig. 3.8 shows the meaning of $\Delta x(m)$, $\Delta x(m-1)$, $\Delta y(n)$ and $\Delta y(n-1)$ defined for the 9 adjacent grid points in the mesh. For the nonuniform mesh, we have $\Delta x(m) \neq \Delta x(m-1)$ and $\Delta y(n) \neq \Delta y(n-1)$ in general. First the various terms in Eq. (3-21a) which is the eigenvalue equation for E_x are discretized as shown by Eqs. (3-22), (3-23) and (3-24).

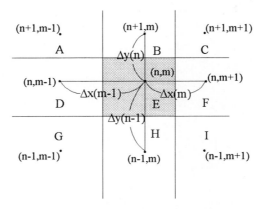

Fig. 3.8. The finite difference scheme with a nonuniform mesh.

1. $\dfrac{\partial}{\partial x}\left[\dfrac{1}{\varepsilon}\dfrac{\partial(\varepsilon E_x)}{\partial x}\right]$

$$\dfrac{\partial}{\partial x}\left[\dfrac{1}{\varepsilon}\dfrac{\partial(\varepsilon E_x)}{\partial x}\right]_{(n,m)} = \dfrac{2}{\Delta x(m-1)+\Delta x(m)} \times$$

$$\left[\dfrac{2}{\varepsilon(n,m+1)+\varepsilon(n,m)}\left(\dfrac{\varepsilon(n,m+1)E_x(n,m+1)-\varepsilon(n,m)E_x(n,m)}{\Delta x(m)}\right) - \dfrac{2}{\varepsilon(n,m-1)+\varepsilon(n,m)}\left(\dfrac{\varepsilon(n,m)E_x(n,m)-\varepsilon(n,m-1)E_x(n,m-1)}{\Delta x(m-1)}\right)\right]$$

$$= \dfrac{2}{(\Delta x(m)+\Delta x(m-1))\Delta x(m)}\dfrac{2\varepsilon(n,m+1)}{\varepsilon(n,m+1)+\varepsilon(n,m)}E_x(n,m+1) -$$

$$\left[\dfrac{2}{(\Delta x(m)+\Delta x(m-1))\Delta x(m)}\dfrac{2\varepsilon(n,m)}{\varepsilon(n,m+1)+\varepsilon(n,m)} + \dfrac{2}{(\Delta x(m)+\Delta x(m-1))\Delta x(m-1)}\dfrac{2\varepsilon(n,m)}{\varepsilon(n,m-1)+\varepsilon(n,m)}\right]E_x(n,m) +$$

$$\dfrac{2}{(\Delta x(m)+\Delta x(m-1))\Delta x(m-1)}\dfrac{2\varepsilon(n,m-1)}{\varepsilon(n,m-1)+\varepsilon(n,m)}E_x(n,m-1) \quad (3\text{-}22)$$

2. $\dfrac{\partial^2 E_x}{\partial y^2}$

$$\left(\dfrac{\partial^2 E_x}{\partial y^2}\right)_{(n,m)} = \dfrac{2E_x(n+1,m)}{\Delta y(n)(\Delta y(n)+\Delta y(n-1))} - \dfrac{2E_x(n,m)}{\Delta y(n)\Delta y(n-1)} + \dfrac{2E_x(n-1,m)}{\Delta y(n-1)(\Delta y(n)+\Delta y(n-1))} \quad (3\text{-}23)$$

3. $\dfrac{\partial}{\partial x}\left[\dfrac{1}{\varepsilon}\dfrac{\partial(\varepsilon E_y)}{\partial y}\right] - \dfrac{\partial^2 E_y}{\partial x \partial y}$

$$\left(\frac{\partial}{\partial x}\left[\frac{1}{\varepsilon}\frac{\partial(\varepsilon E_y)}{\partial y}\right]-\frac{\partial^2 E_y}{\partial x \partial y}\right)_{(n,m)} = \frac{1}{\Delta x(m)+\Delta x(m-1)} \times$$

$$\left[\frac{1}{\varepsilon(n,m+1)}\left(\frac{\varepsilon(n+1,m+1)E_y(n+1,m+1)-\varepsilon(n-1,m+1)E_y(n-1,m+1)}{\Delta y(n)+\Delta y(n-1)}\right)-\right.$$

$$\left.\frac{1}{\varepsilon(n,m-1)}\left(\frac{\varepsilon(n+1,m-1)E_y(n+1,m-1)-\varepsilon(n-1,m-1)E_y(n-1,m-1)}{\Delta y(n)+\Delta y(n-1)}\right)\right]-$$

$$\frac{1}{\Delta x(m)+\Delta x(m-1)}\left[\left(\frac{E_y(n+1,m+1)-E_y(n-1,m+1)}{\Delta y(n)+\Delta y(n-1)}\right)-\right.$$

$$\left.\left(\frac{E_y(n+1,m-1)-E_y(n-1,m-1)}{\Delta y(n)+\Delta y(n-1)}\right)\right]$$

$$= \frac{1}{(\Delta x(m)+\Delta x(m-1))(\Delta y(n)+\Delta y(n-1))}\left(\frac{\varepsilon(n+1,m+1)}{\varepsilon(n,m+1)}-1\right)E_y(n+1,m+1)-$$

$$\frac{1}{(\Delta x(m)+\Delta x(m-1))(\Delta y(n)+\Delta y(n-1))}\left(\frac{\varepsilon(n-1,m+1)}{\varepsilon(n,m+1)}-1\right)E_y(n-1,m+1)-$$

$$\frac{1}{(\Delta x(m)+\Delta x(m-1))(\Delta y(n)+\Delta y(n-1))}\left(\frac{\varepsilon(n+1,m-1)}{\varepsilon(n,m-1)}-1\right)E_y(n+1,m-1)+$$

$$\frac{1}{(\Delta x(m)+\Delta x(m-1))(\Delta y(n)+\Delta y(n-1))}\left(\frac{\varepsilon(n-1,m-1)}{\varepsilon(n,m-1)}-1\right)E_y(n-1,m-1)$$

(3-24)

The various terms in Eq. (3-21b) which is the eigen-value equation for E_y are discretized as shown by Eqs. (3-25), (3-26) and (3-27).

4. $\quad\dfrac{\partial}{\partial y}\left[\dfrac{1}{\varepsilon}\dfrac{\partial(\varepsilon E_y)}{\partial y}\right]$

$$\frac{\partial}{\partial y}\left[\frac{1}{\varepsilon}\frac{\partial(\varepsilon E_y)}{\partial y}\right]_{(n,m)} = \frac{2}{\Delta y(n-1)+\Delta y(n)} \times$$

$$\left[\frac{2}{\varepsilon(n+1,m)+\varepsilon(n,m)}\left(\frac{\varepsilon(n+1,m)E_y(n+1,m)-\varepsilon(n,m)E_y(n,m)}{\Delta y(n)}\right)-\right.$$

$$\left.\frac{2}{\varepsilon(n-1,m)+\varepsilon(n,m)}\left(\frac{\varepsilon(n,m)E_y(n,m)-\varepsilon(n-1,m)E_y(n-1,m)}{\Delta y(n-1)}\right)\right]$$

$$= \frac{2}{(\Delta y(n)+\Delta y(n-1))\Delta y(n)} \frac{2\varepsilon(n+1,m)}{\varepsilon(n+1,m)+\varepsilon(n,m)} E_y(n+1,m) -$$

$$\left[\frac{2}{(\Delta y(n)+\Delta y(n-1))\Delta y(n)} \frac{2\varepsilon(n,m)}{\varepsilon(n+1,m)+\varepsilon(n,m)} + \right.$$

$$\left. \frac{2}{(\Delta y(n)+\Delta y(n-1))\Delta y(n-1)} \frac{2\varepsilon(n,m)}{\varepsilon(n-1,m)+\varepsilon(n,m)} \right] E_y(n,m) +$$

$$\frac{2}{(\Delta y(n)+\Delta y(n-1))\Delta y(n-1)} \frac{2\varepsilon(n-1,m)}{\varepsilon(n-1,m)+\varepsilon(n,m)} E_y(n-1,m)$$

(3-25)

5. $\dfrac{\partial^2 E_y}{\partial x^2}$

$$\left(\frac{\partial^2 E_y}{\partial x^2}\right)_{(n,m)} = \frac{2E_y(n,m+1)}{\Delta x(m)(\Delta x(m)+\Delta x(m-1))} - \frac{2E_y(n,m)}{\Delta x(m)\Delta x(m-1)} +$$

$$\frac{2E_y(n,m-1)}{\Delta x(m-1)(\Delta x(m)+\Delta x(m-1))}$$

(3-26)

6. $\dfrac{\partial}{\partial y}\left[\dfrac{1}{\varepsilon}\dfrac{\partial(\varepsilon E_x)}{\partial x}\right] - \dfrac{\partial^2 E_x}{\partial y \partial x}$

$$\left(\frac{\partial}{\partial y}\left[\frac{1}{\varepsilon}\frac{\partial(\varepsilon E_x)}{\partial x}\right] - \frac{\partial^2 E_x}{\partial y \partial x}\right)_{(n,m)} = \frac{1}{\Delta y(n)+\Delta y(n-1)} \times$$

$$\left[\frac{1}{\varepsilon(n+1,m)}\left(\frac{\varepsilon(n+1,m+1)E_x(n+1,m+1)-\varepsilon(n+1,m-1)E_x(n+1,m-1)}{\Delta x(m)+\Delta x(m-1)}\right) - \right.$$

$$\left. \frac{1}{\varepsilon(n-1,m)}\left(\frac{\varepsilon(n-1,m+1)E_x(n-1,m+1)-\varepsilon(n-1,m-1)E_x(n-1,m-1)}{\Delta x(m)+\Delta x(m-1)}\right) \right] -$$

$$\frac{1}{\Delta y(n)+\Delta y(n-1)}\left[\left(\frac{E_x(n+1,m+1)-E_x(n+1,m-1)}{\Delta x(m)+\Delta x(m-1)}\right) - \right.$$

$$\left. \left(\frac{E_x(n-1,m+1)-E_x(n-1,m-1)}{\Delta x(m)+\Delta x(m-1)}\right)\right]$$

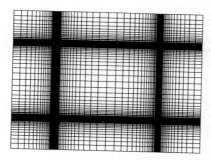

Fig. 3.9. The nonuniform grid used to calculate the modes in the rectangular waveguide.

$$
\begin{aligned}
= & \frac{1}{(\Delta y(n)+\Delta y(n-1))(\Delta x(m)+\Delta x(m-1))}\left(\frac{\varepsilon(n+1,m+1)}{\varepsilon(n+1,m)}-1\right)E_x(n+1,m+1) \\
& -\frac{1}{(\Delta y(n)+\Delta y(n-1))(\Delta x(m)+\Delta x(m-1))}\left(\frac{\varepsilon(n+1,m-1)}{\varepsilon(n+1,m)}-1\right)E_x(n+1,m-1) \\
& -\frac{1}{(\Delta y(n)+\Delta y(n-1))(\Delta x(m)+\Delta x(m-1))}\left(\frac{\varepsilon(n-1,m+1)}{\varepsilon(n-1,m)}-1\right)E_x(n-1,m+1) \\
& +\frac{1}{(\Delta y(n)+\Delta y(n-1))(\Delta x(m)+\Delta x(m-1))}\left(\frac{\varepsilon(n-1,m-1)}{\varepsilon(n-1,m)}-1\right)E_x(n-1,m-1)
\end{aligned}
$$

(3-27)

We used two numerical methods to solve the discretized equations; the alternating direction-implicit (ADI) iterative method and the shifted inverse power method.[19,20] The shifted inverse power method is simpler for solving an eigenvalue problem. The following results were obtained by using the shifted inverse power method. By writing the two finite difference equations at each node point of the calculation window, the following eigenvalue matrix equation is obtained.

$$AE=\beta^2 E, \tag{3-28}$$

where A is asymmetric banded matrix and E is an eigenvector as the following.

$$E=[E_x(1), E_y(1), E_x(2), E_y(2), \cdots\cdots, E_x(N), E_y(N)], \tag{3-29}$$

where N is a total number of nodes. The mode calculation of the rectangular waveguide is carried out on the nonuniform mesh shown schematically in Fig. 3.9, where the mesh size closer to the dielectric boundary is continuously decreased, so that the smallest size at the boundary is more than 1000 times smaller than the mesh size at the waveguide center.

Numerical Examples:

The calculated rectangular waveguides are made up of a high-refractive-index semiconductor core ($n_g=3.4$) surrounded by a low refractive-index cladding ($n_s=1$). The normalized spontaneous emission rates into the guided modes at a/b aspect ratio 2 are shown in Fig. 3.10, where the normalized spontaneous emission rates are expressed as a function of the normalized width of the rectangular waveguide given by a/λ_n, where $\lambda_n = \lambda / \sqrt{n_g^2 - n_s^2}$ with n_g being the refractive index of the guiding region and n_s being the refractive index of the cladding region surrounding the waveguide. The figure shows the normalized spontaneous emission rates of the dominant five modes TE_{00}, TM_{00}, TE_{10}, TM_{10} and TE_{20} in the range of the normalized width shown. The emission rate is the total emission into both the +z and −z propagating directions for each mode. The two largest curves show the normalized spontaneous emission rates from the x-dipole and z-dipole, respectively, emitting into the TE_{00} mode. Their magnitude at the leading-edge is close to 1.2. We see that the y-dipole does not emit into the lowest-order TE_{00} mode. In this figure, we can see that there is a range of optimum waveguide width that gives a single or nearly single mode guiding and high spontaneous emission rate into the lowest-order TE_{00} mode. We define the upper limit of the range of optimum waveguide width as when the width becomes large enough that the emission into other modes is half of the total emission into the TE_{00} mode. (Note that there are two dipoles (x and z) that emit into the TE_{00} mode. Hence this width is at a point when the emission from the x-dipole into TE_{00} mode is about equal to the emission from the y-dipole into the TM_{00}.) We define the lower limit of that as the width when the spontaneous emission rate into the TE_{00} mode drops to half of its maximum value. The optimum waveguide width for the single-mode guiding at the aspect ratio 2 is ranged from 0.32 μm to 0.51 μm. Fig. 3.11 shows the total spontaneous emission

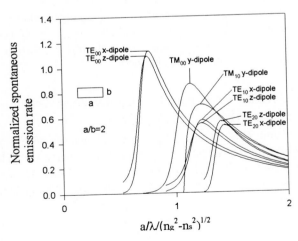

Fig. 3.10. Normalized spontaneous emission rates into TE_{00}, TM_{00}, TE_{10}, TM_{10} and TE_{20} at a/b=2.

rate from the x-dipole. We see that if the normalized width is larger than 1, then the total spontaneous emission rate will oscillate around the value of the bulk medium.

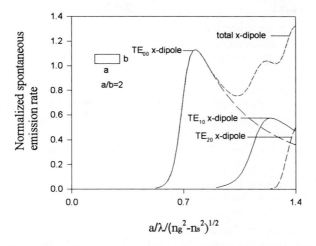

Fig. 3.11. Normalized spontaneous emission rates from x-dipole into TE_{00}, TE_{10} and TE_{20} at a/b= 2. The total spontaneous emission rate from x-dipole is also given.

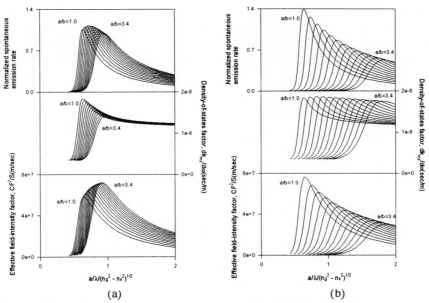

Fig. 3.12. Normalized spontaneous emission rates from x-dipole into TM_{00} mode (a) and y-dipole into TM_{00} mode (b) as aspect ratio is changed from 1.0 to 3.4. The increment of the aspect ratio is 0.2. The normalized spontaneous emission rates are decomposed into the density-of-states factor and the effective-field-intensity factor.

Fig. 3.12 (a) and (b) respectively shows the case of x-dipole emitting into the TE_{00} mode and y-dipole emitting into the TM_{00} mode. Here, the emission rates from z-dipole into the TE_{00} mode are not shown because they are proportional to those from x-dipole. We see that the aspect ratio 1.6 gives the maximum spontaneous emission rate from the x-dipole into the TE_{00} mode, and the aspect ratio 1.0 gives the maximum spontaneous emission rate from the y-dipole into the TM_{00} mode. The spontaneous emission rate curves are decomposed into two factors, the effective field-intensity factor and the density-of-states factor, as shown in Fig. 3.12 (a) and (b), respectively. The characteristics of the maximum point change can be explained by a careful comparison of the effective field-intensity factor and the density-of-states factor. In the case of the spontaneous emission rates from the x-dipole into the TE_{00} mode, the maximum point of the spontaneous emission rate is increased and then decreased as shown by changing the a/b ratio from 1.0 to 3.4. In contrast, the maximum point of the normalized spontaneous curves of the TM_{00} mode constantly decreases with an increase in the a/b ratio as shown in Fig 3.12 (b). We can say that the normalized spontaneous emission rate from the y-dipole into the TM_{00} mode is rapidly suppressed as the aspect ratio increases. In addition to the suppression, the maximum point shifts more toward higher normalized width compared with the shift of the maximum point in TE_{00} mode. As a result of the characteristic change of the maximum points of TE_{00} and TM_{00} modes, increasing the aspect ratio until TM_{00} mode remains the second mode increases the ranges of optimum waveguide width defined above.

Fig. 3.13 shows the increase in the optimum range by comparing the two aspect ratios 2 and 3. Even though the maximum point of the TE_{00} curve at the aspect ratio 3 becomes smaller, we see that the optimum range increases. Therefore, when we consider both the magnitude of maximum spontaneous emission rate and the range of optimum waveguide width, the aspect ratio 2 can be considered as a reasonable optimum width even though the aspect ratio 1.6 gives the highest normalized spontaneous

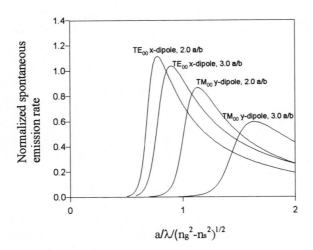

Fig. 3.13. Normalized spontaneous emission rates from x-dipole into TE_{00} mode and y-dipole into TM_{00} mode at a/b=2.0 and 3.0 respectively.

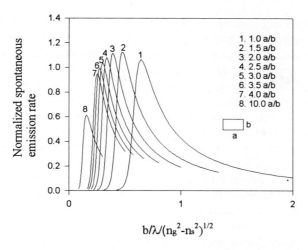

Fig. 3.14. Normalized spontaneous emission rates from x-dipole into TE_{00} mode as functions of normalized height.

emission rate from the x-dipole into the TE_{00} mode.

Fig. 3.14 shows the spontaneous emission rates from the x-dipole into the TE_{00} mode as functions of the normalized height. When the aspect ratio is larger than 10, the emission from the x-dipole approaches that of planar waveguide. The peak value of number 8 curve is smaller than that of the planar waveguide because this emission includes only one mode. However, the peak location is around that of the planar waveguide comparing with Fig. 3.3.

3.5 Realization of Dielectric Photonic-Wire Lasers

Using an appropriate waveguide design, we can take advantage of the modification of spontaneous emission in manipulating laser properties such as the lasing thresholds, the population inversion, the laser modulation, and the carrier dynamics. Specifically in determining the lasing threshold, the fraction of spontaneous emission channeled into the lasing mode called the spontaneous-emission coupling efficiency β is an important factor. A large β value can increase the effective gain of the cavity, making it possible to achieve lasing in a small cavity and to attain low lasing threshold.[21,22,23,24,25] To increase the β value, we need to make a strongly guiding laser-waveguide, which decreases the amount of the emission rates into the radiation modes. This can be accomplished by increasing the refractive index difference between the waveguide core and the cladding.

For semiconductor lasers, the waveguides can be made up of a high-refractive-index semiconductor core (n_g=3.4) surrounded by a low-refractive-index cladding (n_s=1 or 1.5). Compared with the conventional laser cavity, the waveguide structure gives higher optical confinement and lower emission rates into the radiation modes. In addition to the reduction of the emission rates into the radiation modes, the normalized spontaneous emission rates into the guided modes will also increase with increasing refractive index difference. This is because a reduction in the single mode waveguide dimension leads to an enhancement in the mode effective field-intensity factor. Therefore, the use of the strongly guiding waveguides is a major part to a large β value. With the refractive index difference, we can optimize the waveguide structure by varying the aspect ratio. In the last section, we show the optimum width for the single mode guiding in the semiconductor waveguide (n_g=3.4) surrounded by a low refractive-index air cladding (n_s=1). At the optimum aspect ratio of 2, the optimum width would be between 0.32 μm to 0.51 μm as shown in Fig. 3.10. The maximum spontaneous emission rate into the lowest frequency mode is obtained at 0.38 μm waveguide width. We can show that the optimum width is not much different for the case of SiO_2 cladding (n_s=1.5). Using the emission rates into the radiation modes calculated for cylindrical waveguides,[26] we can estimate the β value at 0.19 μm height and 0.38 μm width. From our estimation, for a ring cavity with bi-directional lasing into the lowest-order TE_{00} mode, the β value can reach around 0.35 (70% divided by two directions). This is much higher than the β value of the usual semiconductor laser cavities, typically with $β \cong 10^{-5}$. It is also higher than the microdisk lasers with $β \cong$ 0.1, which can be achieved only with a disk diameter smaller than 3 μm. We call such a tiny ring laser that gives high β value a photonic-wire laser.

Based on the calculation of spontaneous emission rate, we realized the photonic-wire laser using InGaAsP/InGaAs epitaxial layers grown by molecular beam epitaxy on an InP substrate. The epitaxial layers form a 0.19 μm thick InGaAsP/InGaAs laser structure. Within the structure, three 10 nm quantum wells ($In_{0.53}Ga_{0.47}As$) were separated by 100 nm barriers ($In_{0.84}Ga_{0.16}As_{0.33}P_{0.67}$). They were sandwiched by two 70 nm $In_{0.84}Ga_{0.16}As_{0.33}P_{0.67}$ layers on both sides.

Fig. 3.15. Scanning electron microscope image of a 4.5 μm diameter photonic-wire ring laser with 0.4 μm waveguide.

To fabricate the laser, we used nanofabrication techniques involving electron-beam (E-beam) lithography, reactive ion etching (RIE), plasma-enhanced chemical vapor deposition (PECVD) and bonding-etching techniques. First, a 80 nm thick SiO_2 film is deposited on the wafer via plasma-enhanced chemical-vapor deposition (PECVD). E-beam lithography was used to write the ring laser pattern on PMMA coated on the SiO_2 layer. Subsequently RIE with CHF_3 etchant gas was used to transfer the patterns down to the SiO_2 film, and the PMMA was removed. RIE with a gas mixture of methane, hydrogen, and argon was then used to etch the rings down vertically into the InP substrate. To place the thin ring laser structure on a low-refractive-index material, the substrate was removed via the following technique. The RIE etched chip is deposited with 0.75 μm thick SiO_2 using PECVD. A piece of GaAs substrate covered with 0.75 μm thick SiO_2 was then prepared using PECVD. The two samples were SiO_2 face-to-face bonded together with acrylic. Finally, highly selective etchant (1:1, $HCl:H_3PO_4$) was used to remove the InP substrate. Fig. 3.15 shows the photonic-wire ring laser fabricated with 4.5 μm ring diameter and 0.4 μm waveguide width (the waveguide thickness is 0.19 μm).[27]

The photonic-wire ring lasers fabricated were optically pumped with a 514 nm argon-ion laser modulated with 1% duty cycle in a vacuum chamber at 85 K. The pump light was focused to a spot size covering the whole ring laser cavity. The scattered light from the laser structure was collected by an objective lens. The emission spectrum is detected and measured with a spectrometer, a liquid-nitrogen-cooled germanium detector, and a lock-in amplifier. The typical emission spectra of the photonic-wire laser with a waveguide width of 0.4 μm are shown in Fig. 3.16, indicating lasing

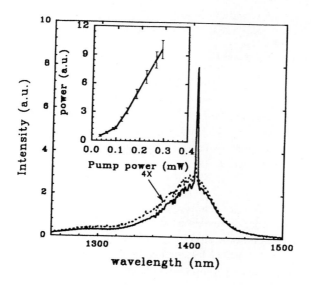

Fig. 3.16. Spectra of a 4.5 μm photonic-wire ring laser with 0.4 μm waveguide. The solid line and the dashed line were measured above 1.5 threshold and near threshold, respectively. Inset shows the measured lasing power as a function of peak pump power.

at 1403 nm. The dashed curve is the spectrum at around threshold where the peak pump power absorbed by the laser is approximately 95 mW. The lasing power as a function the peak pump power is shown in the inset of Fig. 3.16. Its spectral linewidth at 1.5 threshold was measured to be about 0.5 nm with a spectrum analyzer resolution of 0.1 nm. The cavity volume of this laser is approximately 0.27 μm^3, which was among the smallest semiconductor laser cavity ever realized. If the ring diameter is reduced further, it is possible to achieve a cavity volume of lass than 0.1 μm^3 at an optical waveguide of 1.5 μm wavelength.

In order to obtain light output, we have fabricated a waveguide adjacent to the ring laser as shown in Fig. 3.17(a), where the ring laser has 10 μm diameter and 0.45 μm width. Output light at the ends of the U-shape waveguide is imaged using a infrared camera as shown in Fig. 3.17(b). We can see the two bright light spots scattered from the two ends of the U-shape waveguide.[28]

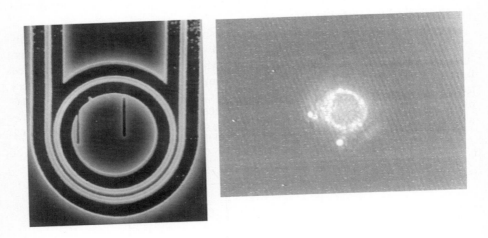

(a)   (b)

Fig. 3.17. (a) Scanning electron microscope image of a waveguide coupled photonic-wire ring laser fabricated with a 10-μm ring diameter. (b) The infrared image of a photonic-wire ring laser at 1.5 threshold pump power. We can see a faint ring pattern and two bright emitting spots at the ends of the U-shape waveguide.

Another type of photonic-wire lasers can be realized by using a linear cavity as depicted in Fig. 3.18. In order to realize a microcavity cavity, strong feedback can be achieved by using a 1-D photonic structures to form the cavity mirrors. The 1-D photonic bandgap structure is formed by etching tiny periodic holes through a 0.8 μm thick and 0.45 μm wide beam of semiconductor suspended in air.[29]

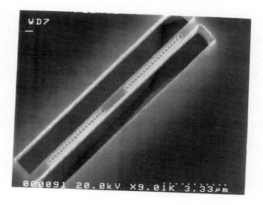

Fig. 3.18. Scanning electron microscope image of a linear photonic wire cavity.

3.6 Smallest Possible Dielectric Cavity?

The smallest dielectric microcavity may be realized with above-discussed dielectric channel waveguide having a length forming a half-wavelength microcavity. This length, denoted as ℓ_c, would be dependent on the velocity of propagation for the guided mode. As discussed above, in order to achieve a large β value, it is desirable to have $a \approx 0.7\lambda/\sqrt{n_g^2 - n_s^2}$ and $b = a/2$. To form a half-wavelength microcavity, we must have $\ell_c = \lambda/(2n_{\text{eff}})$, where n_{eff} is the propagating constant for the guided mode. For $n_g = 3.4$, $n_s = 1$, and $\lambda = 1.5 \mu m$ we will have $a = 0.323 \mu m$ and $b = 0.162 \mu m$. In this case, the propagation constant is given by: $n_{\text{eff}} = 1.64$. The physical volume of the cavity would then be given by $V_{cav}^{physical} = 0.7 \text{x} 0.35 \lambda^3 /(2n_{\text{eff}}(n_g^2 - n_s^2))$. The sinusoidal variation of the mode will give a mode width approximately equal to half the physical width (i.e. half its zero point width). It turns out that the dielectric guided mode will have sinusoidal variation along both side "a" and "b". There will also be sinusoidal field amplitude variation along the cavity length. Hence the mode volume is given approximately by the cavity volume multiply by $(\frac{1}{2})^3$, giving $V_{cav}^{mode} = 0.7 \text{x} 0.35 \lambda^3 /(16 n_{\text{eff}}(n_g^2 - n_s^2))$. The smallest possible microcavity, it turns out, is achieved with $a = 0.323 \mu m$, $b = 0.162 \mu m$, $\ell_c = 0.457 \mu m$, and $V_{cav}^{mode} = 0.00298 \mu m^3$ (This value is close to that of a cavity formed with a cubic half-wavelength box for which: $V_{cubic\ cav}^{mode} = (\frac{\lambda}{2n})^3 * (\frac{1}{2})^3 \approx 0.00134 \mu m^3$). This is close to the metallic cavity case with filled with a medium with a refractive index of 3.4 for which we have $V_{cav}^{mode} = 0.0035 \mu m^3$. Note that we have assumed the "cavity mirrors" to be formed by dielectric-air interfaces, which are of course not very highly reflecting. The mirror reflectivity or the cavity Q can be increased by using photonic-bandgap structures or dielectric coatings to form mirrors. In that case, the effective cavity size will be larger. Another way to achieve high cavity Q is to form a photonic-wire ring cavity as described above. The smallest ring cavity that can be achieved for $\lambda = 1.5\mu m$ (with $n_g = 3.4$, $n_s = 1$), is with a ring diameter of $d = 1\mu m$ as radiation loss will become dominating at smaller than $1\mu m$ diameter. Taking the above parameter for the ring waveguide dimensions gives a waveguide mode area of $A_{waveguide}^{mode} = abx(\frac{1}{2})^2 = \frac{a^2}{2}x(\frac{1}{2})^2 \approx \frac{(0.7\lambda/\sqrt{n_g^2-n_s^2})^2}{2}x(\frac{1}{2})^2$ (note the $(\frac{1}{2})^2$ to covert physical area to mode area). This gives the smallest ring cavity mode volume of $V_{cav}^{mode} = \pi d\, A_{waveguide}^{mode} = 0.041\mu m^3$.

3.7 Summary: The Emergence of Nanophotonic Devices and Systems

From the discussion in this chapter, we see that the use of low-dimensional photonic structures, microcavity structures, and photonic bandgap structures can allow us to realize high-efficiency photonic devices with much smaller physical sizes than the current devices. These devices have critical device dimensions smaller than $0.5\mu m$ and we may refer to them as nanophotonic devices or as sub-wavelength scale devices. We believe that the combined use of low-dimensional photonic and electronic structures will allow us to realize functional photonic devices and systems that may have much lower operating power and faster modulation than those of current devices and may be advantages in certain applications such as that require complex transformation of optical signals. This technology may eventually allow us to realize some types of optical computer or digital optical signal processing system on a chip.

References

[1] See Seng T. Ho et al, in *Optical Processes in Microcavities*, p. 339, edited by Richard K. Chang and Anthony J. Campillo, World Scientific, Singapore, 1996, and references therein.
[2] R. E. Slusher, A. F. J. Levi, U. Mohideen, S. L. McCall, S. J. Pearton, and R. A. Logan, *Appl. Phys. Lett.* **63**, 1310 (1993).
[3] J. P. Zhang, D. Y. Chu, S. L. Wu, W. G. Bi, C. W. Tu, R. C. Tiberio, and S. T. Ho, *Phys. Rev. Lett.* **75**, 2678 (1995).
[4] J. P. Zhang, D. Y. Chu, S. L. Wu, W. G. Bi, R. C. Tiberio, R. M. Joseph, A. Taflove, C. W. Tu, and S. T. Ho, *IEEE Photon. Tech. Lett.* **8**, 491 (1996).
[5] S. T. Ho, *Nuclear Physics B (Proc. Suppl.)* **6**, 306 (1989).
[6] D. Marcuse, *Principles of quantum electronics*, Academic, New York, 1980.
[7] J. J. Sakurai, *Advanced Quantum Mechanics*, Addison-Wesley, Reading, 1982.
[8] Amnon Yariv, *Quantum Electronics*, John Wiley & Sons, New York, 1988.

[9] W. H. Louisell, *Quantum Statistical Properties of Radiation*, Wiley, New York, 1973.

[10] Stuart D. Brorson and Peter M. W. Skovgaard, p. 77, same as in Ref. 1; I. Abram, I. Robert, and R. Kuszelewicz, *IEEE J. Quan. Electron.* **QE-34**, 71 (1998).

[11] E. M. Purcell, *Phys. Rev.* **69** (1946) 681, and A. J. Campillo et al, p. 167, same as in Ref. 1.

[12] S. T. Ho and P. Kumar, *J. Opt. Soc. Amer. B*, **9**, 1620 (1993).

[13] Landau & Lifshitz, *The Classical Theory of Fields* (Pergamon, New York, 1975).

[14] G. L. J. A. Rikken and Y. A. R. R. Kessener, *Phys. Rev. Lett.*, **74**, 880 (1995).

[15] A. Yariv, *Optical Electronics* (Holt, Rinehart and Winston, New York, 4th Ed., (1985).

[16] M. S. Stern, *IEE Proc.-J*, 135, 56 (1988).

[17] C. L Xu, W.P. Huang, M.S. Stern and S. K. Chaudhuri, *IEE Proc.-J*, **141**, 281 (1994).

[18] G. Ronald Hadley and R. E. Smith, *J. Light. Tech.*, **13**, 465 (995).

[19] C. F. Gerald and P. O. Wheatley, *Applied Numerical Analysis*, (Addison-Wesley, Inc., 1984).

[20] J. H. Wilkinson, *The Algebraic Eigenvalue Problem*, (Clarendon Press, Oxford, 1965).

[21] Y. Yamamoto, *Coherence, amplification, and quantum effects in semiconductor lasers*, (John Wiley & Sons, Inc., New York, 1991).

[22] T. Baba, T. Hamano, F. Koyama, and K. Iga, *IEEE J. Quan. Electron.* **QE-27**, 1347 (1991).

[23] H. Yokoyama and S. D. Brorson, *J. Appl. Phys.* **66**, 4801 (1989).

[24] According to quantum mechanics, the spontaneous emission rate in any field mode is equal to its stimulation rate with one photon. An increase in the spontaneous emission coupling efficiency can lead to an increase in stimulated emission and hence gain, which reduces lasing threshold.

[25] Y. Suematsu and K. Furuya, *Trans. IEICE Japan* **60**, 467 (1977).

[26] D. Y. Chu, and S. T. Ho, Or. *Opt. Soc. Amer. B.*, **10**, 381 (1993).

[27] J. P. Zhang, D. Y. Chu, S. L. Wu, W. G. Bi, C. W. Tu, R. C. Tiberio, and S. T. Ho, *Phys. Rev. Lett.*, **75**, 2678 (1995).

[28] J. P. Zhang, D. Y. Chu, S. L. Wu, W. G. Bi, C. W. Tu, R. C. Tiberio, and S. T. Ho, *IEEE Photon. Tech. Lett.*, **8**, 968 (1996).

[29] J. P. Zhang, D. Y. Chu, S. L. Wu, W. G. Bi, C. W. Tu, R. C. Tiberio, R. M. Joseph, A. Taflove, and S. T. Ho, *IEEE Photon. Tech. Lett.*, **8**, 491 (1996).

Cavity QED - where's the Q?

C. J. Hood, T. W. Lynn, M. S. Chapman[2], H. Mabuchi, J. Ye, and H. J. Kimble

[1] Norman Bridge Laboratory of Physics, California Institute of Technology, Pasadena CA 91125, USA
[2] Georgia Tech, Atlanta GA, USA

Abstract. We discuss recent experiments in cavity QED with strongly coupled single atoms, studied one at a time, in real time. Particular emphasis is placed on defining what is, and is not "quantum" about particular parameter regimes of the system.

1 Cavity QED: why use single atoms?

The quest to create and explore interesting quantum states has been one of the primary driving forces in experimental efforts in cavity quantum electrodynamics (QED). A rich array of quantum effects have been predicted theoretically over the last 35 years, since the first work by Jaynes and Cummings (1963), but conditions in which they can be experimentally observed have proved difficult to achieve: to realize the simple Jaynes-Cummings Hamiltonian we are required to be dealing with *single quanta*, and be in the regime of *explicit strong coupling* for which the coherent evolution rate of these single quanta dominates any dissipation in the system. That is, we require $g_0 > \beta \equiv \max[\Gamma, T^{-1}]$, where g_0 is the rate of coherent, reversible, evolution, T is the interaction time and Γ is the set of decoherence rates for the system.

Additionally, to be able to experimentally probe and manipulate a single, strongly coupled atom the "optical information" per atomic transit (to be defined in Section 3) must be large, so that a meaningful signal can be extracted from a single atom in *real-time*. While many quantum optics experiments have been carried out at the single atom level (Thompson, 1992; Brune, 1996; Childs, 1996; Walther 1998a,b), the desired combination of strong coupling and high-information has only recently been realized, using laser-cooled atoms in cavity QED (Mabuchi, 1996a; Hood, 1998).

In contrast, most experiments in cavity QED (and all in semiconductor microcavities) have been in the "weak coupling" regime, for which the semiclassical Maxwell-Bloch equations describe the *structural* properties (eigenstructure) of the coupled system equally well as the full quantum master equation. That is, both theories predict the same level structure for the coupled atom-cavity "molecule". In this regime quantum effects are relegated to remain generally small perturbations to the system *dynamics* such as the nonclassical noise fluctuations of a squeezed output cavity field. Examples of the weak coupling regime include a laser well above threshold, an optical parametric oscillator (OPO), and cavity QED with an ensemble of individually weakly-coupled atoms.

When discussing quantum effects in microcavities and in cavity QED it therefore becomes crucial to be able to parametrize in which of these regimes a given experiment falls, how the simple one-atom Hamiltonian is correctly extended to the case of multiple atoms or emitters, and in what instances the collective coupling of many atoms (each weakly coupled) will still give rise to interesting quantum phenomena. After a brief review of the one-atom cavity QED theory in Section 2.1, this parametrization will be developed in Sections 2.2 and 2.3.

Experimentally, our desire to achieve explicit strong coupling of single atoms has driven us to change tack experimentally, moving from thermal beams as our atom source to laser-cooled trapped atoms, allowing us to work with single atoms, one at a time, in real time. Results from these single-atom experiments will be discussed in Section 3 below.

2 Theory

2.1 Single-atom theory

We initially consider the ideal cavity QED system of a single atom (in this case cesium) coupled to the TEM_{00} longitudinal mode of a high finesse optical cavity, shown schematically in Fig. 1, with curved mirrors providing transverse confinement of the mode. The coherent atom-field coupling rate is given by

$$g_0 = d\sqrt{\frac{\hbar\omega}{2\varepsilon_0 V_m}}, \qquad (1)$$

where d is the atomic dipole matrix element and ω the transition frequency. By making the cavity mode volume V_m small, the magnitude of g_0 can be increased. The rate of dissipation is set by $\gamma_\perp$, the atomic dipole decay rate and κ, the rate of decay of the cavity field. It should be noted that photon decay via κ does not necessarily lead to decoherence in the system, as these output photons can be measured, processed, or even used as the input to another cavity system, maintaining coherences with the atom-cavity (Cirac, 1997; van Enk, 1997; Mabuchi, 1996b). Finally, looking ahead to the experiments of Section 3 with cold atoms we introduce the transit time T of an atom through the cavity mode, and require for explicit strong coupling that $g_0 > (\gamma_\perp, \kappa, T^{-1})$.

In the limit of negligible dissipation, the system is described by the Jaynes-Cummings Hamiltonian,

$$H = \hbar\omega\hat{a}^\dagger\hat{a} + \hbar\omega\frac{\hat{\sigma}_z}{2} + \hbar g_0(\hat{a}\hat{\sigma}_+ + \hat{a}^\dagger\hat{\sigma}_-), \qquad (2)$$

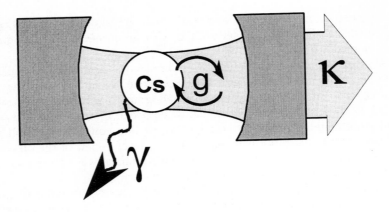

Fig. 1. Schematic: Single-atom cavity QED.

where $(\hat{a}, \hat{a}^\dagger)$ are the field annihilation and creation operators, $(\hat{\sigma}_+, \hat{\sigma}_-)$ the atomic raising and lowering operators (and hence $\hat{\sigma}_z$ the atomic inversion), ω the coincident frequency of the atomic transition and cavity field. This Hamiltonian can be simply diagonalized at n system excitations to find eigenstates $|\pm\rangle = 1/\sqrt{2}(|g,n\rangle \pm |e,n-1\rangle)$, with (g,e) here denoting the atomic ground and excited states. These states represent the atom and cavity equally sharing an excitation, and their corresponding energy levels are shifted by $\pm\sqrt{n}\hbar g_0$.

In the presence of dissipation and allowing for detunings, the Jaynes Cummings theory is extended to give a set of Heisenberg equations of motion for the system operators:

$$\dot{\hat{a}} = -\kappa(1+i\Theta)\hat{a} + g_0\hat{\sigma}_- + \varepsilon e^{-i(\omega_p - \omega_0)t} \quad (3)$$

$$\dot{\hat{\sigma}}_- = -\gamma_\perp(1+i\Delta)\hat{\sigma}_- + g_0\hat{a}\hat{\sigma}_z \quad (4)$$

$$\dot{\hat{\sigma}}_z = -\gamma_\parallel(\hat{\sigma}_z + 1) - 2g_0(\hat{a}^\dagger\hat{\sigma}_- + \hat{a}\hat{\sigma}_+) \quad (5)$$

Here ε is a driving field of frequency ω_p, $\Theta = (\omega_c - \omega_0)/\kappa$ is the cavity detuning and $\Delta = (\omega_a - \omega_0)/\gamma_\parallel$ the atomic detuning from a reference frequency ω_0 which defines the rotating frame for these equations.

In general these operator equations can only be solved by numerical integration; however, in the very restricted case of *weak excitation* (paramatrization of what defines "weak" will be discussed in Section 2.3), we can solve for the system eigenvalues simply by noting that there is never more than one excitation in the system, so that $\langle \hat{a}\hat{\sigma}_z \rangle = -\langle \hat{a} \rangle$ and $\langle \hat{\sigma}_z \rangle = -1$, and the semiclassical Maxwell-Bloch equations are recovered. That is, for measurements probing the *structure* of the system (such as measurement of transmission or fluorescence spectra), a semiclassical approximation is valid for weak excitation, and we therefore only expect to be able to distinguish intrinsically quantum effects in these structural measurements at higher excitation strengths.

That a semiclassical formalism correctly predicts the atom-cavity level structure for weak driving fields should not be taken as a statement that the system contains no interesting quantum mechanics: studying the *dynamics* of a weakly driven, strongly coupled system reveals nonclassicality, such as photon antibunching of the output cavity field (Rempe,1991).

When small amounts of dissipation are included, we note that by varying the cavity parameters κ and g_0, the response of the atom-cavity system to a probe field can be changed qualitatively. For example, if the cavity decay κ is large, we find that the atom and cavity retain their distinct identities, with decay rates *modified* by their coupling. In particular, we can define a "1-D atom" regime by $\kappa > g_0^2/\kappa > \gamma$, in which the atomic decay to the cavity mode is at rate g_0^2/κ and we have an effectively "1-D atom" interacting preferentially with the cavity mode (Turchette, 1995a). In one experiment in this regime, this enhancement corresponds to a Purcell factor of 0.67 (Turchette, 1995b). Other cavity geometries can also lead to an enhanced Purcell factor (Heinzen, 1987; Morin, 1994; Childs, 1996). Note also that since the enhanced atomic linewidth $\gamma + g_0^2/\kappa$ is less than κ, the transmission spectrum is a two-peaked structure (the empty-cavity Lorentzian with an "absorption dip" of the enhanced atomic linewidth superposed).

From the perspective of observing strongly quantum mechanical effects, a more interesting regime is that of *explicit strong coupling*, defined by $g_0 \gg (\kappa, \gamma)$. In this limit the atom-cavity must be considered as a composite coupled system, with structure and dynamics approaching those predicted by the Jaynes-Cummings Hamiltonian. For weak excitation, the familiar vacuum-Rabi spectrum is observed in transmission, while for stronger driving fields the nonlinear response follows from the higher-lying states of the Jaynes-Cummings ladder, directly observed in the microwave domain of cavity QED (Brune, 1996) and reflected in nonlinear transmission measurements in optical cavity QED (Hood, 1998; Mabuchi, 1998a).

2.2 Many Atom Theory

Returning to the case of a lossless interaction, the extension of the Jaynes-Cummings ladder of states to the case of N atoms is summarized in Figure 2 (Tavis and Cummings, 1968; Varada, 1987), where the level structure of the atom-cavity system is plotted as a function of the number of interacting atoms. For one atom the structure is the familiar Jaynes-Cummings ladder, with the n-excitation levels split by $\pm\hbar g_0\sqrt{n}$ in energy.

Moving to larger atom numbers (N), we notice that the magnitude of the vacuum-Rabi splitting scales as $\sqrt{N}$, and are hence led to define an *effective* coupling strength $g_{eff} = g_0\sqrt{N}$. Making this identification suggests that N atoms might act as one *effective* atom of coupling strength g_{eff}, following

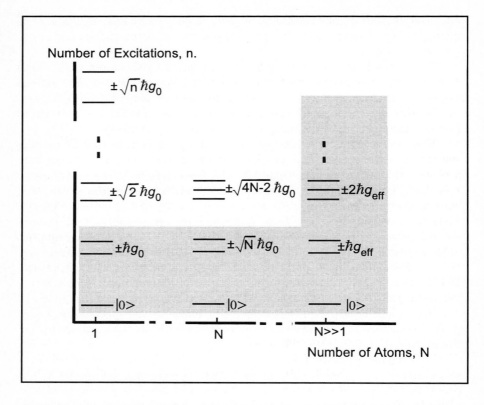

Fig. 2. Energy eigenvalues of the atom-cavity system as a function of number of excitations, n, and number of intracavity atoms, N. In the shaded region semiclassical Maxwell-Bloch theory correctly predicts these eigenvalues.

a Jaynes-Cummings Hamiltonian. Looking more closely at the higher-lying excitations shows that this is clearly *not* the case. The 2-excitation splitting is $\pm\hbar g_0\sqrt{4N-2}$, which for $N \gg 1$ tends to the energy eigenvalues expected for *classical* coupled oscillators, $\pm 2\hbar g_{eff}$.

In the limit of large atom number $N \gg 1$, it therefore follows that semiclassical theory yields the correct values for the structure of the system irrespective of drive strength. Additionally, the quantum and semiclassical descriptions only deviate for $n \geq 2$ excitations, so it should not be surprising that semiclassical theory is also valid in the regime of weak excitation, as was already mentioned.

2.3 Delineating parameter regions: Critical atom and photon numbers

To better parametrize exactly where semiclassical theory is valid we introduce two dimensionless parameters describing the atom-cavity: the critical atom number, $N_0 = (2\kappa\gamma_\perp)/g_0^2$ and critical photon number $n_0 = \gamma_\perp^2/2g_0^2$; the number of quanta (atoms or photons) required to significantly alter the atom-cavity response. Most quantum systems have large critical parameters, for example a typical laser has a threshold photon number $\sqrt{n_0} \simeq 10^3 - 10^4$, indicating that adding or removing one photon has a negligible effect. Similarly, for cavity QED systems with large critical atom numbers, many atoms are required to have an effect on the intracavity light intensity, and accordingly the effect of a single atom is small. In these situations, the equations governing the system can be expanded in the small parameters (n_0^{-1}, N_0^{-1}), and the description of the system reduces again to a semiclassical prediction of the structure, with quantum noise fluctuations about these values (Carmichael, 1993). Note that this expansion is valid for any field strength, and provides the more precise definition of the semiclassical regime, that is $(n_0, N_0) \gg 1$.

As the critical parameters are reduced, we move toward a regime where individual quanta have a profound effect on the system, so the semiclassical approximation above is no longer valid. We can expect to see uniquely quantum effects appearing in the system structure. Our experiments in optical cavity QED have marked a steady progression from the semiclassical to the quantum regime: from $(n_0, N_0) \simeq 10,000$ in 1981 to our latest experiments with $(n_0, N_0) \simeq 0.001$.

In the intermediate regime of $(n_0, N_0) \simeq 1$ there is some disparity between the quantum and semiclassical theories; however, even for one experiment with $(n_0, N_0) = (0.02, 0.9)$, this difference could not be resolved experimentally (Turchette, 1995a). In contrast, for systems with *small* critical parameters $(n_0, N_0) \ll 1$, marked differences between the theories can readily be observed. One experiment with $(n_0, N_0) = (0.0002, 0.015)$, described in Section 3, clearly demonstrates such a difference (Hood, 1998).

Finally, we note that the "weak" excitation regime, for which the system response is linear and a semiclassical approximation is valid, can now be parametrized by reference to the saturation photon number, with "weak-field" meaning $n \ll n_0$. Note that for our current parameters of $n_0 \simeq 10^{-4}$, we are not experimentally in the limit of weak excitation, as field strengths of $10^{-2} < n < 1$ are required for acceptable signal-to-noise ratio.

2.4 Additional considerations with distributions of atoms

In cavity QED experiments which use a thermal atomic beam as an atom source, there is by nature a spatial distribution of atoms within the finite cavity mode, and a temporal variation of this distribution. Although these atoms may be collectively strongly coupled in the sense of $g_{eff} = g_0\sqrt{N}$ being large (Thompson, 1992; Childs, 1996), quantum effects are made difficult to observe by the presence of many atoms as discussed in Section 2.2, even when the effective atom number in the cavity mode is less than one. In addition to the degradation in "quantumness" when moving from single atoms to N atoms, if the *distribution* of atoms varies spatially and temporally there are additional complications. In particular, if $\psi(r)$ is the mode function of the cavity field, then the effective atom number $N_{eff} = \sum_{i=1}^{N} |\psi_i|^2$ gives rise to effective coupling $g_{eff} = g_0\sqrt{N_{eff}}$ for the single photon excitation, however for excitations of 2 photons the splitting is modified to $\pm\hbar g_0 \sqrt{4N_{eff} - 2M/N_{eff}}$ where $M = \sum_{i=1}^{N} |\psi_i|^4$ (Thompson, 1998). That is, various atomic distributions of the same N_{eff} will each produce different structure in the nonlinear spectrum.

Also, temporal variations in the atomic distribution lead to a time variation of g_{eff}, and a resulting averaging of output spectra over the experimental detection time. Intrinsic to thermal beams as an atom source, this effect has led to the inability of optical cavity QED experiments to directly and unambiguously resolve the higher-lying states of the Jaynes-Cummings ladder (Thompson, 1998), and has been our principal motivation in moving to laser cooling as a source of single atoms.

3 Experiments

3.1 Cold atoms in cavity QED - the single atom source

Over the last few years our group has pioneered the use of laser-cooled atoms in cavity QED (Mabuchi, 1996a ; Hood, 1998), enabling experiments involving single atoms, one at a time. Previous experiments using atomic beams required averaging over many atom transits to extract any useful data. More quantitatively, the optical information (Kimble,1997) (the number of photons providing our signal) obtained for a single atom traversing the cavity in time T is given by

$$I = \frac{g_0^2 T}{\beta}, \beta = max(\kappa, \gamma, T^{-1}), \qquad (6)$$

and our parameters yield $I \simeq 54000\pi$, compared to $I \simeq \pi$ for atomic beam experiments. This vast increase in signal allows for the real-time monitoring and manipulation of single atoms as they traverse the cavity mode. It has been shown that this measurement strategy can in fact approach the standard quantum limit for the measurement of atomic position, i.e., the point at which we extract the maximum amount of information allowed by the principles of quantum mechanics (Mabuchi, 1998b).

Our experimental procedure is as follows: a cloud of $\simeq 10^4$ cesium atoms is collected in a magneto-optical trap (MOT), situated 5mm above the cavity mode. The MOT is then switched off, and the atoms allowed to fall under gravity, with collimation by the narrow gap between the cavity mirrors ensuring that few atoms pass through the cavity mode. Figure 3 shows the cloud of atoms 35ms after the MOT is switched off, falling toward the cavity mirrors.

Fig. 3. A cloud of cesium atoms falling to the cavity, 35ms after the MOT is switched off.

3.2 Experiments with explicit strong coupling

The cavity for these experiments, of length 10.1μm, is comprised of two 10cm radius of curvature mirrors of transmission $\simeq 16$ parts per million, giving a finesse of 180,000 and leading to the set of parameters $(g_0, \kappa, \gamma_\perp, T^{-1})/2\pi = (120, 40, 2.6, 0.002)$MHz, where the transit time T through the cavity waist of 15μm is $T \simeq 75\mu$s. The resulting critical parameters $(n_0, N_0) = (0.0002, 0.015)$ put us clearly in the regime of explicit strong coupling for a single atom and photon. In these experiments we wish to investigate the quantum nature of the system.

We continuously probe transmission of the atom-cavity system as cold atoms fall through the cavity mode one by one, with intracavity field strengths corresponding to an energy of $\simeq 1$ photon on average. The strong atom-cavity coupling g_0 gives rise to a dressing of the system eigenstates and a corresponding splitting in the energy levels. For a single, optimally coupled atom the transmission spectrum (solid curve, Figure 4(c)) is similar to the familiar two-peaked vacuum-Rabi spectrum, but is modified by our drive strength of $n \gg n_0$. Optimal coupling is only achieved when the atom is at the point of maximum field strength (at the center of the Gaussian cavity mode). Thus as an atom falls through the cavity the coupling evolves from $g = 0$, with an empty-cavity Lorentzian transmission spectrum (dashed curve, Figure 4(c)) to $g = g_0$ then back to $g = 0$. If the atom does not pass close to an antinode of the cavity standing-wave, a smaller $g_{max} < g_0$ will be achieved.

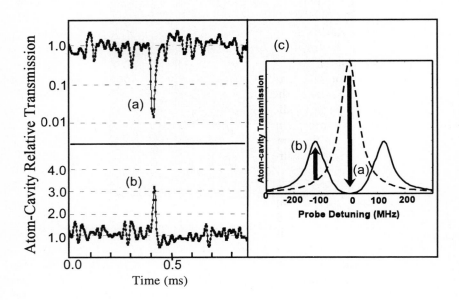

Fig. 4. Transmission for each of two probe beams simultaneously illuminating the atom-cavity as a function of time for a single atom traversing the cavity mode. For (a) the probe detuning is -20 MHz, for (b) -100 MHz. The change in the atom-cavity spectrum which gives rise to these time traces is shown in (c). For these traces the atomic and cavity resonances are coincident.

For an atom traversing the cavity mode (in $\simeq 75\mu s$) the position-dependent coupling gives rise to a time-dependent probe transmission (Rempe,1995; Doherty 1997). For two probe fields of fixed detuning (indicated by the arrow positions of Figure 4(c)) simultaneously illuminating the atom-cavity system and being detected in transmission with heterodyne detection, the real-time atom-cavity transmission is shown in Figure 4(a,b). Close to resonance (Figure 4(a)), transmission drops as an atom enters the cavity and the spectrum shifts from the empty-cavity Lorentzian response to the vacuum-Rabi spectrum. Transmission regularly drops by a factor of 100 at this detuning. For the same atom during the same transit but for a probe of detuning $-g_0$ (Figure 4(b)), the transmission correspondingly rises. For the data of Figure 4, the two probe fields are applied simultaneously, leading to a fundamental decrease in signal to noise over single-probe measurements due to the tradeoff between reduction in shot noise and saturation of the atom-cavity response (Hood, 1998). Detection is by balanced heterodyne at 100kHz bandwidth, with the demodulated output digitized at 500ks/s.

By recording atom transits such as those of Figure 4, the entire transmission spectrum can be mapped out (Hood, 1998), but here we wish to focus on exploring more explicitly the quantum nature of the system in a strongly nonlinear regime. To do this we record resonant transits (as in Figure 4(a)) as the intracavity photon number is increased, to map out the saturation behavior of the system. A quantum system respecting a Jaynes-Cummings ladder of states is predicted to saturate differently than the corresponding semiclassical system, due to the different structure for high-lying excitations. For our parameters the predictions differ by an order of magnitude, with the semiclassical case predicting bistability. The data of (Hood, 1998) shows strong agreement with the quantum master equation calculation, confirming the underlying quantum nature of the atom-field coupling and our ability to access it experimentally.

Beyond simply measuring strong coupling, one experiment in progress in our group is exploiting it to trap single atoms using the vacuum-Rabi splitting (Haroche, 1991; Doherty, 1998). The lower dressed state of the atom-cavity system (with one photon excitation) has an energy minimum at the center of the cavity mode, forming a bound state of atom and cavity sharing a photon. The spatial dependence of the cavity mode (Gaussian transverse distribution, standing-wave along the cavity axis) therefore creates a series of trapping pseudo-potential wells for a red-detuned probe field (for times $\gg \kappa^{-1}$ as photons enter and exit the cavity) of depth $\hbar g_0 p_-$ where p_- is the occupation probability of the lower dressed state. For our parameters this gives a depth of $\simeq 7$mK, so with an initial atomic temperature of 15mK and a fall of only 2mm to the cavity, single atoms have low enough energy to be confined in this bound state. To date trap times of up to 300μs have been observed, a 4-fold increase over the freefall transit time.

4 Conclusion and Prospects

The ability to measure and manipulate strongly coupled single atoms one by one has tremendous possibilities, both from a perspective of being able to experimentally explore fundamental quantum mechanics, and also for its potential application to the fabrication of quantum devices for quantum computing (Pellizzari, 1995; Turchette, 1995b), quantum state preparation (Parkins, 1993; Law, 1997), and quantum communications (Cirac, 1997; van Enk, 1997). For each of these endeavors a critical first step has been to demonstrate the intrinsic quantum behavior of the system, stemming from the very simple underlying Jaynes-Cummings Hamiltonian governing the system: that is, demonstrating the "Q" in cavity QED.

In the world of solid-state devices and microcavity physics, the underlying structure is, as we have seen in other lectures in these proceedings, a far more complex and subtle thing. With the eventual hope of building truly quantum solid state devices, this raises several challenges. Firstly, characterization of the dissipative rates in the system is in itself difficult, while controlling and reducing these decoherence mechanisms poses additional technical problems. In addition, for explicit strong coupling realization of genuinely "single-particle" couplings within the multi-particle system are required, with coupling rates exceeding the dissipative rates for the system. Given the rapid progress to date in microcavity physics it seems merely a matter of time until these current challenges are overcome, and the knowledge accumulated and methods developed in cavity QED experiments can be utilized to demonstrate some "Q" in *micro*-cavity QED.

This work has been supported by DARPA via the QUIC Institute administered by ARO, by the NSF, and by the ONR.

References

1. Brune, M., et. al. (1996); Phys. Rev. Lett **76**, 1800.
2. Carmichael, H. J., (1993); *Quantum Statistical Techniques in Quantum Optics*, Springer.
3. Childs, J. J., An, K., Otteson, M. S., Dasari, R. R., Feld, M. S., (1996); Phys. Rev. Lett. **77**, 2901.
4. Cirac, J. I., Zoller, P., Kimble, H. J., Mabuchi, H., (1997); Phys. Rev. Lett **78**, 3221.
5. Doherty, A. C., Parkins, A. S., Tan, S. M., Walls, D. F., (1997); Phys. Rev. A **56**, 833.
6. Doherty, A. C., Parkins, A. S., Tan, S. M., Walls, D. F., (1998); Phys. Rev. A **57**, 4804.
7. van Enk, S. J., Cirac, J. I., Zoller, P., Kimble, H. J., Mabuchi, H.,(1997); J. Mod Opt **44**, 1727.
8. Haroche, S., Brune, M., Raimond, J. M., (1991); Europhys. Lett. **14**, 19.
9. Heinzen, D. J., Childs, J. L., Thomas, J., Feld, M. S., (1987); Phys. Rev. Lett. **58**, 1321.
10. Hood, C. J., Chapman, M. S., Lynn, T. W., Kimble, H. J., (1998); Phys. Rev. Lett. **80**, 4157.
11. Jaynes, E. T., Cummings, F. W. (1963); Proc. IEEE **51**, 89.
12. Kimble H. J., (1997); Phil. Trans. A **355**, 2327.
13. Law, C. K., Kimble, H. J., (1997); Quantum Semiclass. Opt. **44**, 2067.
14. Mabuchi, H., Turchette, Q. A., Chapman, M. S., Kimble, H. J., (1996a); Opt. Lett. **21**, 1393.
15. Mabuchi, H., Zoller, P., (1996b); Phys. Rev. Lett. **76**, 3108.
16. Mabuchi, H., Ye, J., Kimble, H. J., (1998a); submitted to Appl. Phys. B.

17. Mabuchi, H., (1998b); Phys. Rev. A **58**, 123.
18. Morin, S. E., Yu, C. C., Mossberg, T. W., (1994); Phys. Rev. Lett. **73**, 1489.
19. Parkins, A. S., Marte, P., Zoller, P., Kimble, H. J., (1993); Phys. Rev. Lett. **71**, 3095.
20. Parkins, A. S., (1995); unpublished notes.
21. Pellizzari, T., Gardiner, S., Cirac, C., and Zoller, P., (1995); Phys. Rev. Lett. **75**, 3788.
22. Rempe, G., et. al. (1991); Phys. Rev. Lett **67**, 1727.
23. Rempe, G., (1995); Appl. Phys. **B60**, 233.
24. Thompson, R. J., Rempe, G., and Kimble, H. J., (1992); Phys. Rev. Lett. **68**, 1132.
25. Thompson, R. J., Turchette, Q. A., Carnal, O., and Kimble, H. J., (1998); Phys. Rev. A **57**, 3084.
26. Turchette, Q. A., Thompson, R. J., Kimble, H. J., (1995); App. Phys. B **60** S1-S10.
27. Turchette, Q. A., Hood, C. J., Lange, W., Mabuchi, H., Kimble, H. J., (1995); Phys. Rev. Lett. **75**, 4710.
28. Varada, G. V., Kumar, M. S., Ararwal, G. S., (1987); Opt. Comm **62** 328.
29. Walther, H., (1998a); P. Roy. Soc. A **454**, 431.
30. Walther, H., (1998b); Phys. Scr. **T76**, 138.

Quantum Optics in Semiconductors

Atac Imamoḡlu[1]

University of California, Santa Barbara, CA 93106, USA

One of the ultimate goals of applied quantum optics is the control of light generation at the single photon level. Until recently, the majority of quantum optics research concentrated on achieving this goal using atom-field interactions where relatively simple theoretical models such as Jaynes-Cummings Hamiltonian provide an accurate description of dynamics. In contrast, photon-semiconductor system is inherently more difficult to model primarily due to the importance of Coulomb interactions. Nevertheless, the underlying rich physics and the possibility of practical applications in optoelectronics provides a strong motivation to study quantum optical phenomena in semiconductors.

In this chapter, we will start by a brief review of quantum statistical properties of light and introduce a particular nonclassical state which we refer to as *heralded single photons*. We will then discuss methods to realize quantum noise suppression and heralded single photons in semiconductors using Pauli exclusion and Coulomb interactions. In the last section, we will concentrate on quantum statistical effects associated with bound electron-hole pairs (i.e. excitons) and briefly discuss their optical signatures.

1 Nonclassical states of light

In the semiclassical theory of light-matter interaction, the electromagnetic field is treated as a classical variable that is driven by the mean polarization of the medium. In the full-quantum theory on the other hand, the electromagnetic field is quantized independently of the matter field; the electric and magnetic fields in this case are operators acting on the Hilbert space associated with the electromagnetic field. Despite the drastic difference in formulation, the predictions of the semiclassical and quantum theories agree in all *linear problems* or cases which can be explained by first order (single-photon) coherence functions [1].

On the other hand, the semiclassical and quantum theories predict drastically different results for the two-photon correlation experiments such as the Hanbury-Brown-Twiss (HBT) experiment. The most important quantity to measure in this context is the normalized second-order coherence function $g^{(2)}(\tau)$ defined as [1,2]

$$g^{(2)}(\tau) = \frac{\langle \hat{\mathbf{E}}^{(-)}(t)\hat{\mathbf{E}}^{(-)}(t+\tau)\hat{\mathbf{E}}^{(+)}(t+\tau)\hat{\mathbf{E}}^{(+)}(t)\rangle}{\langle \hat{\mathbf{E}}^{(-)}(t)\hat{\mathbf{E}}^{(+)}(t)\rangle \langle \hat{\mathbf{E}}^{(-)}(t+\tau)\hat{\mathbf{E}}^{(+)}(t+\tau)\rangle} , \qquad (1)$$

where

$$\hat{E}^{(+)}(\mathbf{r},t) = \sum_k i \left(\frac{\hbar\omega_k}{2\epsilon_0 L^3}\right)^{1/2} \cdot e^{i\mathbf{k}\cdot\mathbf{r}} \cdot \hat{a}_k(t) \qquad (2)$$

is the positive-frequency component of the electric field operator and $\hat{E}^{(-)} = (\hat{E}^{(+)})^\dagger$. For a single-mode electromagnetic field, the expression in Eq. (1) simplifies to

$$g^{(2)}(\tau) = \frac{\langle \hat{a}^\dagger(t)\hat{a}^\dagger(t+\tau)\hat{a}(t+\tau)\hat{a}(t)\rangle}{\langle \hat{a}^\dagger(t)\hat{a}(t)\rangle\langle \hat{a}^\dagger(t+\tau)\hat{a}(t+\tau)\rangle} \qquad (3)$$

We now evaluate $g^{(2)}(\tau)$ for particular quantum states of light, that are of interest. The first case that we consider is the thermal state. The density operator for single-mode thermal light is

$$\hat{\rho}_t = (1 - exp[-\frac{\hbar\omega}{k_B T_R}]) \sum_n exp[-\frac{n\hbar\omega}{k_B T_R}]|n\rangle\langle n| \quad , \qquad (4)$$

where T_R is the temperature of the radiation field mode at frequency ω. Since there is no nontrivial time-dependence, the second-order correlation function may be easily evaluated to find

$$g^{(2)}(\tau) = g^{(2)}(0) = 2 \quad . \qquad (5)$$

In order to see the physics more clearly, it is useful to generalize this result to a multimode chaotic light. If we assume that the state of the field is described by the Gaussian distribution of the frequencies ω in the density operator [3], then we can factorize the numerator in (1) and obtain

$$g^{(2)}(\tau) = 1 + |g^{(1)}(\tau)|^2 \quad . \qquad (6)$$

Since $|g^{(1)}(\tau)| = 1$ for any single-mode light, Eq. (5) follows Eq. (6). If the field has a Lorentzian spectrum (with width γ), then $g^{(2)}(\tau) = 1 + exp[-\gamma|\tau|]$. If the light intensities at two time instants are uncorrelated, we obtain $g^{(2)}(\tau) = 1$; this result follows the definition of $g^{(2)}(\tau)$ directly. For thermal light, we see that within the field correlation time γ^{-1}, the detection of a photon makes a second subsequent detection event very likely: This phenomenon is referred to as *photon bunching* and has been demonstrated experimentally in the original HBT experiment.

The bunching effect that we have just obtained for thermal light is a signature of the bosonic nature of photons. Alternatively, we will find a similar bunching effect for all thermal bosonic fields, including for example ultracold sodium-23 and rubidium-87 atoms. Thermal bosons in general and photons in particular tend to be *noisy* with large amplitude fluctuations.

When we consider a quasi-classical or coherent state of light, the calculation of $g^{(2)}(\tau)$ is particularly simple: we find that for a single-mode field $g^{(2)}(\tau) = 1$. The same result would also apply to a classical single-mode light; this is not unexpected since we know that the coherent state is the quantum state of light that most closely resembles a classical field. Photon detection events for a coherent state are completely uncorrelated; detection of a photon at $\tau = 0$ gives us no information regarding the detection time of a second photon. The photocount distribution obtained for a coherent state is Poissonian, which is another indication of the lack of correlation: if we have a coherent state with a mean photon number $\bar{n}$, then the variance in the detected photon number will be $\Delta n^2 = \bar{n}$.

Before proceeding, we discuss some of the inequalities satisifed by the second order correlation function in classical theory. One can use Cauchy's inequality to show that

$$g^{(2)}_{\text{classical}}(0) \geq 1 \tag{7}$$

$$g^{(2)}_{\text{classical}}(0) \geq g^{(2)}_{\text{classical}}(\tau) \quad . \tag{8}$$

Both thermal and coherent states of light satisfy 7 and 8. However, the quantum second-order correlation function $g^{(2)}(\tau)$ can in general violate both of these inequalities. More specifically, it is possible to have

$$0 \leq g^{(2)}(0) < 1 \quad , \tag{9}$$

for certain quantum states of light [1,2]. Since these states violate the classical inequalities, they are termed *nonclassical states of light*. We reiterate that there are no classical fluctuations that would give Eq. (9). This range of values of $g^{(2)}(0)$ implies that photons are *antibunched*; that is detection of a photon makes a subsequent detection event less likely. In most cases of interest, the second inequality is violated along with the first one, as $g^{(2)}(\infty) = 1$ for all finite bandwidth electromagnetic fields.

It is also important to point out that a nonclassical state satisfying Eq. (9) will in general have less intensity noise (smaller photon number fluctuations) than the ideal classical coherent light source. The fact that quantum light can be more *quiet* provides a practical motivation to pursue the generation of such states of light. To quantify the relation between the photon number fluctuations and the nonclassical nature of light, we can show that [2]

$$\Delta n^2 - \bar{n} = \frac{\bar{n}^2}{T^2} \int_{-T}^{T} d\tau \, (T - |\tau|) \, [g^{(2)}(\tau) - 1] \quad , \tag{10}$$

where T is the counting time interval. For the special case of a single-mode field, Eq. (10) can be rewritten as

$$g^{(2)}(0) = 1 + \frac{\Delta n^2 - \bar{n}}{\bar{n}^2} \quad . \tag{11}$$

The prototypical example of a purely quantum state of light is the single-mode photon number eigenstate with n photons $|n\rangle$, for which we obtain

$$g^{(2)}(0) = \frac{n-1}{n} < 1 \quad . \tag{12}$$

It is easy to understand the origin of photon antibunching in a photon number eigenstate: The total number of photons is known and the detection of one makes it less likely to detect other photons. Particularly, for $|n\rangle = |1\rangle$, $g^{(2)}(0) = 0$, as there are no photons left to detect after the first one. In contrast, for $n \gg 1$, the antibunching effect is small. On the other hand, for all values of n the photon number state has no intensity noise, i.e. $\Delta n^2 = 0$. We conclude that for a single-mode photon-number state, there is practically no information on the arrival time of the photons (weak antibunching), even though the total number that will eventually be detected is known precisely.

The ultimate quantum control of photon generation would imply that a single (or a well-defined number) of photons are generated at arbitrarily short time intervals, with a deterministic dwell time between successive photon generation events. In such a case, one has complete information about the total number of photons as well as the generation/detection time of photons. Such a heralded single-photon state [4,5] is best characterized by its second-order coherence function $g^{(2)}(\tau)$: heralded single-photon state exhibits strong antibunching ($g^{(2)}(\tau) \simeq 0$) as there is a deterministic dwell-time (T_{ac}) in between successive photon detection events. In addition, $g^{(2)}(\tau = T_{ac}) > 1$ indicating that at a dwell-time after the first photon detection, the probability of detecting a second photon is much more likely than in the case of coherent photons. This is clearly a nonclassical state of light with possible applications in *quantum optoelectronics*.

Within the last 4 years, there have been several proposals for realizing such a herlded single-photon state [4-6]. In the following section, we will detail a possible implementation in mesoscopic p-i-n junctions that rely on coulomb blockade and quantum confinement effects.

To conclude this section, we will discuss the $g^{(2)}_{fermi}(\tau)$ for fermionic fields qualitatively. We can define an analogous two-time second-order coherence function for a (spin-polarized) fermionic system with field operator $\hat{\psi}(r,t)$ as

$$g^{(2)}_{fermi}(\tau) = \frac{\langle \hat{\psi}^\dagger(t)\hat{\psi}^\dagger(t+\tau)\hat{\psi}(t+\tau)\hat{\psi}(t)\rangle}{\langle \hat{\psi}^\dagger(t)\hat{\psi}(t)\rangle \langle \hat{\psi}^\dagger(t+\tau)\hat{\psi}(t+\tau)\rangle} , \qquad (13)$$

where we have suppressed the spatial dependence of the field operator for the sake of direct comparison with the photon correlation function, even though spatial correlation function is more important in practice. We can consider $g^{(2)}_{fermi}(\tau)$ for fermions with and without repulsive interactions: in both cases $g^{(2)}_{fermi}(0) = 0$, indicating that fermions are antibunched. For non-interacting fermions, the antibunching is purely due to Pauli-exclusion principle and is referred to as an *exchange hole*. The presence of repulsive interactions enhances the antibunching; for example even with electrons of opposite spin, we expect to see antibunching due to mutual Coulomb repulsion: this is referred to as a *Coulomb hole* [7]. In either case, we see that fermions are inherently *quiet*.

The second-order coherence for fermions that we have discussed gives us a clue as to how we can go about generating nonclassical light; more specifically, if we can force photons to *follow* the fermions that generate them, then we could hope to obtain quantum statistical properties for photons that resemble those of fermions. This is in a sense the underlying idea behind the devices discussed in the following section.

2 Quantum optics in mesoscopic systems

In order to control light generation at the single photon level, it is essential to manipulate and suppress the quantum noise that accompanies the photon generation process. Pump noise is crucial in this context as it ultimately determines the noise characteristics of the generated light. In this chapter, we concentrate on the problem of sub-Poissonian or *quiet light* generation from semiconductor p-n junctions. Pump noise associated with electron injection across the junction in this system can be suppressed by either Coulomb interactions or phase-space filling. As we shall see, the nature of the generated nonclassical light depends to a large extent whether the junction is macroscopic or mesoscopic; in the macroscopic limit we obtain number-squeezed light whereas in the mesoscopic limit the generated light can best be described as a regulated single-photon pump or heralded single-photons.

In the first subsection, we will consider a constant current driven macroscopic junction and discuss the roles of current noise and *macroscopic Coulomb blockade* effect on photon-number squeezing. In the second subsection, we will concentrate on mesoscopic junctions and the generation of heralded single photons.

2.1 Quantum noise suppression in macroscopic junctions

As we have already indicated, pump noise plays a fundamental role in determining the noise characteristics of lasers that operate well above threshold [8]. Let's consider a p-n junction laser diode where the junction current is predominantly due to electrons: we will note that for above-threshold operation the stimulated photon emission is the dominant recombination mechanism. This in turn implies that practically each electron injected from the n-type layer into the p-type layer will recombine by emitting a single laser photon. The total number of photons that leave the cavity for long observation times will then be necessarily equal to the number of injected electrons. If in addition, we have a situation where the stimulated emission rate is much faster than the characteristic rates associated with electron injection, the photon statistics will closely follow that of injected electrons. Equivalently, the low-frequency noise spectrum [8] of electrons and photons will be identical. A similar consideration is also valid for light emitting diodes (LED) in the bad-cavity limit, where the dominant recombination mechanism is via spontaneous emission into a single cavity mode. The Heisenberg-Langevin description of the laser dynamics confirm this qualitatitive explanation [8,9].

Since single photons will follow single electrons as we have argued above, we need to understand the possible noise suppression mechanisms for the electron stream in a p-n junction in order to predict the photon statistics of the generated photons. Our primary emphasis should therefore be on the current noise in the circuit. It is well known that the current noise in a macroscopic conductor under a finite bias is independent of the bias voltage and is equal to the Johnson-Nyquist thermal noise. In other words, a *good resistor* does not exhibit partition noise and the observed current noise (due to Johnson-Nyquist noise) is much lower than the shot noise level. Recently, there has been several attempts to explain this well-known phenomenon using microscopic models [10,11]; it appears that the self-feedback provided by the Pauli exclusion principle on dissipative electron transport is responsible for the current noise suppression. Since Pauli exclusion is not as effective for elastic scattering (pure dephasing) processes, it is predicted that shot-noise suppression should be dominant only in dissipative resistors [10].

The current noise in the circuit that drives the p-n junction should therefore depend on whether or not the total circuit resistance seen by the junction is smaller than the differential resistance of the junction. Typically for laser diodes, the differential resistance $R_d = kT/eI \leq 1\Omega$ (where I is the junction current and T is the temperature), implying that even with very small source resistances, the junction will be effectively driven by a constant-current source. As we have argued above, the pump noise of the laser in this case will be well below the shot noise level, which in turn implies that the generated light field will be sub-Poissonian. This prediction have been confirmed by experiments and noise suppression in excess of 10 dB below shot noise has been observed [9]. If on the other hand, the junction resistance is larger than the source resistance, then the junction will be driven by a constant-voltage source: the electron injection across the junction in this case will be random, resulting in a shot-noise limited pump noise. The photons generated by a constant voltage driven macroscopic junction have Poissonian statistics.

Suppression of shot noise in semiconductor p-n junction lasers driven by a high-impedance constant-current source provides us with the simplest source of nonclassical light. The value of the second order correlation function for the light generated by such a constant-current driven macroscopic *pn* junction is below but very close to the Poisson limit, indicating that anti-correlation between successive photon emission events are very small and that practically no information on the photon emission times exist: The anti-correlations only become important for a large number of photons.

The suppression of the driving current shot-noise in a macroscopic conductor connected to a large resistor is an important factor in the observation of the described effects. However, a constant-current source alone does not dictate the correlations between successive injection events. To understand these correlations, we need to consider the influence of charging effects on the individual electron injection across the junction. To this end, we will introduce the concept of Coulomb blockade in the following subsection and after that, we will analyze the transition from a "macroscopic regulation" of many electrons in the *squeezing* regime to the strict regulation of individual electron injection events in the Coulomb blockade regime [12].

2.2 Coulomb blockade of electron injection

Coulomb interactions and phase-space filling (PSF) are the two primary effects that enrich the physics of semiconductors. When the size of the semiconductor becomes small in all 3 dimensions, both of these effects become more prominent. When the size is on the order of 100 Å, size quantization or equivalently PSF is the dominant physical mechanism determining the nonlinear response of the material. For semiconductors with dimensions $0.1\mu - 1\mu$, it is the Coulomb interactions that dominate the transport and even optical properties; this is the mesoscopic limit or the *Coulomb blockade* regime [13].

Let's consider a n-I-n semiconductor junction where I denotes a wide-bandgap undoped semiconductor: when a bias V_0 is applied between the leads, the electrons will tunnel from the emitter side to the collector. The junction can be modelled as having a tunneling resistor R_t and a capacitance C_d. The Capacitance is given by

$$C_d = \varepsilon \frac{A_{eff}}{L_i} \quad , \tag{14}$$

where ε and L_i are the dielectric constant and the length of the insulator layer, respectively. A_{eff} is the effective area of the junction. The resistance R_t can be written in terms of C_d and the tunneling rate of electrons Γ_t

$$R_t = \frac{1}{\Gamma_t C_d} \quad . \tag{15}$$

Coulomb blockade regime can be roughly defined as the parameter range where the single electron charging energy, given by e^2/C_d exceeds all the other relevant energy scales, such as the characteristic energy-scale for thermal fluctuations (kT) and the broadening arising from the tunneling process itself ($\hbar\Gamma_t$). The latter requirement can be stated namely as

$$R_t > R_Q = \frac{h}{2e^2} \quad , \tag{16}$$

where R_Q is commonly referred to as the *quantum unit of resistance* [13]. By using semiclassical arguments, we can deduce that in the limit

$$\frac{e^2}{C_d} \gg kT, \hbar\Gamma_t \quad , \tag{17}$$

single electron charging can induce large correlations between successive electron injection events. More specifically, electron injection through the barrier by tunneling or across the barrier by thermionic emission can be strongly inhibited by an earlier injection event provided that the circuit recovery time is long compared to the tunneling time [13]. The last condition is satisifed only if

$$R_s \gg R_t \quad , \tag{18}$$

where R_s is the (frequency independent) source resistance seen by the junction. When Eq. (19) is valid, the junction is said to be driven by a *constant current source* with $I \simeq V_{dc}/R_s$, where V_{dc} is the applied voltage. Physically, restoring the missing electron (and hence the pre-injection junction voltage) requires a time $\tau_i = e/I$, leading to a dead time between the injection events as both tunneling (in the WKB limit) and thermionic emission rates depend exponentially on the junction voltage.

These simple and interesting predictions of the Coulomb blockade theory for a single constant-current driven mesosocopic junction have not been observed experimentally due to the effects of the electromagnetic environment, which effectively shunts out the source resistance [14]. Due to these restrictions, the experimental demonstration of Coulomb blockade phenomenon was carried out in double-barrier junctions that are of n-I-i-I-n type [15]. In these experiments, electrons in the n-type regions are degenerate and the capacitances C_d of each junction are approximately equal. Provided that $e^2/C_d \gg kT$, the electron injection will only occur if the voltage drop across the first junction exceeds $e/2C_d$. Following the electron injection event, the energy of the *Coulomb island* is increased by e^2/C_d, while the emitter-collector voltage remains fixed due to very short circuit recovery time (constant voltage operation). The injected electron can only leave by tunneling out into the collector region. However, before this happens, injection of another electron from the emitter is inhibited: the electron injection events across the junction are therefore antibunched but not regulated [16].

We note that in order to investigate the signatures of Coulomb blockade in photon statistics, one should consider p-n or p-i-n type semiconductor junctions [12,16]. We proceed by first considering charging effects in macroscopic p-n junctions.

2.3 Macroscopic Coulomb blockade

In this subsection, we will show that the junction capacitance (C_{dep}) and the operating temperature (T) determine the transition from the macroscopic to mesoscopic regime through the ratio $r = e^2/(kTC_{dep})$ of the single electron charging energy to the characteristic energy of the thermal fluctuations. In the macroscopic and high temperature limit ($r \ll 1$), the electron injection process is sub-Poissonian with variance given by $1/r$. On the other hand, for $r > 1$ (mesoscopic and low temperature limit), the individual injection process is regulated, so that a nonstochastic spike appears in the noise spectrum at the single electron charging frequency I/e, with a squeezed background noise. We also note that for measurement times (T_{meas}) short compared to the thermionic emission time τ_{te}, the injection process is Poissonian, even with an ideal constant current source. The relative magnitude of the three fundamental time-scales: τ_i, τ_{te}, and T_{meas} completely determine the noise characteristics of the carrier injection. Once again, provided that the radiative recombination process is fast compared to these time scales, the same noise properties will be transcribed to the generated light field.

To illustrate this connection, we consider a $p-i-n$ $AlGaAs-GaAs$ heterojunction driven by an ideal constant current source. The carrier transport in such a junction occurs by thermionic emission of electrons from the n-type $AlGaAs$ layer into the p-type $GaAs$ layer, across an undoped (i) $AlGaAs$ section. The rate of thermionic emission of electrons is given by

$$\kappa_{te}(t) = \frac{A_{eff} T^2 A^*}{e} \, exp[\frac{e}{kT}(V_j(t) - V_{bi} - \frac{e}{2C_d})] \, . \tag{19}$$

Here, A^* is the Richardson's constant [12]. Under ideal constant current operation ($dI/dt = 0$), the time dependence of the thermionic emission is exponential

$$\kappa_{te}(t) = \kappa_{te}(0) \, exp[\frac{t}{\tau_{te}} - r\, n_e(t)] \, , \tag{20}$$

where

$$\tau_{te} = \frac{kT\, C_{dep}}{e^2}\frac{e}{I} = \frac{1}{r}\frac{e}{I} \, . \tag{21}$$

The time-constant τ_{te} as defined in Eq.(22) gives the time-scale over which the thermionic emission rate changes appreciably and is termed "thermionic emission time". In Eq.(21), $n_e(t)$ denotes the number of thermionically emitted electrons in time interval $(0,t)$. The $exp[-r\, n_e(t)]$ term can be regarded as providing a "feedback mechanism": Emission of an electron results in a decrease in the thermionic emission rate and makes a second emission event less likely. This feedback is at the heart of Coulomb blockade regulation where the decrease in the emission rate is strong enough to strictly block a second emission event for another single electron charging time e/I. It is also the physical origin of macroscopic squeezing where the decrease in the emission rate is small so that only the regulation of many electrons is possible.

We can show analytically that in the limit of $r \ll 1$ (i.e. the macroscopic regime), the probability for observing n_e electron injection events in an observation time T_{meas} is [12]

$$P(n_e, T_{meas}) \simeq \frac{1}{N(r,\bar{n}_e)} \frac{1}{n_e!} \bar{n}_e^{n_e} \, exp[-\bar{n}_e] \, exp[-\frac{r}{2}(n_e - \bar{n}_e)^2] \, , \tag{22}$$

where $\bar{n}_e = T_{meas} I/e$ and $N(r,\bar{n}_e)$ is the normalization factor. If $\bar{n}_e\, r << 1$ (i.e. $\tau_{te} >> T_{meas}$), then the last term in Eq.(23) can be neglected and one obtains the usual Poisson distribution with $\Delta n_e = \sqrt{\bar{n}_e}$; since the observation time is much shorter than τ_{te}, we do not expect to see the effects of the feedback resulting from a decreased thermionic emission rate. If on the other hand, $\bar{n}_e r \gg 1$ (i.e. $\tau_{te} \ll T_{meas}$), then the n_e dependence is predominantly determined by the last term (i.e. n_e is Gaussian distributed). The standard deviation in this limit is

$$\Delta n_e = \frac{1}{\sqrt{r}} = \sqrt{\frac{kTC_{dep}}{e^2}} \ . \tag{23}$$

Δn_e given in Eq.(24) can be considered as a *fundamental noise limit for macroscopic squeezing* in constant-current driven $p-i-n$ heterojunctions. Even in the limit $r \ll 1$ where emission or tunneling of a single electron creates a very small voltage drop, the combined effect of many electrons is sufficient to control and regulate the electron emission to within several Δn_e. Finally, note that Δn_e obtained from Eq. (24) does not depend on T_{meas}, even though $\bar{n}_e$ does.

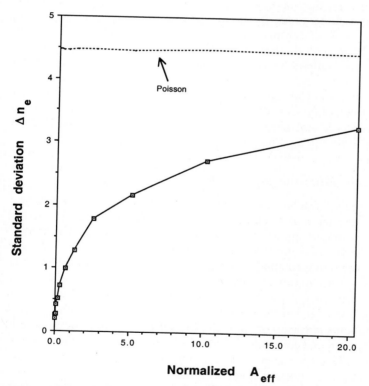

Fig. 1. Δn_e as a function of the junction area

To understand the transition from the mesoscopic ($r > 1$) to macroscopic regime, we consider the dependence of Δn_e on the effective junction area.

Figure 1 shows the results of a classical Monte Carlo simulation [12]. For the chosen (fixed) measurement time $T_{meas} = 20\,e/I$, the expectation value $\bar{n}_e$ of the thermionically emitted electrons is 20. We observe that for very large junction areas ($\tau_{te} > T_{meas}$, or $A_{eff} > 20$), the value of Δn_e is very close to the Poisson value of 4.47: In this limit the electron injection events have practically no effect on the thermionic emission rate. As a result, for the chosen observation time, we have a *random point process* with a constant rate, even though the junction is driven by a *perfect constant-current source*. For junction areas that give $e/I < \tau_{te} < T_{meas}$ ($1 < A_{eff} < 20$), in Fig. 1, Δn_e is approximately proportional to the square root of the area (or C_{dep}) and is clearly below the Poisson limit. Finally, for $e/I > \tau_{te}$ (i.e. $A_{eff} < 1$), Δn_e decreases very sharply: This is the Coulomb-blockade regime where the individual electron injection events become deterministic as the single electron charging energy e^2/C_{dep} exceeds kT.

The ratio r of the single electron charging energy to the characteristic energy of the thermal fluctuations determine the extent of the correlations between the injected electrons: In the mesoscopic limit ($kTC_{dep}/e^2 << 1$), each electron is aware of the previous one due to the significant change in the junction voltage that the last thermionic emission (or tunneling) event created: This is the Coulomb blockade regime discussed above. In the macroscopic limit ($kTC_{dep}/e^2 >> 1$), individual thermionic emission events practically have no effect on the expected injection time of the next electron. A large number of emission events however, do have a combined affect on κ_{te} and it is this feedback that keeps the standard deviation below the Poisson limit.

2.4 Mesoscopic junctions and Coulomb blockade

Even though there are fundamental problems in the realization of a constant current driven submicron-scale single-barrier p-i-n junction, mesoscopic domain provides us unique opportunities for generation of nonclassical light.

First, we consider a constant-voltage driven mesoscopic $P - I_p - i - I_n - N$ junction where the undoped small band-gap i-region acts as the active layer where radiative recombination of injected electrons and holes take place. The capacitance of both $P - I_p - i$ and $i - I_n - N$ regions are assumed to be small enough that both hole and electron injection across the corresponding layers (by either thermionic emission or tunneling) is subject to Coulomb blockade effects. Even if the quantum confinement effects are negligible; the electron and hole injection into the i-layer in such a junction will be correlated. More specifically an electron injection event will increase the energy of the i-layer by e^2/C_{i-I_n-N} thereby enhancing the hole injection rate. Since the injection of a second electron is prohibited until a hole is injected, we expect to find an injection sequence that consists of alternating charges. It is also clear that if we only consider the electron stream, we will find antibunching due to Coulomb blockade. If radiative recombination follows electron and hole injection in a short time-scale, then this antibunching will be transcribed to the generated photons. In this case, the generated light source is antibunched; however, there is no regulation of the photon emission times. This is an example of nonclassical (antibunched) light generation purely due to Coulomb interactions.

If the i-layer of the $P - I_p - i - I_n - N$ junction is a true quantum dot ($L \leq 100$ Å) with strong quantum confinement effects, then we expect to see strong PSF effects. For simplicity, we can consider a junction where charging effects are negligible: due to strong quantum confinement, there will only be two electron and two hole states that are available for radiative recombination at the lowest transition energy of the quantum dot. Even if the injection rate is constant, once two electron-hole pairs are injected, further injection events are prohibited as there are no available final states due to PSF. If one detects only right (or left) hand circularly polarized photons, then only a single electron-hole pair state will contribute to emission: such a system is clearly analogous to a single atom and the generated photons will be antibunched. This is an example of nonclassical (antibunched) light generation that is purely due to PSF effects.

As we have discussed earlier, it is very desirable to go beyond antibunching and regulate the generated photon stream. We next consider an ac-voltage driven mesoscopic junction that achieves this goal: The energy-band diagram of the mesoscopic $P - I_p - i - I_n - N$ $AlGaAs - GaAs$ heterojunction that we analyze is illustrated in Figure 2. If the junction voltage $V_j(t)$ is well below the built-in potential V_{bi} ($V_{bi} - V_j(t) \gg kT$), the carrier transport in such a structure takes place by resonant tunneling of electrons and holes through the undoped I_n and $I_p - AlGaAs$ barrier layers, respectively. The injected electron-hole pairs then recombine radiatively in the $i - GaAs$ layer. We assume that the width of the $i - GaAs$ Coulomb island is small enough that the energy separation of the quantized subbands well exceed the single electron (hole) charging energy and that resonant tunneling into a single conduction (valence) subband need to be considered. The resonant tunneling of an electron or a hole is allowed only when the junction voltage is such that

$$E_{fn} - e^2/2C_{ni} \geq E_{res,e} \geq E_{nc} - e^2/2C_{ni} \quad (electrons) \;, \qquad (24)$$

and

$$E_{fp} + e^2/2C_{pi} \leq E_{res,h} \leq E_{pv} + e^2/2C_{pi} \quad (holes) \;. \qquad (25)$$

Here, $E_{res,e}$ ($E_{res,h}$) is the energy of the electron (hole) resonant subband of the $i - GaAs$ quantum well (or dot); E_{nc} and E_{pv} are the energies of the conduction and valence bands in the $n-$ and $p-$ type layers, respectively; and, E_{fn} and E_{fp} are the Fermi energies in the corresponding layers. C_{ni} and C_{pi} are the capacitances of the $n - i_n - i$ and $p - i_p - i$ regions, respectively. The energies in Eq. (1) are determined by the applied junction voltage $V_j(t) = V_o + v(t)$, where $v(t) = 0$ ($0 \leq t < T_{ac}/2$); and $v(t) = \Delta V$ ($T_{ac}/2 \leq t < T_{ac} = $

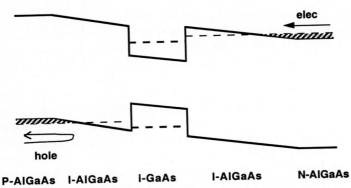

P-AlGaAs I-AlGaAs i-GaAs I-AlGaAs N-AlGaAs

Fig. 2. Energy-band diagram of a mesoscopic $P - I_p - i - I_n - N$ junction

f_{ac}^{-1}). The impurity concentrations on both n and p sides should be small enough that $E_{fn} - E_{nc} \simeq e^2/C_{ni}$ and $E_{pv} - E_{fp} \simeq e^2/C_{pi}$, since we want to be able to turn the tunneling of a particular carrier on and off by applying a voltage pulse whose magnitude is on the order of (but larger than) the single-charge charging energy. Finally, we assume that the Al concentrations in the two barrier regions are chosen independently so as to guarantee that peak electron and hole resonant tunneling occur at (significantly) different values of the applied junction voltage.

We choose the dc-bias voltage (V_o) so that the electron tunneling is resonantly enhanced when $v(t) = 0$. The applied square voltage pulses ($v(t) = \Delta V$) enable resonant hole tunneling, while blocking electron tunneling due to the violation of the inequality in Eq. (25), i.e. by quantum confinement. A second tunneling event of the same carrier during the time interval where the junction voltage remains unchanged, is blocked by Coulomb blockade. Therefore, only one electron and one hole can tunnel into the $i - GaAs$ layer within a single cycle of the applied ac-voltage. Assuming that the radiative recombination occurs in a time-scale short compared to the period of the ac-voltage, a single photon is generated in each cycle with a probability approaching unity. If in addition, the heterostructure is embedded in a photonic band-gap structure, then the photons are spontaneously emitted predominantly into a single mode of the radiation field [4].

Figure 3 shows the junction dynamics obtained using the classical Monte-Carlo method. The period of the ac-voltage is such that both the electron and hole tunneling occurs with very high probability during the time-intervals in which they are allowed. The photon emission events follow hole tunneling in a very short time interval. We can consider the ratio of the period of the photo-emission events to the jitter in the single-photon generation time as a *quality factor* Q for the generated single photon stream. For a source with a time-independent generation rate, this ratio is unity (Poisson limit). In the proposed device, the quality factor is given by the ratio of T_{ac} to

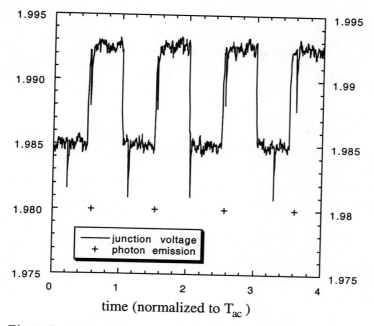

Fig. 3. Junction voltage as a function of time for the mesoscopic $P-I_p-i-I_n-N$ junction

τ_{rad}, provided that the peak hole tunneling rate satisfies $\Gamma_{tunn,h} > \Gamma_{rad}$. For the parameters of Fig. 3, $Q = 30$. If the temperature is kept low enough to avoid secondary tunneling events in a single cycle, one can increase the quality factor by increasing T_{ac}.

The output light field generated by such a junction consists of optical pulses that contain one-and-only-one photon. As we have discussed earlier, these heralded single photons in a given observation-time window form a special class of *multimode number-states*: they have well defined number and emission time information, which is achieved at the expense of increased phase and energy uncertainty [5].

To summarize, we have seen that both Coulomb interactions and PSF effects in mesoscopic p-i-n junctions offer a wealth of opportunities for the generation of nonclassical states of light. Considering the trend in miniaturization in optoelectronic devices, one would expect the physics that we discussed to become relevant for practical devices in the near future.

3 The exciton boser

As we have seen, the quantum statistical nature of electrons and holes can play a crucial role in determining the nonclassical properties of the generated light. On the other hand, it is well known that under weak optical excitation, the absorption and emission spectrum of semiconductors is dominated by excitonic features: an exciton is a bound electron-hole pair that appears due to the Coulomb attraction between the electrons and holes. We can define the exciton operator as [7]

$$\hat{C}_{\nu,k} = \sum_p \varphi_\nu(p) \hat{h}_{\frac{k}{2}-p} \hat{e}_{\frac{k}{2}+p} \quad, \tag{26}$$

where $\hat{e}_k$ and $\hat{h}_k$ denote electron and hole annihilation operators with wavevector k, respectively. Provided that $\varphi_\nu(p)$ is the spatial Fourier transform of the ν'th eigenstate solution of the Wannier equation, the exciton operator $\hat{C}_{\nu,k}$ diagonalizes the single interacting electron-hole pair Hamiltonian. As a result, we expect that at *ultra-low densities* excitons will be the relevant quasi-particles describing the optical properties of the semiconductor. It is also important to note that excitons have the remarkable feature of a giant oscillator strength, which results in ultra-short radiative recombination lifetimes ($\simeq 30psec$) for quantum-well excitons and large normal-mode splitting for cavity-polaritons [17].

3.1 Excitons as bosons

Despite their importance in fundamental spectroscopy of semiconductors, excitons are of relatively little importance from a device perspective. This is predominantly due to the fact that ideal excitons at low densities do not exhibit stimulated photon emission and therefore cannot be used to make coherent light sources based on this gain mechanism. This is due to the quasi-bosonic nature of low-density excitons: if we calculate the commutation relation for $\hat{C}_{\nu,k}$ and its Hermitian conjugate, we find [7]

$$[\hat{C}_{\nu,k}, \hat{C}^\dagger_{\mu,k'}] \simeq \delta_{\nu,\mu} \delta_{p,p'} - \sum_p |\varphi_\nu(p)|^2 (n_{ep} + n_{hp}) \quad, \tag{27}$$

where n_{ep} and n_{hp} denote the electron and hole occupancy, respectively. From Eq. (28) we observe that we can treat the *exciton annihilation* operators $\hat{C}_{\nu,k}$ as *bosonic* provided that the (electron and hole) carrier density N satisfies

$$N a_B^d \ll 1 \quad. \tag{28}$$

Eq. (29) thereby defines the ultra-low density limit for which the treatment of excitons as *composite bosons* is justified; we will assume that this condition holds in our discussion of excitons.

The bosonic nature of excitons that inhibit stimulated photon emission could on the other hand allow for condensation of excitons into a single state: if it can be realized, such an exciton condensate will correspond to a *macroscopic polarization* that forms *spontaneously*, i.e. without an external coherent source. If a macroscopic polarization forms in a semiconductor, coherent light will be generated by spontanoeus emission without any need for stimulated photon emission or a cavity mode.

There has been considerable interest in the realization of exciton BEC over the last 20 years and so far two groups have reported experimental evidence [18,19]. If one assumes an ideal dissipation-free exciton gas with no dynamic screening, one can show that the ground-state of the optically excited semiconductor is given by [20]

$$|\Psi_g\rangle = |\Psi_{BCS}\rangle = \prod_k [u(k) + v(k)\hat{e}_k^\dagger \hat{h}_{-k}^\dagger]|0\rangle \quad , \qquad (29)$$

where $u(k)$ and $v(k)$ are subject to the normalization condition $|u(k)|^2 + |v(k)|^2 = 1$, and are determined by variational techniques. In the low density limit, one can show that $u(k) \simeq 1, \forall k$ and $v(k) = \sqrt{N_{exc}}\varphi_{1s}(k)$ where N_{exc} is the total number of electron-hole pairs in the system. We can also show that in this limit

$$|\Psi_g\rangle \simeq exp[\alpha \hat{C}^\dagger]|0\rangle \quad , \qquad (30)$$

which describes a coherent state of ground-state excitons. One of the fascinating aspects of exciton condensation problem is that the composite-boson density is tunable by the optical excitation strength which in principle allows us to explore both the low-density limit where one should obtain a BEC of excitons and the high density limit where under special circumstances, a BCS-type state can form. Both of these limits are governed by the same ground-state wavefunction given in Eq. (30).

Of particular interest in the context of quantum optics is exciton condensation in a QW structure. The signature of BEC in such a structure is the generation of bi-directional and all-orders coherent light. Since the ground-state of the system is the zero in-plane momentum ($k_\perp = 0$) state, the photons generated by annihilating excitons should conserve the transverse momentum and propagate along $\pm z$ direction. The quantum statistics of the spontaneously generated light also follow those of excitons, in the ultra-low density limit [21]. Even though true condensation is not possible in a homogenoeus two-dimensional system, intentional or unintentional (weak) exciton trapping in the QW plane will allow for BEC in a (finite) QW system [22].

A distinguishing feature of QW excitons however is the finite lifetime of the ground-state excitons due to radiative recombination. In the presence of this dissipation mechanism, a more appopriate description of the condensation process would be the analog of a laser or *boser*, as excitons in this case will remain out-of-equilibrium even in steady-state [21]. From a laser perspective, the condensation process will require the presence of a gain (or cooling) process that can be stimulated by the occupancy of the final state. In the case of excitons, electron-hole-phonon coupling provides a natural mechanism for the system to cool down to its ground-state. In the ultra-low density limit, quantum Boltzman equations predict that a macroscopic fraction of excitons indeed accumulate in the ground-state, provided that the cooling rate is at least comparable to the dissipation rate [22]. The strength of the bosonic enhancement of scattering (i.e. gain) however, strongly depends on the density of excitons. The dissipation (or heating mechanism) that needs to be overcome is given by the finite radiative lifetime which strongly restricts the observability of exciton condensation.

As the previous discussion suggests, the stimulated scattering of excitons is the crucial process that should allow for the realization of a nonequilibrium condensate, or equivalently an *exciton boser*. Since an exciton consists of two (relatively) weakly coupled fermions, it is important to understand the limits in which bosonic enhancement of scattering can be observed.

3.2 Stimulated scattering of composite bosons

It is well known that scattering of identical particles are strongly affected by their quantum statistical nature. In the case of electrons interacting with a phonon reservoir for example, the scattering rate is reduced by the occupancy of the final state, which is a direct signature of the Pauli exclusion principle. For bosonic particles on the other hand, there is an enhancement of scattering by the final-state occupancy, which is termed as *stimulated scattering* or Bose enhancement of scattering. Since we have seen that excitons in the ultra-low density limit can be treated as bosonic particles, we would expect similar stimulation effects for exciton scattering. If for example, we consider relaxation of an excited excitonic state denoted by $\hat{C}_k$ into the ground-state $\hat{C}_0$, we obtain a quantum Boltzman equation

$$\frac{d<\hat{C}_k^\dagger \hat{C}_k>}{dt} = \Gamma_{scatt} <\hat{C}_k^\dagger \hat{C}_k> [1+ <\hat{C}_0^\dagger \hat{C}_0>] \quad , \tag{31}$$

where we assume a zero temperature phonon reservoir ($\bar{n}_{phonon} = 0$) and ideal (bosonic) excitons. Even though we expect Eq. (32) to hold for all particles that can be treated as bosons, a natural question to ask in the case of excitons is the compatibility of stimulated scattering with the Pauli-exclusion of the fermionic particles that form the composite boson. For example, as the composite-boson density increases, we expect the stimulated scattering to be suppressed by Pauli exclusion [23].

In this subsection, we derive an expression for the statistical enhancement factor of composite boson scattering that is valid for all densities. Even though we concentrate on electron-hole-pair phonon interaction in the Born-Markov limit, our basic result can be used for a large class of composite bosons and for arbitrary dimensionality, provided that the Bardeen-Cooper-Schrieffer (BCS)-type ground-state of the system is predetermined [20].

The starting point of this analysis is the interaction of a specific many-body electron-hole state with the phonon reservoir, in the fermion basis. The corresponding interaction Hamiltonian is [23]

$$\hat{H}_{int} = \hbar \sum_{k,q}[g_{e-ph}(q)\hat{e}^{\dagger}_{k+q}\hat{e}_k + g_{h-ph}(q)\hat{h}^{\dagger}_{k+q}\hat{h}_k](\hat{b}_q + \hat{b}^{\dagger}_{-q}) \tag{32}$$

$g_{e-ph}(q)$ and $g_{e-ph}(q)$ are the corresponding electron-phonon and hole-phonon interaction coefficients. The spin index and the vector nature of the momenta are suppressed for simplicity. The same set of variables could also be used to describe electron-hole-phonon system under large magnetic fields.

We proceed by assuming that the initial many-body electron-hole state state can be written as [20]

$$|\Psi_{in}\rangle = \hat{C}^{\dagger}_{\nu,K}|\tilde{\Psi}_{BCS}\rangle = \hat{C}^{\dagger}_{\nu,K}\prod_{k}[\tilde{u}(k) + \tilde{v}(k)\hat{e}^{\dagger}_{k}\hat{h}^{\dagger}_{-k}]|0\rangle \tag{33}$$

The final (electronic) state of the scattering process is simply $|\Psi_{BCS}\rangle = \prod_{k}[u(k) + v(k)\hat{e}^{\dagger}_{k}\hat{h}^{\dagger}_{-k}]|0\rangle$ where $\sum_{p}|v(p)|^2 = N+1$ is the mean number of electron-hole pairs. In the following discussion, we will assume that $N \gg 1$, so that the difference between $v(p)$ ($u(p)$) and $\tilde{v}(p)$ ($\tilde{u}(p)$) is negligible $\forall p$.

The next step is to obtain the scattering rate in the Born-Markov approximation using the Fermi's Golden Rule. For simplicity, we will assume that the lattice is at zero temperature and only consider the spontaneous phonon processes. We then obtain the scattering rate

$$W_s = 2\pi \sum_{q}|g_{e-ph}(q)M(q) + g_{h-ph}(q)M(-q)|^2 \delta(\omega_{in} - \omega_{fin} - \omega_q), \tag{34}$$

where

$$M(q) = \sum_{k}\varphi^{*}_{\nu}(k + \frac{q}{2})u^{*}(k)v(k)[1 - |v(k+q)|^2] \quad . \tag{35}$$

Here, $\hbar\omega_{in}$ and $\hbar\omega_{fin}$ denote the the energies of the initial and final many-body eigenstates, respectively. ω_q is the frequency of the emitted phonon. We reiterate that Eq. (35) is derived using the interaction Hamiltonian of Eq. (33) and no assumption regarding the *bosonic character* of electron-hole pairs were made. The product $u^{*}(k)v(k)$ is proportional to the electron-hole pair wave-function. In the Hartree-Fock approximation, $u^{*}(k)v(k)$ can be determined from the semiconductor Bloch equation (SBE) for the polarization term. As we have already seen, for ultra-low density excitons, this equation reduces to the Wannier equation in real space and yields the 1s exciton wave-function $\varphi_{1s}(p)$. In the high density limit ($Na_B^d/L^d > 1$), it is the counterpart of superconducting gap equation. In many cases however, Hartree-Fock approximation is not valid and the calculation of the pair wave-function would require the inclusion of screening and scattering terms.

The factor $[1 - |v(k+q)|^2]$ in Eq. (36) gives the correction to the electron-hole-phonon scattering arising from the fact that the presence of a BCS ground-state with a large number of composite bosons modify the commutation relation of high-momentum electron-hole pairs as well. To the extent that $K \gg k_F$ where k_F is the Fermi wave-vector, the contribution of this term is negligible. If we in addition assume $K \sim q \gg \pi/a_B$, we can also neglect the k-dependence of $\varphi_\nu^*(k + \frac{q}{2})$, provided that φ_ν is a Hydrogenic wavefunction. In this limit, we obtain

$$M(q) \simeq \varphi_\nu^*(\frac{q}{2}) \sum_k u^*(k)v(k) \quad , \tag{36}$$

and

$$W_s \simeq 2\pi \sum_q |G(q)|^2 . I . \delta(\omega_{in} - \omega_{fin} - \omega_q) \quad , \tag{37}$$

where

$$G(q) = g_{e-ph}(q)\varphi_\nu^*(\frac{q}{2}) + g_{h-ph}(q)\varphi_\nu^*(\frac{-q}{2}) \tag{38}$$

and

$$I = |\sum_k u^*(k)v(k)|^2 \quad . \tag{39}$$

The expression for I given in Eq. (40) contains the statistical enhancement factor for the scattering of a phonon by a many-body composite boson (electron-hole pair) system. In the low density limit where $v(k) \simeq \sqrt{N}\varphi_{1s}(k)$, we obtain $I \propto N$ for all composite bosons, as expected. In this case ($u(k) \sim 1, \forall k$), bosonic enhancement arises from a constructive interference of the contributions from all partially occupied pair states. In the opposite high density limit, the qualitative nature of saturation and Pauli blocking of statistical enhancement factor strongly depends on the particular BCS-state (i.e. the coherence factors $u(k), v(k)$). Here, only the states around the Fermi level for which $u^*(k)v(k) \neq 0$ contribute to I. Equivalently for this latter case, the electron-hole pairs with $k \ll k_F$ have exhausted the phase-space avaliable for them ($v(k) \simeq 1$) and can no longer participate in stimulated scattering. Physically, this is due to the fact that the mean separation of the electron-hole pairs is less than their size, which makes the Pauli exclusion dominant.

Eq. (40) shows that the stimulated scattering explicitly depends on the overlap $u^*(k)v(k)$. Therefore it is the coherence between the electron-hole pair states that results in bosonic enhancement. Conversely, if the ground-state of the many-body system is an electron-hole plasma state where $u^*(k)v(k) = 0, \forall k$, there is no final-state stimulation at any electron-hole pair density. We remark that even though we assume a BCS-state in our analysis, the assumption of a well-defined condensate phase should not be relevant for the bosonic enhancement factor.

Next, we consider the special case of two-dimensional (2D) magnetoexcitons. It has been shown that in the strong magnetic field limit where the magnetic length $a_0 = \sqrt{\hbar/eB}$ is much smaller than a_B, the magnetoexcitons become ideal non-interacting bosons [24,25]. More specifically, Paquet et al. [25] have shown that the single-particle wave-function remains unchanged for all occupancies of the lowest exciton-band. For this system we have

$$v(k) = v = \sqrt{2\pi a_o^2 N/L^2} \quad , \tag{40}$$

where L is the transverse size of the 2D structure. The evaluation of the stimulated scattering contribution is then straightforward:

$$I = N\left(1 - \frac{2\pi a_0^2 N}{L^2}\right)\frac{L^2}{2\pi a_0^2} \quad . \tag{41}$$

The analytical expression given in Eq. (42) is valid for $1 \leq N \leq \frac{L^2}{2\pi a_0^2}$. In the low density limit ($N \ll \frac{L^2}{2\pi a_0^2}$), magnetoexcitons behave as ideal bosons ($I \propto N$). The total scattering rate peaks at $N_{max} = N_M = \frac{L^2}{4\pi a_0^2}$ where only half of the magnetoexcitons contribute to stimulation. For $N > N_M$, stimulated scattering rate into the ground-state starts to decrease and goes to zero as $N \to 2N_M$; at this occupancy, all the underlying electron and hole fermionic phase-space is exhausted (i.e. the first electron and hole Landau levels are full) and it is not possible to create another ground-state magnetoexciton.

We therefore see that stimulated scattering of excitons can indeed survive exciton densities far exceeding the more restrictive ultra-low density limit, provided that the electron-hole pairing continues to exist. As we discussed earlier, this stimulated scattering of excitons plays the key role of stimulated emission in an exciton laser or boser and hence is of fundamental importance [21]. Of particular interest is the possibility of observing stimulated scattering of cavity-polaritons, which could lead to a novel form of coherent light source [26].

References

1. R. Loudon 1985 *The Quantum Theory of Light*. Oxford University Press, New-York
2. D. F. Walls and G. J. Milburn (1994) *Quantum Optics*. Springer-Verlag, Berlin
3. C. W. Gardiner (1991) *Quantum Noise*. Springer-Verlag, Berlin
4. A. Imamoglu and Y. Yamamoto (1994) Turnstile device for heralded single photons: Coulomb blockade of electron and hole tunneling in quantum confined p-i-n heterojunctions. Phys. Rev. Lett. **72**, 210–213
5. A. Imamoglu, H. Schmidt, G. Woods, and M. Deutsch (1997) Strongly interacting photons in a nonlinear cavity. Phys. Rev. Lett., **79**, 1467–1470
6. C. K. Law and H. J. Kimble (1997) Deterministic generation of a bit-stream of single-photon pulses. J. of Modern Optics **44**, 2067–2074
7. H. Haug and S. W. Koch (1993) *Quantum Theory of the Optical and Electronic Properties of Semiconductors*. World Scientific, Singapore

8. Y. Yamamoto, S. Machida, and O. Nilsson (1991) in *Coherence, Amplification and Quantum Effects in Semiconductor Lasers*, ed. Y. Yamamoto. John Wiley, NY
9. W. H. Richardson, S. Machida, and Y. Yamamoto (1991) Squeezed photon-number and sub-Poissonian electrical partition noise ina semiconductor laser. Phys. Rev. Lett. **66**, 2867–2870
10. A. Shimizu and M. Ueda (1992) Effects of dephasing and dissipation on quantum noise in conductors. Phys. Rev. Lett. **69**, 1403–1406
11. C. W. J. Beenakker and M. Buttiker (1992) Suppression of shot noise in metallic diffusive conductors. Phys. Rev. B **46**, 1889–1892
12. A. Imamoglu and Y. Yamamoto (1993) Noise suppression in semiconductor p-i-n junctions: Transition from macroscopic squeezing to mesoscopic Coulomb blockade of electron emission processes. Phys. Rev. Lett. **70**, 3327–3330
13. D. V. Averin and K. K. Likharev (1986) Coulomb blockade of single-electron tunneling, and coherent oscillations in small tunnel junctions. J. Low Temp. Phys. **62**, 345–373
14. M. H. Devoret, D. Esteve, H. Grabert, G. -L. Ingold, H. Pothier, and C. Urbina (1990) Effect of the electromagnetic environment on the Coulomb blockade in ultrasmall tunnel junctions. Phys. Rev. Lett. **64**, 1824–1827
15. L. P. Kouwenhoven, A. T. Johnson, N. C. van der Vaart, C. J. P. M. Harmans, and C. T. Foxon (1991) Quantized current in a quantum-dot turnstile using oscillating tunnel barriers. Phys. Rev. Lett. **67**, 1626–1629
16. A. Imamoglu, Y. Yamamoto, and P. Solomon (1992) Single-electron thermionic emission oscillations in pn microjunctions. Phys. Rev. B **46**, 9555–9563
17. C. Weisbuch, M. Nishioka, A. Ishikawa, and Y. Arakawa (1992) Observation of the coupled exciton-photon mode splitting in a semiconductor quantum microcavity. Phys. Rev. Lett. **69**, 3314–3317
18. J.-L. Lin and J. P. Wolfe (1993) Bose-Einstein Condensation of Paraexcitons in Stressed Cu_2O. Phys. Rev. Lett. **71**, 1222–1225
19. L. V. Butov, A. Zrenner, G. Abstreiter, G. Bohm, and G. Weimann (1994) Condensation of Indirect Excitons in Coupled AlAs/GaAs Quantum Wells. Phys. Rev. Lett. **73**, 304–307
20. C. Comte and P. Nozieres (1982) Exciton Bose condensation: the ground state of an electron-hole gas. J. Physique **43**, 1069–1081
21. A. Imamoglu and. R. J. Ram (1996) Quantum dynamics of exciton lasers. Phys. Lett. A **214**, 193–198
22. W. Zhao, P. Stenius, and A. Imamoglu (1997) Kinetics of condensation in trapped exciton gases. Phys. Rev. B **56**, 5306–5315
23. A. Imamoglu (1998) Phase-space filling and stimulated scattering of composite bosons. Phys. Rev. B **57**, R4195–R4197
24. I. V. Lerner and Yu. E. Lozovik (1981) Two dimensional electron-hole system in a strong magnetic field as an almost ideal exciton gas. Zh. Eksp. Teor. Fiz. **80**, 1488-1495 (1981) [Sov. Phys.-JETP **53**, 763–770]
25. D. Paquet, T. M. Rice, and K. Ueda (1985) Two dimensional electron-hole fluid in a strong perpendicular magnetic field: exciton Bose condensate or maximum density two-dimensional droplet. Phys. Rev. B **32**, 5208–5221
26. A. Imamoglu, R. J. Ram, S. Pau, and Y. Yamamoto (1996) Nonequilibrium condensates and lasers without inversion: Exciton polariton lasers. Phys. Rev. A **53**, 4250–4253

Semiconductor Microcavities, Quantum Boxes and the Purcell Effect

Jean-Michel Gérard and Bruno Gayral

France Telecom/CNET/DTD/CDP, 196 avenue Henri Ravera,
92220 Bagneux, FRANCE

Abstract : We discuss the recent observation of a strong enhancement of the spontaneous emission rate (Purcell effect) for self-assembled InAs/GaAs quantum boxes inserted in GaAs-based pillar microcavities or microdisks.

1 Introduction

As proposed by Purcell fifty years ago[1], the spontaneous emission (SE) rate of a dipole can be tailored by using a cavity to modify the dipole-field coupling and the density of available photon modes. Cavity Quantum Electrodynamics (CQED) has provided a solid theoretical basis as well as a spectacular experimental support to this revolutionary concept[2-3], and has been since 1990 a major source of inspiration for the research activity on solid-state optical microcavities. An ability to enhance the SE rate (Purcell effect) of a solid-state emitter in the weak coupling regime would open major novel avenues for physics and engineering, such as the fabrication of high frequency and/or high efficiency light-emitting diodes.

Until very recently, attempts to observe the Purcell effect in solid-state microcavities have been somewhat disappointing. High quality planar cavities can be produced by layer-by-layer deposition techniques, but such cavities entail only minor modifications of the SE rate, as predicted theoretically[4] and observed for rare-earth atoms[5], semiconductor quantum wells[6] (QWs) or quantum boxes[7] (QBs) in the weak coupling regime. Hopefully, progress in microfabrication has allowed a three-dimensional (3D) engineering of the refractive index on the wavelength scale, and the fabrication of several kinds of solid-state microcavities providing a strong three-dimensional photon confinement. Many of these have been for years good enough to generate a potentially strong Purcell effect (see section III), but the lack of an appropriate emitter has been a major hindrance to its clear observation. Firstly, most available solid-state emitters (bulk semiconductor or QW, rare earth atoms...) are indeed spectrally much broader than the resonant modes of these cavities, which weakens considerably the magnitude of the Purcell effect. Secondly, non-radiative carrier recombination at the microcavity sidewalls may also dominate, for bulk material or QWs, the intrinsic modification of the carrier lifetime[8].

Using self-assembled InAs/GaAs QBs drastically changes this state of affairs, due to a unique combination of assets[9,10]. Firstly, such nanometer-scale QBs support well-separated discrete electronic states and exhibit a single narrow emission line[11,12] (<<

kT), which permits to exploit the full potentialities of high Q cavities (up to $Q\sim10000$ typically). Standard InAs QBs, emitting in the 0.9-1.1µm range depending of their size, are also defect-free and display a high radiative quantum yield ($\eta\sim1$). They capture and trap charge carriers very efficiently, which prevents their diffusion toward non-radiative recombination centers[9,13]. As a result, this ensures that any modification of the SE rate observed for QBs in 3D cavities, is assigned to intrinsic cavity effects. Finally self-assembled QB arrays also offer practical advantages, such as an easy insertion within semiconductor cavities during their epitaxy, and an absorption level which is most often low enough to avoid any significant degradation of the cavity optical quality[14]. It should be noted however that these various advantages are rather fragile. Firstly, a raise of the temperature -above 100K for the shallower QBs- entails carrier thermoemission and results to a decrease of η. Secondly, novel emission lines and an overall broadening of the spectrum of single QBs appear under high excitation conditions, i.e. when more than one electron-hole pair are injected in the QB, due to the strong Coulomb interaction between the trapped carriers[15,16]. These effects generate only minor constraints when performing CQED experiments on QBs, but might be a more severe impediment to the room-temperature operation of devices based on CQED effects on QBs.

Nevertheless, this unique combination of assets has permitted a very clear observation of the Purcell effect in 1997 for InAs QBs embedded in pillar microresonators[10] and more recently in microdisks. We present in this paper our experimental results in section IV, and show that a good quantitative estimate of the magnitude of the Purcell effect can be obtained from simple considerations. We emphasize in particular the usefulness of an estimate of Purcell's cavity figure of merit F_p in this context. However, various expressions can be found for F_p in the litterature, which differ by a factor as large as 10. For the sake of clarity F_p we thus first derive explicitly in section II the original expression of Purcell's factor F_p and precise its physical meaning. We then evaluate in section III F_p for different 3D microcavities in order to compare their respective assets for SE control. We will finally discuss in section V potential developpements of these studies, ranging from the fabrication of revolutionary single-mode emitters emitting photons one-by-one in a deterministic way, to the search of other CQED effects such as strong coupling on single QBs.

2 Purcell's factor revisited

2.1 SE in a strongly damped single-mode microcavity[17]

We consider a single localized emitter, initially in its excited state and resonantly coupled to a single empty mode of a microcavity. If the cavity were perfect, the system would experience a Rabi oscillation at the angular frequency Ω. In practical cases, cavity relaxation must be included in order to get a realistic description of the SE process in the cavity. It is well known that when the cavity losses are large enough ($4\Omega<\omega/Q$, where Q is the mode quality factor and ω the emitter's angular frequency), the oscillatory behavior is lost. In this so-called « weak coupling » or « strong damping » regime, the evolution of the emitter to its ground state is exponential, as when the emitter is in the free space, but occurs at a different rate. In the strong damping limit, ($\Omega \ll \omega/Q$), the SE is characterized by an exponential decay whose rate is given by[17]:

$$\frac{1}{\tau} = \frac{4\Omega^2 Q}{\omega} \quad (1)$$

This value is to be compared with the SE rate of this emitter when it is imbedded in a transparent homogeneous medium of refractive index n:

$$\frac{1}{\tau_{free}} = \frac{|\vec{d}|^2 \omega^3 n}{3\pi\varepsilon_0 \hbar c^3} \quad (2)$$

where $\vec{d}$ is the electric dipole of the localized emitter.

In order to compare both rates, let us introduce some notations and recall the expression of the Rabi angular frequency Ω in a general case. Field quantization leads to the following expression for the electric field operator for the cavity mode (see Ho and Fabre lectures):

$$\vec{\hat{E}}(\vec{r},t) = i\, \varepsilon_{max}\, \vec{f}(\vec{r})\hat{a}(t) + h.c. \quad (3)$$

where h.c. means hermitian conjugate, $\hat{a}$ is the photon creation operator and $\vec{f}$ the mode spatial function. $\vec{f}$ is a complex vector which describes the local field polarization and relative field amplitude ; it obeys Maxwell equations and is normalized so that its norm is unity at the antinode of the electric field. The numerical prefactor ε_{max}, which is often named in a somewhat improper way « maximum field per photon » can be estimated by expressing that the vacuum-field energy is, for each mode, $\hbar\omega/2$:

$$\varepsilon_{max} = \sqrt{\frac{\hbar\omega}{2\varepsilon_0 n^2 V_{eff}}}, \quad \text{where} \quad V_{eff} = \frac{1}{n^2}\iiint_{\vec{r}} n(\vec{r})^2\, |\vec{f}(\vec{r})|\, d^3\vec{r} \quad (4)$$

In this expression, n is the refractive index at the field maximum and V_{eff} the effective cavity volume, which describes how efficiently the cavity concentrates the electromagnetic field in a restricted space. More precisely, V_{eff} is the volume of an hypothetic cavity, defined by Born-Von Karman periodic boundary conditions, which would provide the same maximum field per photon than the cavity under study.

For an electric-dipole transition and within the standard rotating wave approximation, the atom-field interaction hamiltonian becomes :

$$\hat{H}_{int} = -i\,\varepsilon_{max}\, \vec{d}.\vec{f}(\vec{r}_e)|g\rangle\langle e|\hat{a}\ +\ h.c \quad (5)$$

where |g> and |e> design the ground and excited states of the two-level system and $\vec{r}_e$ the position of the localized emitter. For a perfect (lossless) cavity, its Rabi frequency would thus be:

$$\hbar\Omega = \left|\varepsilon_{max}\, \vec{d}.\vec{f}(\vec{r}_e)\right| \quad (6)$$

Using these expressions, we see that the emitter's SE rate is enhanced (or inhibited) in the cavity with respect to its value for an homogeneous surrounding material by a factor :

$$\frac{\tau_{free}}{\tau_{cav}} = \frac{3Q(\lambda_c/n)^3}{4\pi^2 V_{eff}} \cdot \frac{|\vec{d}\cdot\vec{f}(\vec{r}_e)|^2}{|\vec{d}|^2} \quad (7)$$

Whereas the first term is only related to cavity properties (Q, V_{eff}), the second one (which is always smaller than 1) depends on the relative field amplitude at the emitter's location and on the orientation matching of the transition dipole and electric field. In order to find a figure of merit describing the ability of the cavity for SE control, it is convenient to consider the SE rate of an « ideal » emitter, whose properties allow to maximize the magnitude of the Purcell effect. This ideal emitter should be located at a maximum of the electric field, with its dipole aligned with the local electric field. This figure of merit takes the form proposed by Purcell fifty years ago :

$$F_p = \frac{\tau_{free}}{\tau_{cav}} = \frac{3Q(\lambda_c/n)^3}{4\pi^2 V_{eff}} \quad (8)$$

Let us finally note that the strong damping condition can be rewritten as :

$$\Delta\omega_e = \frac{1}{\tau_{cav}} = \frac{4\Omega^2 Q}{\omega} \ll \frac{\omega}{Q} = \Delta\omega_c \quad (9)$$

which shows that the homogeneous emitter's linewidth $\Delta\omega_e$ is much smaller than the linewidth of the cavity mode $\Delta\omega_c$. As discussed by Ho in his lecture, it is then valid to treat the single cavity mode as a continuum, and to apply the Fermi Golden Rule, which was the original approach of Purcell[1]. We will also detail this method here, in order to show that both calculations lead to the same result. This derivation will also highlight the role of a detuning between the emitter and the cavity mode.

2.2 SE rate from the Fermi Golden Rule

For an electric dipole transition, the Fermi Golden Rule can be written as :

$$\frac{1}{\tau} = \frac{2\pi}{\hbar^2}\rho(\omega_e).<|<\vec{d}\cdot\hat{\vec{\varepsilon}}(\vec{r}_e)>|^2> \quad (10)$$

where $\rho(\omega_e)$ the density of photon modes at the emitter's angular frequency ω_e and where the averaging of the squared dipolar matrix element is performed over the various modes seen by the emitter.

The insertion of the radiating dipole inside the cavity will change its SE rate in three ways : the spectral density of modes, the amplitude of the vacuum field and its orientation with respect to the radiating dipole are indeed all modified. We evaluate in the following the resulting change of the SE rate for a cavity supporting a single-mode (angular frequency ω_c, linewidth $\Delta\omega_c$ and quality factor $Q=\omega_c/\Delta\omega_c$). In this case, the mode density seen by the emitter is given by a normalized Lorentzian :

$$\rho_{cav}(\omega) = \frac{2}{\pi\,\Delta\omega_c}\cdot\frac{\Delta\omega_c^{\;2}}{4(\omega-\omega_c)^2 + \Delta\omega_c^{\;2}} \quad \text{and} \quad \rho_{cav}(\omega_c) = \frac{2}{\pi\,\Delta\omega_c} = \frac{2Q}{\pi\omega_c} \quad (11)$$

whereas the « free-space » mode density can be written as:

$$\rho_{free}(\omega) = \frac{\omega^2 V n^3}{\pi^2 c^3} \quad (12)$$

Using a normalization volume V, the field operator for each free-space mode is :

$$\vec{E}(\vec{r},t) = i\,\vec{\varepsilon}\,\sqrt{\frac{\hbar\omega}{2\varepsilon_0 n^2 V}}\,e^{i\vec{k}\vec{r}}\,\hat{a}(t) + h.c. \quad (13)$$

where $\vec{\varepsilon}$ is a unit vector describing the polarisation of the mode.

Using the Fermi Golden Rule for both estimates, we can compare the SE rate of the emitter in the single-mode cavity to the case of an homogeneous surrounding medium :

$$\frac{\tau_{free}}{\tau_{cav}} = \frac{2\pi}{V_{eff} n^3 c^3 \omega_c^2 \Delta\omega_c} \cdot \frac{\Delta\omega_c^2}{4(\omega_e - \omega_c)^2 + \Delta\omega_c^2} \cdot \frac{\xi^2 |\vec{f}(\vec{r}_e)|^2}{1/3} = \frac{3Q(\lambda_c/n)^3}{4\pi^2 V_{eff}} \frac{\Delta\omega_c^2}{4(\omega_e - \omega_c)^2 + \Delta\omega_c^2} \xi^2 |\vec{f}(\vec{r}_e)|^2 \quad (14)$$

where $\xi = |\vec{d}.\vec{f}(\vec{r}_e)| / |\vec{d}|.|\vec{f}(\vec{r}_e)|$ describes the orientation matching of $\vec{d}$ and $\vec{f}(\vec{r}_e)$, and where 1/3 is the averaging factor accounting for the random polarization of free-space modes with respect to the dipole. We thus retrieve the enhancement factor given by Eq. (8) for an emitter perfectly on resonance with the cavity mode ($\omega_e = \omega_c$). Here again, F_p appears as the largest SE rate enhancement which can be induced by the cavity. In order to observe the Purcell effect in its full magnitude, our « ideal » emitter should be well matched with the frequency, spatial distribution, and polarisation of the mode, and should have an emission line much narrower than the cavity mode.

The Purcell factor F_p provides a practical means for comparing different types of microcavities, as it will be done in the next section. As already stated, one should however keep in mind that F_p is a figure of merit for one of the resonant cavity modes alone. In order to interpret experimental data on SE rate enhancement, we first have to take into account the location, detuning and dipole orientation of the emitter(s). If it is coupled to a continuum of leaky modes (which is usual for solid-state cavities) and/or to several confined modes, we will also add the partial SE rates obtained for the various modes in order to get the total SE rate of the emitter.

3 Purcell factor of various 3D solid-state microcavities

For cavities providing a 3D confinement, it is in general possible to tune Q and V_{eff} independently. This is a major advantage over planar cavities, for which SE rate modifications only depend on the effective cavity length[4] (and emitter location). Since 1990, many approaches have been used to obtain a 3D photon confinement. Silica microspheres[18-19], pillar microresonators[8-10,20-28], photonic disks[29-32] and wires[33], 1D and 2D photonic bandgap (PBG) microcavities[34-36], all sustain a discrete set of resonant modes and are likely to display the Purcell effect ($F_p \gg 1$), provided that a « sufficiently ideal » emitter is used. We give in this section a brief overview of this quest for high F_p semiconductor microcavities.

3.1 Pillar microcavities

Micropillars have received much attention since the first fabrication of vertical cavity surface lasers[20,21]. Their interest for spontaneous emission control has been acknowledged as early as 1991[22]. We present in figure 1 a typical micropillar fabricated through the reactive ion etching (RIE) of a GaAs/AlAs λ-cavity resonant around 0.9 μm. A 3D optical confinement is obtained through the combination of the waveguiding along the pillar (due to the high refractive index contrast at the semiconductor/air interface) and of the longitudinal confinement by the distributed Bragg reflectors (DBRs). A good estimate of the energies and field distributions of the resonant modes can be obtained by expressing these as simple linear combinations of the guided modes of a GaAs cylinder[24]. We can in particular estimate in this framework, their effective height ($\sim 2\lambda_c/n$) and area ($\sim \pi R^2/4$), where R is the pillar radius. Therefore the 1 μm diameter pillar shown in figure 1 is able to confine light within an effective volume as small as $\sim 5(\lambda_c/n)^3$.

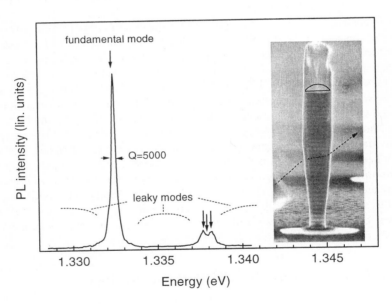

Fig. 1. Typical c.w. PL spectrum, obtained for a 3 μm diameter micropillar containing InAs QBs. The arrows indicate the calculated energies for the resonant modes of the pillar. The noise level for this spectrum is hundred times smaller than the background PL from leaky modes. Insert : SEM displaying a 1μm diameter GaAs/AlAs micropillar.

The resonant modes of 3D microcavities can be easily studied by micro-photoluminescence (mPL), using QB arrays placed in the cavity as a broadband internal light source, as shown for micropillars[9,24] as well as PBG microstructures[35]. These modes contribute to a series of sharp lines in the mPL spectrum (fig 1), which allows to measure their energies and Q. Unlike QWs, the insertion of QBs allows to probe the empty cavity Q, since it introduces an absorption loss which is in general negligible[14].

The way Q evolves when the pillar size is reduced is a major issue in the context of our study. We show in figure 2 a typical result obtained for the fundamental mode on a series of micropillars etched in a high-finesse planar microcavity (Q=5200). For large enough pillars, Q takes a constant value, equal to the planar cavity Q. Below a certain critical diameter, here 3 µm, Q decreases[24,27,28], which indicates a reduction of the escape time of the photons. This trend is due to the increased efficiency of several diffusion/diffraction processes which can scatter confined photons out of the pillar. We have paid much attention for the present pillars to two usually dominant processes: a) unlike previous work[24,25], we etch here the major part of the bottom mirror; this reduces the field intensity at the bottom of the pillar and quenches the diffraction by the finite aperture of the pillar foot; b) an optimization of the RIE process has allowed to improve the smoothness of the semiconductor sidewall, and to limit the efficiency of the related scattering process. As a result, our smallest pillars (d~ 0.8 µm) retain very large Qs (Q>1000). Considering now F_p, we see that the decrease of V_{eff} overcomes the degradation of Q, so that F_p still tends to increase for our smallest pillars. We obtain for a 1µm diameter F_p's as large as 32, which constitutes, to the best of our knowledge, the highest value ever achieved for pillar microcavities.

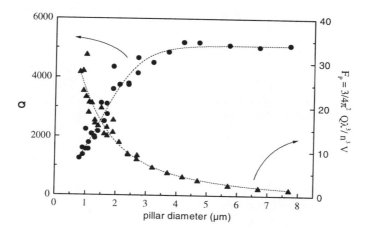

Fig. 2. Plot of the cavity quality factor Q (circles) and Purcell factor F_p (triangles) for the fundamental mode of GaAs/AlAs micropillars as a function of their diameter d.

In practice, the sensitivity of Q on the sidewall morphology is such that reproducibility is still a major subject of concern for pillar diameters below 1.5 µm typically. Further improvements of F_p for the pillar geometry are more likely to come from novel material combinations than from a further reduction of the pillar diameter. In this context, challenging novel opportunities are opened by the selective oxydation of AlAs into low-index AlOx[37-39]. A substitution of GaAs/AlAs DBRs by GaAs/AlOx ones, drastically reduces the spreading of the confined modes into the DBRs, and allows a three-fold decrease of V_{eff}. Previous studies have also shown that high Q values can also be obtained with a small number of DBR periods (e.g. $Q \sim 1000$ with 4 periods for both DBRs[39]). We present in figure 3 some recent results obtained for GaAs/AlOx micropillars based on a GaAs λ-cavity (containing InAs QBs) surrounded by two four-period DBRs. As shown on the SEM micrograph, our present process leads to very rough sidewalls. This explains why we could not observe until now Purcell factors larger than 14 in this system, in spite of the record-high Q's (>3000) obtained for the largest pillars. Further technological developments should nevertheless lead to major improvements in the near future.

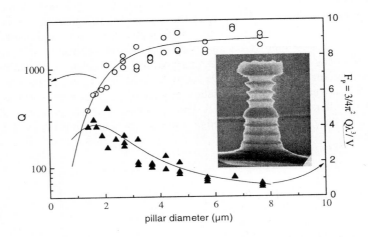

Fig. 3. Plot of the cavity quality factor Q (circles) and Purcell factor F_p (triangles) for the fundamental mode of GaAs/AlOx micropillars as a function of their diameter d.

3.2 Microdisks[29-32]

It is well known that semiconductor microdisks support a series of whispering gallery modes (WGMs), which are guided by total internal reflection –and tightly confined– at the lateral edge of the disk. Microdisks as shown in figure 4 are usually fabricated using a combination of RIE, which defines a vertical GaAs/AlGaAs cylinder, and of a selective wet chemical etching which forms the disk pedestal. Compared to the pillars, for which the resonant modes are partially delocalized in the DBRs and all over the pillar area, this geometry provides a more efficient 3D confinement. For a small radius GaAs disk ($\lambda_c/n < R < 20 \lambda_c/n$), the effective area of the lowest radial quantum number WGM is approximately given by $0.86 \lambda_c^2 (R/\lambda_c)^{3/2}$, where all lengths are expressed in μm. The effective height H_{eff} of the WGM can be evaluated by considering the field distribution of the guided mode of the air-confined GaAs slab. For the 250 nm thick GaAs disk shown on fig 4, H_{eff} is 175nm or $\sim 0.6 \lambda_c/n$ when λ_c is around 1 μm, and the effective volume is of the order of $6 (\lambda_c/n)^3$.

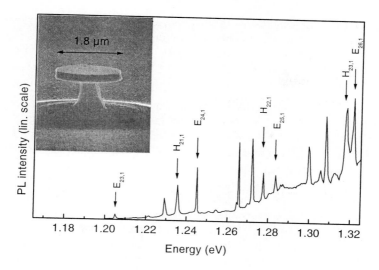

Fig. 4. Typical c.w. PL spectrum, obtained for a 3 μm diameter microdisk containing InAs QBs. The arrows indicate the estimated energies of some whispering gallery modes. Insert : Scanning electron micrograph displaying a 1.8μm diameter GaAs microdisk fabricated using a two-step wet-etching process.

For WGMs, Q is intrinsically limited by radiation losses[29]. In practice however, the diffusion by the roughness of the disk edge is by far the major escape path for confined photons. When the disk contains QWs, their absorption becomes the dominant loss mechanism ; it is then necessary to work at the transparency threshold to study the optical properties of the empty cavity[30,31]. As for pillars, a simpler way for investigating the empty cavity Q is to perform mPL on disks containing QBs.

Figure 4 shows a typical mPL spectrum obtained on a single 3μm diameter microdisk. The sharp lines constitute the WGM contribution to the spectrum, while the broad background due to the emission of non-resonant QBs into the leaky modes. The large number of observable WGM is quite remarkable, since only one or two WGMs (with TE polarization and radial quantum number n_r =1) are clearly observed for QW emitters, and only above the lasing threshold. An attempt to identify these various WGM shows that both TE and TM modes are observed. Within the broad spectral range covered by the QB array, all modes with n_r =1 are observed as well as some higher order WGMs (n_r =2 and probably 3).

The small linewidth of the WGMs highlights the good optical quality of such microcavities. In a first experiment, we have studied microdisks processed using RIE, with Q's of the order of 3000 for the best modes. We have more recently optimized a two-stage wet etching process, which improves considerably the smoothness of the disk edge. Q's close to 12000 have been obtained in this case for 1.8μm diameter microdisks, which is eight times higher than the best previously reported results[31,32]. Q values in the 5000-10000 range are routinely obtained. These large Q's also corresponds to a very high Purcell factor, of the order of 125 for Q=10000. Microdisks appear therefore as excellent candidates for the study and implementation of the Purcell effect in a solid-state microcavity, all the more since their processing is by far more simple and more reproducible than for high-F_p pillars.

3.3 Other 3D microcavities

Several other 3D microcavities have been presented during this summer school. S.T. Ho described the properties of photonic wire lasers[33], based on a ridge waveguide forming a ring resonator. Compared to microdisks, the wire geometry potentially provides a better radial confinement, thus reducing the number of resonant cavity modes for a given diameter. This is the main reason why larger β SE coupling coefficients can be observed on wires (β>0.3) compared to disks (β~0.1-0.2 [30]) for QW emitters. In order to keep a good confinement in the vertical direction, a low-refractive index substrate such as SiO_2 must be used. For the best structures, the mode volume is as low as 3 $(\lambda_c/n)^3$ for a 4 μm diameter. However, the strong lateral confinement is also responsible of the strong sensitivity to sidewall roughness and Q is only close to 300. As a result, we estimate that F_p is of the order of 7 for state-of-the-art photonic wires.

The ultimate approach for obtaining very small cavities is certainly the implementation of photonic bandgap structures (PBGs) as confining material as discussed by Yablonovitch, Joannopoulos and Labilloy. In spite of impressive recent progress, 3D PBGs are not yet mature enough to build optical microcavities. Their development remains an important challenge since they would allow to get rid of the continuum of leaky modes present in other 3D cavities. Until now, only 1D[34,36] or 2D[35] PBG cavities have been fabricated, using a combination of waveguiding and reflection on the PBG to obtain a 3D photon confinement. As an exemple, we present in figure 5 a 2D PBG-microcavity processed within an air-bridge and designed for an operation around 1 μm. As discussed by Yablonovitch in his lecture, defect modes can exhibit an effective volume as small as 0.3 $(\lambda_c/n)^3$, due to their small penetration in the surrounding PBG crystal. The air-bridge geometry or a low refractive-index substrate is in general necessary to minimize the leakage of the confined photons toward the substrate[40]. A cavity Q of 265 has been measured by the MIT group for a Si-based 1D PBG structure on SiO_2, which corresponds to F_p =34 (here V_{eff} = 0.6 $(\lambda_c/n)^3$).[34]

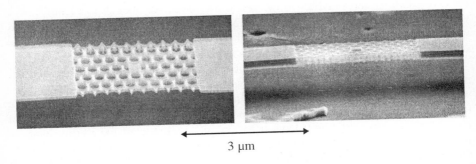

3 μm

Fig. 5. Scanning electron micrograph of an air-bridge microcavity based on a 2D PBG crystal, and designed for an operation around 1 μm.

3.4 Conclusion

We have shown in this section that many approaches allow to build high-F_p 3D semiconductor microcavities. Compared to silica microspheres (V_{eff}~1000 $(\lambda_c/n)^3$, Q~10^9) these structures allow a much stronger photon confinement, but exhibit a much smaller Q. In the prospect of CQED experiments, these 3D solid-state microcavities thus have different advantages. The long-lived WGM of the microspheres are for instance ideal for fabricating very-low threshold lasers[19]. The small mode volume of semiconductor cavities allows to get a large Purcell factor for moderate Q's (100-1000) ; the constraint on the emitter's linewidth is then less severe, which opens the choice of usable emitters for observing the Purcell effect. The same considerations hold obviously within the family of the 3D semiconductor microcavities, when comparing e.g. relatively high-Q microdisks and ultimately small volume PBG structures.

4 Purcell effect for QBs in solid-state microcavities

We present in this section some experimental evidence of the Purcell effect for QBs in micropillars and microdisks.

4.1 Purcell effect for QBs in micropillars.

Single pillars containing InAs QBs have been studied at 8K by time-resolved mPL, using a set-up based on a streak-camera. Experimental conditions have already been detailed elsewhere[10]. Let us simply mention that the excitation conditions are such that

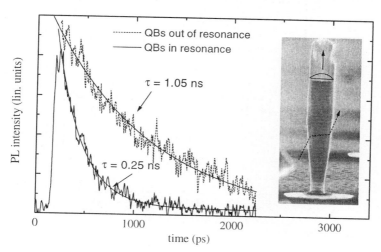

Fig. 6. Time-resolved PL spectra for QBs in the core of the pillar shown in the insert and either placed on resonance (solid line) or out of resonance (dashed line) with the fundamental mode. The dashed curve is a monoexponential fit of the PL decay ; the solid one features the result of our theoretical model for the QBs on resonance.

less than one electron-hole per QB is generated per pulse. This ensures that each QB behaves as a quasi-monochromatic light emitter which is a prerequisite for observing the Purcell effect. Since both resonant and leaky modes are observed on the mPL spectra of the thinner pillars, we can directly compare in a single time-resolved experiment the PL decay rate of QBs either on-resonance or out-of-resonance with one of the cavity modes. Such a comparison is highlighted on figure 6 for QBs emitting around 1.35 eV at 8K, which exhibit a 1.3 ns lifetime when placed in a reference bulk sample of GaAs. We observe a clear shortening of the PL decay (x1/5) when the QBs are on resonance with the fundamental mode of the pillar. On the opposite, out-of-resonance QBs exhibit a behavior quite similar to the reference QBs. This selectivity shows unambiguously that the decay rate shortening is not due to an extrinsic effect, such as non-radiative recombination at the sidewalls, but to the enhancement of the SE of the QBs when on-resonance with one cavity mode.

This effect has been studied for variable pillar diameters (fig. 7). For each pillar under study, the on-resonance lifetime of the QBs, τ_d^{on}, is extracted from a monoexponential fit of the PL decay over the first nanosecond after the pulse. Within experimental accuracy, we observe a smooth regular increase of $1/\tau_d^{on}$ as a function of Purcell's factor F_p. It does not exhibit such a monotonic dependence on Q or d since Q displays large fluctuations from pillar to pillar for a given pillar size in the 1-2 µm range. As expected, the Purcell factor is the relevant microcavity figure of merit for our problem.

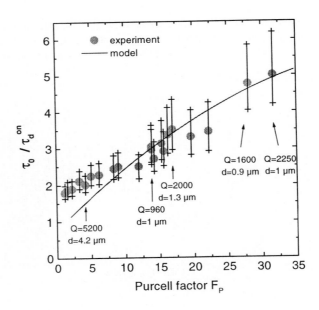

Fig. 7. Experimental (dots) dependence of the PL decay time τ_d^{on} as a function of the Purcell factor F_p. d and Q are indicated for some of the pillars under study. Error bars correspond to a ± 70 ps uncertainty on τ_d^{on}. The solid line shows the result of our calculation of the average lifetime of on-resonance QBs.

The enhancement factor of SE rate is however much smaller than the Purcell factor of the pillar. As stated in section II, only those QBs which are both well matched spectrally with the mode and located close to its antinode (i.e. near the center of the pillar) experience a strong enhancement of their SE rate. According to (14), we need to take into account the random spectral and spatial distributions of the QBs, to explain quantitatively the observed SE enhancement factor ϕ_{exp}. Even for our smallest pillars, the number of QBs on resonance with the fundamental pillar mode is large enough (20) to perform a statistical averaging of the Purcell effect. As reported previously[10], the result of a numerical calculation allows to reproduce properly the time-resolved PL profile (fig. 6) and the dependence of the SE enhancement factor as a function of F_p (fig 7) without any adjustable parameter.

In the context of this school, it is however more interesting to show that simple hand-waving arguments allow to obtain a good first-estimate of ϕ_{exp}. Considering Equ. 14, one notes that the spectral averaging of the Lorentzian factor gives a factor 1/2 (This averaging is necessary since we study the emission into the resonant mode globally due to the limited spectral resolution of our experimental set-up). Furthermore, the QBs are distributed all over the cross-section of the pillar so that we loose (on average) the antinode enhancement for in-plane directions, which introduces typically a factor varying between 1/4 for large diameter pillars and 1/3 for the smaller ones. Finally, the dipole associated to the fundamental optical transition of the QBs is essentially randomly oriented[41], so that $\xi^2 \sim 1/3$. Finally, the fundamental mode is two-fold degenerate, and the SE rate into the leaky modes is of the order of $1/\tau_{free}$. The rate for the SE into each confined mode is thus typically $F_p/18\,\tau_{free}$ and the global SE rate (2 $F_p/18+1)/\tau_{free}$, which gives us $\phi_{exp} \sim 4.6$ for our best micropillars, in good agreement with the experimental result.

4.2 Purcell effect in microdisks

Time-resolved mPL experiments on microdisks have not yet been performed, and might prove to be more tricky than for micropillars due to the poor collection efficiency of the PL emitted from the disks. However, the Purcell effect can be also observed using a simple cw mPL experiment as shown now.

Figure 8 displays PL spectra obtained on a single microdisk (R=1.5 μm) at 8K for various excitation powers P_{ex}. The sample under study is tilted so that the plane of the disk makes a 30° angle with respect to the optical axis of our experiment set-up. This geometry allows to collect more efficiently the emission from WGMs, which is sharply concentrated around the disk plane (~20° FWHM)[29]. When P_{ex} increases, one observes a clear saturation of the background emission for energies below 1.26 eV due to the filling of the QB states. Due to this background saturation, WGM show up much more clearly when we raise P_{ex}. For microdisks containing QWs, a similar behavior is observed due to lasing. For this microdisk (Q~2500, R=1.5 μm, $F_p \sim 18$), simple estimates show that the gain provided by the fundamental transition of the QBs is not large enough to support lasing[42]. As a result, lasing is only observed when enough additional gain is brought by the optical transitions involving excited QB states. This is the case here for the higher energy WGMs (above 1.26 eV), as shown

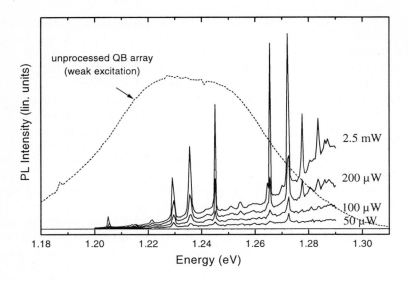

Fig.8. PL spectra obtained for the as-grown (unprocessed) sample, and for a 3μm diameter microdisk containing QBs as a function of the excitation power.

by the observation of a standard « S-like » shape for their input-output curves in log scale.

The low-energy WGMs (e.g. at 1.205 eV) display a more interesting behavior in the context of this work. When one raises the excitation power, their peak intensity first increases linearly and then saturates. This behavior is thus similar to the one of the background, but for the onset of the saturation, which is observed for a much higher (x 5) critical excitation power as shown in figure 9b). Since all QBs are similarly excited in our PL experiment, this observation shows that on-resonance QBs are less subject to state-filling than out-of-resonance QBs, or, in other words, that the radiative lifetime of trapped electron-hole pairs is much smaller for QBs on-resonance with a WGM ($\phi_{exp} \sim 5$). Much larger effects are observed for high-F_p microdisks obtained by wet chemical etching (see figure 9a). For a 1.8 μm diameter disk (Q=5300 and F_p =68 for the WGM under study) we obtain here $\phi_{exp} \sim 14$ for the QB which are on-resonance with this WGM.

Our interpretation of these cw PL data is supported by a simple estimate of ϕ_{exp}. One expects to observe here an average enhancement factor $\phi_{exp} \sim 2F_p/(2.2.3)+1$, where the various corrective factors (from the left to the right) come from the WGM two-fold degeneracy, the spectral and spatial averagings, the random orientation of the QB dipole, and the emission into leaky modes. We get in this way $\phi_{exp} \sim 12.5$ for the high-F_p disk and $\phi_{exp} \sim 4$ for the low-F_p disk, which is, here again, in good qualitative agreement with our experimental observation.

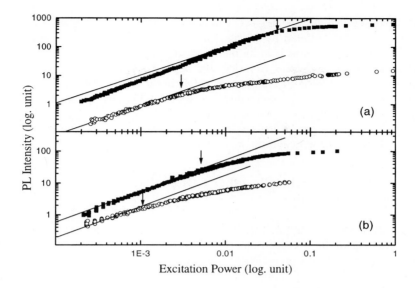

Fig. 9. Input-output curves for two microdisks : (a) two-step wet etching and $R=0.9$ µm ; (b) RIE etching, $R=1.5$ µm . Full squares correspond to a low-energy WGM, and empty dots to reference out of resonance QBs. The arrows mark the onset of the PL intensity saturation.

4.3 Potential impact of collective effects.

Until now, QBs have been treated as independent light emitters. In principle, power dependent collective effects could induce a deviation of the PL decay rate (or saturated PL intensity) from the value we estimate in our framework. For a weak excitation power, emitted photons can be reabsorbed by the QB array, which leads to photon recycling and to an apparent *slowing* of the PL decay. On the opposite, amplification by stimulated emission (ASE) could be observed when more than one electron-hole pair per QBs are injected on average, which would *enhance* the emission rate. It is thus important at this stage to estimate the average photon number for the cavity mode. We focus here on our time-resolved experiment on pillars.

Our pillars contain five QB arrays, with $4\ 10^{10}$ cm^{-2} areal density, so that the smallest cavity under study ($d=1$ µm) contains about 1600 QBs. Since the mode is spectrally very narrow (0.6 meV FWHM) compared to the inhomogeneous distribution of QBs bandgaps (60 meV FWHM), only 15 to 20 QBs are coupled to the cavity mode (This estimate includes the detuning of the mode with respect to the maximum of the QB emission band). For our experimental conditions, we do not saturate the emission from the fundamental optical transition of the QBs, so that each QB emits at most one photon per pulse. Since the escape time of the photons outside the cavity (1ps for $Q=2000$) is much shorter than the average radiative lifetime (250 ps) we see the mean photon number in the cavity is smaller than 0.08, even for short delays after the excitation pulse. It is therefore reasonable to neglect ASE in our experiment. We can as well neglect reabsorption under weak excitation conditions, since the cavity Q is limited by the photon escape rather than absorption by the QB array. The same conclusions hold for our smallest microdisks.

When the pillar surface S is varied, the dependence of the Purcell factor on S compensates the variation of the number of QB in the pillar ; we can easily see then that the maximum average photon number scales as Q. For large pillars (d~ 4 to 8 µm, Q~5000), it is thus of the order of 0.2. The effect of ASE on PL decay curves is still small, but can explain part of the systematic deviation of experimental PL decay rates with respect to our theoretical estimate for the larger pillar diameters (see fig 7).

5 Purcell effect on single QBs, other CQED effects and related applications

We have shown for micropillars and microdisks, that the Purcell effect can lead to a clear shortening of the average radiative lifetime of the QBs which are coupled to a resonant cavity mode. In principle, this global effect could be used to design high-frequency (>1 GHz) LEDs, for instance for board to board or intra-chip optical interconnexions. A major prerequisite for this application however is room-temperature operation. Whether sufficiently deep QBs would display a single narrow emission line at 300K - and experience the Purcell effect in cavities - is therefore a major open question. In the medium term, devices based on the Purcell effect (if any!) will more likely use a single QB as an active medium, as discussed in the following sections.

5.1 Purcell effect on single QBs

Unlike inhomogeneous collections of QBs, a single QB allows to make the best of high F_p microcavities. For our best micropillars (resp. microdisks), a single QB perfectly on-resonance and placed at an antinode of the mode should exhibit a SE rate enhancement (2/3 F_p+1)/τ_{free} ~ 20/τ_{free} or ~60 ps (resp. 80/τ_{free} or 15 ps), where the various terms account respectively for the polarization degeneracy (2), the random dipole orientation (3) and the contribution of the emission into leaky modes (1). For microdisks, SE dynamics would therefore probably be limited under non-resonant excitation by the finite carrier relaxation time (20-30ps).

Due to this strong Purcell effect, a very large fraction of the single QB SE is funnelled into the confined modes. The fundamental mode of circular micropillars and the WGM of microdisks are doubly degenerate. In this case, $\beta = F_p/(2F_p+3)$ ~ 0.5 for each resonant mode. Lifting this degeneracy is thus of obvious importance. Micropillars with elliptical circular cross-section and diameter in the 1-2µm range exhibit a clear splitting of their fundamental pair of modes for moderate excentricities[26]. By tailoring the size of a defect placed at the edge of microdisks, we can also induce a coupling between contra-propagating WGMs and lift their degeneracy. For such single mode microcavities and a single QB on-resonance, β~$F_p/(F_p+3)$ > 0.9 for pillars (0.95 for disks). Observing such a large β would constitute a huge step forward for solid-state microcavities, since the best value reported to date, obtained for photonic wire lasers, is of the order of 0.3.[33]

How can we achieve in practice such an ideal configuration ? Dilute QB arrays with areal density of the order of one QB per μm^2 can be obtained by MBE for deposited quantities of InAs very close to the critical thickness[43,44]. If we fabricate a collection of 1μm diameter micropillars containing on average one QB, we see that one fourth of the pillars typically will contain a single QB located reasonably close to the field antinode. Temperature tuning of the QB bandgap could be used to obtain the resonance with the cavity mode. Assuming a random distribution of the bandgap of the single QBs over a ~50 meV bandwidth, and a 5 meV tuning range (which corresponds to temperatures in the 8K-77K range, for which the QB emission would remain quasi-monochromatic), about one out of 40 pillars would be well suited for studying a single QB on-resonance. This system, which mimics the standard « single atom in a cavity » CQED system, will allow to probe whether other CQED effects - such as strong coupling or photon number state squeezing- can be obtained for solid-state cavities containing a single emitter, as discussed now.

5.2 Strong coupling regime for single QBs ?

The tight photon confinement in semiconductor microcavities corresponds to an extremely strong maximum field per photon. Using eq. (4), we obtain as an estimate for ε_{max} 0.8 10^5, 0.7 10^5 and 2.2 10^5 V/m respectively for our best micropillars and microdisks, and for 1D PBG microcavities[34]. These fields are obviously extremely large, for instance more than one order of magnitude larger than for microspheres[19]. Whether the strong coupling regime can be achieved for a single QBs in such 3D microcavities is quite an interesting question, owing to the richness of CQED developments on strongly-coupled single atoms.

Normal incidence absorption experiments on InAs QB arrays have shown that the oscillator strength per QB is of the order of f~10,[45] in agreement with simple estimates[46]. Therefore, the electric dipole component d_x (or d_y) given by :

$$f = \frac{2m\omega d_x^2}{e^2\hbar} \quad (15)$$

is of the order of 9 10^{-29} C.m, which is also quite larger than typical values for atomic optical transitions (2.4 10^{-29} C.m for the 5s->5p transition of rubidium at 1.59 eV)[19].

According to Eq.(6), the Rabi energy $\hbar\Omega$ of a QB placed at an antinode of the vacuum field is equal to ~50 μeV when $\varepsilon_{max} = 10^5$ V/m. In order to observe the vacuum Rabi splitting for a single QB, this Rabi energy should be larger than the arithmetic average of the emitter and mode linewidth. This goal is clearly out of reach for present pillars and PBG cavities due to their poor cavity Q. On the opposite, future developments on microdisks -especially a reduction of their radius and optimisation of their thickness- will most likely allow to raise $\hbar\Omega$ above 100 μeV, which is comparable to the linewidths of both our best WGM (Q=12000) and the QB emission line. We therefore think that an observation of the strong coupling regime on single InAs QBs, though difficult experimentally, will probably become possible in the future. Other QBs displaying a larger oscillator strength, such as those formed by interface monolayer fluctuations of QWs, might also be very interesting in this context.[47]

Whether this regime can lead to interesting applications remains however an open question. Unlike atoms in CQED experiments, QBs cannot be inserted and extracted at will from the solid-state microcavity. In order to overcome this difficulty we should develop means for controlling the interaction time of the QB with the cavity mode on the time scale of the Rabi oscillation (~1-3ps!). This will be, obviously, a difficult task ! The most promising approach might be the implementation of a χ_3 non-linearity to control optically the refractive index of the cavity material and the mode/emitter detuning.

5.3 Single-photon generation

The fabrication of single-photon generators, able to emit single photon pulses at deterministic times, has been for years a major challenge. Such a source would allow to encode information on the single photon level, and to implement efficiently novel transmission protocols such as quantum cryptography[48]. Sources emitting a regular stream of single photon pulses would also provide a high precision photon-flux standard[49].

Three properties must be combined in order to get a single-photon source. First of all, the nature of the emitter should ensure that photons are emitted one by one. Such an antibunching behavior has been observed for single atoms[50,51] or molecules[52]. In order to use suitably this photon flux, it is highly desirable to place this peculiar emitter inside a single-mode output coupler, i.e. on resonance in a high β (β~1) microcavity. Finally, the active medium should also have a high quantum efficiency (η~1) in order to avoid a random partition between radiative and non-radiative recombination events.

Single QBs in solid-state microcavities open very challenging opportunities in this context, and might allow the fabrication of a compact monolithic single-photon source. Let us note however that the anticorrelation of photon-emission events has not yet been observed for single QBs. In principle, two electron-hole pairs can be placed in the fundamental electronic states of the QB, and nothing prevents them from recombining with an arbitrarily short delay. Due to the strong electronic confinement however, the energy of emitted photons depends significantly of the QB state of charge due to Coulomb interaction[15,16]. Considering a QB which is on resonance - when empty- with a high Q (Q>500) discrete cavity mode, and supposing that several electron-hole pairs are injected in the QB, we see that only the last emitted photon lies within the spectral window defined by the cavity mode. As shown before, nearly all photons are collected by the cavity mode when the QB is on resonance (β~1) thanks to the Purcell effect. Finally, the quantum efficiency of InAs QBs is close to 1 as long as carrier thermoemission can be neglected. Even for small QBs (such as those implemented in micropillars and disks until now for practical reasons) η~1 for temperatures as high as 100K, which might be enough for most applications, and can be improved by using deeper QBs.

Unlike single atoms or molecules, QBs can be excited through a non-resonant pumping, either optical or electrical, which makes a practical implementation much easier. Pumping pulses should be adjusted so that few electron-holes pairs are captured by the QB (with a small probability of having no injected pair). As stated previously, the microcavity containing a single QB should act as a converter of such classical pulses, characterised by poissonian statistics, into single photon pulses well synchronized (within a few nanoseconds, corresponding to the recombination time of the extra carriers of charge) with the excitation pulses.

Coulomb Blockade (CB) of both electron and hole tunneling can also be used[53] to inject exactly one electron and one hole in a QB (obtained e.g. through the lateral patterning of a QW) as proposed few years ago[49]. This scheme has recently been sucessfully implemented in a single-photon turnstile device[53]. Unless nanometer-scale tunnel junctions are used however, this approach will be restricted to very low temperatures (<0.1K for refs 49-53). Using Coulomb interaction within our tiny InAs QBs so as to control the energy of emitted photons, instead of regulating the electron-hole injection, might be much more efficient. The insertion of the single photon source within a microcavity is particularly simple when using QBs, and operation temperatures above 77K seem well within reach.

Acknowledgements

The authors gratefully acknowledge the important experimental contributions of B. Legrand, B. Sermage, G. Ungaro, A. Lemaître (CNET), V. Thierry-Mieg, C. Dupuis, L. Couraud and L. Manin (CNRS-L2M) and E. Costard (Thomson-CSF/LCR). They are also very grateful to L.C. Andreani (U. Pavia), T. Rivera, E. Sagnes, R. Kuszelewicz, I. Abram, J.L. Oudar (CNET), J.Y. Marzin, J. Bloch and R. Planel (L2M) for very fruitful interactions.

References :

[1] E.M. Purcell, Phys.Rev. **69**, 681 (1946)
[2] for an early review, see S. Haroche and D. Kleppner, Phys. Today **42**, 24 (1989) ; see also the lectures by S.T. Ho, C. Fabre and C. Hood in this book.
[3] for a recent review, see *Microcavities and Photonic Bandgaps : Physics and Applications,* C. Weisbuch and J. Rarity eds, NATO ASI series E324, Kluwer, Dordrecht, 1996
[4] G. Björk et al, Phys. Rev. A **44**, 669 (1991) and IEEE J. Quantum Electron.**30**, 2314 (1994) ; S.D. Brorson et al, J. Quantum Electron. **QE26**, 1492 (1990)
[5] A.M. Vredenberg et al, Phys.Rev. Lett. **71**, 517 (1993)
[6] K. Tanaka et al, Phys.Rev. Lett. **74**, 3380 (1995)
[7] L.A. Graham et al, Appl. Phys. Lett. **72**, 1670 (1998)
[8] T. Tezuka et al, Jpn. J. Appl. Phys. **32**, L54 (1993)
[9] See the contribution of J.M. Gérard et al in ref 3
[10] J.M. Gérard et al , Phys. Rev. Lett. **81**, 1110 (1998)
[11] J.Y. Marzin et al, Phys. Rev. Lett. **73**, 716 (1994)
[12] M. Grundmann et al, Phys. Rev. Lett. **74**, 4043 (1995)

[13] J.M. Gérard et al, Appl. Phys. Lett. **68**, 1113 (1996)
[14] Let us consider for instance a GaAs/AlAs planar λ-cavity containing a thin absorbing layer placed at the field antinode ; in this case $Q<m\pi/2\gamma$ due to absorption losses, where m is the effective order of the cavity (here $m\sim 8$) and γ the fraction of the light which is absorbed for one normal-incidence path through the layer. If this layer is a QW, $\gamma\sim 0.01$ and $Q<1200$. Absorption experiments on QB arrays [45] show that $\gamma\sim 0.0006$ for a single layer of InAs QBs with $4\ 10^{10}$ cm^{-2} areal density and 60 meV FWHM, so that $Q<20000$. Since the best GaAs/AlAs cavities have a Q of 12000, the insertion of a layer of QBs has a negligible effect on their optical quality.
[15] L. Landin et al, Science **280**, 262 (1998)
[16] J.M. Gérard et al, proceedings MBEX to be published in J. Crystal Growth
[17] S. Haroche and J.M. Raimond, in « Advances in Atomic and Molecular Physics, vol XX », D. Bates and B. Bederson eds, Academic Press, New York, 1985
[18] V.B. Braginsky et al, Phys. Lett. A **137**, 393 (1989)
[19] L. Collot et al, Europhys. Lett. **23**, 327 (1993) ; V. Sandoghdar et al, Phys. Rev. A **54**, R1777 (1996)
[20] K. Iga et al, IEEE J. Quantum Electron. **24**, 1845 (1988)
[21] J.L. Jewell et al, Appl.Phys. Lett. **55**, 22 (1989)
[22] T. Baba et al, IEEE J. Quantum Electron. **27**, 1347 (1991)
[23] J. Jewell et al, Electron. Lett. **25**, 1123 (1989)
[24] J.M. Gérard et al, Appl. Phys. Lett. **69**, 449 (1996)
[25] J. P. Reithmaier et al, Phys. Rev. Lett. **78**, 378 (1997)
[26] B. Gayral et al, Appl. Phys. Lett.**72**, 1421 (1998)
[27] T. Tezuka et al, J. Appl. Phys. **79**, 2875 (1996)
[28] J.M. Gérard et al, Physica **E2**, 804 (1998) ; T. Rivera et al, Appl. Phys. Lett. **74**, 911 (1999)
[29] S.L. Mc Call et al, Appl. Phys. Lett. **60**, 289 (1992)
[30] R.E. Slusher et al, Appl. Phys. Lett. **63**, 1310 (1993)
[31] U. Mohideen et al, Appl. Phys. Lett. **64**, 1911 (1994)
[32] T. Baba et al, IEEE Photon. Tech. Lett. **9**, 878 (1997)
[33] J.P. Zhang, Phys. Rev. Lett. **75**, 2678 (1995)
[34] J.S. Foresi et al, Nature **390**, 143 (1997)
[35] D. Labilloy et al, Appl. Phys. Lett. **73**, 1314 (1998) ; see also Labilloy's contribution in this volume.
[36] D. Labilloy et al, Electron. Lett. **33**, 1978 (1997)
[37] A.L. Holmes et al, Appl. Phys. Lett. **66**, 2831 (1995)
[38] T.R. Nelson et al, Appl. Phys. Lett. **69**, 3031 (1996)
[39] H.E. Shin et al, Appl. Phys. Lett. **72**, 2205 (1998)
[40] V. Berger et al, J. Appl. Phys. **82**, 5300 (1997)
[41] When collecting the QB PL from the cleaved edge of the sample, we observe similar contributions for TE and TM polarizations (within 10-20%), which shows that the vertical component of the QB dipole is as large as the in-plane components.

[42] For a microdisk, the modal loss coefficient α is related to Q through $Q = 2\pi n_{eff}/\lambda\alpha$. For Q=2500, we get α= 75 cm^{-1}, which is much larger than the estimated saturated modal gain of the three QB arrays (~25 cm^{-1} around 1.2 eV).

[43] J.M. Gérard et al, J. Crystal Growth **150**, 351 (1995)

[44] D. Leonard et al, Phys. Rev. B. **50**, 11687 (1994)

[45] R.J. Warburton et al, Phys. Rev. Lett. **79**, 5282 (1997)

[46] For such QB which provide a strong electronic confinement, the oscillator strength does not depend on the size of the QB and is simply related to the Kane bulk matrix element E_p through the relation :

$$f = \frac{E_p |\langle\psi_e|\psi_h\rangle|^2}{E}$$

where ψ_e and ψ_h design the electron and hole envelope functions. Their square overlap is of the order of 0.8 in our case, E_p is about 18 eV for InAs, and the bandgap energy E 1.2 eV so that we get f~12, in reasonable agreement with the experiment of ref 45.

[47] L.C. Andreani et al, submitted to Phys. Rev. B.

[48] Marand et al, Opt. Lett. **20**, 1695 (1995) and references therein ; see also John Rarity's tutorial in this book.

[49] A. Imamoglu et al, Phys. Rev. Lett. **72**, 210 (1994)

[50] H. J. Kimble et al, Phys. Rev. Lett. **39**, 691 (1977)

[51] P. Grangier et al, Phys. Rev. Lett. **57**, 687 (1986)

[52] T. Basché et al, Phys. Rev. Lett. **69**, 1516 (1992)

[53] J. Kim et al, Nature **397**, 500 (1999)

Single Photon Sources and Applications

John G. Rarity, Stephen C. Kitson, and Paul R. Tapster

DERA Malvern, St Andrews Rd, Malvern, UK WR14 3PS

Abstract. We discuss sources of single photons for quantum information processing. For limited applications the attenuated laser is an adequate source. Further improvement would be obtained from single atom or molecule emission. Collection efficiency into a narrow band and single mode could be improved by photonic bandgap material surrounding the emitter. We discuss a measurement showing light emission from single dye molecules modified by planar cavity. Another source could be time gated single photons created from the parametric downconversion process. We go on to show a simple interference effect between separate single photons that confirms the quantum nature of this source.

1 Introduction

In the spirit of driving our physics from eventual applications we write this paper in the following order. First we introduce the concept of quantum information in optical terms. We show that information can be encoded on single photons using interferometers or polarisation (itself an interference effect). We then show that these interferometric encoding schemes can be used to establish identical random binary numbers (**Keys**) at remote locations in the technique of quantum cryptography. The security of this key distribution scheme is dependent on the interferometric encoding of the bits.

We then describe the possible single photon sources to be used in such schemes. The first, in general use at present, is the attenuated pulsed laser. When the energy per pulse is much less than $\hbar\omega$ (the energy associated with a single photon) most pulses will contain zero photons and a small percentage will contain single photons. However we cannot predict which pulses will contain a photon and a very small percentage of pulses will always contain more than one photon. In quantum cryptography we are limited us to about 0.1 photons per pulse at the transmitter. This limits the maximum bit rate of a system.

A second single photon source is a single atom or molecule. Here we exploit the naturally quantum mechanical nature of the emission process. An atom excited by an optical pulse much shorter than its lifetime can only emit a single photon. The problem then is to direct the single photon emission efficiently into a single mode. Here we describe a demonstration of the principles of a single photon source based on emission from single dye molecules. Of interest in the context of this school is the tailoring of the dye spontaneous emission to the modes of a planar microcavity. In principle we can make a

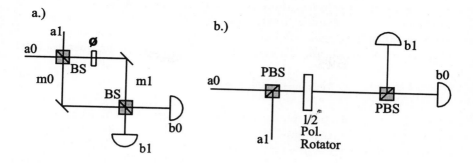

Fig. 1. Single photon encoding schemes using a) Mach.-Zehnder interferometer and b) polarisation

high efficiency single photon source with a microcavity system engineered to capture a high percentage of the spontaneous emission.

The third source of single photon pulses is parametric downconversion. In a parametric downconversion crystal pumped by a suitable short wavelength laser, pairs of long wavelength photons are created simultaneously, travelling in correlated directions, and correlated in energy. Detection of one of the photons can be used to gate its partner thus producing a source rich in time tagged single photons. A problem with this source is the randomness in the emission times of the pairs. This can be overcome by pumping the crystal with ultrashort laser pulses. In this case the time uncertainty of the single photon pulses can be comparable to the inverse of their bandwidth. Such time bandwidth limited pulses are what is required to show interference effects between separate single photon pulses and to build up multi-photon entangled states. We show a simple example of interference between nominally separate sources to illustrate this.

2 Optical Quantum Information Processing

2.1 Single Photon Interference

Take a symmetric Mach-Zehnder interferometer as shown in figure 1. In a simple classical analysis with an input field E_{a0} and associated intensity $I_{a0} = |E_{a0}|^2$ the output intensity will vary as

$$I_{b0/b1} = |E_{b0/b1}|^2 = \frac{1}{2}I_{a0}(1 \pm \cos\phi) \tag{1}$$

This is a linear loss free device as $I_{b0} + I_{b1} = I_{a0}$ for all interferometer phase differences ϕ. Here we are interested in the behaviour of a single quantum incident on the same interferometer. We associate probability amplitudes with the presence of a photon in the input modes and deal with the probability amplitudes in a similar way to the classical fields. In the simplest case the photon is input from one mode ($a0$ say) and represented by the number state $|1>_{a0}$ with unit amplitude. This state is transformed at the first beamsplitter

$$|1>_{a0} \to \frac{1}{\sqrt{2}}[|1>_{m0} + i|1>_{m1}] \qquad (2)$$

noting that the reflection and transmission amplitudes of the beamsplitter are $i/\sqrt{2}, 1/\sqrt{2}$ with the phase change on reflection required for energy conservation. Obviously the presence of the photon in one arm implies an empty mode or vacuum state in the other. Here we specialise to single photon states throughout and thus leave the vacuum implicit. After propagating and incurring a phase delay $e^{i\phi}$ only in the $m1$ arm of the interferometer a similar transformation occurs at the second beamsplitter and the state becomes

$$|1>_{a0} \to \frac{i}{2}(1+e^{i\phi})|1>_{b0} + \frac{1}{2}(1-e^{i\phi})|1>_{b1} \qquad (3)$$

We now identify the probability of detecting a single photon at a particular the interferometer output as the modulus square of the associated probability amplitude

$$P_{b0/b1} = \frac{1}{2}(1 \pm \cos\phi) \qquad (4)$$

which of course is identical to the classical result when we have unit intensity input. Again we see that $P_{b0} + P_{b1} \equiv 1$ the total probability of detecting a photon is unity in this loss free case.

We can now use this system to encode information on a single photon. A phase $\phi = 0$ sends all photons to b0 while $\phi = \pi$ sends all pulses to b1. We can interpret a detection in b0 as a 'zero' and b1 as a '1'. One extension from any classical encoding scheme is the situation where the phase lies between these two extremes. Setting $\phi = \pi/2$ transforms the input state to

$$|1>_{a0} \to \frac{e^{3i\pi/4}}{\sqrt{2}}(|1>_{b0} + |1>_{b1}) \qquad (5)$$

and now our single photon data is in a superposition state of a 1 and a zero. Detection at this point will provide random information as we have a 50% chance of detecting a 1 or a zero.

Information can also be coded on a single photon using two orthogonal polarisations such as vertical and horizontal. The coding can be manipulated simply by rotating the polarisation in a waveplate. We show this to be equivalent to the above scheme by drawing a polarisation interferometer in figure 1b. Now the $m0, m1$-modes are co-linear circular polarised modes and the relative phase between them is varied by rotating the waveplate. If we define θ as the angle between the waveplate fast axis and the polarisation direction of the $a0$ mode then the interferometer transforms

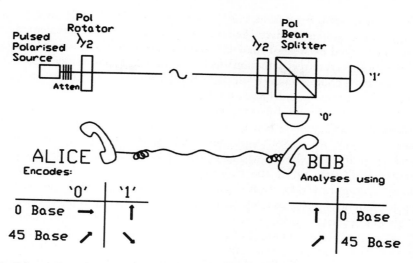

Fig. 2. Schematic of polarisation based cryptography based on the original scheme of Bennett et al (1992)

$$|1>_{a0} \rightarrow (\cos 2\theta \ |1>_{b0} - \sin 2\theta \ |1>_{b1}) \qquad (6)$$

and again we see definite outputs when $2\theta = 0, \pi/2$ and superposition states otherwise.

2.2 Quantum Cryptography

A direct application of this simple single photon coding has been in secure key sharing schemes commonly known as quantum cryptography (Bennett *et al* 1992, Townsend *et al* 1993, Townsend 1994, Rarity *et al* 1994). In most communication applications losses are large and most photons (ie bits) do not arrive. This is not a problem in quantum cryptography because all we want to do is **establish identical random numbers** at two remote locations. The sender encodes a random series of bits on single photon pulses. The receiver analyses all single photons that arrive then communicates back from the remote location (using a conventional communication channel) the times at which these photons have arrived. Only those bits that arrive are then incorporated into the 'key'. Here one sees the first level of security of such a system. Photons are indivisible objects thus if an eavesdropper picks off a small percentage of photons and measures them, they will not reach the receiver and not be included in the key. However, a subtle eaves-dropper can measure the photons then create copies to reinject into the communication channel thus breaching this security. To prevent copies being made the sender has to randomly change the encoding basis. For instance in polarisation based quantum cryptography (figure 2) 0's are randomly encoded with either 0^0 OR

45^0 polarised single photons and 1's are encoded in 90^0 OR 135^0 polarised single photons. The receiver then incorporates a polarising beamsplitter randomly switched between 0^0 and 45^0 measurement bases. As seen above, 100% correlation (error free detection) only occurs when the sender and receiver use the same coding basis (0^0 or 45^0). Sender and receiver must also communicate this measurement basis and discard all received pulses where the send and receive bases were different. After all uncorrelated bits are discarded the transmitter and receiver are left with near identical random bit strings to be used as a key. An eavesdropper must now guess which measurement basis was used. He will choose wrongly 50% of the time and 25% of reinjected photons will turn up at the wrong output. The error rate is estimated by openly comparing a fraction of the key bits which are then discarded. If a large number of errors are detected the sender and receiver must assume that their key security has been compromised and restart the key exchange.

Recent experiments are showing that such key exchange schemes could be used in real communication networks (Townsend 1997) with adequate environmental stability (Muller *et al* 1997) and even in free space systems (Buttler *et al* 1998) aimed at uploading keys to low earth orbit satellites. The main limitation to quantum cryptography is that we cannot amplify the signal to extend the system range as we can in a conventional communication system. Essentially any amplification that did not introduce errors would be a form of quantum cloning and as a result would violate the Heisenberg uncertainty principle. Present systems also rely on weak lasers as a source approximating single photons (see below) and this limits the effective single photon generation efficiency to around 10%. Coupled with detector inefficiencies at longer wavelengths this limits the present maximum range of fibre based systems to 50km (Hughes *et al* 1996).

3 The Weak Laser

To date we have no true single photon sources available off-the-shelf. However for simple applications such as quantum cryptography attenuated laser pulses are adequate. The photocount probability distribution $P(n)$ for an ideal (classical) laser pulse of energy E is given by a Poisson distribution (Loudon 1987)

$$P(n) = \frac{\exp(-\overline{m})}{n!} \cdot \overline{m}^n \qquad (7)$$

with mean photocount $\overline{m} = E/\hbar\omega$ where $\hbar\omega$ is the energy associated with a single quantum. When the mean photon number per pulse is $\overline{m} = 0.1$ the probability of seeing one photon is $P(1) \simeq 0.09$ while the chance of seeing two photons is $P(1) \simeq 0.004$. Thus of those pulses that contain at least one photon only 1 in 20 will contain more than one photon. This is assumed adequate proof against eavesdropping in a quantum cryptography system.

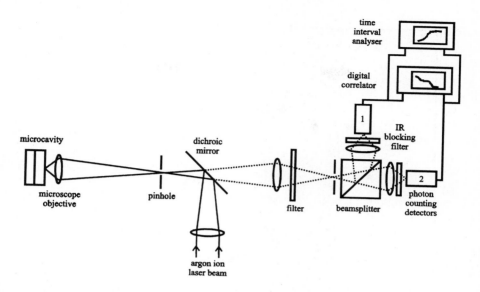

Fig. 3. Schematic of the apparatus used to study the fluorescence fluctuations from a small number of dye molecules in a microcavity

However absolute proof of security is yet to be shown with such weak pulses. The quantum state associated with such pulses is mostly vacuum

$$|\Psi> \simeq \exp(-\frac{\overline{m}}{2}) \left(|vac> + \alpha |1> + \frac{\alpha^2}{\sqrt{2}} |2> \ldots \right) \qquad (8)$$

where the amplitude of the coherent state is α and $|\alpha|^2 = \overline{m}$. When $\overline{m} = 0.1$ we see that the amplitude $|\alpha| \simeq 0.3$ suggesting that coherent methods of measuring the state may provide more information than direct detection (Yuen 1998). Weak pulses also have the disadvantage of starting out only 10% occupied.

4 Single Molecule Sources

A convenient future single photon source may be a single molecule, atom, ion or even single quantum dot (see chapter by J M Gerard). These two level systems cannot emit more than one photon at a time. The detection of a single photon conditions the system to be in the ground state and there is a finite time of order the excited state lifetime before emission can occur again. When we pump such a system with pulses shorter than the excited state lifetime we expect to see single photon emission in a regular train (De Martini 1996).

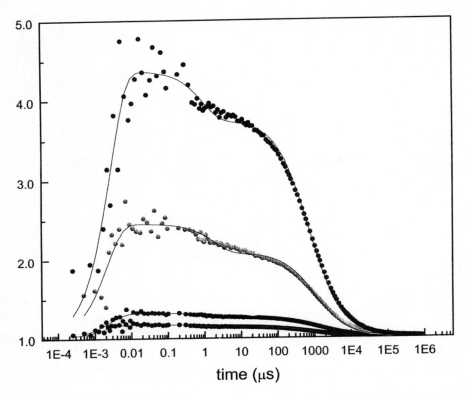

Fig. 4. $g^{(2)}(t)$ versus delay t for a cavity containing a range of concentrations of R6G from 10^{-8} to 10^{-9} molar. As dye concentration decreases the value of $g^{(2)}(t)$ increases. The circles are experimental data and the lines are a fits to equation 10

Here we illustrate this with a system consisting of a dilute dye solution in a thin (40nm) layer confined within a Fabry-Perot cavity (Kitson et al 1998). The cavity serves to increase the efficiency with which the spontaneous emission is collected. The single molecule regime is reached by using very dilute (10^{-9}M) dye solutions and a small illumination volume (around 40nm thick by 6μm diameter). When there are a small average number of dye molecules in the excitation volume the fluorescence signal exhibits strong fluctuations. The fastest fluctuations arise from the mechanism cited above where a dye molecule that has emitted a photon must remain dark for a time comparable to the excited state lifetime. This is commonly known as anti-bunching. On intermediate timescales we see fluctuations arising from passage through the triplet state to the ground state by non-radiative routes. On long timescales there are fluctuations in the number of dye molecules in the measurement volume due to diffusion in the liquid suspension.

The dye used was Rhodamine 6G (R6G) dissolved in propylene carbonate. Propylene carbonate was chosen as the solvent because of its low intrinsic fluorescence and low volatility. R6G is highly efficient and can be excited with the 488nm line from an argon ion laser. The dye solution is placed in a microcavity consisting of two dielectric mirrors, made from alternating layers of silica (n = 1.5) and tantalum pentoxide (n = 2.265). The peak reflectivity of the mirrors was designed to be at a wavelength of 560nm by making the thickness of each layer $\lambda/4n$ (where λ is 560nm and n is the refractive index of the layer). The top layer of each mirror is silica, deliberately grown 20nm thinner than the $\lambda/4n$ condition. The microcavity is formed by placing a drop of the dye solution on one mirror, and then pressing the other one on top. Placing the cavity in a vacuum chamber for around an hour causes the liquid to slowly evaporate, pulling the two mirrors together. The resulting structure is a $\lambda/2n$ thick microcavity with a dye layer, approximately 40nm thick, at the centre. The cavity material is the lower index silica so that the dye layer sits at the antinode of the electric field of the fundamental cavity mode, into which the molecules preferentially emit. This narrows the emission spectrum to match the cavity band width and so increases the efficiency with which the light can be collected through a narrow band pass filter which is used to discriminate the fluorescence from scattered laser light.

Figure 3 is a schematic of the confocal fluorescence microscope used to study the fluorescence from the microcavity. The 488nm light from a CW argon ion laser is focused through a 100μm pinhole and then through a microscope objective (x25, 0.35 numerical aperture) onto the microcavity. The same lens also collects the fluorescence from the dye molecules. The diameter of the region from which the light is collected is defined by the pinhole to be around 6μm. The collected light passes through the dichroic mirror and additional filters are used to remove any remaining laser light. The fluorescence light is divided equally between two avalanche diode single photon counting detectors which are connected to a photocount correlator and to a time interval analyser. Two detectors are used to circumvent the problems associated with the deadtime of the detectors $\sim$1μs, allowing the measurement of time intervals as small as 0.5ns.

The electronics estimates photocount correlation function

$$g^{(2)}(\tau) = \frac{\langle I(t)I(t+\tau)\rangle}{\langle I^2 \rangle} \tag{9}$$

where $I(t_0)$ is the measure intensity at time t_0 and $I(t_0 + t)$ the intensity measure a time t later. Over the time range 100ns to 1s, $g^{(2)}(t)$ is measured using parallel digital correlator which evaluates equation (9) in real time. The system used here contains eight 32 channel correlators with sample times T, 6T, 36T,.....etc. with T = 100ns. The time range from 0.5 to 100ns is covered by a time interval analyser which records a histogram of the time interval between consecutive pairs of photons. At low count rates, so that the average number of photocounts per 100ns sweep is much less than one, the time interval distribution is a good approximation to $g^{(2)}(t)$. In our system the photocount rate is typically 30kHz so that this approximation holds. Combining the data from the time interval analyser and the correlator then gives $g^{(2)}(t)$ over the time range ns to s. Using a simple three level model for the Rhodamine molecule we have evaluated a theoretical form for the correlation function

$$g^{(2)}(t) = 1 + \frac{1}{\langle M \rangle (1 + t/t_d)} \left[1 - (1+a) \exp(-t/t_e) + a \exp(-t/t_t) \right] \quad (10)$$

where t_e is the excited state lifetime, t_t is the triplet state lifetime and t_d the typical diffusion time across the volume. Figure 4 is a plot of $g^{(2)}(t)$ versus τ for time scales ranging from ns to seconds obtained for a cavity containing various concentrations of dye solution with an incident laser power of 30 mW. Solid lines are fitted by equation (10). The curves clearly show the three features that we expect. The initial rise is due to antibunching, the drop at around 1μs is due to triplet state shelving and the drop at around 1ms is due to diffusion. We see that on average there are of order $\langle M \rangle = 0.4$ molecules in the volume for the most dilute sample. The positive slope at zero delay in the correlation function is a clear sign that the emitted light is non-classical. However for a true single photon source the conditional probability of seeing a second photon some time after an emission would be zero and as a result $g^{(2)}(t \rightarrow 0) \equiv 0$. Here the Poisson number fluctuations due to diffusion exactly cancel the short time antibunching. This was also the case in the original anti-bunched light experiments carried out with two level atoms (Kimble et al 1977). The emitted light tends, therefore, to consist of bursts of antibunched photons separated by dark intervals of order μs, due to triplet state shelving, and of order ms due to diffusion in-and-out of the volume. In order to see constant antibunched light we would need to freeze out the diffusional motion and study a single dye molecule. However the short triplet lifetime we measure is a result of triplet state quenching by dissolved oxygen with the subsequent creation of singlet oxygen. From experiments carried out at various laser powers we find clear evidence that the dye bleaches due to oxidisation by the highly reactive singlet oxygen. In ongoing experiments we are investigating dye performance in the presence of other triplet quenching agents in oxygen free solutions. We do see evidence of improved dye lifetime but as yet have not been able to freeze our system and study fixed single molecules. Other workers (Baché et al 1992) have been able to show more stable dyes (with low triplet branching ratios) at cryogenic temperatures.

When we do manage to identify a stable two level system preferably operating at temperatures close to ambient we will also have to address the collection efficiency. The dye emission is naturally much broader band than the cavity bandwidth and as a result any lifetime alterations (Purcell factor) due to the cavity will be negligible. We can estimate the geometric collection efficiency from the ratio of the solid angle subtended by our collecting lens

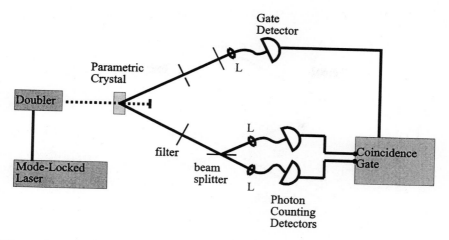

Fig. 5. Gated single photon source based on a pulsed laser (double mode locked Ti Sapphire, 407.5nm) pumping a Beta Barium Borate crystal cut to emit pair photons in the near infra-red (815nm wavelength) The gating detector signals the presence of single photons in the measurement arm. The presence on single photon states can be inferred from a lack of coincidences across the beamsplitter.

in the silica mirror substrate Ω compared with the full 2π steradians that the dye emits into. For our lens of NA 0,35 we find a geometric collection efficiency of 2.5%. To improve on this we need to increase our numerical aperture or to strongly alter the emission pattern by going to a non-planar cavity such as a micro-pillar or confocal arrangement (see chapters in this volume by Gerard et al and Abram et al).

5 Gated Parametric Downconversion

A third source of single photon states exploits the pair photon generation process of parametric downconversion. Detection of one photon of a pair can be used to gate detection of the other. Demonstration experiments showing optical and electronic gating of single photons were performed some time ago in our laboratory (Rarity et al 1987) using bulk non-linear crystals. Although single photon states were detected the counting rates were extremely low due to the poor detectors and other losses in the system. More recently we have been working on a pulsed source (Rarity 1995, Rarity and Tapster 1996, Rarity et al 1997, 1997a, Rarity and Tapster 1998, 1999) where the single photon state is confined to within a 130fs time window by the duration of the pumping pulse (a doubled mode locked Ti-sapphire laser). In this experiment (figure 5) we again used a bulk crystal and then launched the pair photon beams into single mode fibres as would be required for a quantum cryptography source. In the experiment we count photons in fibre coupled

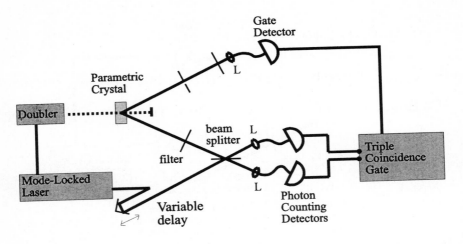

Fig. 6. The experimental apparatus used to demonstrate interference between separate sources. Solid lines indicate light of 815nm wavelength, dashed lines indicate light of 407.5nm wavelength and thick curved solid lines represent optical fibres.

detectors with efficiencies of order $\eta=40\%$. However the singles rates in the detectors are only of order $S=5$ kilocounts per second while the gated single photon source rate is of order $G=1$ kilocount per second.

The gated rate G can be expressed as

$$G = \alpha^2 \eta^2 P r \tag{11}$$

where r is the pulse repetition rate of the pump laser (100Mhz here), P is the number of pair photons created in the single mode per pulse, and α is the loss due to mode matching in the fibres and other filter edge effects. Similarly the singles rates are $S = \alpha \eta P r$. From this we see that $\alpha = 0.5$ and that the gated rate is reduced by a factor of 4 by mode matching losses. Thus given a photodetection in the gate detector there is at least a 50% probability that the corresponding pulse in the other fibre contains a single photon and the probability of two photons in the pulse is vanishingly small. This is a five fold improvement on an attenuated classical source but the overall emission rate of 1 kilocount per second is far too low for cryptography applications. The efficiency of creation of these single spatial mode single photon states could be increased if we were to use a medium engineered such that only two modes are available to the downconverted photons. Obviously this could be done in a suitably designed photonic crystal medium.

In practice second order non-linearity has recently been demonstrated in poled single-mode optical fibres (Kazanski *et al* 1994). If adequate non-linearity can be achieved we expect to be able to have downconversion pairs emitted directly into the fibre mode. There would be small losses in separating the pairs using wavelength selective couplers but we might expect to see effective numbers of photons per gate greater than 0.9.

6 Interference between separate single photon sources

We illustrate the time-bandwidth product limited nature of our pulsed gated source in a simple interference experiment where we mix a gated single photon pulse from our source with a weak laser pulse at a beamsplitter. The experiment is shown in figure 6 (Rarity et al 1997). A frequency-doubled mode-locked laser (407.5nm wavelength) pumps a thin parametric downconversion crystal cut for non-degenerate operation. Signal and idler photons satisfying energy conservation are emitted spontaneously in a broad band cone behind the crystal and apertures are placed to select 815nm wavelength beams from opposite ends of a cone diameter. Detection of an idler (or gate) photon in one beam with time resolution better than the pump pulse separation time essentially localises a single signal (a−mode) photon within a pulse length which in the experiment is around 130fs. As the signal photon so selected is a good approximation to a one-photon state it must have random phase when we measure it with respect to the original near infra-red beam. The two can be thought of as separate sources. Thus we expect to see no first order interference fringes when we mix this single photon source with coherent pulses from the undoubled mode-locked laser in a beamsplitter as shown. However when we reduce the intensity of the coherent pulses to the point where they too approximate single photons we see a strong two-photon interference effect. As seen in equation 2 passage through a 50/50 beamsplitter takes a single photon state to the superposition

$$|1>_{a0} \rightarrow \frac{1}{\sqrt{2}}[|1>_{m0} + i|1>_{m1}] \qquad (12)$$

$$|1>_{a1} \rightarrow \frac{1}{\sqrt{2}}[i|1>_{m0} + |1>_{m1}]$$

And thus when we input simultaneously into both ports we obtain (Fearn and Loudon 1987)

$$|1>_{a0}|1>_{a1} \rightarrow \frac{1}{\sqrt{2}}[|2>_{m0} + i|2>_{m1}] \qquad (13)$$

where all terms in $|1>_{m0}|1>_{m1}$ cancel due to destructive interference arising from the phase change on reflection. Thus coincidences between outputs across the beamsplitter disappear when the two inputs are made indistinguishable from measurements made at the outputs. We find the same effect when one input is replaced by a weak coherent state as it is dominated by vacuum and one photon contributions. However the requirement for indistinguishability can only be met by filtering our pulsed source through narrow filters that effectively produce a coherence length as long or longer than the original pulses. We also need to restrict our observations to a single spatial mode through the use of single mode optical fibres at the detectors. In the experiment (figure 6, Rarity et al 1996, 1997, 1997a) we were able to show a reduction of the coincidence rate to 36% of its value when pulses were distinguishable. This experimental arrangement can be extended to show the non-local inteference effects unique to quantum mechanics (Rarity and Tapster 1997a), to demonstrate quantum teleportation (Bouwmeister et al 1997) and to three photon entanglement (Rarity and Tapster 1999)

7 Conclusions

We have shown that for present applications, namely that of quantum cryptography the weak laser pulse is still the best source of, albeit approximate, single photon states. However progress is being made in the generation of single photon states from single quantum systems. We have shown in principle that single dye molecules could be used as a single photon source but need to improve the efficiency of collection into a single mode. and the dye stability (possible by cooling to cryogenic temperatures. Other sources could be single trapped atoms or ions and a more technologically approachable source could be the single quantum dot coupled to a microcavity. However all these sources will only rival the weak laser when the efficiency of coupling into a single mode exceeds 10%.

The gated parametric source can easily be made more than 50% efficient. However here it is the low, and random, gating rate which limits the application to quantum cryptography. A brighter source of photon pairs is required possibly creating the pairs directly into a single mode fibre or waveguide.

We can extend the single photon encoding via interference to multiple photon encoding in the multi-photon interference experiments that lead on from the above section. Eventually we will be able to create a non-linear element sensitive at the single photon level and be able to build arbitrary entangled states of many photons. Such a system forms a quantum computer capable of performing arbitrary calculations. The extra degree of freedom, when a bit can be in a superposition of two values, may lead to more rapid evaluation of certain difficult problems such as factorisation. However we must remember that we are still a long way from such a system as it has only recently been possible to demonstrate certain simple three photon entanglement effects.

Acknowledgements

The authors acknowledge Claude Weisbuch for his untiring efforts in bringing this school and the ensuing book to fruition.

References

1. Baché Th., Moerner W.E., Orrit M. and Talon H. 1992, *Phys. Rev. Lett.* **69**, 1516
2. Bennett C.H., Bessette F., Brassard G., Salvail L. and Smolin J 1992, *J Cryptology* **5**, 3.
3. Bouwmeester D. Pan J-W. Mattle K. Eibl M. Weinfurter H. and Zeilinger A. 1997, Nature **390**, 575.
4. Buttler W.T. Hughes R.J. Kwiat P.G. Luther G.G. Morgan G.L. Nordholt J.E. Peterson C.G. and SimmonsC.M. 1998, *Phys. Rev. A* **57**, 2379.
5. De Martini F., Di Guiseppe G. and Marrocco M. 1996, *Phys. Rev. Lett.* **76**, 900.
6. Fearn H. and Loudon R. 1987, *Opt. Commun.* **64** 485-490.
7. Hong C.K., Ou Z.Y. and Mandel L. 1987, *Phys. Rev. Lett.* **59**, 2044.
8. Hughes R.J. Luther G.G. Morgan G.L. Nordholt J.E. Peterson C.G. and SimmonsC.M. 1996, *Lecture Notes in Computer Science* **1109**, 329.
9. Kazansky P.G., Dong L. and P St J Russell (1994), *Opt. Lett.* **19**, 701
10. Kimble H.J., Dagenais M., and Mandel L. 1977 *Phys. Rev. Lett.* **39**, 691.
11. Kitson S.C., Jonsson P., Rarity J.G. and Tapster P.R. (1998), *Phys. Rev. A.* **58**, 620.
12. Loudon R. 1983, *The quantum theory of light*, Oxford University Press 2nd edition.
13. Muller A. , Gautier J.D. Gisin N. Huttner B. Tittel W. and Zbinden H. 1997, *Applied Phys. Lett.* **70**, 793
14. Rarity J.G., Tapster P.R. and Jakeman E. 1987 *Opt. Commun.*, **62**, 201.
15. Rarity J.G. Tapster P.R. and Owens P.C.M. 1994, *J. Modern Opt.* **41**, 2435.
16. Rarity J.G. 1995, In *Fundamental Problems in Quantum Theory, Ann. New York Acad. Sci.* **755**, 624.
17. Rarity J.G. and Tapster P.R. 1996, In *Quantum Interferometry II*, F De Martini and Y H Shih eds., Adam Hilger, p122.;
18. Rarity J.G., Tapster P.R. and Loudon R. 1997 *Los Alamos Preprint Server* quant-ph/9702032.
19. Rarity J.G. and Tapster P.R. 1997, *Phil. Trans. R. Soc. Lon.* ,**355** 2267.
20. Rarity J.G. and Tapster P.R. 1997, *J.Mod.Opt.***45** 595
21. Rarity J.G. and Tapster P.R. 1999, *Phys. Rev. A* **59** R35.
22. Townsend P.D. Rarity J.G. and Tapster P.R. 1993, *Electron. Lett.* **29**, 634.
23. Townsend P.D. 1994, *Electron. Lett.* **30**, 809-811.
24. Townsend P.D. 1997, *Nature* **385**, 47-49.
25. Yuen H.P. 1998, *J.Mod.Opt.***8**, 939

Photonic Crystals for Nonlinear Optical Frequency Conversion

Vincent Berger

THOMSON CSF Laboratoire Central de Recherches
Domaine de Corbeville, 91400 ORSAY, FRANCE.

Abstract. Most of the work on photonic crystals has been devoted to the study of the linear optical properties of these refractive index heterostructures. In this lecture, the use of optical heterostructures for nonlinear optics and in particular frequency conversion is reviewed. A key issue for frequency conversion is the possibility of phase matching, that is to compensate for optical dispersion, which results in different phase velocities for light of different frequencies. Various optical heterostructures, which can be used for phase matching nonlinear interactions, are described. A first way consists of engineering form birenfringence in a composite multilayer material. In that case the different waves propagate in the plane of the layers, and phase matching is obtained by making use of the different dispersion relations for two perpendicularly polarized Bloch waves. Generalization of form birefringence phase matching in two-dimensional (2D) photonic crystals is discussed. Quasi phase matching (QPM) is another possibility, which has achieved considerable success for instance in periodically poled $LiNbO_3$: in that case, the propagation is perpendicular to a one dimensional multilayer system of periodic nonlinear susceptibility. QPM can be generalized in 2D: the possibilities offered by frequency conversion in a 2D photonic crystal of $\chi^{(2)}$ are discussed. Interesting perspectives are opened with this kind of 2D nonlinear interaction, described by a nonlinear Bragg law.

1 Introduction

1.1 Photonic crystals need nonlinear optics

In the developpment of the field of photonic band gaps materials, the analogy between electrons and photons has been extensively stressed: the mathematical tools (Floquet-Bloch theorem) and the semantics (conduction and valence band, donors, acceptors) from the well established theory of semiconductors were used for the description of photonic band gap materials[1]. Following this analogy, the most optimistic people compared the promising early years of photonic band gap materials with the revolution of semiconductors in the 50ths. It has also been stressed that important differences exist between electrons and photons, which make this comparison very risky. These differences between electrons and photons are summarized in figure 1. First, the photons are described by a vectorial field whereas the electron wavefonction is scalar. A second difference is the mass of electrons: electrons are usually localized (on atomic potentials) but, in a crystal, the electron wavefunction is delocalized in the array of coupled potential wells. On the opposite, the photon has a delocalized intrinsic nature, and in a photonic crystal (PC), the field can be localized on a defect in the array of refractive index. The most important difference between these two particles is nevertheless the Coulomb interaction. For electrons, the Coulomb interaction makes the calculation of electron states in bulk or quantum heterostructures a very complex problem (see the lecture of Stefan Koch), with phenomena as excitons, band gap renormalization and other collective effects. In the case of photons, the absence of the Coulomb interaction makes the theoretical calculation of, for instance, relation dispersions in crystals, far easier: In the linear approximation of Maxwell's equations, electromagnetics is a single particle problem. But what is an advantage as far as theoretical calulation is concerned is a severe drawback for device purposes: the success of electronic devices (for instance the transistor) is based on the possibility of controlling the motion of electrons with other electrons *via* the Coulomb interaction. Doing the same with photons requires nonlinear optics.

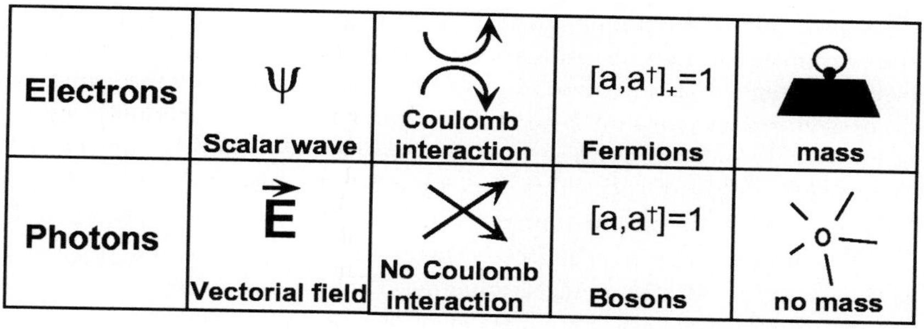

Electrons	ψ Scalar wave	Coulomb interaction	$[a,a^\dagger]_+ = 1$ Fermions	mass
Photons	$\vec{E}$ Vectorial field	No Coulomb interaction	$[a,a^\dagger] = 1$ Bosons	no mass

Fig. 1. Some differences between electrons and photons

In order to make active devices with PCs, we need indeed to control the photons with an external parameter, we need a way to switch the electromagnetic energy from one mode to the other. The electrons change very often their energy in solids: they meet other electrons, phonons, defects... Changing the energy of photons is another matter: this is the field of nonlinear optics. The first point in this introduction is then the following: Due to the absence of the Coulomb interaction, PCs need nonlinear optics, in order to make some active devices. In this paper, we will deal with an important part of nonlinear optics : second-order frequency conversion.

1.2 Nonlinear optics needs photonic crystals

Optical frequency conversion[2,3] by second order nonlinear interaction is a way to obtain coherent light in various spectral regions. The frequency doubling (or second harmonic generation, SHG) process is used for instance to obtain green light from the very efficient near infrared YAG laser, or blue light from semiconductor laser diodes, whereas difference frequency generation (DFG) is the basic process for high power mid infrared sources such as optical parametric oscillators (OPOs). The most famous nonlinear materials used for frequency conversion are KDP, KTP or LiNbO$_3$[3], but several other crystals can be used: AgGaSe$_2$, GaAs, synthetic materials as organic molecules or semiconductor quantum wells (QWs)... In the simple case of plane-wave, collinear SHG, with a non-depleted pump, the second harmonic power scales typically as[4]:

$$\frac{P^{2\omega}}{P^\omega} \propto \frac{\omega^2 \left(\chi^{(2)}\right)^2 L^2}{n^3} P^\omega \frac{\sin^2\left(\frac{\Delta k L}{2}\right)}{\left(\frac{\Delta k L}{2}\right)^2} \tag{1}$$

L is the interaction length of the non linear process, P^ω the pump power at the frequency ω, and n the refractive index of the medium. In Eq. (1), two parameters are of paramount importance, among the different characteristics of nonlinear crystals.

The first one is the nonlinear coefficient $\chi^{(2)}$, which reflects the strength of the nonlinear interaction, and is related to the degree of asymmetry of the electronic potential at the microscopic level. Let us justify in a few words why such an asymmetry gives birth to nonlinear frequency conversion[4]. An incident pump field puts an electron in a superposition of different eigenstates of the potential ; in linear optics this electronic displacement is described by the polarisation, which is proportionnal to the field (*via* the linear susceptibility $\chi^{(1)}$). The polarisation is a source term in Maxwell equation ; this means that the electron radiates an electromagnetic field in response to the incident field, and this is the basic explanation of the refractive index in materials. In nonlinear optics, due to the asymmetry of the potential, the eigenstates are not centrosymmetric and the polarisation is not linearly proportionnal to the incident electric field. The polarisation has still the same temporal period as the incident field, but the nonlinearity has given birth to harmonics in the Fourier spectrum of the polarisation : Second harmonic frequency, in particular, is generated.

The second crucial parameter in Eq. (1) is Δk, which accounts for the possibility to match the phase velocities between the interacting fields at ω and 2ω[2]. Indeed, due to the optical dispersion, the two waves do not travel at the same velocity in the material. This results in a momentum mismatch $\Delta k = k^{2\omega} - 2k^\omega = \frac{2\omega}{c}(n^{2\omega} - n^\omega)$ between the propagating harmonic wave and the non linear polarization, the latter being source of the former. After a distance $L_{coh} = \frac{\pi}{\Delta k}$, called the "coherence length", the nonlinear polarization and the generated wave acquire a phase lag of π, and, due to destructive interference, the second harmonic power decreases for $L > L_{coh}$ (note, in Eq. (1), the $\frac{sinx}{x}$ dependence in L). High SHG efficiencies require small Δk processes. Reaching or approaching $\Delta k = 0$ is called "phase-matching", and can be achieved for example by waves having different polarizations in birefringent crystals. It appears that the phase-matching condition $\Delta k < \frac{\pi}{L}$ is more severe when the crystal length increases. This is because the phase matching condition expresses the photon momentum conservation, which is required with an accuracy inversely proportional to the interaction length. Parameters with some impact in the choice of a nonlinear crystal are the refractive index (note the cubic dependence in Eq. (1)) and the damage threshold for high-power applications. However, the $\chi^{(2)}$ and the possibility of phase-matching are of greatest importance because the nonlinear efficiency varies by orders of magnitude between two different nonlinear crystals (with different $\chi^{(2)}$) or two different temperatures or angles of incidence ($\Delta k = 0$ depends on T and θ). In general, they correspond to the two basic requirements for high-power coherent electromagnetic emission (for instance by an array of antennas): high power emission of individual emitters, and phase coherence between all the emitters, according to the Huyghens-Fresnel principle.

Semiconductors (and especially GaAs) are very interesting materials for nonlinear optics, because the high degree of control of the technology of this material widely used for optoelectronics gives the opportunity to create artificial structures in which these two key features (the nonlinear susceptibility and the phase matching) can be controlled. Thanks to the progress of molecular beam epitaxy or metal organic chemical vapor deposition, different materials such as GaAs, AlGaAs or InGaAs, GaSb or AlGaSb, and a great number of related alloys, can be grown in very thin adjacent layers, with a high control of the interface between the different materials up to the atomic layer. Through an optimization of the widths and compositions of the different layers, it is possible to engineer the energy levels of electrons in semiconductor quantum wells in order to get the desired electronic properties. This has been called band gap engineering or quantum design[5,6]. The energy between different quantized levels can be tuned and also the position of the levels with respect to the barrier potential. The dipole matrix elements describing the strength of the interaction of the material with an electromagnetic field can also be engineered, and this has been used for engineering the $\chi^{(2)}$ in semiconductor heterostructures. A review of this field can be found in [7].

In addition to band gap engineering for the control of the behavior of electrons, semiconductor heterostructure growth and technology have given birth to a great number of structures controlling the motion of photons[8]; and all these structures are interesting for frequency conversion purposes: integrated waveguides [9], Fabry Perot cavities [10,11], or more recent objects as whispering gallery structures [12], pillar microcavities [13], photonic wires [14], air bridges [15] or photonic band gap materials [16–18]. By analogy to band gap engineering, this field can be called "refractive index engineering" [18]. Among all photonic applications, the possibility of phase matching nonlinear interactions will be the subject of this paper. If band gap engineering enables one to engineer the microscopic electronic properties of materials (and in particular $\chi^{(2)}$, it will be shown how refractive index engineering enables one to engineer the macroscopic optical properties of the composite material, in particular to get the phase matching conditions $\Delta k = 0$. In both cases, the success relies on the possibility of realizing materials by design, thanks to growth and processing of materials.

In second order nonlinear optics, the material is fully described by two optical susceptibilities: The linear one ($\chi^{(1)}$) (which is related to the refractive index) and the nonlinear one ($\chi^{(2)}$), which gives the source term for a second harmonic beam. The possibility of building a PC with a nonlinear material can then be divided in two ways, depending on the susceptibility which is periodic. This lecture is thus divided in two parts. In the first one, structures where the linear susceptibility is periodic will be discussed, these structures belong to the family of PCs in the classical sense. In a second part, structures where the nonlinear susceptibility is periodic will be described. This includes QPM structures that have been used for years in nonlinear optics, but new possibilities in 2D PCs of $\chi^{(2)}$ will also be presented.

2 Photonic crystals of $\chi^{(1)}$

2.1 1D photonic crystals: Form birefringence in multilayer heterostructures

As already said in the introduction, a key feature of a nonlinear material is the possibility of phase matching. Phase matching can be achieved by varying the velocity of waves in the nonlinear medium, which implies engineering the refractive indices in the material. In this part, in particular, it will be shown how this is possible in a multilayer stack. Phase-matching is indeed obtained by using a built-in *artificial birefringence* in the new composite multilayer material: the isotropy of bulk GaAs is broken by inserting thin oxidized AlAs (Alox) layers in GaAs [19]. This concept, called form birefringence[20,21], was proposed in 1975 by Van der Ziel[22] for frequency conversion phase-matching. However, the experimental realization of this proposal has been achieved only very recently, due to the lack of a well-suited pair of materials having a high nonlinear coefficient, and high enough refractive index contrast for form birefringence phase matching[19].

An intuitive way to understand the origin of form birefringence is to consider the macroscopic crystal formed by a GaAs/Alox multilayer system. GaAs is a cubic semiconductor of point group $\bar{4}3m$, and therefore it is not birefringent. The presence of thin Alox layers grown on a (100) substrate breaks the symmetry of 3-fold rotation axes and the point group of the composite material becomes $\bar{4}2m$, *i.e.* the same as KDP. This artificial material has the nonlinear properties of GaAs: in particular the same tensorial character and roughly the same nonlinear coefficient, (if we neglect the small zero contribution of thin Alox layers), but the linear optical symmetry of

KDP. Otherwise stated: one takes advantage of the microscopic nature of the nonlinear polarization given by the ionicity of GaAs, and the macroscopic engineering of the refractive index on the scale of the extended electromagnetic wavelengths. It can be noticed that with (111) oriented GaAs, the introduction of Alox layers switches the point group from $\bar{4}3m$ to $3m$, which is identical to that of another nonlinear birefringent material: LiNbO$_3$. These symmetry considerations are illustrated in Fig.2.

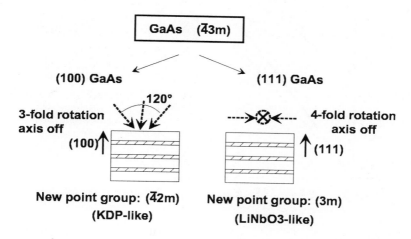

Fig. 2. Point groups of multilayer stacks with GaAs. The point group of GaAs is $\bar{4}3m$, this means that this material is not birefringent. If some layers of an amorphous material are inserted in a (100) GaAs substrate (on the left), the 3-fold rotation axes (dashed arrows) disappear and point group of the composite material is now $\bar{4}2m$, the same as KDP. If the same layers are inserted on a (111) oriented GaAs substrate (on the right), the 4-fold rotation axes (dashed arrows) are killed in that case, and the point group of this other composite material is $3m$, which is that of another nonlinear material: LiNbO$_3$. These are two artificial birefringent nonlinear materials.

In the pioneering paper of Van der Ziel[22], form birefringence was calculated directly from Maxwell's equations in the large wavelength approximation. Another physical explanation can be given in terms of a modal wavefunction approach. Let us consider an infinite periodic multilayer material. Following Joannopoulos[23], for a given wavevector the frequency of an allowed electromagnetic mode in a composite medium increases with the fraction of electric field in the low index material. In fact, as illustrated in Fig. 3, the difference between the Transverse Electric (TE) and Transverse Magnetic (TM) polarizations arises from the continuity equations at the boundaries between the two materials. In the TM polarization, the continuity of the electric displacement forces the electric field to have an high value in the low index material. This mode thus has a higher frequency for a given wavevector.

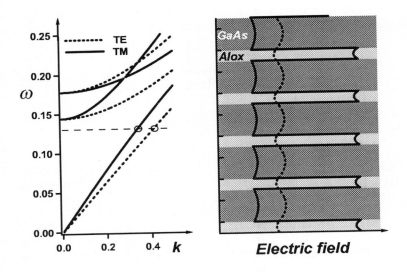

Fig. 3. Dispersion relation for an in-plane propagation in a periodic composite material which consists of 25% of Alox (n≃1.6) and 75% of GaAs (n≃3.5), for TM modes (full line) and TE modes (dotted lines). In this theoretical plot illustrating only the principle of form birefringence, the dispersion of the materials has been neglected. The frequency is given in units of $\frac{2\pi c}{d}$ and the wavevector in units of $\frac{2\pi}{d}$, where d is the period of the multilayer. The physical origin of form birefringence appears in the mode wavefunction, pictured to the left for a frequency $\omega = 0.13 \times \frac{2\pi}{d}$ (they correspond to the open circles in the dispersion relation). These Bloch waves have been calculated using standard periodic multilayer theory[21]. The direction of propagation is perpendicular to the plane of the figure. Due to the continuity of the electric displacement εE normal to the layers, the TM mode (full line) has an important overlap with the low ε layer (Alox, in light gray), and a lower average dielectric constant. The continuous TE electric field (dotted line) has a higher value in GaAs (dark gray), and a higher average dielectric constant. From reference [19].

For large wavelengths (compared to the unit cell), light experiences an effective medium. As a consequence dispersion relations $\omega(k)$ in Fig. 3 are linear near the origin, where form birefringence appears as the difference between slopes (*i.e.* phase velocities) for TE and TM waves. In that case the two dielectric constants of the uniaxial composite material are given by:

$$\varepsilon_{TE} = \alpha_1 \varepsilon_1 + \alpha_2 \varepsilon_2 \tag{2}$$

$$\frac{1}{\varepsilon_{TM}} = \frac{\alpha_1}{\varepsilon_1} + \frac{\alpha_2}{\varepsilon_2} \qquad (3)$$

where α_i and ε_i are the filling factors ($\alpha_1 + \alpha_2 = 1$) and the dielectric constant of the two constitutive materials, respectively. These equations are analogous to electrical series and parallel capacitors: The charge equality $C_1 V_1 = C_2 V_2$ between series capacitors is in fact the static limit of the electric displacement continuity relation for TM waves ($\varepsilon_1 E_1 = \varepsilon_2 E_2$), and the bias equality $V_1 = V_2$ for parallel capacitors is equivalent to the electric field continuity for TE waves ($E_1 = E_2$).

In the dispersion diagram of Fig. 4, a simple picture of phase matching is given, energy and momentum conservation corresponding to the equality between the vectors associated with the nonlinear interaction. This is usual in solid state physics, for instance in the description of optical transitions in semiconductors, Raman or Brillouin scattering, acousto-optics, etc...

It appears from Eqs.(1) and (2) that form birefringence ($\sqrt{\varepsilon_{TE}} - \sqrt{\varepsilon_{TM}}$) increases with the refractive index contrast between the two materials in the multilayer, as for photonic band gap effects in PCs[25,18]. Although a GaAs/AlGaAs multilayer structure was formerly proposed for phase matching[22], the refractive index contrast between GaAs (n$\simeq$3.5) and AlAs (n$\simeq$2.9) is too low to provide the birefringence required to compensate for the dispersion. This is the reason why thin film layers of Alox (n$\simeq$1.6) in GaAs have been used to get sufficient form birefringence. Alox results from selective oxidation at 400$-$500°C in a water vapour atmosphere of AlAs layers embedded in GaAs. This technology of AlAs oxidation has emerged in the early 90's[26], and since then Alox has led to breakthroughs in the fields of semiconductor lasers[27] and Bragg mirrors[28] -again thanks to its refractive index contrast with GaAs-.

To illustrate experimentally the efficiency of form birefringence phase matching, a DFG interaction was first demonstrated. An example of a structure used for DFG is presented in Fig. 5, along with the three modes involved in the nonlinear interaction. The details of the experiments can be found in [19,29,30]. In the structure depicted in Fig. 5[31], a form birefringence $n(\text{TE}) - n(\text{TM}) = 0.154$ has been measured for a wavelength of 1.06 μm, and even higher birefringences, of about 0.2, have been obtained with different samples. Such birefringences are sufficient to phase match mid-infrared generation between 3 μm and 10 μm by DFG from two near-infrared beams. Note that by increasing the width of Alox layers (as in the example of Fig. 3), much higher birefringences up to 0.65 could be achieved, in principle.

The DFG process (1.035 μm, TM) $-$ (1.32 μm, TE) $\mapsto$ (4.8 μm, TE) is represented schematically in Fig. 5. The experimental set up is simple: Two near infrared beams (a tunable Ti:Sa and a YAG laser at 1.32 μm) are simultaneously end fire coupled into the waveguide by a microscope objective. The output light is collected by an achromatic reflecting microscope objective, and the mid-Infrared DFG signal is finally measured by an 77K InSb detector.

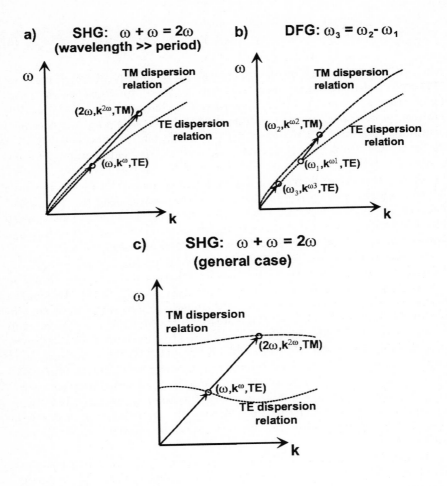

Fig. 4. Phase matching is nothing but a coherent transfer of energy between different modes $(\omega, \mathbf{k})$, with momentum and energy conservation. For every kind of phase matching, this can be illustrated by a simple picture in the dispersion relation. In our case of form birefringent phase matching: a) for SHG in the long wavelength approximation (near the origin of the dispersion diagram). The first arrow represents the mode $(\omega, k^\omega, \mathbf{TE})$ and the second one the transition to the mode $(2\omega, k^{2\omega}, \mathbf{TM})$. Phase matching means that these two arrows are identical. b) for DFG the same kind of schematic picture applies. c) in the general case, phase matching between two different bands in the multilayer structure. Only the two bands of interest have been represented. This case is also a generalization of modal phase matching that has been demonstrated in waveguides. In that case, a careful look at the overlapp nonlinear integral between the different modes has to be considered, in addition to phase matching [24].

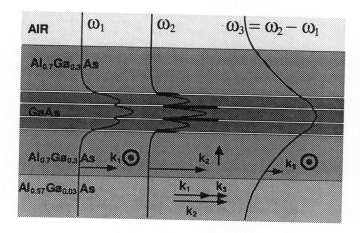

Fig. 5. DFG process in the GaAs/Alox multilayer structure, in which form birefringence phase matching was first demonstrated. Three periods of the composite material GaAs(325 nm)/Alox(40 nm) constitute the core of the waveguide. The birefringence of the composite material was engineered to compensate for the dispersion arising from both the natural dispersion in bulk GaAs and the optical confinement dispersion in the waveguide. The sample was grown by molecular beam epitaxy on a GaAs (100) substrate and consists of: 2800nm $Al_{0.97}Ga_{0.03}As$; 1500nm $Al_{0.70}Ga_{0.30}As$ (waveguide cladding layers), three periods of birefringent composite material (40 nm Alox; 325nm GaAs)×3 and 40 nm Alox; 1500nm$Al_{0.70}Ga_{0.30}As$ and a final 30 nm GaAs cap layer. The oxidation process is described in detail in reference [31]. The three modes involved in the DFG process are pictured together with their polarization (↑ for TM, ⊙ for TE). The higher overlap of the TM mode with the low refractive index Alox layers is apparent, which is the origin of form birefringence. The arrows recall the "phase matching" momentum conservation. From reference [19].

A typical infrared output is shown in Fig. 6 versus the Ti:Sa wavelength. This function has the expected $\left(\frac{\sin x}{x}\right)^2$ shape, which is clear evidence of phase matching. This kind of experiment has been the first achievement of perfect phase matching with a cubic nonlinear material, and also the first realization of Van der Ziel's 22-year-old proposal.

Typical mid-IR output powers of 120 nW were obtained for 0.4 mW and 17 mW of Nd:YAG and Ti:Sa pump powers, respectively[30]. By increasing pump powers and reducing scattering losses originating from processing[30], this result can easily be pushed into the μW range, which is an interesting power level for mid-infrared spectroscopic applications. It is worth mentioning that wavelengths up to 5.3 μm at room temperature, and 5.6 μm at a waveguide temperature of 150°C have been generated[30].

The birefringence of the composite structure is sufficient not only to phase match DFG in the mid infrared, but also for SHG around 1.55 μm, recently

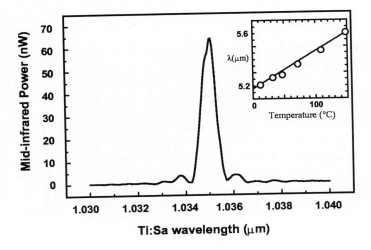

Fig. 6. Mid-Infrared DFG signal as a function of the Ti:Sa laser wavelength. The pump powers were 0.2 mW and 1.6 mW for the YAG and the Ti:Sa lasers, respectively. This function has the well known shape of a phase matching resonance: It is a $\left(\frac{\sin x}{x}\right)^2$ function, characteristic of a momentum conservation in a given interaction length.

demonstrated[32]. In this respect, two examples deserve a mention because of their great application interest: continuously tunable mid-infrared compact sources are desirable for pollutant detection in the molecular fingerprint region or for process monitoring; and a 1.55 μm signal can be wavelength-shifted by mixing it with a 0.75 μm pump, which is a function required in wavelength division multiplexing.

Among the processing issues of the GaAs/Alox system, of critical importance is the accurate characterization of near- and mid-infrared losses, due to absorption and scattering on the ridge inhomogeneities introduced during the etching process. These parameters are crucial for the successful utilisation of this composite material in frequency conversion devices, and have been analyzed in ref.[30], where rather high mid IR losses (of the order of 50 cm^{-1}) were mainly attributed to ridge sidewall scattering. Experiments are underway to reduce losses by improving the fabrication process and by an optimization of the geometrical parameters of the ridge.

A lot of perspectives are offered by form birefringence phase-matching in multilayers. Let us cite some of them:

Simultaneous frequency conversions

Form birefringence is an example of building an artificial structure with the desired optical properties, that is an illustration of refractive index engineering. It is obvious that the large number of degrees of freedom in the design of the structures enables to design waveguides with several simultaneous constraints. For instance it is possible to design a guide in which the doubling process $\omega + \omega \to 2\omega$ and the sum frequency (SFG) process $\omega + 2\omega \to 3\omega$ are simultaneously phase matched. Note that since two polarizations exist in the waveguide, modal phase matching (*i.e.* between guided modes of different order[24]) is necessary to phase match more than two waves. This novel third harmonic generation (THG) scheme considerably simplifies the usual THG method of realizing first SHG and then SFG in two different crystals. The possibility of phase matching these two nonlinear processes simultaneously, making the three waves ω, 2ω and 3ω, which obey to a new system of nonlinear coupled equations, interact simultaneously. This constitutes a new type of nonlinear interaction.

QWs as nonlinear material

Form birefringence phase matching has been demonstrated using bulk GaAs as nonlinear material. On the other hand, the interest of GaAs, and especially molecular beam epitaxy on GaAs, is that it is also the material for QW's. As already pointed out in the introduction, band gap engineering of multiple QW's has lead to the demonstration of the highest $\chi^{(2)}$ ever reported[7]. In principle, using intersubband-based $\chi^{(2)}$ inside such phase-matched structures is readily feasible by growing asymmetric QW's in the core of a GaAs/Alox structure. Due to the huge nonlinear susceptibilities demonstrated in these asymmetric QW structures [7], this may lead to highly efficient devices. For such a purpose, deep QW's for high energy intersubband transitions have to be choosen (see for example ref. [33]), to avoid the absorption range of Alox. A careful calculation of the nonlinear efficiency, taking into account the selection rules for transitions between confined electronic levels in semiconductor heterostructures, has to be considered.

Optical parametric oscillator on a GaAs chip

Once DFG has been demonstrated, it seems obvious that parametric fluorescence can be observed in the same structures. From this, to obtain an OPO with these waveguides is just a matter of reduction of losses in the mid-IR. A lot of processing work is necessary to reach this purpose, especially a precise measure of optical losses in oxidized AlAs. This is a crucial point, very difficult to measure with a good accuracy since this material exists only in thin layers and small surfaces.

Non-critical phase matching

Another consequence of the large number of degrees of freedom in the design of the structures is the possibility to choose, among all the possible phase-matched structures, the one with the widest phase matching resonance. Relaxing the phase matching condition is practically important to reduce the device sensitivity to *e.g.* temperature and pump wavelength. A preliminary calculation shows that to achieve a non critical phase matching for the DFG process $\omega_3 = \omega_1 - \omega_2$ (that is to vanish the phase mismatch at the first order around the resonance), one has to minimize the quantity

$$F(n_1(\omega), n_3(\omega)) = n_3 - n_1 + \omega_3 \frac{dn_3}{d\omega} - \omega_1 \frac{dn_1}{d\omega} \qquad (4)$$

, among all the possible structures that are phase matched (*i.e.* with $n_3\omega_3 = n_1\omega_1 - n_2\omega_2$). Again, this is a refractive index engineering problem, with two constraints (the phase matching and its non-critical character) instead of one.

Another way to get a relaxation of the phase matching condition: Insertion of optical amplification in the nonlinear structure

As previously stated, one of the major advantages of GaAs/Alox composite nonlinear material is its fabrication on a GaAs substrate, fully compatible with the GaAs technology, with the possibility of integration with the GaAs-based devices developed in the field of optoelectronics: in particular the very efficient GaAs-based QW lasers, which represent the major item of the semiconductor laser market. The introduction of QWs with population inversion and gain in a phase-matched nonlinear waveguide has useful properties. In particular, it was shown[34] that merging optical amplification and three-wave mixing can strongly relax the phase-matching condition for DFG.

Self-pumped optical parametric oscillator: A QW laser OPO

The introduction of inverted QWs in the nonlinear structure would be of great interest if the optical gain could be pushed above lasing threshold. In such case, the structure would be basically a semiconductor laser, with intracavity phase-matched parametric fluorescence. If losses at the parametric frequencies are lower than the parametric gain, the structure is an OPO which provides its own pump internally. The full OPO system, including laser pump, nonlinear material and cavity, would be integrated in the same cavity on a semiconductor chip. This would represent the smallest OPO system ever realized, and also an alternative way to extend the range of room temperature c.w. semiconductor lasers in spectral regions where they are not yet available (*e.g.* between 3 and 4 μm).

Several problems need to be solved, however, before the realization of such a device : losses have to be reduced (this is a processing issue) ; since the lowest wavelength in the nonlinear interaction must be TM polarized, the QWs have to lase in the TM polarization, which is possible but not usual.

Another relevant issue is the competition between QW gain and the losses constituted by the parametric fluorescence, which alter the spectral gain and hence the laser wavelength. Such situations have been studied by Khurgin[35]. Finally, for electrical pumping, the insulator character of Alox also represents a serious problem.

2.2 2D photonic crystals of $\chi^{(1)}$: Form birefringence in arrays of cylindric air holes

The composite material GaAs/AlOx bears a considerable drawback: it only exists in thin layers (a few microns), due to its fabrication by molecular beam epitaxy. Considering the great success met by periodically-poled Lithium Niobate as it became technologically available in thick samples rather than waveguides[36], it would be very interesting to obtain a thick form-birefringent material. One could focus greater energy onto the structure and obtain higher nonlinear efficiencies. In this respect, the study of form birefringence in 2D materials may represent an alternative way. The modeling of form birefringence in a 2D set of air cylinders in GaAs, for waves propagating in the plane perpendicular to the cylinders, ressembles the calculation of the dispersion relations in photonic band gap materials. Form birefringence appears again, in the long wavelength approximation, as the difference of slopes between the TE and TM dispersion relations, which have been calculated for years now[25]. In our case, however, the photonic band gaps of the material have to be avoided, since we want light to propagate, and the dielectric structure is designed only for phase matching purposes. Calculations show that it is possible for instance to phase match the SHG process of the CO_2 laser radiation at 10.6 μm with air holes in GaAs, of diameter 0.2 μm, located on a triangular lattice of period 3 μm. Such a material is very difficult to obtain with the state-of-the-art of GaAs technology (especially in large thicknesses, lets say 100 μm), but very impressive structures have been demonstrated in macroporous Silicon[37], with these typical dimensions. Silicon has unfortunately no $\chi^{(2)}$; however phase matching in a 2D form birefringent material could also be demonstrated for a third order process, such as third harmonic generation of a CO_2 laser. Calculations of sum-frequency phase matching in 2D photonic lattices can be found in [38].

3 Photonic crystals of $\chi^{(2)}$

3.1 1D photonic crystals: Quasi phase matching

Since we are dealing with second-order optical processes, the materials are described by two susceptibilities $\chi^{(1)}$ and $\chi^{(2)}$. They are then two ways to make a PC; with a periodic $\chi^{(1)}$ or $\chi^{(2)}$, respectively. The first part was devoted to the description of PCs of $\chi^{(1)}$, this second part will describe the possibilities of phase matching in structures with a periodic $\chi^{(2)}$.

We focus now on PCs of $\chi^{(2)}$, where the linear susceptibility is supposed to be homogenous. 1D crystals (*e.g.* multilayers ($\chi^{(2)}, -\chi^{(2)}$)) will be first presented. These structures are well known in the field of nonlinear optics and are very efficient materials. Note that the light is propagating perpendicularly to the layers and not in the plane of the layers, on the opposite to form birefringence phase matching presented in the first part. The purpose is anyway the same: phase matching the nonlinear interaction. This particular scheme is called quasi phase matching (QPM).

QPM was first proposed in a seminal paper of Bloembergen and co-workers[2], to solve the problem of phase mismatch. The principle of QPM is the following: We know that due to the dispersion, the phase mismatch between the interacting waves reduces the useful interaction length for frequency conversion to the so-called "coherence length", which is inversely proportional to the dispersion and equal to $L_{coh} = \lambda_0^\omega/4 \left(n^{2\omega} - n^\omega\right)$ for SHG of a wavelength λ_0^ω in vacuum. QPM consists of reversing the sign of the nonlinear susceptibility of the material every coherence length. The consequence is the change of sign of the nonlinear polarization. This switches also the sign of the generated harmonic wave, and this change of sign exactly compensates for the destructive interference coming from the dispersive propagation. The constructive build-up of the generated wave occurs then on the entire length of the QPM structure, increasing the overall energy conversion. The principle of QPM is presented on figure 7, and an excellent review about this topics can be found in [39], with a lot of references about different materials that have been used for QPM.

Various demonstrations of QPM have been performed, with different materials. Among them, Periodically Poled Lithium Niobate (PPLN)[40,41] or periodic poled KTP[42] have become some of the most attractive nonlinear materials for OPOs. QPM structures using semiconductors have also been demonstrated: stacks of GaAs plates have been used, obtained by wafer bonding[43]. In that case, several substrates of thickness one coherence length (for example 108 μm for frequency doubling at 10.6 μm) are bonded, two successive substrates having opposite $\chi^{(2)}$ orientations. GaAs waveguides with periodic (100) and (-100) oriented zones were also recently demonstrated[44].

Let us stress the fact that a lot of analogies can be found between Bragg mirrors or more generally multilayer structures used for linear optics (linear phase matching) and QPM structures used for nonlinear phase matching. It was shown that in periodic multilayer structures the reflections bands depend on the *Fourier transform of the $\chi^{(1)}$ function*, and this was used for the formulation of a Fourier transform method for optical multilayer design[45]. For instance, the width of the nth gap of 1D photonic band gap material is proportional to the value of the fundamental spatial component of the Fourier expansion of the dielectric constant in the multilayer. That is why the even order gaps of a perfectly balanced Bragg mirror ($\frac{\lambda}{4},\frac{\lambda}{4}$) vanish[21]. From the point of view of QPM materials, the fundamental equation of the evolution of the harmonic field in a $\chi^{(2)}$ material is given by[4]:

$$\frac{dE^{2\omega}}{dz} \propto (E^\omega)^2 \; \chi^{(2)}(z) \; exp(i\Delta kz) \tag{5}$$

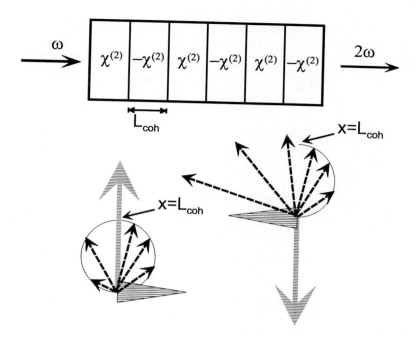

Fig. 7. Quasi phase matching principle: The QPM structure (top) is a periodic structure of $\chi^{(2)}$, as proposed in the pioneering reference[2]. The waves are travelling perpendicularly to the layers and the thickness of the $\chi^{(2)}$ domains is equal to one coherence length of the nonlinear process. The basic principle of QPM is explained in the schematic picture. The thick striped arrows represent the nonlinear polarization, fixed in the scheme rotating at $2k^\omega$. It creates a SH field in quadrature (striped triangles at the origin). The thin dashed black arrows represent the SH field, which is dephased with respect to the polarization during the dispersive propagation. After one coherence length, the SH field decreases back to zero, because the direction of creation of the second harmonic wave (triangle) is opposite to the harmonic wave itself: a destructive interference occurs (left picture). Right picture: QPM consists of changing the sign of the polarization (by reversing $\chi^{(2)}$) after L_{coh}. The sign of the nonlinear polarisation is changed and the interference is still constructive.

From this equation it appears that the harmonic field is the *Fourier transform* of the $\chi^{(2)}$ function[39]:

$$E^{2\omega} \propto \int (E^\omega)^2 \, \chi^{(2)}(z) \, exp(i\Delta k z) dz \tag{6}$$

The QPM principle is very clear from this expression: If the $\chi^{(2)}$ function has a component $exp(-i\Delta k z)$, the harmonic field can grow constructively on a large scale. This exactly what happens in a QPM material, since $\chi^{(2)}$ is periodic with a period equal to $2L_{coh}$, where the coherence length $L_{coh} = \frac{\pi}{\Delta k}$.

Following these observations and this analogy between the Fourier transforms of the $\chi^{(1)}$ and $\chi^{(2)}$ functions, a lot of interesting variations from the QPM principle have been demonstrated: Broad band QPM structures have been realized (*i.e.* with a large tuning curve resonance) by making a structure with deviations from perfect periodicity[46]. By means of a variation of the period of the QPM grating, Arbore *et al.* have demonstrated the possibility of SH pulses that are stretched or compressed relative to the input pulses at the fondamental frequency[47]. This is analoguous to the possibilities of pulse compression in Bragg mirrors obtained by engineering the dispersion of the reflection spectrum. Wedged QPM structures with a variation of the $\chi^{(2)}$ period on the surface of the sample have also been proposed, in analogy with "tunable" wedged interference filters[43].

Another analogy between Bragg heterostructures and QPM structures will be studied in the last section of this lecture: QPM in a 2D crystal of $\chi^{(2)}$, which is the a generalization of QPM in two dimensions, as photonic band gap materials are the generalization of Bragg mirrors in several dimensions.

3.2 2D photonic crystals of $\chi^{(2)}$

This last part will focus on the possibilities offered by a 2D PC of $\chi^{(2)}$. Such a structure presents a space-independent linear dielectric constant, but has a periodic second order nonlinear coefficient in a 2D plane. Figure 8 shows schematically an example of the structure under study: a 2D triangular lattice of cylinders with nonlinear susceptibility tensor $(-\chi^{(2)})$ in a medium of nonlinear susceptibility $\chi^{(2)}$.

In the same way as the 1D case of a $\chi^{(1)}$ crystal is the Bragg mirror, the 1D case of a $\chi^{(2)}$ crystal is the QPM structure presented in the previous section. 2D crystals of $\chi^{(2)}$ are then a generalization in two dimensions of QPM structures in the same way as photonic band gap materials are a generalization in two dimensions of Bragg mirrors.

The first point addressed in this paper concerns the possibility of realizing a 2D or 3D PC of $\chi^{(2)}$. In 1D QPM structures, for GaAs waveguides as for PPLN, the 1D periodicity of the nonlinear susceptibility is defined by the design of a metallic grating. In the case of GaAs waveguides, the grating is used as a mask for a reactive ion etching step[44], and in the case of PPLN, the grating is an electrode for ferroelectric domain reversal. Though these techniques are very different, they both use a metallic grating, defined by electron-beam lithography, which defines the pattern of the QPM structure. Both techniques can be generalized to the 2D structure presented in Fig.??. One only has to change the metallic grating into a metallic honeycomb mask during the technological process. In the case of PPLN, it is necessary to choose a connex area for the metal (white area in figure 8), in order to apply easily the voltage on the whole pattern. We conclude that 2D PCs of $\chi^{(2)}$ are easy to obtain as a generalization of 1D QPM technology.

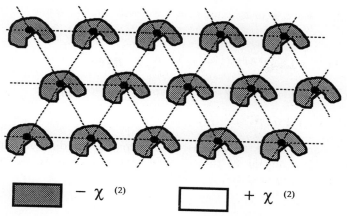

Fig. 8. Schematic picture of a 2D crystal of $\chi^{(2)}$. The material presents a translation invariance perpendicular to the figure, and is invariant by translation in a 2D lattice (here a triangular lattice). The linear susceptibility is constant in the whole material but the sign of the second order susceptibilty $\chi^{(2)}$ presents a given pattern in the unit cell.

Conversely, the fabrication of 3D crystals of $\chi^{(2)}$ seems to be very tricky. One may imagine a complicated multistep technology with bondings, etchings and regrowth of GaAs, resulting in a 3D stack of domain reversals. Such a process, theoretically possible, is however far beyond the state-of-the-art of GaAs technology. For this reason, this paper will consider mainly 2D $\chi^{(2)}$ crystals. However, it is clear that a hypothetical 3D structure would present analogous properties to those described here, and the assumption of a 2D structure in the following is not a loss of generality.

Let us assume that a plane wave at the frequency ω propagates in the transverse plane of a 2D $\chi^{(2)}$ crystal, that is perpendicularly to the translation axis of the cylinders, of arbitrary section. Let us recall that the linear dielectric constant is constant in the whole structure. This ensures that multiple reflections, leading to PBG effects, are not present. In this 2D structure, the problem can be considered as scalar[48], which simplifies the notations. For instance, in the case of a 2D PPLN crystal, fundamental and harmonic waves are TM polarized, *i.e.* with the electric field in the translational direction. Although they are constant in space, the linear dielectric constants are assumed to be different at ω and 2ω, this dispersion being the source of phase mismatch. An efficient SHG process in the $\chi^{(2)}$ crystal is obtained if a quasi-plane wave at the harmonic frequency is observed to increase at a large scale, compared to the coherence length L_{coh} and to the $\chi^{(2)}$ period order.

By quasi-plane wave, it is assumed that if we write the harmonic electric field as:

$$\mathcal{E}^{2\omega}(\mathbf{r},t) = \frac{1}{2}E^{2\omega}(\mathbf{r}) \, exp[i\left(2\omega t - \mathbf{k}^{2\omega}\mathbf{r}\right)] + c.c. \qquad (7)$$

then the classical slow varying envelope approximation applies:

$$\mathbf{k}^{2\omega}.\nabla\left(E^{2\omega}(\mathbf{r})\right) \gg \nabla^2 E^{2\omega}(\mathbf{r}) \qquad (8)$$

In these equations, $\mathbf{r} \equiv (x,y)$ is the 2D spatial coordinate. Under this assumption, the evolution of the SH field amplitude can be written as a function of the pump field and the second order coefficient $\chi^{(2)}(\mathbf{r})$:

$$\mathbf{k}^{2\omega}.\nabla\left(E^{2\omega}(\mathbf{r})\right) = $$
$$-2i\frac{\omega^2}{c^2}\left(E^\omega\right)^2 \chi^{(2)}(\mathbf{r}) \, exp[i\left(\mathbf{k}^{2\omega} - 2\mathbf{k}^\omega\right)\mathbf{r}] \qquad (9)$$

This equation is a simple generalization in two dimensions of the 1D harmonic field evolution equation[4], where the derivative has been replaced by a gradient and $(E^\omega)^2$ is assumed to be constant. The nonlinear susceptibility can be written as a Fourier series:

$$\chi^{(2)}(\mathbf{r}) = \sum_{\mathbf{G} \in RL} \kappa_\mathbf{G} \cdot exp\left(-i\mathbf{G}.\mathbf{r}\right) \qquad (10)$$

where the sum is extended over the whole 2D reciprocal lattice (RL)[1]. Inserting this expresion in Eq.(9), the increase of the SH field appears to be related to a sum of $exp[i\left(\mathbf{k}^{2\omega} - 2\mathbf{k}^\omega - \mathbf{G}\right)\mathbf{r}]$. The QPM condition appears then as the expression of the momentum conservation:

$$\mathbf{k}^{2\omega} - 2\mathbf{k}^\omega - \mathbf{G} = 0 \qquad (11)$$

For 1D QPM, the phase mismatch can be compensated in a structure of period d if it is equal to a multiple of the fundamental spatial frequency of the structure $\frac{2\pi}{d}$[39]. In contrast to this, QPM in a 2D PC of $\chi^{(2)}$ involves a momentum taken in a the 2D RL. The possibilities of QPM are not only six-fold degenerate (thanks to the symmetry of the triangular lattice), but new QPM orders appear in the 2D crystal which are not multiples of the fundamental QPM process, opposite to the 1D situation. Two examples of 2D

---

[1] In the case of the honeycomb lattice of cylinders, following [49], $\kappa_\mathbf{G} = 4f\frac{J_1(|\mathbf{G}|R)}{|\mathbf{G}|R} \times \chi^{(2)}$, where f is the filling factor of the circle in the Wigner-Seitz cell, and J_1 the first Bessel function. For a radius R of cylinders equal to 0.38 times the period, we have $\kappa_\mathbf{G} = 0.13\chi^{(2)}$ for a [1,0] order process. This can be compared to $\frac{2}{\pi}\chi^{(2)}$ in the case of a 1D first order QPM. Higher [1,0] Fourier coefficients can be obtained with a triangular lattice of hexagons instead of cylinders.

QPM processes are shown in Fig.2 9: the fundamental process, which involves the shortest possible **G** vector, and a 2D QPM process with a momentum transfer $\sqrt{3}$ times greater, which is impossible in a 1D structure. The 2D QPM order can be labeled with two integer co-ordinates, given in the $(\mathbf{G_1}, \mathbf{G_2})$ basis of the RL. In Fig.9 for instance, 2D QPM processes of orders [1,0] and [1,1] are represented. In the case of a unit cell invariant by the symmetry operations of the Bravais lattice (as for instance a triangular lattice of circular patterns), it is obvious that points in a 30° sector ΓM-ΓK form a complete set of QPM schemes, all other points in the RL playing the same role for a reason of symmetry. This means that $[x,y] \in N^2$ orders with $y \geq 0$ and $x - \sqrt{3}y \geq 0$ represent all the 2D QPM processes in these structures. In the case of an asymmetric unit cell (as for example the graphite-like structure[50]), the RL and the QPM schemes are the same as above. However, the related conversion efficiency depends of the Fourier coefficient of Eq.(10), which depends on the shape of the $\chi^{(2)}$ pattern at the unit cell level, and is generally not the same for different vectors of equal modulus in the RL.

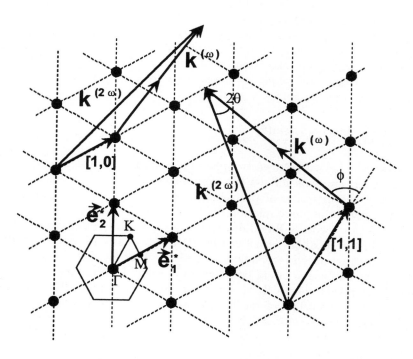

Fig. 9. Reciprocal lattice of the structure of Fig. 8, with the 2D QPM processes of order [1,0] and [1,1] shown schematically. The efficiency of the nonlinear process is proportional to the corresponding 2D Fourier series coefficient, which depends on the unit cell filling factor, and is not represented here. The first Brillouin zone with the usual Γ, M, and K points, is represented on the left.

A particularly interesting case occurs when $\mathbf{k}^{2\omega}$ and $\mathbf{k}^{\omega}$ are colinear, because the interaction length is not limited by the walk-off between pump and harmonic waves. Such a process is obtained when the phase mismatch is equal to the modulus of a vector in the RL. For instance, in the case of a structure with a triangular lattice of period d, the phase mismatches that can be compensated are equal to $\sqrt{x^2 + y^2 + xy} \times \frac{2\pi}{d}$, $[x,y] \in N^2$. They are then belonging to the series $(1, \sqrt{3}, 2, \sqrt{7}, 3, 2\sqrt{3}...)$. This has to be compared to the series $(1, 3, 5, 7, ...)$ which is obtained in the usual 1D QPM process.

Using some trigonometry, Fig.9 leads to:

$$\lambda^{2\omega} = \frac{2\pi}{|\mathbf{G}|} \sqrt{\left(1 - \frac{n^{\omega}}{n^{2\omega}}\right)^2 + 4\frac{n^{\omega}}{n^{2\omega}} sin^2\theta} \qquad (12)$$

where $\lambda^{2\omega}$ is the SH wavelength inside the material and 2θ the walk-off angle between $\mathbf{k}^{2\omega}$ and $\mathbf{k}^{\omega}$. More generally, this equation gives the direction of coherent radiation at the wavelength $\lambda^{2\omega}$ for a phased array of nonlinear dipoles having a phase relation fixed by the propagation of the pump. Eq. (12) appears then as a *nonlinear Bragg law*, and is a generalization for nonlinear optics of the Bragg law. It gives the direction of resonant scattering at the wavelength $\lambda^{2\omega}$ of a plane wave with vector $\mathbf{k}^{\omega}$ by a set of nonlinear dipoles. If the medium has no dispersion, $n^{\omega} = n^{2\omega}$ and Eq. 12 is reduced to the well known Bragg law, which expresses the resonant scattering direction by a periodic set of scatterers:

$$\lambda = \frac{4\pi}{|\mathbf{G}|} \sin(\theta) = 2d\sin(\theta) \qquad (13)$$

where d is the period between two planes of scatterers. In the case $n^{\omega} = n^{2\omega}$, the nonlinear emission follows the same behavior as a linear scattering: In both cases the direction of propagation is given by the Huyghens-Fresnel principle, given the phase relation between the scatterers.

The analogy with X-Ray diffraction by crystals is useful for understanding the different possibilities offered by 2D QPM. Figure 10 shows a modified Ewald construction corresponding to Eq.(12). This figure follows the same principle as the usual Ewald construction, except for the fact that the radius of the Ewald sphere $|\mathbf{k}^{2\omega}|$ is greater than the distance $2|\mathbf{k}^{\omega}|$ between its center and the origin of the RL. As in the case of X ray diffraction, for a given pump wavevector $\mathbf{k}^{\omega}$, there is in general no reciprocal vector $|\mathbf{k}^{2\omega}|$ on the Ewald sphere. This means that the 2D QPM is a resonant process, an "accident", which can be obtained by varying either the angle of propagation of the pump, or the wavelength. It is interesting to note that for specific angles and wavelengths several points can be located simultaneously on the Ewald sphere. In such a case of multiple resonance, SH beams can be generated simultaneously in different directions in the plane, in a similar way as the linear diffraction in several order beams by a diffraction grating.

These different beams will present anticorrelation noise properties which can be useful in quantum optics experiments.

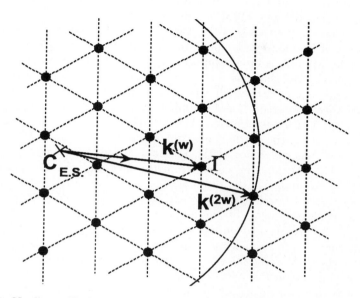

Fig. 10. Nonlinear Ewald construction: The center of the Ewald sphere is located $2k^\omega$ away from the origin of the RL and the radius of the sphere is $k^{2\omega}$. The main difference with the usual Ewald construction is that the Ewald sphere does not contain the origin of the RL. If a point of the RL is located on the Ewald sphere, phase matching occurs for the SHG process.

The new possibilities offered by PCs of $\chi^{(2)}$ can be classified into two categories: First, for a given pump frequency ω, a phase matched direction in the lattice is required for an efficient SHG. In that case, by changing the angle of incidence in the structure, the Ewald sphere crosses for some angle a point of the RL. For this direction, phase matching occurs resonantly in a very similar way as a Bragg resonance in a rotating crystal X-ray diffraction experiment. The walk-off of the nonlinear interaction corresponding to this resonance is given by the nonlinear Bragg law (12). For a unit cell having the same symmetry as the crystal, as explained before several directions of propagation are equivalent. This can be used for ring cavity purposes. At variance with previous ring cavity nonlinear optics experiments[51], a ring cavity (having the shape of a hexagon for instance) can be designed so that the constructive interaction occurs along in the entire intracavity optical path, increasing the final efficiency. The lower efficiency coming from structure factor is overcome by the increase of the interaction length on the entire cavity round trip in the 2D case.

A second application is the search of the phase-matched SHG spectrum, for a given direction of propagation. The X-ray analog of this kind of experiment is the broad band X-Ray diffraction analysis used in the Laue method[52]. The different QPM resonances are found from the scheme of Fig.10 by changing the radius of the Ewald sphere. It is obvious that several resonances will be found; this opens the possibility of multiple wavelength generation by SHG. It is interesting to compare this phenomenon with multiple wavelength SHG that has been recently obtained in a quasi periodic 1D Fibonacci optical superlattice[53]. Multiple resonances were observed in the QPM SHG spectrum arising from the different reciprocal vectors $G_{m,n}$ of the quasi periodic optical superlattice. The 2D RL indexing of the quasi periodic 1D structure is the fundamental difference from the usual 1D periodic structure, and this difference is the reason for multiwavelength frequency conversion. The 2D indexing comes from the fact that the 1D quasi periodic lattice is nothing but the projection of a 2D periodic crystal on a 1D axis. This follows the well known geometrical construction of quasicrystalline structures. The experiments of Zhu and co-workers[53] are thus a projection of a 2D $\chi^{(2)}$ PC experiment on one particular direction of propagation. 1D QPM experiments are also such a projection but the difference is the following: in the 1D QPM case, the propagation is on such an axis that the $\chi^{(2)}$ function is periodic on this axis, whereas in the quasi periodic structure the projection is done on an axis with an irrational slope. All these cases are contained in the 2D PC of $\chi^{(2)}$, and can be obtained by changing the angle of propagation in the 2D plane of the material.

This kind of photonic crystal of $\chi^{(2)}$ is an example of a more general possibility, which consists of patterning a $(\chi^{(2)}, -\chi^{(2)})$ function in a nonlinear crystal. Holography consists of patterning a material with a $\chi^{(1)}$ grating. By illuminating the hologram with light, an image can be created by linear diffraction in this grating. Here, we have presented nonlinear optical frequency conversion as a diffraction process. By analogy with holography, the nonlinear diffraction of a pump wave can be controlled by the pattern of $\chi^{(2)}$ in the nonlinear material, in such a way that the second harmonic beam is generated with a desired spatial shape. This field can be referred to as nonlinear holography, and opens the way to a lot of exciting perspectives.

4 Conclusion

Semiconductor heterostructures form a system where both linear and nonlinear optical properties can be engineered at will. In this paper, it has been shown how phase matching for nonlinear frequency conversion processes can be achieved in photonic crystals, with periodic linear or nonlinear susceptibilities. These heterostructures make possible frequency conversion in optically

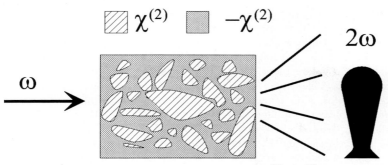

Fig. 11. Nonlinear holography: The pattern of $(\chi^{(2)}, -\chi^{(2)})$ in the nonlinear material has been designed in such a way that the SH generated wave has the desired spatial shape

isotropic materials, for which classical birefringent phase matching cannot be achieved. Among all the perspectives opened by these heterostructures, the realization of a micro optical parametric oscillator in a GaAs based material is a major challenge.

References

1. E. Yablonovitch, "Inhibited Spontaneous Emission in Solid-State Physics and Electronics," Phys. Rev. Lett. **58,** 2059 (1987).
2. J. A. Armstrong, N. Bloembergen, J. Ducuing, and P. S. Pershan, "Interactions between light waves in a non linear dielectric," Phys. Rev. **127,** 1918–1939 (1962).
3. M. M. Fejer, "Nonlinear optical frequency conversion," Phys. Today **May 1994,** 25–32 (1994).
4. A. Yariv, *Quantum electronics* (John Wiley & sons, New York, 1989).
5. F. Capasso, Science **235,** 172 (1987).
6. F. Capasso, J. Faist, and C. Sirtori, "Mesoscopic phenomena in semiconductor nanostructures by quantum design," J. of Math. Phys. **37,** 4775–4792 (1996).
7. V. Berger, "Frequency conversion in semiconductor heterostructures," In *Advanced Photonics with Second-Order Optically Nonlinear processes*, A. Boardman, ed., (Kluwer, 1998).
8. *Microcavities and Photonic Band Gaps: Physics and Applications*, J. Rarity and C. Weisbuch, eds., (Kluwer Academic Publishers, Dordrecht, 1996).
9. T. Tamir, *Guided-Wave Optoelectronics* (Springer-Verlag, 1990).
10. J. L. Jewell, J. P. Harbison, A. Scherer, Y. H. Lee, and L. T. Florez, "Vertical-Cavity Surface-Emitting Lasers: Design, Growth, Fabrication, Characterization," IEEE J. of Quant. Elec. **27,** 1332 (1991).
11. R. P. Stanley, R. Houdré, U. Oesterle, M. Gailhanou, and M. Ilegems, "Ultra-high finesse microcavity with distributed Bragg reflectors," Appl. Phys. Lett. **65,** 1883 (1994).
12. U. Mohideen, W. S. Hobson, S. J. Pearton, F. Ren, and R. E. Slusher, "GaAs/AlGaAs microdisks lasers," Appl. Phys. Lett. **64,** 1911 (1994).

13. J. M. Gérard, D. Barrier, J. Y. Marzin, R. Kuszelewicz, L. Manin, E. Costard, V. Thierry-Mieg, and T. Rivera, "Quantum boxes as active probes for photonic microstructures: The pillar microcavity case," Appl. Phys. Lett. **69**, 449 (1996).
14. J. Zhang, D. Y. Chu, S. L. Wu, S. T. Ho, W. G. Bi, C. W. Tu, and R. C. Tiberio, "Photonic-Wire Laser," Phys. Rev. Lett. **75**, 2678 (1995).
15. J. S. Foresi, P. R. Villeneuve, J. Ferrera, E. R. Thoen, G. Steinmeyer, S. Fan, J. D. Joannopoulos, L. C. Kimerling, H. I. Smith, and E. P. Ippen, "Photonic-Band-Gap Microcavities in Optical Waveguides," Nature **390**, 143 (1997).
16. D. Labilloy, H. Benisty, C. Weisbuch, V. Bardinal, T. Krauss, R. Houdré, U. Oesterle, D. Cassagne, and C. Jouanin, "Quantitative measurement of transmission, reflexion and diffraction of two-dimensional photonic bandgap structures at near-infrared wavelengths," Phys. Rev. Lett. **79**, 4147 (1997).
17. C. C. Cheng, A. Sherer, V. Arbet-Engels, and E. Yablonovitch, "Lithographic band gap tuning in photonic band gap crystals," J. Vac. Sci. Technol. B **14**, 4110 (1996).
18. V. Berger, "From photonic band gaps to refractive index engineering," Optical Materials **11**, 131 (1999).
19. A. Fiore, V. Berger, E. Rosencher, P. Bravetti, and J. Nagle, "Phase matching using an isotropic nonlinear material," Nature **391**, 463 (1998).
20. M. Born and E. Wolf, *Principle of Optics* (Pergamon Press, Oxford, 1980).
21. P. Yeh, *Optical Waves in Layered media* (John Wiley & Sons, New York, 1988).
22. J. V. der Ziel, "Phase-matched harmonic generation in a laminar structure with wave propagation in the plane of the layers," Appl. Phys. Lett. **26**, 60 (1975).
23. J. D. Joannopoulos, P. R. Villeneuve, and S. Fan, "Photonic crystals : putting a new twist on light," Nature **386**, 143 (1997).
24. M. Jäger, G. I. Stegeman, M. C. Flipse, M. Diemeer, and G. Möhlmann, "Modal dispersion phase matching over 7 mm length in overdamped polymeric channel waveguides," Appl. Phys. Lett. **69**, 4139 (1996).
25. J. D. Joannopoulos, R. D. Meade, and J. N. Winn, *Photonic Crystals* (Princeton University Press, Princeton, 1995).
26. J. M. Dallesasse, J. N. Holonyak, A. R. Sugg, T. A. Richard, and N. El-Zein, "Hydrolysation Oxidation of AlGaAs-AlAs-GaAs Quantum Well Heterostructures and Superlattices," Appl. Phys. Lett. **57**, 2844 (1990).
27. D. L. Huffaker, D. G. Deppe, and K. Kumar, "Native oxide defined ring contact for low-threshold vertical cavity lasers," Appl. Phys. Lett. **65**, 97 (1994).
28. M. H. MacDougal, H. Zao, P. D. Dapkus, M. Ziari, and W. H. Steier, "Wide-Bandwidth Distributed Bragg Reflectors Using Oxide/GaAs Multilayers," Electr. Lett. **30**, 1147 (1994).
29. A. Fiore, V. Berger, E. Rosencher, P. Bravetti, N. Laurent, and J. Nagle, "Phase-matched mid-IR difference frequency generation in GaAs-based waveguides," Appl. Phys. Lett. **71**, 3622 (1997).
30. P. Bravetti, A. Fiore, V. Berger, E. Rosencher, J. Nagle, and O. Gauthier-Lafaye, "5.2 - 5.6 microns tunable source by frequency conversion in a GaAs based waveguide," Optics Lett. **23**, 331 (1998).
31. A. Fiore, V. Berger, E. Rosencher, S. Crouzy, N. Laurent, and J. Nagle, "$\Delta n=0.22$ birefringence measurement by surface emitting second harmonic generation in selectively oxidized GaAs/AlAs optical waveguides," Appl. Phys. Lett. **71**, 2587 (1997).

32. A. Fiore, S. Janz, L. Delobel, P. van der Meer, P. Bravetti, V. Berger, E. Rosencher, and J. Nagle, "Second harmonic generation at $\lambda = 1.06\mu$m in GaAs based waveguides using birefringence phase matching," Appl. Phys. Lett. **72**, 2942 (1998).
33. H. C. Chui, E. L. Martinet, G. L. Woods, M. M. Fejer, J. S. Harris, C. A. Rella, B. I. Richman, and H. A. Schwettman, "Doubly resonant second harmonic generation of 2.0μm light in coupled InGaAs/AlAs quantum wells," Appl. Phys. Lett. **64**, 3365 (1994).
34. G. Leo and E. Rosencher, "Analysis of optically amplified mid-IR parametric generation in AlGaAs waveguides," Opt. Lett. p. accepted for publication (1998).
35. J. Khurgin, E. Rosencher, and Y. J. Ding, "Analysis of all-semiconductor intracavity optical parametric oscillators," J. of Opt. Soc. Am. B **15**, 1726 (1998).
36. G. Dixon, Laser Focus World **May**, 105 (1997).
37. U. Grüning, V. Lehmann, S. Ottow, and K. Busch, "Macroporous silicon with a complete two-dimensional photonic band gap centered at 5μm.," Appl. Phys. Lett. **68**, 747 (1996).
38. K. Sakoda and K. Ohtaka, "Sum frequency generation in a two-dimensional photonic lattice," Phys. Rev. B **54**, 5742 (1996).
39. M. M. Fejer, G. A. Magel, D. H. Jundt, and R. L. Byer, "Quasi-Phase-Matched Second Harmonic Generation : Tuning and Tolerances," IEEE J. of Quant. Elec. **28**, 2631 (1992).
40. L. E. Myers, G. D. Miller, R. C. Eckardt, M. M. Fejer, R. L. Byer, and W. R. Bosenberg, "Quasi-Phase-Matched 1.064-μm-pumped optical parametric oscillator in bulk periodically poled LiNbO$_3$," Opt. Lett. **20**, 52 (1995).
41. L. E. Myers, R. C. Eckardt, M. M. Fejer, R. L. Byer, W. R. Bosenberg, and J. W. Pierce, "Quasi-Phase-Matched optical parametric oscillators in bulk periodically poled LiNbO$_3$," J. of Opt. Soc. Am. B **12**, 2102 (1995).
42. H. Karlsson and F. Laurell, "Electric field poling of flux grown KTiOPO$_4$," Appl. Phys. Lett. **71**, 3474 (1997).
43. L. A. Gordon, G. L. Woods, R. C. Eckardt, R. R. Route, R. S. Feigelson, M. M. Fejer, and R. L. Byer, "Diffusion-bonded stacked GaAs for quasi-phase matched second-harmonic generation of a carbon dioxide laser," Electron. Lett. **29**, 1942 (1993).
44. S. J. B. Yoo, C. Caneau, R. Bhat, M. A. Koza, A. Rajhel, and N. Antoniades, "Wavelength conversion by difference frequency generation in AlGaAs waveguides with periodic domain inversion achieved by wafer bonding," Appl. Phys. Lett. **68**, 2609 (1996).
45. J. A. Dobrowolski and D. Lowe, "Optical thin film synthesis program based on the use of Fourier transforms," Appl. Optics. **17**, 3039 (1978).
46. M. L. Bortz, M. Fujimura, and M. M. Fejer, "Increase acceptance bandwidth for quasi-phase matched second harmonic generation in LiNbO$_3$ waveguides," Electron. Let. **30**, 34 (1994).
47. M. A. Arbore, A. Galvanauskas, D. Harter, M. H. Chou, and M. M. Fejer, "Engineerable compression of ultrashort pulses by use of second-harmonic generation in chirped-period-poled lithium niobate," Optics Letters **22**, 1341 (1997).
48. P. R. Villeneuve and M. Piché, "Photonic band gaps in two-dimensional square and hexagonal lattices," Phys. Rev. B **46**, 4969 (1992).
49. V. Berger, O. Gauthier-Lafaye, and E. Costard, "Photonic band gaps and holography," J. of Appl. Phys. **82**, 60 (1997).

50. D. Cassagne, C. Jouanin, and D. Bertho, "Hexagonal photonic band gap structures," Phys. Rev. B **53**, 7134 (1996).
51. W. J. Kozlovsky, C. D. Nabors, and R. L. Byer, "Efficient second harmonic generation of a diode-laser-pumped CW Nd:YAG laser using monolithic MgO:LiNbO3 external resonant cavities," IEEE J. of Quant. Electron. **24**, 913–919 (1988).
52. C. Kittel, *Introduction to Solid State Physics* (John Wiley & sons, New York, 1976).
53. S. N. Zhu, Y. Y. Zhu, Y. Q. Qin, H. F. Wang, C. Z. Ge, and N. B. Ming, "Experimental Realization of Second Harmonic Generation in a Fibonacci Optical Superlattice of $LiTaO_3$," Phys. Rev. Lett. **78**, 2752 (1997).

Physics of Light Extraction Efficiency in Planar Microcavity Light-Emitting Diodes

H. Benisty

Laboratoire de Physique de la Matière Condensée, UMR 7643 du CNRS, Ecole Polytechnique, 91128, Palaiseau cedex, France

Abstract. We use the modifications of spatial mode distributions in planar microcavities to address the issue of light outcoupling in devices such as LEDs. Whereas detailed calculations are needed to fully model and optimize the structures, we show that a simple picture can be obtained that clarifies the main physical effects of the microcavity. In particular, it is shown that the cavity order m_c is the chief parameter in determining the extraction efficiency according to $\eta=1/m_c$ under some reasonable, simplifying assumptions. When the source in the cavity has a linewidth, light redistribution among both angles and wavelengths occurs and can either be a limiting factor for the overall extraction efficiency or a positive one in case of spectral narrowing requirements. In extraction-optimized systems, one gains a factor of up to ten on the single-face extraction as well as brightness (radiance).

1 The Issue of Light Extraction from High-Index Solids

Although semiconductor laser diodes achieve a number of desirable characteristics for light emitters, including a high efficiency, adverse factors that include their cost, difficulty of fabrication (varying with wavelength range), threshold behavior, speckled spot, eye-damage risks, large thermal sensitivity, etc., limit their use. As a result, the three-times larger market share in compound semiconductor light sources belongs to the much cheaper, simpler and robust light-emitting diodes (LEDs)[1] used by billions in displays, infrared free-air transmission or fibre transmission in local area networks (LANs), while LED-based board-to-board interconnects are also envisioned. Physicists have nevertheless devoted less efforts to LEDs based on spontaneous emission than to the more fascinating lasers based on stimulated emission. As will be seen, microcavity effects offer a good opportunity to apply physics-based concepts to the improvement of LEDs[2,3].

Viewed as electro-optic converters, LEDs lie well behind lasers (when these can be made!) in terms of light extraction efficiency: Geometric optics tells us that the high index (n~3) of the semiconductor parallelepiped (Fig.1a) containing the p-n junction in which light is generated is unfortunately a severe hindrance to light extraction from a single planar face: Assuming air as the outside medium for simplicity ($n_{out}=1$), this leads to a small critical angle $\theta_c = \sin^{-1}(n_{out}/n^{-1}) \approx n^{-1} \approx 20°$ beyond which total internal reflection takes place, so that the internal solid angle coupled to rays that propagate in air, $\Omega=2\pi(1-\cos(\theta_c))$, represents only a few percents of the 4π steradian total space

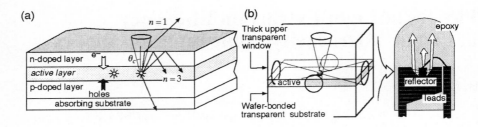

Fig.1 (a) The fate of internal light in a semiconductor: mostly total internal reflection. (b) Side collection in "high-brightness" LEDs requires a thick transparent overlayer and, ideally, substrate lift-off followed by bonding to a nonepitaxial transparent conducting substrate.

solid angle. A simple approximation ($\cos\theta_c \approx 1 - 1/2n^2$) leads to the fraction $\eta = \Omega/4\pi \approx 1/4n^2$, which is thus the extraction efficiency at a single planar face for an isotropic emitter in the medium, typically only 2-3%. This isotropy is a valid approximation in bulk active regions of standard Double Heterostructure LEDs. For quantum-well–based active regions, the dominant e-hh recombination corresponds to electric dipoles in the junction plane, which are somewhat better coupled to the outside. In the following, we neglect such dipole orientation and polarization effects for simplicity. What we will aim at by means of planar microcavity effects is to widely redistribute the directions of emitted photons so that emission into the outside-coupled directions are enhanced to the expense of other ones.

In practice, industrial solutions to the low efficiency issue make use of light collection through the sides of the semiconductor block as illustrated in Fig1.b [1], employing transparent "window" layers (or a thick dielectric guide) above and below the light-emitting layer, this latter step implying a costly removal of the substrate in the frequent case in which it is absorbing, as is, e.g., GaAs at wavelengths below 900 nm. These so-called "high-brightness LEDs" are however not "brighter" at the chip level since the increase in optical power P is accompanied by an equivalent increase in the source area S so that the radiance $dP/dSd\Omega$ (a conserved quantity in optics that limits subsequent beam transformations) remains basically unchanged, setting a limit to fibre-coupled transfer efficiency for example.

Other ways to raise the extraction efficiency rest on the principle of randomizing the photon direction until a favorable angle $\theta < \theta_c$ is found. For this purpose, one may use an optically rough interface[4] or reabsorption-emission cycles, the so-called "photon recycling" phenomenon [5]. These means add demand to the absorption length, the internal quantum efficiency etc. Note that photon recycling phenomena are not exclusive of microcavity effects since both can be combined to still increase device performances.

2 Photons in Two Dimensions

2.1 Modes and Light Emission

In the other lectures of this Volume (by Savona, Fabre, Ho, etc.) the modifications of photonic modal density in microcavities of various dimensionalities have been discussed. In particular, spontaneous emission is described as the transfer of a quantum of energy from an oscillating dipole (of an atom or an electron-hole pair) to modes of the electromagnetic field. In the quantum picture, vacuum mode fluctuations are responsible for the spontaneous emission process. The main effect of planar microcavities we are concerned with is that the amplitude of some modes at the emitter location are much larger than the others, thereby favoring emission in privileged modes, i.e. directions, an anisotropy that can benefit to light extraction. In terms of the Fermi Golden Rule, the total density of final photon states $\rho(\omega)$ at the dipole frequency ω is not necessarily much changed (see e.g. Ref.[6, 7]), but the squared matrix element M^2 of some modes is much larger than that associated with other modes.

The perfect two-dimensional cavity of thickness L, index n, uncoupled to the outside world is a limit case where modes are rigorously discretized along the normal of the cavity, according to $k_z^m = m\pi/L$, with a field amplitude $E \propto \sin(k_z^m z)$ at the location z of the dipole (Fig.2a). Along the cavity plane, the in-plane wavevector $\mathbf{k}_{//}$ is a good quasi-continuous variable (in the "large box" limit) so that, denoting $k = n\omega/c$

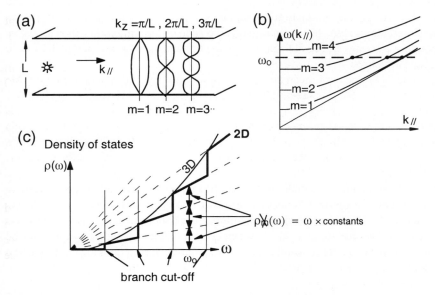

Fig.2(a) The planar ideal cavity and its mode profiles m=1,2,3,... ; (b) the dispersion relation of branches m=1,2,3... ; (c) the density of scalar photons states in 2D (bold line), 3D (thin line) and the branch individual contributions, the same for all m (dahsed lines).

(c=light velocity), the angle θ_m of a mode at frequency ω can be conveniently defined as $\cos\theta_m = k_z^m/k = m\pi c/nL\omega$. Modes are grouped by branches indexed by m, each branch having the simple dispersion relation $\omega=(c/n)[(k_z^m)^2+k_{//}^2]^{1/2}=\omega_m(k_{//})$ (Fig.2b). They have a cut-off frequency $\omega_m(0)$. We will assume in the following that we work above the cut-off of the cavity, $\omega > \omega_1(0)$ so that emission couples at least to one mode in our scalar picture. Here, M^2 is simply zero for all directions except the discrete θ_m's.

A very important feature of these branches (which we will call also modes for simplicity) in our scalar picture is that their density-of-states (DOS) $\rho_m(\omega)=dN/d\omega=(dN/dk_{//}) \times (dk_{//}/d\omega)$ is independent of the branch and simply proportional to ω, reading $\rho_m(\omega)=\rho_1(\omega)=A\omega n^2$, as a basic calculation shows. The total DOS $\rho_{2D}(\omega)$ is the sum of the individual $\rho_m(\omega)$ up to $m=m_c$, the largest integer such that $\omega_m(0)<\omega$, and can be simply written $\rho_{2D}(\omega)=m_c A\omega n^2$. We term m_c the cavity order, a number which depends on the frequency (wavelength) of interest (It is also the number of half-wavelengths ($\lambda/2n$) that fully fit into the cavity).

The resulting $\rho_{2D}(\omega)$ is illustrated in Fig.2c. In this figure, the dashed lines that prolong the bold segments illustrate the weight of each branch. When a cut is made at a given ω, one readily sees that all the m_c branches have the same weight, $1/m_c$. One also sees the relationship between $\rho_{2D}(\omega)$ and the familiar 3D photon DOS $\rho_{3D}(\omega) \sim n^3\omega^2$ (thin line).

It is clear from the comparison that, except below or around the cavity cut-off $\omega_1(0)$ where scalar approximations are much too drastic, the total DOS may not be changed by large factors in a *planar* cavity: the ratio $\rho_{2D}(\omega)/\rho_{3D}(\omega)$ (a kind of "Purcell factor") does not shift from unity by more than a quantity of order $1/m_c$ (see e.g. Brorson's work [8,9] for mode-counting arguments when taking polarization into account). Hence, no important lifetime changes are expected for emitters in planar cavities of some thickness. The main effect is the *redistribution among discrete angles* corresponding to cones in the 3D space (modes, branches), with an identical DOS in each discrete direction.

An equal DOS does not mean an equal emission however: the matrix element M^2 in the Fermi Golden Rule is essentially the squared field amplitude $E^2 \propto \sin^2(k_z^m z)$. The quantity $\zeta_m(z)=2\sin^2(k_z^m z)$ is also known as the antinode factor. It tells us whether the emitter stands at a node of mode m ($\zeta_m(z)=0$, no coupling) or an antinode ($\zeta_m(z)=2$, optimal coupling), or in between. For thick enough emitting layers, $k_z^m z$ will span more than π within the layer, and the z-averaged antinode factor $<\zeta_m(z)>_z$ tends towards unity.

The emission is thus essentially redistributed among the m_c branches according to ζ_m factors. It is unfortunately impossible to favor one ζ_m to the expense of all others as soon as $m_c \geq 3$. One may check by examining the canonical case $z=L/2$ that upon averaging on m, one has $<\zeta_m>_m =1$. This remains true in most other useful cases (the limit case $z<<\lambda/4$ is more subtle, as all ζ_m vanish like $z^2(m/m_c)^2$, but often unrealistic), so that we may write when summing over the m_c modes in which emission takes place : $\Sigma\zeta_j \approx m_c$.

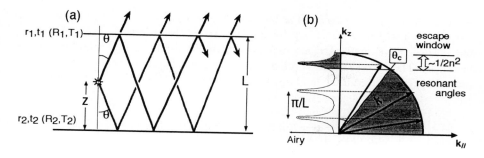

Fig.3 (a) An emitter radiating two waves in a planar cavity that emerge after undergoing multiple-beam interferences. Both series interefere in the far-field.(b) Scheme in the $k_{//}$-k_z plane to quantify emission in the direction θ with the escape cone; on the left is the Airy factor as a function of k_z.

The fraction of emission η_j radiated in a given mode j now simply reads :

$$\eta_j = \frac{\zeta_j}{\sum \zeta_m} \approx \frac{\zeta_j}{m_c} = \frac{\text{antinode factor}}{\text{cavity order}} \quad . \tag{1}$$

This fraction is at most $2/m_c$. To discuss the trends on extraction efficiency, we need to couple this cavity to the outside world. We will see that most of the ingredients defined above still hold and that extraction is enhanced whenever a single emitted mode lies in the angular range $[0, \theta_c]$.

3 The Fabry-Perot Cavity with Few Modes

3.1 Lossless Mirrors with Finite Transmission

Coupling of a strongly resonant cavity mode to the outside continuum is treated quantum-mechanically by introducing "quasi modes" in Savona's lecture. Here, we retain a simpler classical ansatz introduced by Kastler [10] to address the modification of spontaneous emission for atoms emitting into a Fabry-Perot cavity (Fig.3a).

The source emits rays that interfere before being collected in the far-field, provisionally in a medium of the same index as the cavity so as to avoid guided modes. Two series of rays add in a given outside direction (above the cavity, $z \to +\infty$) : One is initially emitted at θ ($k_z = k \cos\theta$), and the other at $\pi-\theta$, ($k_z = -k \cos\theta$). Each undergoes multiple-beam interference that classically leads to the well-known Airy function, Airy $= |1 - r_1 r_2 \exp(2i\phi)|^{-2}$ whose properties are conveniently cast into the cavity quality factor Q that measures the maximum of this function and the peak relative width via $Q^{-1} \sim \Delta\omega/\omega$. The far-field intensity E^2 normalized to the no-cavity case E_o^2 is just $(E/E_o)^2 = T_1 \times$Airy for the first series (Notations t_1, T_1, r_1, $R_1 = 1 - T_1$, and t_2 etc.) have their obvious meaning from Fig.3, 2ϕ is the round-trip phase $2\phi = 2k_zL$). The second series is reflected on the bottom mirror, hence a phase factor $r_2\exp(2ik_zz)$; the overall far-field intensity thus reads:

$$\left|\frac{E}{E_o}\right|^2 = \frac{T_1}{\left|1 - r_1r_2\exp(2i\phi)\right|^2} \times 2 \times \tfrac{1}{2}\left|1 + r_2\exp(2ik_zz)\right|^2. \tag{2}$$

Noting that the right-hand factor may be written in the limit $r_2 = -1$ $\tfrac{1}{2}\left|1 - \exp(2ik_zz)\right|^2 = 2\left|\tfrac{1}{2}[\exp(-ik_zz) - \exp(ik_zz)]\right|^2$, it is seen to retrieve the antinode factor $\zeta(z) = 2\sin^2(k_z z)$.

The left hand Airy factor peaks for 2ϕ multiple of 2π (assume $r_1r_2 > 0$): in the limit $r_1r_2 = 1$, one retrieves rigorous discrete modes $m = 1, 2, ...$ up to the cavity order $m = m_c$, the mode with the largest allowed k_z (keep in mind that $k_z < k = n\omega/c$ since our emitter has a given frequency!). Equation (2) provides us with a convenient way of dealing with an out-coupled planar cavity with the same concepts as those of the uncoupled cavity: modes (say, resonant angles), characteristic of the cavity, and antinode factors which tell where the emitter lies in each mode profile.

Note that the Airy factor plays its enhancement/inhibition role whatever ζ. If an emitter is located close to the node of a resonant mode ($\zeta \to 0$), but emission is monitored in the very direction of this mode (where the Airy factor is maximum), a fair amount of light can result from the compensation between both factors. For example, if one wants to suppress emission in a resonant mode *below the no-cavity level* by proper location of the emitter, one requires an accuracy on the emitter position z much better than λ/Qn. This is to be contrasted with the same kind of "inhibition" experiment carried out for a single mirror $R_2 \approx 1$ ($R_1 = 0$), in which case diminishing emission below the same no-cavity reference level "only" amounts to requiring $\zeta \ll 1$, i.e. an accuracy much better than $\lambda/2n$.

The scheme of Fig.3b is useful to quantify emission from a source of given k in a given solid angle (we assume $R_2 = 1$ to collect light only on one side): For this purpose, we note that the elemental solid angle (ring) subtended by $d\theta$ around θ, which reads $d\Omega = 2\pi \times d(\cos\theta)$, is proportional to $dk_z = k \times d(\cos\theta)$. This means that the amount of emission in a given cone such as the escape cone $[0, \theta_c]$ of interest for extraction efficiency is given within a constant factor $k/2\pi$ by the shaded integral below the Airy function on the left between $k_z = k\cos\theta_c$ and $k_z = k$. The quarter circle of radius k is helpful in visualizing the associated angles. This picture helps us finding what are the reasonable requirements on the Airy function that will provide extraction efficiency enhancement at the smallest cost.

Finally, one may be surprised that in this scheme, an angle-independent reflectivity is used (Airy peaks have the same width). Actually, beyond θ_c, guided modes (and, in DBR systems, leaky modes) can occur as the reflectivity is a strongly angle-dependent

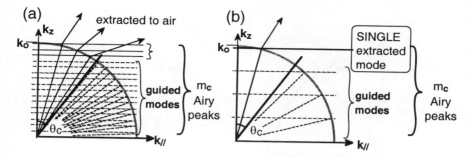

Fig.4 (a) Macro-cavity with $m_c \gg 2n^2$; (b) micro-cavity with $m_c \ll 2n^2$ and a single mode in the escape cone.

quantity. However, we will not need the detailed contribution of all modes beyond θ_c but only their sum. Given the basic fact that, within simplifying assumptions on the ζ's, these modes have the same weight (the DOS argument), we need not know about their detail. This statement is further confirmed by exact simulations [3].

3.2 Micro- Versus Macro-Cavity

Let us first assume that the cavity Q is large enough for the Airy peaks to be narrower than the escape window [k $\cos\theta_c$, k]. The story of the extraction efficiency then becomes a mere mode-counting issue for which one parameter is the cavity order m_c and the other the relative width of the escape window $(k - k\cos\theta_c)/k = (1 - \cos\theta_c) \approx 1/2n^2$. Given the equally spaced Airy resonances, $1/m_c$ in relative terms and their basically identical weight, if $m_c \gg 2n^2$, there are many Fabry-Perot modes in the escape window, but still much more below (Fig.4a), in a ratio (hence an extraction efficiency η) which is asymptotically the relative size of the escape window $1/2n^2$: we just gained a factor of two over the no-cavity case ($\eta = 1/4n^2$) due to the back mirror. Of course some other phenomena among which spectral narrowing and brightness enhancement (see below) already takes place as emission is strongly redistributed into the Fabry-Perot rings [10].

But extraction efficiency η is not significantly enhanced. If, on the contrary, we have few modes, $m_c < 2n^2$ (cavity order less than ~20 in a typical semiconductor), we have at most one mode in the escape window while all the ($m_c - 1$) others are below (Fig.4b). *This is the microcavity regime.* The extraction efficiency η is then given by eqn. (1) where ζ_j has to be the antinode factor of the only outcoupled Fabry-Perot mode ($\zeta_j \equiv \zeta_{out}$). Even for a thick source layer for which $\langle\zeta\rangle_z \approx 1$, we have $\eta \approx 1/m_c$, i.e. we retain most of the micro-cavity effect even with a random emitter location. Below $m_c = 2$, of course, most approximations made become too drastic (starting with the meaning of an average !) and (1) is not valid. But for many useful cases, DBR mirrors

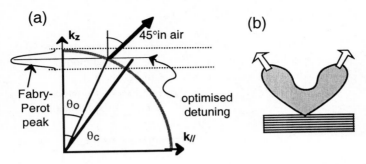

Fig.5 (a) Optimal position and sufficiently narrow width of the Airy peak in the escape window ; (b) corresponding emission pattern peaking at oblique angles of the detuned microcavity.

increase the cavity order m_c up to values of at least 3, for which (1) is again valid. The formula $\eta \approx \zeta_{out}/m_c$ for a micro-cavity is the chief result of this simplified approach that emphasizes the benefit of a small cavity order.

3.3 Monochromatic Extraction: Which Mirrors?

We now discuss the peak central position and width of the outgoing mode. Firstly, since the peak has some width, it should be centered in the escape window in terms of k_z to maximize outgoing light (Fig.5a): denoting θ_o the resonant angle, this means $\cos\theta_o \approx 1-1/4n^2$, that is $\theta_o \approx \theta_c/\sqrt{2}$. Translated to air by Snell's law, this gives an incidence of 45°. An optimized microcavity LED will not have its emission maximum at normal angle, which would truncate one half of an Airy peak ; in other words, there has to be a sizable "detuning" to oblique angles (Fig.5b).

Secondly, what is the cavity Q, or in other terms the R_1R_2 product, needed to achieve $\eta \approx \zeta_{out}/m_c$? For this, one may use an analytically integrable Lorentzian approximation of the Airy function around resonance to see that about 80% of the peak area is contained within those points at one tenth of the peak value. Applying this criterion in the case $r_2=1$ [2], one finds a very simple result:

$$R_1 \geq 1-m_c/n^2 \quad . \tag{3}$$

This result could have been partly predicted using the relative peak width formula $1/Q=1/Fm_c$ where the cavity finesse F basically scales like $(1-R_1)^{-1}$ in our approach. Fitting most of the peak into the escape window means $1/Q < 1/2n^2$, i.e., taking the inverses, $Fm_c < 2n^2$, which gives a form akin to eqn. (3) by rearranging the factors. Note that in most practical cases, R_1 needs not be larger than 50–80%. This is to be contrasted with reflectivities in excess of 99.5% required for vertical laser operation (VCSELs) for which the issue is not where a photon is redirected, but how to compensate the small gain per round-trip intrinsic to a VCSEL geometry.

3.4 Real-World Microcavity-LEDs: Metallic and DBR Mirrors

We could not, in the present simplified framework, take losses into account. In general, one cannot neglect losses of metal mirrors at optical frequencies, typically in the 20% range (but reaching in extreme cases between 3% to 70%) [2, 3]. Let us just say that they will limit the cavity Q and that they are one more reason to avoid too high reflectivity mirrors which would cause photons to be absorbed during the extra round-trips instead of being extracted : in many cases, the point of vanishing return occurs when losses per round-trip are of the same order as the output mirror transmission [2].

The other type of mirrors also used are DBR stacks of semiconductors that can be made reasonably lossless and conductive. However, they increase the effective cavity size as explained in Savona's notes. In half-wavelengths units, this increase is $n/2\Delta n$ per DBR mirror, where Δn is the DBR index step, and n, their average index, is most often close to the cavity index due to material epitaxy considerations. The cavity order to be used for a microcavity with two similar DBR mirrors is now $m_c = m_o + n/\Delta n$, where m_o is the order of the bare cavity (for example, $m_o = 2$ for the so-called "lambda-cavity" of many DBR-containing microcavities)[2, 3]. One might be surprised that such a simple treatment is sufficient to treat the many complexities of DBR stacks (leaky Bloch modes, tunneling, total reflexion effects, etc.) In the simplest terms, the reason why such a simplicity is possible is that the power in all these other modes remain quite constant and only the power in the resonant outcoupled mode has to be evaluated, a power given by the area below the Airy peak when plotted vs. k_z. This area (more exactly its inverse) is indeed measured by the increased cavity order.

This increased order is a penalty for extraction efficiency, so that one may wonder whether it compensates the lower losses. In practice, detailed analysis shows that it is rather optimum to use a hybrid cavity with a metal mirror on one side and a DBR mirror on the other side.

The IMEC group in Gent has achieved within the ESPRIT SMILES Project impressive performances at 980 nm: $\eta \approx 23\%$ total conversion efficiency [11-13]. Our simplified estimate for their system ($\Delta n=0.5$, $n=3.5$, $m_o=2$) is $m_c = 2 + 3.5/2/0.5 = 5.5$ and, with $\zeta=2$, we predict up to $2/m_c \approx 35\%$ extraction efficiency in a lossless, monochromatic case, and with 100% internal QE, all optimistic assumptions that mostly explain the discrepancy, with some room for photon recycling as well [14]. Work at other wavelengths is pursued actively , in particular in the ESPRIT SMILED project.

For these various systems, the crucial DBR parameter is its materials index contrast, as it dictates the penalizing increase in cavity order. In this respect, the advent of "AlOx"-containing DBR (oxidized AlAs) with a step $\Delta n \approx 2$ instead of $\Delta n \approx 0.5$ for AlAs/GaAs certainly enhances the performances of some of the GaAs LED families. One should however note that for index contrast $\Delta n > 1$, the simple formulation $n/\Delta n$ given above for the DBRs penetration does not hold any more, especially in our case of interest, when the variation in k_z stems from an angular variation. Detailed simulations [3] give the exact enhancement as less than what the naive $n/\Delta n$ formulation predicts. On the other side of the index contrast spectrum, many material families (quaternaries at 1.3 and 1.5 μm, visible 630-550 nm AlGaInP LEDs, ...[1]) feature rather low Δn's, of the order of 0.2, which are a severe hindrance in going to the microcavity regime $m_c < 2n^2$.

4 The Impact of Source Linewidth

In the ideal uncoupled planar cavity of §2, where modes are fully discrete, each wavelength λ is emitted at a different angle θ following the dispersion relation $\omega(k_{//}) = \omega(k \sin\theta)$ with $k = n2\pi/\lambda$. The finite reflectivity of actual microcavities allows an angular spread for each wavelength, given for internal angles θ by $\Delta(\cos\theta) = \Delta k_z/k \sim Q^{-1}$.

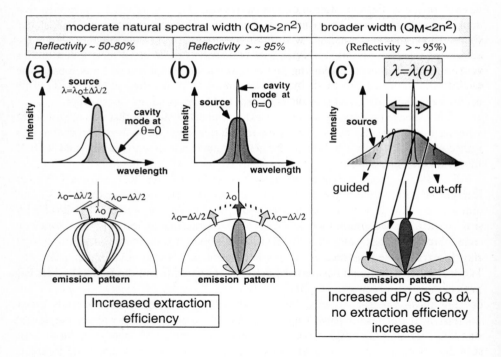

Fig.6 Spectral effects (a) quasi-monochromatic case $Q \ll Q_M$, light output is defined chiefly by the cavity characteritics; (b) polychromatic case $Q \gg Q_M$, spectral narrowing occurs and different wavelengths emerge in different directions. (c) for $Q_M < 2n^2 < Q$, some wavelengths are not allowed to escape, an intrinsic limit of the planar cavity.

Consequently, there are two regimes : if the material's quality factor Q_M (the inverse of its intrinsic spread in emitting wavelength $\Delta\lambda/\lambda = Q_M^{-1}$, a terminology used e.g. in Yablonovitch's lecture) is much larger than the cavity Q (Fig.6a), we are in the quasi-monochromatic regime discussed above and η is effectively ζ_{out}/m_c if losses are neglected.

Conversely, the material may be substantially polychromatic, $Q_M << Q$. In this case, the cavity redistributes the emitted wavelengths at different angles as discussed for the ideal cavity. In a given direction, a spectrum of width $Q^{-1} << Q_M^{-1}$ is collected and spectral narrowing occurs [15] (Fig.6b). However, it may very well be that the large Q used to achieve this effect has degraded the overall (angle-integrated) extraction efficiency due to, e.g., losses. In addition, if one looks at the radiance with a wavelength-integrating detector, there is also no more gain above $Q=Q_M$ since what is gained in spectral power density is lost in spectral narrowing. Therefore, the regime $Q>>Q_M$ is only of interest when one needs spectral narrowing, for example to limit chromatic dispersion at high bit rates [15]

A last issue that is critical in long-wavelength system is that below $Q_M \approx 2n^2 \sim 20$, the source is so broad that a microcavity with a Q of the same order or larger (as would result from the "minimal" choice of Eqn.(3)) redistributes wavelengths in angles that necessarily go beyond the escape cone (Fig.6c): a planar cavity cannot provide simultaneous resonant escape to wavelengths differing by more than $1/2n^2$ in relative terms. It is then necessary to revisit our approach [2,3]. From a practical point of view, the epoxy used in packaging provides some relief to this linewidth limit since escape to epoxy is accounted for by replacing $1/2n^2$ by $n_{epoxy}^2/2n^2$: the linewidth limit is then increased by a factor of more than two ($n_{epoxy} \approx 1.5$).

5 Conclusion

It might be thought that the effect on light emission of a 2D microcavity of moderate order is negligible as the overall DOS is almost unchanged (no "van Hove singularities" in $\rho_{2D}(\omega)$), and therefore the lifetime also. However, the effect of redistribution among emission *directions* does lead to a major increase of extraction efficiency, at least one order of magnitude, as demonstrated by Gent's team [11-13]. The simple picture drawn here gives the basic physics of this increase, explaining the threshold cavity order and the fundamental $1/m_c$ behaviour of extraction efficiency beyond this threshold.

Transfer of microcavity concepts to industrial LEDs is in principle easy as the standard planar technology is retained, in contrast with a number of alternative solutions such as outcoupling gratings or roughened interfaces. In many existing semiconductor systems, the material properties allow mirrors with sufficiently small penetration length to attain at least 10–20 % extraction efficiency.

Radiance (brightness) is also increased in numerous cases, which certainly constitutes an advantage for coupling to fibers. Quantifying the exact advantage for classes of applications in which the spectra and/or the exact solid-angle to couple to are important is beyond the scope of these short notes.

However, let us remark that the above picture of Fabry-Perot modes also tells us that there are some limits to planar cavities for these more demanding requirements. Let us examine, for example, coupling to fibres with numerical aperture 0.5-0.2. In such a case, an escape cone much narrower than 2π is required, which can be translated as if the material index n were multiplied by typical factors 4-25. Conditions detailed above for the microcavity regime to appear then become much more stringent: consider the case $Q_M \sim 10$ and the linewidth limit set by the modified factor $2n^2 \times 25$ of the order of 500 instead of 20 : the planar cavity cannot redirect such a large spectrum in such a narrow cone, with a discrepancy of a factor 50! In this view, one should consider in the future what added benefit could then be drawn from the peculiar DOS of 1D and 0D photon systems (Photonic Band Gap, Photonic wire...), an approach which challenges both technology and physics.

Nevertheless, while planar microcavities are basically well-understood, it does not mean that nothing is to be done : implementation of the concepts to various material/wavelength situations and the search for the physical limits of planar cavities should well lead us into major achievements, both in the market place and in the new physics areas opened by high efficiency emitters such as single photon generators.

The present work is supported by the ESPRIT Basic research Project SMILED.

References

1. High-Brightness Light-Emitting Diodes, G. B. Stringfellow, and M. G. Craford, eds (San Diego, Academic Press, 1997) 48.
2. H. Benisty, H. De Neve, and C. Weisbuch, "Impact of planar microcavity effects on light extraction: I. basic concepts and analytical trends," IEEE J. Quantum Electron. **34**, 1612-1631 (1998).
3. H. Benisty, H. De Neve, and C. Weisbuch, "Impact of planar microcavity effects on light extraction: II. selected exact simulations and role of photon recycling," IEEE Quantum Electron. **34**, 1632-1643 (1998).
4. I. Schnitzer, E. Yablonovitch, C. Caneau, T. J. Gmitter, and A. Scherer, "30-Percent External Quantum Efficiency from Surface Textured, Thin-Film Light-Emitting Diodes," Appl. Phys. Lett. **63**, 2174-2176 (1993).
5. I. Schnitzer, E. Yablonovitch, C. Caneau, and T. J. Gmitter, "Ultrahigh Spontaneous Emission Quantum Efficiency, 99.7 % Internally and 72 % Externally, from AlGaAs/GaAs/AlGaAs Double Heterostructures," Appl. Phys. Lett. **62**, 131-13 (1993).
6. G. Björk, S. Machida, Y. Yamamoto, and K. Igeta, "Modification of spontaneous emission rate in planar dielectric microcavity structures," Phys. Rev. A **44**, 669-68 (1991).

7. G. Björk, "On the Spontaneous Lifetime Change in an Ideal Planar Microcavity - Transition from a Mode Continuum to Quantized Modes," IEEE J. Quantum Electron. **QE 30**, 2314-2318 (1994).

8. S. D. Brorson, H. Yokoyama, and E. Ippen, "Spontaneous emission rate alteration in optical waveguide structures," IEEE J. Quantum Electron. **26**, 1492-1499 (1990).

9. S. D. Brorson, in *Spontaneous emission and laser oscillation in microcavities,* ed. H. Yokoyama, and K. Ujihara. Laser and Optical Science and Technology Series, (Boca Raton, CRC Press, 1995) 151-188.

10. A. Kastler, "Atomes à l'intérieur d'un interféromètre Perot-Fabry," Applied Optics **1**, 17-24 (1962).

11. J. Blondelle, H. De Neve, G. Borghs, P. Vandaele, P. Demeester, and R. Baets, "High efficiency (>20%) microcavity LEDs", (IEE Topical Meeting, London, 1996).

12. J. Blondelle, *"Realisation of high-efficiency substrate-emitting InGaAs/(Al)GaAs microcavity LEDs by means of MOCVD"* Ph. D Thesis. University of Gent, 1997.

13. H. De Neve, J. Blondelle, P. Vandaele, P. Demeester, R. Baets, and G. Borghs, "Planar substrate-emitting microcavity light emitting diodes with 20% external QE", (San Jose, California: 1997), 74-84.

14. H. De Neve, J. Blondelle, P. Vandaele, P. Demeester, R. Baets, and G. Borghs, "Recycling of guided mode light emission in planar microcavity light emitting diodes," Appl. Phys. Lett. **70**, 799-801 (1997).

15. N. E. J. Hunt, et al., in *Confined Electrons and Photons*, ed. E. Burstein, and C. Weisbuch, (New-York, Plenum Press, 1995) 703-714.

Measuring the Optical Properties of Two-Dimensional Photonic Crystals in the Near Infrared

D. Labilloy[a], H. Benisty[a], C. Weisbuch[a], T.F. Krauss[b], C.J.M. Smith[b], R.M. De La Rue[b], D. Cassagne[c], C. Jouanin[c], R. Houdré[d], U. Oesterle[d]

a) Laboratoire de Physique de la Matière Condensée, Ecole Polytechnique, 91128 Palaiseau France
b) Glasgow University, Department of Electronics and Electrical Engineering, Glasgow University, Glasgow, G12 8LT - United Kingdom
c) Groupe d'Etude des Semiconducteurs, Université Montpellier II CC074, place E. Bataillon, 34095 Montpellier Cedex 05 - France
d) Institut de Micro et Opto-électronique, Ecole Polytechnique Fédérale de Lausanne, CH-1015 Lausanne - Switzerland

Abstract. We present measurements of two-dimensional photonic crystals in a waveguided geometry, using photoluminescence from emitters inserted in the guiding heterostructure as an internal light source. A complete set of measurements is given, including quantitative evaluation of the transmission, reflection and also diffraction coefficients of the samples. Their behaviour is shown to follow mostly the pure 2D theory. Capitalizing on the measured properties, we fabricated one-dimensional cavities. The cavity modes are probed through transmission measurement. The measured quality factor leads to an estimation of the reflectivity of the mirror of the order of 95%. We also designed and fabricated disk cavities surrounded by circular Bragg mirrors. The resonances are probed by exciting the photoluminescence of quantum dots placed inside the cavity. Resonances with quality factors up to 650 are found corresponding to the confined Quasi-Radial Modes.

1 Introduction

The concept of Photonic Crystals (PCs) raised a lot of interest in the elapsed decade because these crystals, when properly tailored, have the unique property of creating a zero density of states (DOS) over a range of energies called the photonic bandgap [1,2]. This property could open the way to the much desired goal of controlling the spontaneous emission of an emitter if it is placed into a cavity surrounded by such a photonic crystal [3]. The simplest consequence of the zero DOS is the vanishing transmission at the bandgap energies, at which light propagation is forbidden.

However, three-dimensional (3-D) PCs have proven very difficult to fabricate at visible or near infrared wavelengths i.e. at typical useful sizes of a few hundred nanometers [4]. On the contrary, 1-D and 2-D PC are within reach using present microelectronics technology [5,6]. There is however a price to pay, as 3-D structures are needed for a full control of light in the whole k-space. An alternative way of controlling an already large fraction of the 3-D solid-angle is to combine 1-D and 2-D

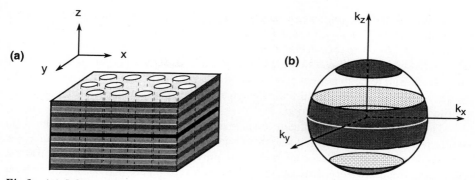

Fig.1 - *(a) Schematics of a hybrid microcavity surrounded vertically by 1-D planar Bragg mirrors and laterally by 2-D photonic crystal (b) Controlled solid angle is shaded: The two polar calots are controlled by the 1-D vertical microcavity, while the equatorial belt is controlled by the lateral 2-D structure.*

approaches: a planar microcavity would provide the vertical confinement in a given acceptance cone, while a 2-D PC would then control the lateral propagation of light, e.g. in guided modes (Fig. 1).

Planar microcavities are already the subject of numerous studies [7]. In the following, we will rather focus on 2-D PCs. We will present quantitative measurements of the optical properties of 2-D PCs, and test the validity of the purely 2D theory on real samples, in the presence of a waveguide in the third dimension. We will then present results on cavities, confined with 1-D and 2-D-like photonic bandgap materials, capitalizing on the measured basic parameters to design proper cavities.

2   Designing 2-D photonic crystals

Many calculations have been performed on 2-D PC systems, due to their simplicity [8,9]. We will first describe the theoretical optimum crystals and then describe to what extent practical samples should be inspired by those.

A favourite candidate to achieve large gaps is a structure made of a triangular array of air holes in a dielectric matrix of high index n (typically n=3-4 in semiconductors) [10]. In these 2-D systems, the eigenmodes are either TE or TM polarized, and a complete gap can appear only if these two polarizations are inhibited. Calculations [11] show that this happens in the triangular structure for large air-filling factors f_{air} (fraction of air per unit cell) above 60%, the largest gap being obtained for $f_{air} \approx 78\%$. (Fig. 2). At these values, the air cylinders almost touch and are separated only by small veins of dielectric.

Being less stringent, bandgaps for either polarization arise in the triangular structure, at values of a/λ (a being the period of the lattice) in the range 0.2–0.6 which means that crystal periods a should be around 200–600 nm to control propagation of light at λ =1 μm.

Experimentally, samples are realized using PMMA masks, defined by e-beam lithography and subsequently transferred on a GaAs substrate by two Reactive Ion Etching steps [6]. With this technique, or similar etching methods, period and diameter in the hundreds of nanometer range are quite feasible, as far as 2-D periodicity (only) is concerned. But, to be in perfect agreement with 2-D model, these structures should be made "infinitely" long in the third dimension and propagating waves should have a flat profile along this infinite direction.

This is of course very difficult to achieve in practice: the mentioned technology hardly produces patterns with a [depth vs width] aspect ratio of 10. In particular, the optimal configuration of large air fraction requires dielectric veins less than 100 nm thickness which are very difficult to realize with reasonable depths, in the present state-of-the-art. Light also needs to be confined in or around the plane of periodicity, e.g. by means of a standard GaAlAs/GaAs waveguide, so that the light field rather has a guided mode profile in the third direction. This profile is not taken into account in

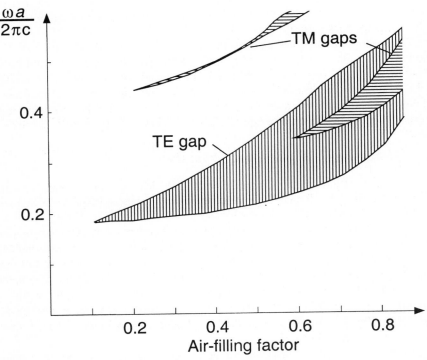

Fig. 2 Bandgap map for a lattice of triangular air holes ($\varepsilon=1$) in a matrix of dielectric constant $\varepsilon=13.6$. The striped areas indicate the position and width of the omnidirectional bandgaps, for each polarization TE and TM, as a function of the fraction of air f_{air}. They overlap only in the high f_{air} region.

most calculations and one should ensure that waveguide and PCs are compatible, i.e. that light mainly remains guided upon interaction with the PC. In particular, it should be noticed that the presence of the air holes is an interruption of the horizontal waveguide. If these holes are too large, the guiding property may be lost and light might be strongly scattered into air and substrate when traversing a hole [12,13].

This scattering would be deeply detrimental to practical applications and should thus be avoided. A way to obtain this is to limit the amount of air in the structure to a moderate value, typically 0.2 to 0.3 in our experiments [14]. This value also decreases the stringent aspect ratio of the horizontal structure, making veins larger, so that depths of 1 µm can be reached by technology, bringing our samples even closer to a 2-D model [6]. The price to pay is to obtain a bandgap only for TE polarization (see Fig. 2). This is acceptable as many emitters are TE polarized, in particular the strained QWs we use [15]. Our goal was then to demonstrate that strong photonic effects still exist for guided light and despite the moderate hole depth ($\approx 3\lambda/n$ in our case).

3 Description of Experiments

To probe PC effects on light propagating in a waveguide configuration, the photonic structures are etched through a 250 nm wide GaAs monomode waveguide with large AlGaAs barriers. We use photoluminescence (PL) as a built-in light source. PL is obtained from the focused spot of a red laser diode (678 nm) which excites 3 InGaAs strained QWs embedded in the waveguide. Because of the presence of the guiding heterostructure, part of the PL propagates parallel to the surface as a guided mode and interacts with the photonic structure under study. Guided light then exits the sample through a cleaved facet of the sample where it is collected with a microscope equipped with polarizers (Fig. 3) [14].

Two perpendicular CCD cameras provide images of the surface and cleaved edge of the sample, in order to detect emitted light. The image formed on either of the CCD sensors is partly projected onto a multimode fiber of diameter ϕ by means of a beam splitter. The optical fiber then feeds an Optical Multichannel Analyzer, thus performing a localized spectral analysis of a known disk in the image of the CCD sensor, with resolution ϕ/γ, where γ is the magnification of the optical line [16].

The emitted PL reaches the microscope through three different channels, propagating through air, through the substrate, or propagating in the waveguide. These three channels appear clearly separated in the microscope camera, e.g. when imaging the cleaved facet of the sample (Fig 4), so that the optical fiber allows to spatially select the guided mode of interest only, getting rid of the two unwanted channels where light is not affected by the structure.

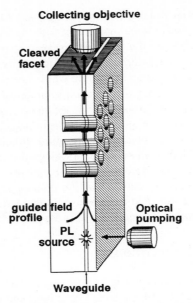

Fig. 3 Schematics of the experiment

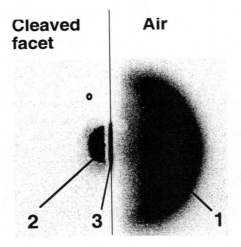

Fig. 4 Image of the cleaved facet in the microscope. Guided light (3) appears as a thick vertical line and can be spatially separated from other contributions (through air (1) and through substrate (2)) by means of an optical fiber. The black circle represents the equivalent diameter of the fiber. Warning: this picture has been contrast-enhanced to make all 3 signals clearly apparent. The guided signal is usually much brighter than the others.

The available light intensity of this built-in probe $I_1(\lambda)$ when exciting the as-grown heterostructure may be written as:

$$I_1(\lambda) = I_0(\lambda) \times \phi/d \times \exp[-\alpha(\lambda)d]T_{GaAs/air} \quad ,$$

where the first factor $I_0(\lambda)$ is the initial PL intensity emitted into the guided mode. The second factor is the part of the guided signal collected by the fiber, with d the distance from excitation spot to cleaved edge. The exponential term represents the intrinsic absorption of the guided mode on distance d, and the last factor is the intensity transmission coefficient at the interface with air.

When exciting PL at such a position that the collected guided mode runs through a PC structure, the material section of thickness d' between the parallel cleaved edge and the PC pattern has a special role: it forms a slab bounded with partially reflecting boundaries — the cleaved facet and the pattern (Fig. 5) [16].

Therefore, guided waves undergo in-plane multiple-beam interferences and the collected intensity becomes $I_2(\lambda) = T_{FP}(\lambda) \times I_1(\lambda)$, where $T_{FP}(\lambda)$ is the well-known Fabry-Perot transmission [17]:

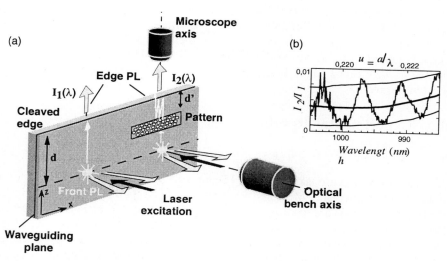

Fig. 5 (a) Multiple-beam interferences between the cleaved edge and the PC boundary. The measured intensity is $I_2(\lambda)$. A reference spectrum $I_1(\lambda)$ is also measured in an on-etched area. (b) Spectral oscillations appear in the ratio. The mean value (thick line) gives the transmission, while the fringe amplitude (between the two light lines) is proportional to the square root of the reflectivity of the PC.

$$T_{FP}(\lambda) = \left| \frac{t}{1 - rr_2 \exp(2i\Phi)\exp[-\alpha d']} \right|^2$$

t is the amplitude transmission and r the amplitude reflection coefficient of the photonic crystal and r_2 the amplitude reflection coefficient of the cleaved edge; $2\Phi = 4\pi d' n_{eff}/\lambda$ is the round-trip phase and $\exp(-\alpha d')$ the absorption both at normal incidence. Except if $\alpha d' \gg 1$, spectral oscillations (Fig. 5 (b)) appear in the ratio $T_{FP}(\lambda)$. They allow to extract $T(\lambda) = |t|^2$ from the mean value and on the range where $I_1(\lambda)$ is non-zero $R(\lambda) = |r|^2$ from the fringe amplitude (proportional to $\sqrt{R(\lambda)}$).

Due to reabsorption of the guided mode by quantum well excitons, the probed wavelength range is usually quite narrow (about 20 nm wide only) in the case of QWs, so that only a small part of the band structure can be checked with one sample. Since the spectral response of the photonic structure scales with the size of the lattice, we fabricated photonic lattices with 7 different periods at a constant air-filling factor, ranging from $a=180$ nm to $a=360$ nm, in order to probe a large range of $u = a/\lambda$ from 0.18 to 0.4. It should also be noted that due to refraction at the semiconductor/air interface (ratio of refraction indices about 3.5) combined to the limited numerical aperture of the objective (NA=0.4), guided light collected in the microscope has a maximum internal angle of 6.5°, so that the photonic structure is probed in a quasi-directional way, at almost normal incidence. The different crystal axis, were probed by fabricating two types of pattern, with either the ΓM or the ΓK principal crystallographic axis of the Brillouin zone, aligned along the probing beam (i.e. normal to the cleaved edge). Finally, a polarizer in the microscope selects the TE or the TM beams.

4   Experimental Results

4.1   Quantum Well Samples

All subsequent data are taken through 15 unit cells thick quasi-infinite (30 μm-long) slabs of photonic crystal. Measured transmission $T(u)$ (points) and their derivatives $\partial T(u)/\partial u$ (arrows) are reported for all 7 samples and the 2 directions of propagation ΓM and ΓK [16]. Lines are guides to the eye. They are shown only for the TE case where a gap is expected (Fig. 6). For comparison, we show the theoretically predicted transmission (using Transfer Matrix Method) of a triangular array of infinitely deep air cylinders in a uniform dielectric, next to the experimental data. Parameters used in the calculation were $f=28.5\%$, consistent with experimental values and dielectric constant $\varepsilon=10.2$, somewhat lower than the effective index of the waveguide.

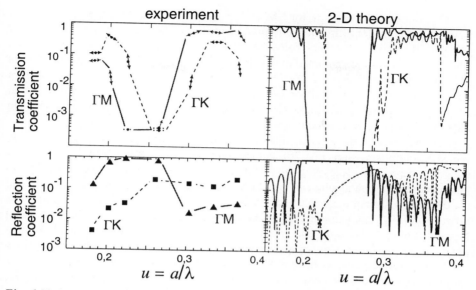

Fig. 6 *Measured (left) and calculated (right) transmission (top) and reflectivity (bottom) for a slab of PC with 15 rows thickness*

The general behavior of intensities as well as derivatives is very consistent with the calculation. In particular, one can note the relative positions of ΓM and ΓK curves and the two overlapping stop-bands around $u=0.25$, going down to the noise level. Clear falling band-edges appear at $u=0.2$ for ΓM and $u=0.23$ for ΓK. In the pass-window between $u=0.3$ and $u=0.35$, transmission in excess of 50% is observed not far from the theoretical value. The contrast between pass- and stop-windows exceeds three orders of magnitude.

A crucial test that would ensure that waveguided light has only in-plane interaction with the photonic crystal is that low transmission spectral regions coincide with high reflection ones. Reflection data obtained from fringe visibility are shown in the bottom frames. Unlike transmission data, points only are displayed, not derivatives, because there are not enough fringes in the 20 nm spectral window. Again, a very satisfying agreement is found. The highest reflectivity of R>80% is obtained for TE polarization propagating along ΓM, coinciding with the low transmission window.

Reflection and transmission alone, however, do not tell the whole PBG story: one can see that both in theory and experiments, the low transmission window along ΓK does *not* coincide with a high specular reflectivity. If only transmission and reflection were allowed for guided light, one could conclude that guided light is lost out of the waveguide. But in fact, due to the periodic nature of PBG patterns, the in-plane interaction may also be largely diffractive. As discussed by Sakoda [18], plane waves may be diffracted at angles predicted by the standard ruled grating formulae. In our case, guided waves may be redirected in the Bragg direction given by the surface period of the 30 μm-long PBG slab: a for ΓM and $\sqrt{3}a$ for ΓK, still propagating as guided waves. Conditions for diffraction at normal incidence are then $u \geq (n_{eff})^{-1}$ for ΓM and $u \geq \sqrt{3}^{-1} n_{eff}^{-1}$ for ΓK. Only below these cut-offs should one observe $R+T=1$. Above, four beams are diffracted in first-Bragg orders at angle θ, two with efficiencies η_R and two with efficiencies η_T (Fig. 7 (a)). This is not a loss mechanism and does not preclude the use of PBG for spontaneous emission control: lossless interaction with the structure now reads $R+T+D=1$ where $D=2\eta_R+2\eta_T$ is the diffracted power for unit incident power.

Experimentally, the in-plane diffraction is detected in the geometry shown in Fig. 7(b & c). In transmission (Fig. 7b), guided light also appears at a point B, away from the direct beam in A. Recalling the ~6° directional selectivity achieved by our setup, light occuring at such a point can only have been redirected by the lattice from oblique incidence θ to normal incidence and is therefore an unambiguous signature of in-plane diffraction [19]. The same holds for the reflection geometry of Fig. 7c. The direct beam is seen at A', but light also emerges at B'. In both cases, one measures at B or B' the diffraction efficiency at oblique incidence θ if the reference is taken at a distance d'' from the edge, equal to the total diffracted light path ($d''=SC+CB$ in Fig. 7). The result is, from calculation (and it can be shown by time reversal symmetry arguments)

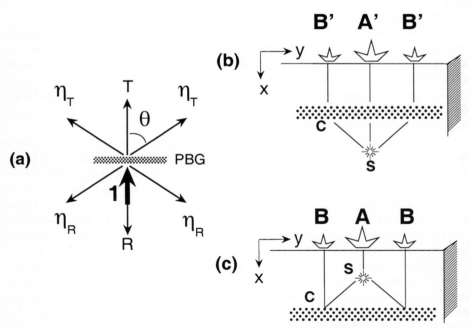

Fig. 7 (a) Incoming guided light can be directed in any of the Bragg directions, i.e. it can either be transmitted, reflected or diffracted forward and backward. (b) Visualization of forward diffraction in our setup. (c) same for backward diffraction.

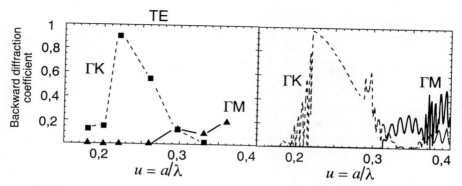

Fig.8 *Measurements and calculations of backward diffraction from our samples*

that a reciprocity rule holds for oblique to normal incidence and normal to oblique incidence diffraction efficiencies, so that the measured efficiencies in our experiment are the same η_R and η_T of Fig. 4a.

The set of measurements on all samples is shown in Fig. 8 for TE polarization in the backward diffraction geometry and compared to calculated curves for 2 η_R, using the same fitting parameters as above [19]. One can clearly see that along ΓM, no diffraction occurs below the cut-off value. Along ΓK, diffraction is likely everywhere in the displayed u range, in particular, in the bandgap between $u=0.22$ and $u=0.28$. The maximum value reaches unity in theory and is measured at 90% for $u=0.21$. Diffraction efficiencies in the forward geometry were also measured, but they are not displayed here, as efficiencies (predicted and measured) are all below 30%.

In summary, diffraction phenomena are an integral part of photonic bandgap concepts if the outside medium is just the unetched dielectric matrix, a quite canonical case indeed: the periodic nature of photonic crystals results in the fact that light whose propagation is forbidden is not necessarily specularly reflected by the crystal. Backward diffraction is the other possible channel, which can be predominant in the bandgap region. As a consequence, photonic crystals cannot always be considered as perfect specular mirrors in the bandgap.

Experimentally, losses outside the waveguide achieved by the photonic structures can be estimated through the value of $\Sigma = R + T + D$: light not directed in one of the three channels — transmission, specular reflection or diffraction,— is most certainly lost out of the waveguide. The high values achieved for R and D in the bandgap in agreement with theory, drive Σ to values near unity. Out-of-plane losses L estimated through $L = 1 - \Sigma$ are thus weak so that applications of photonic crystals as integrated optics elements can already be envisioned with present samples. Finally, the consistency of the three coefficients R, T, D with the perfect 2-D theory (no waveguide, infinite holes) shows that, in the moderate f_{air} limit, the main trends of the 2D picture holds in the deep-etched guide configuration despite the lack of waveguiding in the holes and their finite height.

4.2 QDs Samples

Previous measurements show that the main trends of the perfect 2-D picture holds in our samples, e.g. the presence of a TE bandgap as well as the occurence of diffraction. In particular, it is of practical interest for future devices to note that the bandgap width obtained with $f_{air} \approx 30\%$ is large enough to obtain an inhibition of the propagation of the whole TE spectrum emitted by the sample. This is the case e.g. in our 4th sample with $a=260$ nm. However, the use of QWs emitters limits the collected data to discrete points along the $u = a/\lambda$ scale whereas a continuum spectrum would provide much more detailed information essential to the design of future devices.

Let us remind that the available probe light intensity is:

$$I_1(\lambda) = I_0(\lambda) \times \phi/d \times \exp[-\alpha(\lambda)d] T_{GaAs/air}$$

The probe spectral width is thus limited both by $I_0(\lambda)$ and the reabsorption spectrum $\alpha(\lambda)$. In particular, in quantum well samples, the spectral width of $I_0(\lambda)$ can reach 100 nm under strong optical excitation, whereas the exciton strong reabsorption reduces the "useful" of $I_1(\lambda)$ width to 20 nm only (Fig. 9), in the low energy tail of $I_0(\lambda)$.

InAs self-organized quantum dots (QDs) overcome these two limitations [20, 21]: a deliberately large inhomogeneous dot size distribution translates into a broadened $I_0(\lambda)$ [22], as well as into a weak reabsorption compared to QWs (in other words, the oscillator strength is distributed over a broader spectrum), so that $I_1(\lambda)$ and $I_0(\lambda)$ have comparable widths over 100 nm (Fig. 9). Measurements from samples with 10% varying PC periods thus have spectra which overlap in the a/λ scale. Spectra obtained from our 7 samples then give a complete image of the transmission at the bandgap and far into its boundaries [23]. As an example, we present in Fig. 10 a set of data obtained for TE polarization, along ΓK orientation, when the number of rows of the structure varies from 3 to 15. Instead of having discrete data points as in Fig. 6, spectra now completely overlap for a given number of rows, within small stitching errors, confirming the good reproducibility of the air-filling factor in our fabrication technology. The bandgap clearly builds up when the number of rows in the sample increases confirming by the way that the PBG effects overcome largely the losses. One can see in particular that the steepness of the band edge is already maximum for N=12, which appears as an optimum number of rows, in term of passwindow / stopband contrast.

The data for samples with a large number of rows also show oscillations of the transmission in the "pass-window". These oscillations are DBR-like oscillations due to interferences of Bloch modes reflected at the PC boundaries in the structure. They also prove the good quality of the samples and allow to determine the band structure of the sample [23]. A more detailed comparison with theory can now be performed. QWs samples showed that the 2D model holds in its globality, QDs samples can now show the detailed intrinsic effects of real samples which includes possible limitations of the 2-D model.

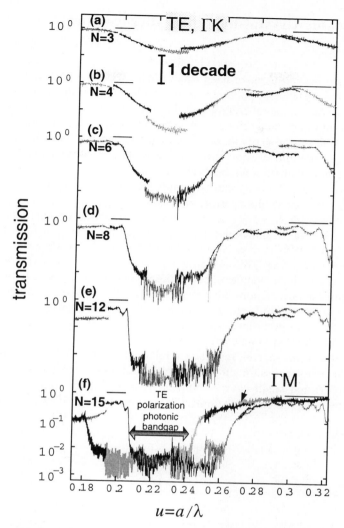

Fig. 10 Transmission curves as in Fig. 5, using InAs QDs layers as emitters in the waveguide, so that spectra from successive samples overlap, giving rise to a continuous spectrum. Transmission measurements show the building up of the TE bandgap along ΓK when the number of PC rows increases from 3 to 15 rows. The final omnidirectional bandgap is visualized for 15 rows by adding the transmission curve along ΓM.

5 Photonic-Crystal-Bounded Microcavities

5.1 1-D Cavity

Once the basic properties have been established, PCs can be used as constitutive elements of more elaborate photonic structures. We remind that these 2-D PCs are expected to perform a lateral control of light propagating in a planar waveguide, and ultimately to surround small cavities, where control of spontaneous emission could be achieved. A first step towards these goals is the demonstration of horizontal cavity effects, using photonic crystal mirrors. They offer a good insight on reflector performance, through the finesse and peak transmission of cavity modes [24].

A horizontal 1-D cavity can be simply designed by etching 2 slabs of PC, a few hundreds of nanometers apart, through the waveguide. A simple transmission measurement through the cavity, as described before, allows to determine the cavity resonances [25]. Fig.11 shows the obtained spectra for cavities with a large 3 µm spacing between the PC mirrors. The number of rows of each mirror increases from 3 to 9. Measurements using QWs as emitters, clearly show a transmission peak, whose finesse increases with the number of rows, clearly demonstrating the cavity effect. The small superimposed oscillations are due to parasitic interferences between the cleaved edge and the pattern (similar to those used to measure reflectivity of single slabs). Fabry-Perot fits were performed and give a maximum reflectivity of 80% and losses estimated to 11% for the sample with 9 rows.

However a 3 µm spacing results in a relatively high-order cavity ($m > 20$). To affect spontaneous emission directionality, small cavity orders are desired (see Benisty's notes) and small cavity volumes are required for lifetime changes (see Gérard's notes). Fig. 12 shows the transmission measurement on a cavity as small as

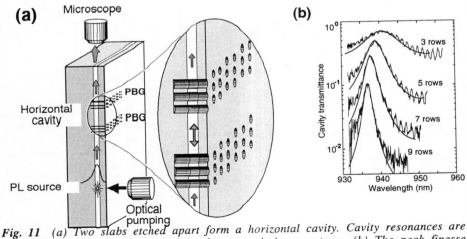

Fig. 11 (a) Two slabs etched apart form a horizontal cavity. Cavity resonances are detected through peaks observed in the transmission spectrum. (b) The peak finesse increases with the number of rows, indicating the reflectivity of the boundaries also increases

70 nm width [26]. It was carefully designed according to passive reflectivity measurements: QDs were chosen as the active material for a broader spectrum. The periodicity was chosen as 220 nm with f_{air} still around 30% and the orientation of the photonic crystal was chosen along ΓM, so that no in-plane diffraction is allowed and the expected reflectivity is high as the QDs emission is inside in the expected bandgap. The cavities have only 4-row PC mirrors on each side. As a comparison, the transmission of a 8-row mirror (no cavity) was found to be around 1%, while the cavity transmission exhibit a clear peak with maximum transmission around 40% (Fig. 12). Half width is 8 nm, which gives a quality factor of 125. This quality factor translates into a finesse of 63 (the cavity order is close to 2) and a mirror reflectivity of 95%, so that the estimated losses are as weak as 4% per mirror, including absorption losses within the cavity, due to the presence of the QDs. As a result, losses through out-of-plane scattering are as weak as a few per cent in our samples. Even smaller values could be expected for future devices by providing greater optical confinement in the out-of-plane direction, by means of a vertical microcavity or by using oxidized AlAs in the confining layers, both solutions being compatible with optoelectronics technology.

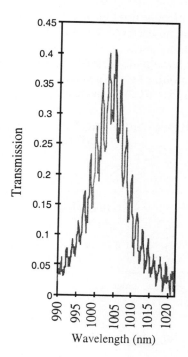

Fig. 12 Peak transmission through a 70 nm-wide horizontal cavity. The Q is 125 for a cavity order equal to 2, e.g. the mirror reflectivity is over 95%. Note the small period FP fringes due to the cavity reflectivity, that are superimposed to the cavity resonance,.

5.2 2-D Cavity

The extension to cavities surrounded by 2-D photonic crystals is at hand, but their interpretation promises to be awkward, as soon as the size is small enough for the structured boundary to play a predominant role in detailed mode shape and frequencies [27]. We therefore investigated first simpler circular microdisk cavities surrounded by mirrors inspired by the photonic crystal approach, but which can be considered neither 1-D nor 2-D [28].

Modes in microdisks of radius R are at wavelengths $\lambda_{n,m}$ characterized by a radial number n (number of nodes along a radius) and an azimuthal number m (number of nodes along a circle) [29]. The well-known Whispering-Gallery Modes (WGMs) have $m \gg n$, which means in k-space representation that they have almost grazing incidence on the circle boundary: they are well confined by total internal reflection at a semiconductor/air interface (Fig. 13a). High-Q modes as well as lasers with very low thresholds have been demonstrated in flat GaAs disks surrounded by air [30]. On the opposite end, Quasi-Radial Modes (QRMs) have $n \gg m$, which means in a ray-tracing model that they have almost normal incidence on the circular boundary: in simple disks, they are only weakly confined by the 30% reflectivity of the interface (Fig. 13b). Some kind of PC mirrors are thus required to confine those modes, the simplest of which is a circular Bragg mirror, made of concentric air trenches [31].

We etched such mirrors through our heterostructure, using QDs as emitters (Fig. 14a) [32]. Gaining on our experience of moderate amount of air in 2-D PC, we designed a 4th-order Bragg mirror, with narrow air-trenches (about 70 nm wide). The center 3 µm diameter waveguide disk forms the cavity. The PL of QDs is excited inside the resonator, in order for the guided spontaneous emission to probe the disk

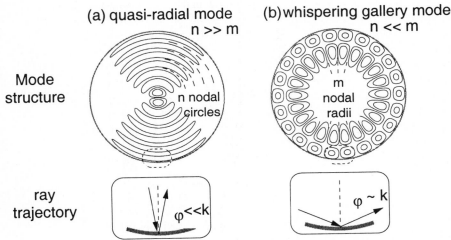

Fig.13 (a) Whispering-Gallery Mode and (b) Quasi-Radial Mode in a circular microcavity.

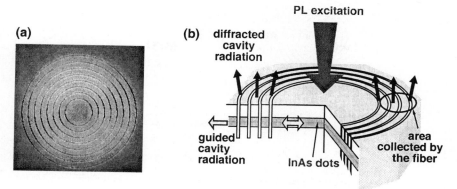

Fig. 14 (a) *Micrograph of the cavity (the central unetched area is 3 μm diameter) surrounded by the circular Bragg mirror made of concentric deeply etched trenches.* (b) *The PL of QDs is excited inside the cavity. The resonances build up and are observed through light leakage in the 4th order grating*

resonance. Not only do QDs have a broad spectrum, but they also efficiently reduce carrier diffusion by trapping electron-hole pairs, so that the intensity of PL is still large, despite the close presence of recombination centers at the etched boundaries. The disk resonances build up in the cavity and are primarily outcoupled to the guided mode, where they slightly leak to the air through second order Bragg diffraction (4π phase-difference towards the air and the substrate), similarly to surface emission mechanism of second-order DBRs (2π phase difference towards the air). Light collected from the Bragg grating thus allows to observe the horizontal resonances (Fig. 14b).

Spectra are shown in Fig. 15 for structures with gratings of variable periods. The main observation is that many sharp peaks of width $\Delta\lambda=1.5-5$ nm show up in the spectra in clusters of one to four, while light collected from simple mesas (shown in inset) exhibit only very broad features. The shift in period results in a shift in the expected stop-band. The stop-bands calculated in the simplified case of 1-D grating are indicated as thick black lines in the graphs: the sharpest multiple clusters are located into the stop-bands, while there are almost no marked feature outside them, confirming that the peaks are indeed the signature of confinement due to the grating.

Using a 2-D model with metallic boundary on an effective radius of 3.7 μm, we could further identify the peaks as QRMs, as expected. The best Q's we could measure are up to 650, which results in a reflectivity close to 90% at normal incidence. These novel mirrors inspired by the photonic bandgap approach prove a very efficient confinement of modes, still in the small air fraction limit. Useful implementation such as for the change of spontaneous emission time, can thus be envisioned for smaller cavities.

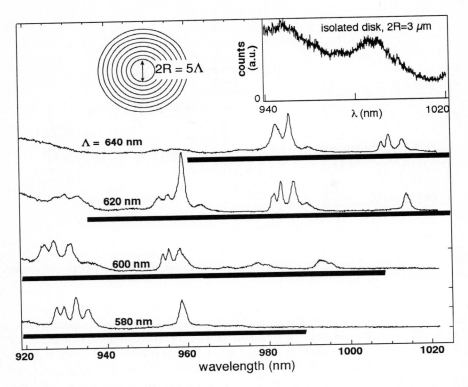

Fig. 15 Light is collected from cavities with various pitches Λ and central diameter equal to 5 times the pitch. Clusters of peaks appear in light collected from the grating area, mainly at wavelengths within the grating stopbands, calculated in a 1-D model and indicated by thick black lines. Inset: Collected spectrum from a simple mesa shows only smooth features.

6. Conclusion

This set of quantitative measurements amply demonstrates that the 2-D PCs which can presently be fabricated can already control guided light propagation through reflection, transmission and diffraction, although it was feared that this would be prevented by scattering into the substrate. Those promising results are further confirmed by the high quality of 1-D and 2-D microcavities bounded by PC structures. The next steps are clearly to demonstrate microcavity effects such as lifetime changes (Purcell effect), enhancement of LED efficiency and applications in integrated optics. Let us elaborate on the latter. While some application of PC's have been predicted, and sometimes demonstrated in the microwave regime [33,34], the present experiments show that a number of building blocks necessary for integrated optics can be achieved with 2D PC's, some being even quite original compared to the usual integrated optics toolbox [35].

We gratefully acknowledge the support of European Community through Long Term Research contract ESPRIT SMILED n°24997. TFK is supported by a Royal Society Research Fellowship. CJMS is supported by an EPSRC case with BT. We also thank the Nanoelectronics research Centre at Glasgow University for technical support.

References

1. E. Yablonovitch, "Inhibited Spontaneous Emission in Solid-State Physics and Electronics," *Phys. Rev. Lett.*, vol. 58, pp. 2059-2062, 1987.
2. J. D. Joannopoulos, R. D. Meade, and J. N. Winn, *Photonic Crystals, Molding the Flow of Light*. Princeton, NJ: Princeton University Press, 1995.
3. E. M. Purcell, "Spontaneous emission probabilities at radio frequencies," *Phys. Rev.*, vol. 69, pp. 681, 1946.
4. C. C. Cheng, A. Scherer, V. Arbet-Engels, and E. Yablonovitch, "Lithographic band gap tuning in photonic band gap crystals," *J. Vac. Sci. Technol. B*, vol. 14, pp. 4110, 1996.
5. T. Krauss, Y. P. Song, S. Thoms, C. D. W. Wilkinson, and R. M. Delarue, "Fabrication of 2-D photonic bandgap structures in GaAs/AlGaAs," *Electronics Letters*, vol. 30, pp. 1444, 1994.
6. T. F. Krauss, R. M. De La Rue, and S. Brand, "Two-dimensional photonic-bandgap structures operating at near-infrared wavelengths," *Nature*, vol. 383, pp. 699-702, 1996.
7. For references, see in this volume the various lecture notes on microcavities.
8. A. A. Maradudin and A. R. McGurn, "Photonic band structure of a truncated, two-dimensional, periodic dielectric medium," *J. Opt. Soc. Am. B*, vol. 10, pp. 307, 1993.
9. D. Cassagne, C. Jouanin, and D. Bertho, "Photonic band gaps in two-dimensional graphite structure," *Phys. Rev. B*, vol. 52, pp. R2217, 1995.
10. R. D. Meade, K. D. Brommer, A. M. Rappe, and J. D. Joannopoulos, "Existence of a photonic band gap in two dimensions," *Appl. Phys. Lett.*, vol. 61, pp. 495, 1992.
11. J. M. Gérard, A. Izraël, J. Y. Marzin, R. Padjen, and F. R. Ladan, "Photonic bandgap of two-dimensional dielectric crystals," *Solid-State Electronics*, vol. 37, pp. 1341, 1994.
12. V. Berger, I. Pavel, E. Ducloux, and F. Lafon, "Finite-element Maxwell's equations modelling of etched air/dielectric Bragg mirrors," *J. Appl. Phys.*, vol. 82, pp. 5300-5304, 1997.
13. B. D'Urso, O. Painter, J. O'Brien, T. Tombrello, A. Yariv, and A. Scherer, "Modal reflectivity in finite-depth two-dimensional photonic crystal microcavities," *J. Opt. Soc. Am. B*, vol. 15, pp. 1155-1159, 1998.

14. D. Labilloy, H. Benisty, C. Weisbuch, T. F. Krauss, R. Houdré, and U. Oesterle, "Use of guided spontaneous emission of a semiconductor to probe the optical properties of two-dimensional photonic crystals," *Appl. Phys. Lett.*, vol. 71, pp. 738-740, 1997.
15. J.-Y. Marzin, M. N. Charasse, and B. Sermage, "Optical investigation of a new type of valence-band configuration in InxGal-xAs-GaAs strained superlattices," *Physical Review B*, vol. 31, pp. 8298-8301, 1985.
16. D. Labilloy, H. Benisty, C. Weisbuch, T. F. Krauss, R. M. De La Rue, V. Bardinal, R. Houdré, U. Oesterle, D. Cassagne, and C. Jouanin, "Quantitative measurement of transmission, reflection and diffraction of two-dimensional photonic bandgap structures at near-infrared wavelengths," *Phys. Rev. Lett.*, vol. 79, pp. 4147-4150, 1997.
17. M. Born and E. Wolf, *Principles of Optics*. Oxford: Pergamon Press, 1980.
18. K. Sakoda, "Transmittance and Bragg reflectivity of two-dimensional photonic lattices," *Phys. Rev. B*, vol. 52, pp. 8992-9002, 1995.
19. D. Labilloy, H. Benisty, C. Weisbuch, T. F. Krauss, D. Cassagne, C. Jouanin, R. Houdré, U. Oesterle, and V. Bardinal, "Diffraction efficiency and guided light control by two-dimensional photonic-band-gap lattices," *IEEE J. Quantum Electron.*, to be published, 1999.
20. L. Goldstein, F. Glas, J. Y. Marzin, M. N. Charasse, and G. Le Roux, "Growth by molecular beam epitaxy and characterization of InAs/GaAs strained-layer superlattices," *Appl. Phys. Lett.*, vol. 47, pp. 1099-1101, 1985.
21. J. Y. Marzin, J. M. Gérard, A. Izraël, and D. Barrier, "Photoluminescence of single InAs quantum dots obtained by self-organized growth on GaAs," *Phys. Rev. Lett.*, vol. 73, pp. 716-719, 1994.
22. J. M. Gérard, J. Y. Marzin, G. Zimmermann, A. Ponchet, O. Cabrol, D. Barrier, B. Jusserand, and B. Sermage, "InAs/GaAs quantum boxes obtained by self-organized growth: intrinsic electronic properties and applications.," *Solid State Electronics*, vol. 40, pp. 807-814, 1996.
23. D. Labilloy, H. Benisty, C. Weisbuch, C. J. M. Smith, T. F. Krauss, R. Houdré, and U. Oesterle, "Finely resolved transmission spectra and band structure of two-dimensional photonic crystals using InAs quantum dots emission," *Phys Rev. B*, vol. 59, pp. 1649-1652, 1999.
24. U. Oesterle, R. P. Stanley, R. Houdré, M. Gailhanou, and M. Ilegems, "Molecular beam epitaxy of an ultrahigh finesse microcavity," *J. Cryst. Growth*, vol. 150, pp. 1313-1317, 1995.
25. D. Labilloy, H. Benisty, C. Weisbuch, T. F. Krauss, V. Bardinal, and U. Oesterle, "Demonstration of a cavity mode between two-dimensional photonic-crystal mirrors," *Electronics Letters*, vol. 33, pp. 1978-1980, 1997.
26. C. J. M. Smith, T. F. Krauss, R. D. L. Rue, D. Labilloy, H. Benisty, C. Weisbuch, U. Oesterle, and R. Houdré, "In-plane microcavity resonators with

two-dimensional photonic bandgap mirrors," *EE Proc.-Optoelectron., Special issue on Photonic Crystals*, vol. 145, pp. 373-378, 1998.

27. C. J. M. Smith, T. F. Krauss, R. D. L. Rue, D. Labilloy, H. Benisty, C. Weisbuch, U. Oesterle, and R. Houdré, "Near-infrared microcavities confined by two-dimensional photonic bandgap crystals," *Electron. Lett.*, vol. 35, pp. 228-230, 1999.

28. D. Labilloy, H. Benisty, C. Weisbuch, T. F. Krauss, C. J. M. Smith, R. Houdré, and U. Oesterle, "High-finesse disk microcavity based on a circular Bragg reflector," *Appl. Phys. Lett.*, vol. 73, pp. 1314-1316, 1998.

29. R. E. Slusher, A. F. J. Levi, U. Mohideen, S. L. McCall, S. J. Pearton, and R. A. Logan, "Threshold characteristics of semiconductor microdisk lasers," *Appl. Phys. Lett.*, vol. 63, 1993.

30. S. L. McCall, A. F. J. Levi, R. E. Slusher, S. J. Pearton, and R. A. Logan, "Whispering-gallery mode microdisk lasers," *Appl. Phys. Lett.*, vol. 60, pp. 289, 1992.

31. A. A. Tovar and G. H. Clark, "Concentric-circle-grating, surface emitting laser beam propoagation in complex optical systems," *J. Opt. Soc. Am. A*, vol. 14, pp. 3333-3340, 1997.

32. J. M. Gérard, D. Barrier, J.-Y. Marzin, R. Kuszelewicz, L. Manin, E. Costard, V. Thierry-Mieg, and T. Rivera, "Quantum boxes as active probes for photonic microstructures: the pillar microcavity case," *Appl. Phys. Lett.*, vol. 69, pp. 449, 1996.

33. S.-Y. Lin, V. M. Hietala, L. Wang, and E. D. Jones, "Highly-dispersive photonic band-gap prism," *Optics Letters*, vol. 21, pp. 1771-1773, 1996.

34. S. Y. Lin, E. Chow, V. Hietala, P. R. Villeneuve, and J. D. Joannopoulos, "Experimental demonstration of guiding and bending of electromagnetic waves in a photonic crystal," *Science*, vol. 282, pp. 274-276, 1998.

35. R. D. Meade, A. Deveny, J. D. Joannopoulos, O. L. Alerhand, D. A. Smith, and K. Kash, "Novel applications of photonic band gap materials: Low-loss bends and high Q cavities," *J. Appl. Phys.*, vol. 75, pp. 4753-4755, 1994.

Limitations to optical communications

J.E.Midwinter

University College London, Torrington Place, London WC1E 7JE

LECTURE I - Links without amplifiers

1 Background

The origins of optical fibre communications as seen today trace back to two key developments. The first was that of the laser in 1960 which made available for the first time a coherent optical source at a carrier frequency in the range of 10^{14} to 10^{15} Hz. This soon sparked discussion of the communication properties of light, noting, for example, that a mere 1% bandwidth would offer in the range 1 to 10 THz of spectrum space, an almost unimaginable amount compared to the norms for electrical communication of the time. (One notes that the entire radio, microwave and millimetre wave range embraces 300 GHz).

Early work on optical communications was largely aimed at exploiting this huge potential, usually using free space or gas as the transmission medium, until the paper published by Kao & Hockham in 1966 proposed the use of a glass fibre dielectric waveguide and linked it with much more modest albeit realistic potential application. Coming from a telecommunications laboratory, they had in mind the transport of data at rates of perhaps 140 Mbit/s, the highest rate then commonly used in telecommunications, coupled with a desire to achieve 2 km repeater (source to detector) separation which implied achieving an attenuation of less than about 20 dB/km. This was a technologically challenging target, since fibre-optic guides of the time had attenuations that were more typically 1000 dB/km. As a result, an intensive study was launched of the loss mechanisms in glasses and means to overcome. Today, glasses based predominantly upon the oxide of silicon (SiO_2) and known as silica dominate almost all thinking on the subject, just as chips based upon silicon dominate most electronic design.

In passing it is worth noting the data rates commonly found in telecommunications networks today. The single telephone channel is digitised to produce a 64 kbit/s data stream. 30 of these are electronically multiplexed together to form a basic "building block" level of 2.048 Mbit/s . Increasingly, data is also carried in the form of ATM cells, short packets of data each consisting of 53 bytes (424 bits) containing 48 bytes of message data and 5 bytes of header giving source and destination data. Fibre transmission systems are typically found carrying data at the rates listed below :

Data Rate (MBit/sec)	No. ATM cells per second (approx)	No of telephony channels (approx)
155	365 thousand	2000
1200	2.8 million	15000
10,000	23 million	130000

Table 1. Typical data rates found in fibre telecommunications transmission systems (per fibre)

2 Types of Fibre

Two types of fibre are in widespread use today, namely the single-mode and graded-index designs. Both are of circular cross section and feature a guiding core of higher index glass surrounded by a cladding of lower refractive index. Their general characteristics are summarised below.

Type	Outside diameter (mm)	Core Diameter (mm)	Peak Core Index	Cladding Index (approx)
Graded Index	0.125	0.05	1.46	1.45
Single Mode	0.125	0.008	1.454	1.45

Table 2. Typical Fibre Characteristics

For simple thinking purposes, guidance can be considered to occur by Total Internal Reflection at the core cladding interface but a proper analysis involves the solving of Maxwell's equation to solve for the guided modes of the waveguide structure.

The single mode fibre is characterised by having a V value of :

$$V = \frac{2\pi a}{\lambda}\sqrt{n_1^2 - n_2^2} \leq 2.404$$

where a is the core radius, n_1 the core index, n_2 the cladding index and λ the wavelength of the light guided. It should be noted that such fibres actually guide two modes having degenerate spatial mode patterns but with orthogonal polarisations. The mode pattern looks very much alike a Gaussian spot in cross section.

The graded-index fibre has a refractive-index in its core which varies parabolically with radius. Because its core is much larger than that of the single mode fibre, it guides many modes and in general, these travel at different speeds through the structure. As a result, pulse spread by multipath or multimode dispersion. A popular theoretical model for studying this is the "alpha" profile model which models the index variation as :

$$n(r) = n_o\left[1 - 2\Delta\left(\frac{r}{a}\right)^\alpha\right]^{0.5} \quad 0 < r \leq a$$

$$n(r) = n_o[1 - 2\Delta] \quad r > a$$

For reasons we will discuss below, profiles having alpha approximately equal to 2 are strongly favoured.

The overwhelming attraction of the graded-index fibre design over the single mode designs is its larger core, making jointing and launching much easier and hence cheaper. As a result, it is to this day used for cost sensitive intermediate performance applications, such as FDDI links on campuses.

2.1 Multipath or Multimode Dispersion

The single mode fibre, being characterised by only one spatial mode, allows light to travel by only one pathway through its length and accordingly, pulse spreading arising from the multiple paths that exist in multimode fibres cannot occur. The Graded Index fibre, because both its core and index difference are much larger, has V >> 2.404 and accordingly supports many modes. (Very approximately, the number of modes M in a graded index fibre will be given by $M = V^2/4$). Since different modes travel at different speeds, pulse spreading occurs since light inevitably travels in all modes on a long fibre because of mode coupling.

Using the alpha profile model above, the wave equation has been solved analytically [3] to derive expressions for the pulse spreading arising in such a structure and typical results are given below in Fig. 1 in units of ns/km. This shows that close to an alpha value of 2, there is a sharp minimum in pulse spreading implying that an index profile close to a parabolically varying one is optimum. Production graded index fibre uses such a profile and typically achieves multimode dispersion values of about 1 ns/km, to within a factor of 3 either way. The problems of production control make it very difficult to hold the profile in the minimum and thus the spread arises from variations in the profile produced away from the optimum.

Fibre Type	Multipath dispersion
Graded Index	0.5 to 5 ns/km
Single Mode	0 ns/km

Table 3. Typical fibre multipath dispersion values

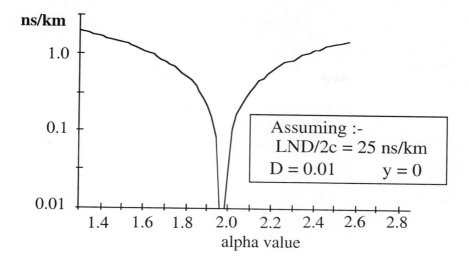

Fig. 1. Variation of multipath dispersion with alpha profile parameter

2.2 Material Dispersion

A completely separate dispersion in fibres arises from the dispersive properties of the fibre material and the fact in a dispersive glass, different wavelengths of light travel at different speeds. Given that the refractive index is n and phase velocity $v_p = c/n$, we know that pulses travel at the group velocity where $v_g = c/N$ and $N = n - \lambda\, dn/d\lambda$. The transit time through a length L of fibre for a pulse is therefore :

$$\tau_{transit} = \frac{LN}{c}$$

and the spread in transit times arising from a spread in wavelength of $\delta\lambda$ must be given by :

$$\delta\tau_{transit} = \tau_{spread} = \frac{L}{c}\frac{dN}{d\lambda}\delta\lambda = \frac{L}{c}\frac{d^2n}{d\lambda^2}\delta\lambda$$

An examination of the form of the refractive index versus wavelength curve for silica shows that there is a point of inflection at about 1300 nm wavelength implying that this effect must go to zero at that wavelength. Evaluation of the effect yields the curve below where the material dispersion is plotted in Fig. 2. in practical units of picoseconds of pulse spreading per kilometre travelled per nanometre source line width (e.g. ps/(nm.km))

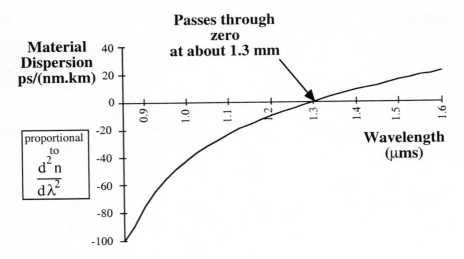

Fig. 2. Variation of material dispersion with wavelength in silica fibre.

This curve makes abundantly clear that there is large benefit to working at 1300 nm wavelength as opposed to the 850-900 nm range characteristics of GaAs based sources that was exploited for the first production systems.

2.3 Fibre Attenuation

The attenuation properties of silica have been studied extensively with the result that the fundamental loss mechanisms associated with the material are now fairly well tabulated. There are three primary ones. The Ultra Violet region is dominated by the electronic band edge absorption of the SiO_2 but this falls away very rapidly as one moves into the visible and is considered negligible in the range of interest for the fibre systems use. Throughout the visible region, there is also Rayleigh Scattering which arises from the random density fluctuations of the silica glass as well as composition fluctuations when the glass is doped, as in the core, to change its refractive index. Typically this takes a value of 1 dB/km at a wavelength of 1 micron (1000 nms) and extrapolates as λ^{-4}. Finally, there is a very intense dipole absorption arising from the vibration of the Si–O bond which has its fundamental frequency at about $3.3 \cdot 10^{13}$ Hz or 9 microns (9000 nm). At this wavelength, silica (and all oxide glasses) are "jet black". The wings of this absorption give rise to a rapidly increasing absorption seen in fibres starting at about 1600 nm and rising towards longer wavelengths. These effects are shown below together with the resultant curve, with the attenuation plotted in units of dB/km, showing a clear and fundamental minimum in the region of 1550 nm.

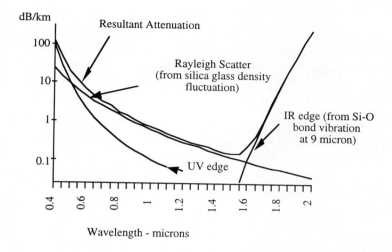

Fig. 3. Typical constituent effects in silica fibre attenuation versus wavelength.

These effects set rather solid limits to the attenuation values that can be achieved in a silica based fibre as follows :

Wavelength (nm)	Minimum (dB/km)	Typical Range (dB/km)
850-900	1.9	3 - 10
1300	0.35	0.35 - 0.6
1550	0.17	0.17 - 0.25

Table 4. Typical fibre attenuation values

To achieve lower losses than these, one must move to a different materials systems. Crystal quartz has vastly lower Rayleigh scattering but is hardly a serious candidate for fibre manufacture, many non-oxide glasses offer superior IR transmission and hence the potential of lower attenuation at longer wavelengths but are also extremely intractable materials to work with. As a result and because of other reasons that will emerge later, silica is firmly anchored as the fibre medium of choice and is not expected to be challenged in the foreseeable future.

Taking the three wavebands of interests, we can summarise our conclusions so far as :

- 850 to 900 nm.
Allows use of low cost devices (GaAs sources, Si detectors) but characterised by high material dispersion (circa 100 ps/(nm.km), and high attenuation (3-10 dB/km). Ideal for cheap modest performance links using graded-index fibres.

- 1300 nm
Attractive combination of zero material dispersion and lower attenuation (0.35 - 0.6 dB/km) coupled with availability of sources and detectors developed for this wavelength have established as the wavelength region of first choice for most land based telecommunications applications, using both single mode and graded-index fibres.

- 1500 nm
Offers the ultimately lowest attenuation but carries a penalty of higher material dispersion. However, in single mode fibres, this can be overcome and is now the preferred wavelength for very high performance systems.

We note in passing that there are other effects that can easily increase the attenuation of fibres above the minimum values quoted. In the 1300-1500 nm region, the most notable is the presence of O-H ions in the glass giving rise to a fundamental dipole absorption at 2.8 microns and a second harmonic absorption at 1.4 microns. The resulting transmission spectrum for a typical modern fibre is thus more like that shown below in Fig. 4.

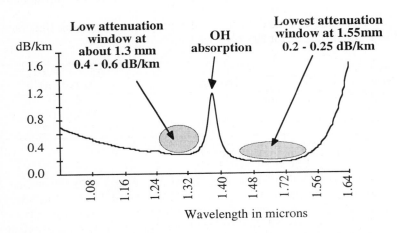

Fig. 4. Overall attenuation seen in a typical single mode fibre.

3 Simple Fibre System Models

A simple approach to establishing what performance a given combination of components is capable of is to do two calculations, one concerning power budget and the other concerning dispersion budget. In reality, these are linked in a much more complex model that involves calculating the exact size and shape of the eye-diagram at the receiver as well as the noise level and hence the error rate but a good indication of performance is obtained using the simple approach. The power budget model is illustrated schematically below in Fig. 5.

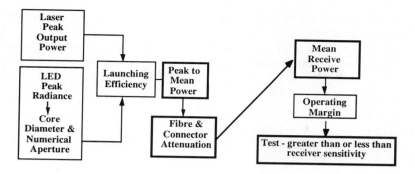

Fig. 5. Power budget model for a simple fibre system.

It involves establishing the mean power launched into the fibre, the mean power required by the receiver to achieve an acceptable error rate at the desired bit rate and hence establishing what power ratio can be dropped across the fibre. Hence, knowing the fibre attenuation, an upper length is established for the transmitter to receiver distance.

Then a calculation must be done to explore the pulse spreading properties to ensure that the pulse is not completely scrambled over the length. This is illustrated schematically in Fig. 6.

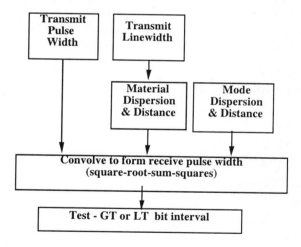

Fig.6. Dispersion budget calculation.

To make life simple, it is common to assume that the transmit pulse width is half the bit interval, to calculate the pulse spreading due to material dispersion and to multipath dispersion if a multimode fibre is used, then convolve these together using square-root-sum-squares and then to test whether the resultant and greater or less than the bit interval. Applying the equality to this condition sets an approximate upper distance limit from dispersion considerations.

The result of such calculations generates two curves, one power budget derived and one dispersion budget derived, that take the form shown below in Fig. 7.

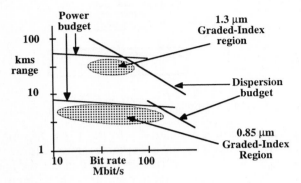

Fig. 7. Typical results of the power and dispersion budget calculations for 0.85 and 1.3 micron graded-index operation.

One should note that the exact position of the limit curves evidently depends on the exact values chosen for the various devices modelled so that the above curves (for graded index fibres) indicate typical operating regions and typical limiting regions. By very careful selection of components, it might be possible to do better.

The above curves show that moving to 1300 nm has markedly increased range because of the lower attenuation but bit rate is still seriously

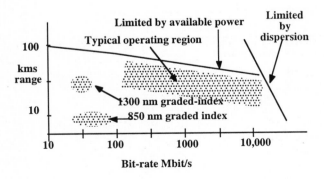

Fig. 8. Typical power and dispersion budget calculations for 1.3 micron single mode fibre operation.

constrained because of the multi-path dispersion expected in the graded index fibre. Moving to single mode fibre avoids this problem with the result shown in Fig. 8.

These curves indicate very clearly why 1300 nm single-mode systems have been the standard deployment for inland telecommunications use throughout the world for the last decade, with systems operating at bit rates of 155, 650, 1200, 2400 and now 10,000 Mbit/s.

4 Receiver Sensitivity

The above system curves were derived using an assumed receiver sensitivity of order 2500 photons (mean) per bit as a crude approximation to what is routinely achieved using APD or PIN detectors with low noise amplifiers. This is far higher than is required. For example, if we had a noiseless receiver, then we could ask how many photons/bit would be required on grounds of (photon) Shot Noise.

Assume that photons arrive with a Poisson distribution. When a ZERO is sent, no photons are sent and none will arrive since there is no spontaneous emission at these wavelengths. When a ONE is sent, assume that on average m photons arrive. How large must m be to achieve 10^{-9} error rate. Poisson statistics tells us that :

$$P(n,m) = \frac{m^n}{n!} \exp(-m)$$

and hence the probability that n arrive when m were expected. Given our noiseless receiver, when 1 or more photons arrive, we will assume correctly a ONE was sent and when 0 photons arrive we will assume a ZERO was

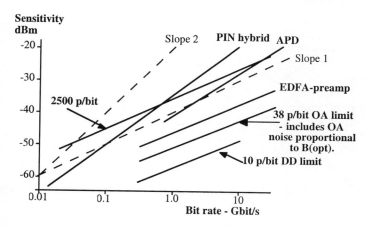

Fig. 9. Typical receiver sensitivity values for various types of receiver.

sent but in so doing will make an occasional error since P(0,m) is finite. The question is now - what value should m be for $P(0,m) = 10^{-9}$ to which the answer is m=21. Thus a mean arrival rate of 10.5 photons could suffice, some 24 dB better than that assumed in our simple model.

Much of this gap has been closed by recent developments in optical amplifiers to be discussed later. As a result, the incoming optical signal can be amplified (with an EDFA - see next Chapter) so that, on detection, it is much larger compared to the receiver noise. Using this approach, a theoretical sensitivity of about 38 photons/bit has been predicted and of order 100 photons per bit achieved. Typical sensitivity curves are shown above in Fig. 9.

5 Increased Range - Dispersion Shifted Fibre

In our discussion of single mode fibre above, the range allowed between transmitter and receiver is still limited by the power budget although the dispersion budget allows very high bit rates to be transmitted. Form the attenuation curves, moving to 1550 nm operation is obviously beneficial but, as noted in the discussion on material dispersion, this takes one away from the favourable zero of material dispersion at about 1300 nm wavelength.

However, waveguide analysis of the single mode fibre structure shows that there is another dispersion mechanism present that adds to the material dispersion and that arises from the physical structure of the guide itself. The signs are such that this "waveguide dispersion" terms adds to the material dispersion in such a way as to shift the zero of dispersion towards longer wavelength. Designing a guide specifically to achieve large waveguide dispersion therefore allows the designer to shift the effective zero of dispersion to 1550 nm. Such fibre is known as Dispersion Shifted (Single Mode) Fibre and is referred to as DSF. The typical dispersion curves for such a fibre are shown below in Fig. 10.

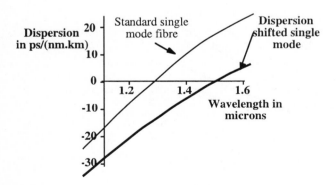

Fig. 10. Typical dispersion curves for standard and dispersion shifted single mode fibres.

The result is a fibre with near ideal transmission properties, low attenuation of circa 0.2 dB/km coupled with zero dispersion both at a wavelength of about 1550 nm. Not surprisingly, this has attracted great interest and usage in very long haul transmission systems such as those that run undersea. Typical operating windows are illustrated for this fibre below in Fig. 11

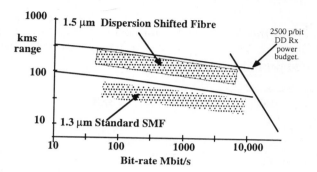

Fig.11. Results of dispersion and power budget calculations for DSF fibre versus standard single mode fibre.

6 Fundamental limits to dispersion

In what we have said so far, we have implied that dispersion is controlled by the source line width interacting with the material dispersion (or in the case of single mode, the overall dispersion) but we have said nothing about what the source line width might be. If we use an LED source, the source line width is typically 30-50 nm which corresponds to some 4000-6000 GHz and thus has nothing whatsoever to do with the modulation being applied. Simple cleaved-cavity semiconductor lasers can easily emit light over a range of 5-10 nm or 700-1400 GHz, similarly unconnected to the data modulation but purely reflecting the overmoded character of the laser cavity.

However, modern Distributed Feedback (DFB) Lasers emit light in a single reasonably stable line with perhaps a natural line width of 100 MHz to 1 GHz so that, once modulated by high speed data, the effective line width can easily be controlled by the data rather than the source. Under these conditions, one show that the maximum bit rate allowed is roughly given by (in practical units for easy evaluation):

$$B_{max}(bit/sec) = \frac{2.9 * 10^{14}}{\lambda(nm)} \sqrt{\frac{1}{L(km)D_{mat}(ps/(nm.km))}}$$

If we plot this result for various values of fibre dispersion and for 1550 nm operation, we obtain the following results shown in Fig. 12.

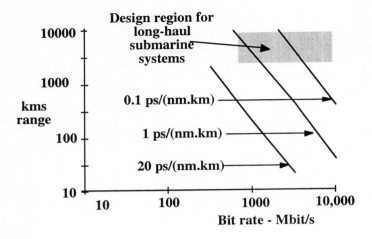

Fig. 12. Ultimate limits set by dispersion

In the short term, this suggests that dispersion should be little problem provided that careful control of the source line width is exercised. However, as we shall see later, the development of fibre Amplifiers now means that the system length can be increased almost indefinitely so far as the power budget is concerned so that distances of 3000-10000 kms have become of real engineering interest (corresponding to Trans-Atlantic and Trans-Pacific types of system). Here it is apparent dispersion will be significant factor and extreme care will have to be exercised to retain data free of distortion.

6.1 Polarisation Mode Dispersion (PMD)

Another linear mechanism that can also lead to pulse broadening arises from the fact that the single mode fibre actually carries two modes of orthogonal polarisation state. When the fibre is perfectly circularly symmetric, the two modes are degenerate but any departure from that leads to splitting of the degeneracy and hence another pulse spreading effect. To give some feel for the scale of the problem, a PMD value of $1ps/\sqrt{km}$ would arise from any one of the following :
  - 0.7% core ellipticity
  - 4 gm.wt./cm lateral stress
  - 3 cm radius bend

It needs little imagination to see that the first two in particular could occur very easily. The results of such PMD limitations are estimated on the curves below in Fig. 13 to give a feel for the levels needed in advanced system design. Once again, we have an effect that can very easily become serious for long haul trans-oceanic systems but is not likely to be serious for short haul inland systems.

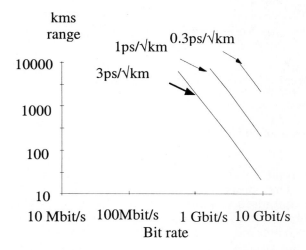

Fig. 13. Implications of various PMD values for systems.

LECTURE II - Amplifiers and Non-Linear effects

7   The Erbium Doped Fibre Amplifier (EDFA)

We saw in Lecture I that the range of a fibre link was limited by the minimum attenuation possible in a (silica) fibre coupled with the practical launch power and receiver sensitivity and that, whilst all of these could be tweaked and optimised, other than operating at the minimum attenuation wavelength, we ran into a wall of rapidly increasingly difficulty somewhere in the 100-200 km range region. The addition of the optical amplifier as a means to boost the optical power in the fibre has largely solved this problem. Early studies were based upon the use of semiconductor laser chips operated below threshold as amplifiers but the development of the EDFA has essentially removed them from contention except in very special cases.

The EDFA is based upon a short length of single mode fibre whose core has been doped with the rare-earth Erbium ion. This can be optically pumped at about 980 nm or 1480 nm to produce gain in a window centred on about 1550 nm. The configuration of such an amplifier is shown schematically below in Fig. 14.

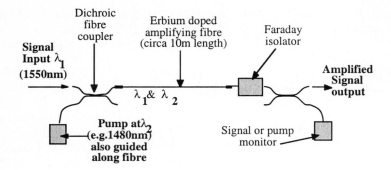

Fig. 14. Schematic layout of an EDFA.

The whole assembly typically is contained within a box the size of a lap-top computer. The pump laser is a specially produced semiconductor laser optimised to emit at the appropriate wavelength but because the absorption lines are fairly broad, this does not have to be controlled with great precision nor does the emission line width. The optical lifetime of the inverted Erbium ion is measured in many milliseconds with the result that this does a superb job in averaging out the rapid intensity fluctuations common in most semiconductor lasers. Gain values of 10-30 dB are obtained so that placing one after every 100km of 0.2dB/km fibre operating with a gain of 20dB provides a loss-free data-pipe section that can be repeated ad-infinitum.

With such high gains possible, great care must be taken to avoid unwanted reflections and to stop the build-up of an unwanted backward propagating wave so a Faraday Isolator is frequently found within the overall package. The inputs (signal + pump) are combined using a dichroic fibre coupler which allows the pump wavelength to cross over and the signal to bypass so that the whole assembly involves single mode splices rather than the lossy and reflective joints found in interfacing fibres to semiconductor lasers.

In practice, however, such links cannot be extended indefinitely since although the power levels can be restored, there is an inevitable slow build up of (optical spontaneous emission) noise that eventually will drown the signal. The noise power at the output of a single amplifier is :

$$P_n = 2 n_{sp} h\nu (G-1) B_{opt}.$$

In this equation, the term n_{sp} is the spontaneous noise factor and is typically $1 < n_{sp} < 4$ for the EDFA but much larger for semiconductor amplifiers, G is the gain and B_{opt} is the full optical bandwidth in Hz. In the case of the EDFA, this is typically 50 nm or 7000 GHz so that unless something is done to restrict it by means of very narrow optical line filters, a great deal of unnecessary noise can be accumulated in transit. However, by careful design, amplified fibre systems are now operating and carrying data in the Gbit/s range over trans-oceanic distance (3,000-10,000 kms).

Notice the implications of this statement for the system power budget. A 10,000 km fibre link of 0.2dB/km fibre has an insertion loss (without amplification) of 2000 dB or a transmission of 10^{-200}. The amplification required is 2000 dB implying some 100 amplifiers spaced by 100 km intervals each providing 20 dB gain. Notice also that any small departure from perfect gain equalisation across the signal band will lead to massive errors overall. For example, a 0.5 dB gain mismatch repeated across 100 amplifiers implies a 50 dB gain error or a signal power that is in error by a factor of 10^5. Such numbers highlight the extreme care that is required in designing such systems both in terms of controlling signal and gain levels very precisely and also in controlling the spectral bandwidth and hence noise build-up along the link. But it is also very important to notice that, since the signals now stay in the optical domain throughout the link rather than just between regenerators or amplifiers, the effective link length for dispersion calculation purposes is now the overall link length, not the amplifier section length. Thus, when we considered the performance of 10,000 km links in our dispersion discussion earlier, there was a good reason for doing so and we see also that the link design for such systems must pay very close attention to the issues of source wavelength, spectral spread and fibre dispersion properties.

8 Wavelength Division Multiplexing (WDM)

One single number about the EDFA brings home the scale of the possibility it might offer. The gain spectrum for an EDFA is shown schematically below in Fig. 15.

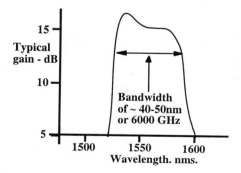

Fig. 15. Schematic gain spectrum for an EDFA.

For comparison purposes, one should note that the entire electrical spectrum from DC to millimetre wave embraces about 300 GHz yet within each amplified fibre, there is perhaps 10-20 times that much available for exploitation. This has stimulated great interest and discussion on how it can be exploited.

The first point to note is that with electronic circuits running into increasing difficulty in operating in excess of perhaps 10-20 Gbit/s, it seems fanciful to imagine filling this space by electronically multiplexing data to ever higher rates, particularly into the multi-Terabit range. Increasingly, the favoured route is to use multiple optical carriers each of which will be modulated at perhaps 10 Gbit/s. This is known as Wavelength Division Multiplexing (WDM). Whilst there is still debate on exactly how far this approach can be extended, there seems to be a convergence on 100 GHz (optical) as the carrier spacing for the first WDM and already systems with 16 carriers are carrying traffic. The optical spectral spreads this gives rise to are shown below in Table 5.

No Carriers	Carrier Separation 100 GHz	Optical Spread (nms)
8	100	5.25
16	100	11.25
32	100	23.25
64	100	47.25

Table 5. Total spectral spread for different WDM systems.

The precise choice of carrier spacing involves many factors. An obvious one is that any one carrier plus its modulation must not spectrally overlap that of an adjoining carrier and that a sufficient spectral window must be left between them to allow one to be filtered from another. But we shall see shortly that other more complex factors enter the design process involving the control of non-linear cross talk between carriers in the same fibre.

Notice that the technology above implicitly assumes that optical sources and filters of high stability are available. A stability of carrier frequency of 1 GHz centred at a wavelength of 1500 nm implies frequency stability of 1:200,000 which in turn means exceptional control of the key components.

9 Non-Linear Effects in Fibre

As interest in exploiting WDM has grown, so also has the realisation that the design optimisation of fibre systems to suppress non-linear interactions will be a key factor in their success. Some simple numbers illustrate the problem. Consider 50 carriers each with 1mW of power in a single mode fibre having a typical core area of 50 square-microns. The power density in the core is 100 kW/cm^2, far more than enough to melt fire-brick! Yet this power density could be maintained over the best part of 10,000 kms or 10^{13} wavelengths so that only a very small non-linearity could be expected to be case for concern. In practice, all materials are known to suffer from the Intensity Dependent Refractive Effect whereby we find that the refractive index is given by an expression of the form :

$$n_{total} = n_0 + n_2 E^2$$

Here the term n_0 is the normal linear refractive index and the term n_2 is the intensity dependent refractive index (IDRI) and is coupled with the optical electrical field E. The second term is small but in our situation, where E can be large, the result is significant.

The IDRI arises from the Kerr non-linearity or third order polarisability in the material. This is written as :

$$P_i(\omega_1) = \varepsilon_0 \chi_{ijkl}(\omega_1, \omega_2, \omega_3, \omega_4) E_j(\omega_2) E_k(\omega_3) E_l(\omega_4)$$

where P is dipole polarisation in the material at frequency 1, the term χ is the 3rd order polarisability tensor and the three terms in E represent the electric fields of up to three different driving waves. In this more general form, the frequencies are related by the relationship :

$$\omega_1 = \omega_2 \pm \omega_3 \pm \omega_4$$

out of which a number of different interactions are possible :

i. $\omega_1 = \omega_2 + \omega_3 + \omega_4$ or $3\omega_1 = \omega_1 + \omega_1 + \omega_1$

This is sum frequency generation or 3rd harmonic generation if all the driving waves are the same frequency. This is not significant in fibres.

ii. $\omega_1 = \omega_2 - \omega_3 + \omega_4$

Here, if the driving waves are closely spaced as in a WDM system, the new frequency 1 can coincide with or be close to them. This is known as Four Wave Mixing (FWM) in fibres.

iii. $\omega_1 = \omega_1 - \omega_1 + \omega_1$

This is the classic IDRI effect in which a single driving wave produces a polarisation at the same frequency which shows up as a modified refractive index. In fibres, this effect is known as Self Phase Modulation (SPM)

iv. $\omega_1 = \omega_2 - \omega_2 + \omega_1$

This is a very similar case except that two waves interact to generate a polarisation at the frequency of one of them proportional to the intensity of the other. This effect is know as Cross Phase Modulation, (CPM).
We will now examine each of these in a little more detail.

9.1 Self Phase Modulation (SPM)

This effect describes the effect of the IDRI being excited by a wave that modulates itself. The simple IDRI expression suffices so that we can write :

$$n_{total} = n_0 + n_2 E^2$$

which, if the electric field were constant, would simply lead to change in the phase length of the fibre. However, because the wave is modulated, E changes with time and hence we can write :

$$\delta v = -(Ln(2)/\lambda)\frac{d(E^2)}{dt}$$

A frequency shift is induced in the wave proportional to the rate of change of intensity. Schematically, we can represent this as shown below in Fig. 16.

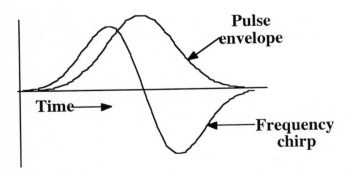

Fig. 16. Chirp induced by IDRI on the carrier frequency of a pulse.

The carrier wave underlying the pulse envelope is chirped as shown. This change in carrier frequency during the pulse then interacts with the linear dispersion of the fibre to change the pulse shape. This can lead to a variety of effects ranging from pulse compression through exact cancellation of the linear dispersion (soliton propagation) to excess pulse spreading or pulse break up and the formation of chaotic pulse streams.

The design of links to minimise this effect concentrates on two criteria :
- minimising the power level in the fibre (subject to Signal/Noise ratio considerations)
- maintaining low overall dispersion in the link as well as low dispersion along the link to minimise the effect the induced chirp has on the pulse shape.

9.2 Cross Phase Modulation

Here the requisite expression for the refractive index in the fibre is given by :

$$n_{total} = n_0 + n_2 \left[\left| E^2(\omega_m) \right| + 2 \sum_{n \neq m}^{N} \left| E^2(\omega_n) \right| \right]$$

where there are N carriers and we focus on their effect on carrier m. The expression includes the term of the SPM but also the new CPM terms via the summation. Now, in addition to SPM, the carrier is phase modulated and hence chirped by each of the other carriers present. Noting that N-1 might be large and there is a factor of 2 present in the CPM term, we might reasonably deduce this effect is of more concern, to the tune of 2(N-1). In practice this is true although there is a simple way of reducing CPM. By designing the link so that it has HIGH DISPERSION, the carriers involved slide past each other over short distances compared to the distance required for the chirp to change a given pulse shape. In that case, following Any one point on our carrier m and its data stream, any other pulse stream is sliding past and respectively chirping that section of carrier up and down again as it passes, leaving the original largely unchanged.

Evidently to design this properly requires proper modelling of the non-linear interaction which, specifically, involves solving the Non-Linear Schrödinger equation for the full set of coupled waves. In general, this can only be done numerically.

However, our discussion has highlighted an apparent non-sense in that we seek a link specification that offers simultaneously both high and low dispersion. This apparent impossibility can be achieved in practice once careful attention is given to the relevant interaction lengths involved. For example, during the 1/e distance during which the power decays in the fibre, one might ensure that high dispersion operates so that carriers slide past each other in terms of a few bit intervals to cancel the CPM effect whereas on the distances associated with the power restoration, one might aim to achieve low dispersion. Using special fibres this is possible. Two such link designs are shown below.

In the first, standard fibre having a zero of dispersion at 1300 nm is used at 1550 nm and hence shows a high dispersion of order 20 ps/(nm.km). However, before each amplifier is inserted, a length of fibre having a higher dispersion of the opposite sign is inserted, typically say of -100 ps/(nm.km). Such fibres are known as Dispersion Compensating fibres (DCF). The resultant dispersion map is shown below (Fig. 17) and implies that during propagation, pulses will stretch out and then recompress as they traverse each amplifier section.

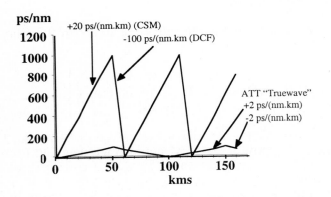

Fig. 17. Dispersion compensation using different fibre designs.

But more importantly, during this process, adjacent carriers will be racing past each other. Noting that 100 GHz carrier separation corresponds to 0.75 nm, then with 20 ps/(nm.km), adjacent carriers will slide past each other at a rate of 15 ps/km or over a 1/e length of around 20 km by 300 ps. or 3 bit intervals at 10 Gbit/s.

Also shown on the plot is another fibre design known as the Truewave fibre in which alternative amplifier sections are constructed of the lengths of fibre have the same numerical dispersion (modulus) but of alternating sign. Typically, fibre of +/- 2ps/(nm.km) is used which, on a 50 km amplifier section implies as slide distance of $0.75 \times 2 \times 50 = 75$ ps for adjoining carriers. This is claimed to be sufficient by ATT for high bit rate (10Gbit/s+ systems). Note here that alternative sections will see the relative sign of d/dt change as the adjoining carrier slides alternatively forwards and backwards past the same point on the observed pulse.

We will note also in passing that the large dispersion criterion need to minimise CPM also minimise Four Wave Mixing (FWM) even more effectively so that the same design criteria largely meet both situations although in detail the mechanism for suppression is quite different and is controlled by lack of phase-matching in the travelling wave interaction.

Finally, we should note that the perfect dispersion compensation shown in the figure above can only occur at a single carrier wavelength in practice because of higher order dispersion terms. In reality, a dispersion compensation plot for several wavelengths look like that shown below in Fig. 18, further complicating the design process.

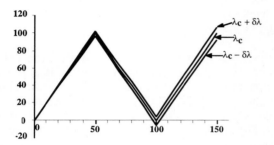

Fig.18. Dispersion compensation map for several wavelength channels

9.3 Soliton Propagation

Soliton propagation is the special case of SPM where the self induced chirp interacts with the linear dispersion in such a way as to exactly cancel the ordinary dispersion that arises from the modulation line width with the result that the pulse propagates without change in shape. This can only happen when exactly the right shape pulse is present for the linear dispersion value operating. When one solves the Non-Linear Schrödinger Equation, one finds that the optimum pulse shape is that of a sech(t/τ) pulse in which the peak pulse power is linked to the FWHM pulse width as follows :

$$P_{peak} = 1340 \, D / T^2 \, mW$$

where :
 D is the dispersion in ps/(nm.km)
 T is the FWHM pulse width in picoseconds.

If we insert numbers into this expression, then we find the following for example :

D = 0.1
T = 10 ps
P = 1.34 mW

However, the above numbers assume that there is no attenuation in the fibre. When attenuation is present, it is found that by launching pulses with roughly twice this power, soliton propagation still pertains. Such solitons are described as "Average Solitons". A further condition for this to work is that the amplifiers in the system must be closely spaced compared to the "soliton distance" which is given by :

$Z_{period} = 0.42\ T^2/D$ kms

or

$Z_{period} = 562.8\ /P_{peak}$

Another characteristics of solitons is that they attract each other through the same non-linear effect. In general, to prevent soliton pulses coalescing, they must be spaced by 7-10 times their pulse width. Thus, for 10 Gbit/s data streams, pulse of only 10-15 ps duration are required. Summarising the above we see that as the bit rate increases, the peak power must increase (as $1/T^2$) and the amplifier spacing must decrease as T^2. Together, these fundamental relations seems to impose a rather tough practical constraint upon the bit rate per carrier in a soliton system in the region of 10 Gbit/s.

9.4 Noise in soliton systems

Just as with a linear system, in a soliton system with amplifiers, noise will build up from the spontaneous emission of the amplifiers. This leads to a new effect known as "Gordon Haus" jitter after its discovers. The effect of the growing noise signal interacting with the soliton pulses is similar to that of the attraction of two pulses for each other. The randomly varying noise power pulls the soliton and introduces a random jitter in its time position within the pulse stream. In time, this can lead to break up of the data stream and the distances over which this is shown to occur are typical of those for undersea transoceanic systems. However, there is a very clever fix for this problem known as the sliding filter.

We have already made clear that in an amplified system, optical matched filters are required, matched to the signal spectrum, to suppress noise everywhere except in the signal channel where nothing can be done about it. However, in the soliton system, one can do much better. By misaligning successive filters by a small amount, typically by an average amount of 10 MHz/km, the signal spectrum associated with the soliton pulse is distorted but regenerated by the non-linear interaction to follow the filter spectrum whereas the noise spectrum, in the absence of

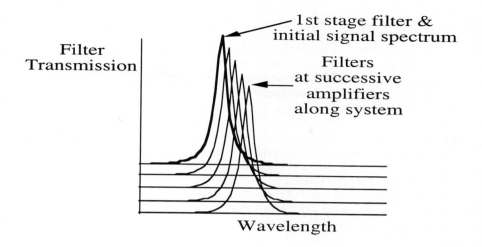

Fig. 19. Schematic of sliding filter spectra on a long link.

the non-linear dragging, cannot follow and is blocked. Note that on a 10,000km system, this slide rate corresponds to a total slide of 1000 GHz so that for linear signals, the system is completely non-transparent from end to end yet for the non-linearly dragged soliton, it offers data transparency. The sliding filter concept is illustrated above in Fig. 19.

9.5 Power stabilisation in Soliton Systems

The non-linear interaction also offers another intriguing possibility. As we have commented already, in long amplified systems power stabilisation is important but in soliton systems, it takes on an additional meaning because of the constraints of the soliton pulse. However, if we look back to the formula for the peak power of the soliton, we recall that :

$$P_{peak} = 1340 \, D / T^2 \, mW$$

so that the pulse width and peak power are intimately linked. Assume then a long transmission system with filters after each amplifier matched to the desired signal spectrum (linked with the value of T). Assume further that the gain is too large. We can now envisage the following sequence of events :

- gain too large
- pulse peak power too large
- pulse shortens or compresses to make T smaller
- pulse spectrum broadens as T shortens
- spectral width increases with respect to filter window
- power lost in filter increases
- peak power decreases

Evidently the possibility exists of a passive power stabilisation by exploiting the soliton effect and this also has been demonstrated. To date, soliton propagation using sliding filter noise reduction has been demonstrated in the laboratory over distances of order 1Mkm at a data rate equivalent to about 10Gbit/s by allowing pulses to make multiple rotations round a long fibre loop with amplifiers etc. No soliton system appears to be in use as yet since linear WDM systems have so far proved capable of meeting the need.

10 Stimulated Brillouin Scattering (SBS)

A separate non-linear scattering effect involves the scattering of light from longitudinal acoustic waves in the fibre. This is seen as backward scattering shifted in frequency to longer wavelength by about 10-12 GHz. The scattering line width is about 20 MHz and the threshold for the onset of stimulated scattering is given by :

$$P_B^{th} = \frac{42 A_e}{g_B L_e}\left(1 + \frac{\delta v_s}{\delta v_B}\right)$$

Here the subscripts s and B refer to the line widths of the source and the Brillouin scattering, A_e is the effective area of the fibre core, L_e is the effective length of interaction (typically 15–20km) and g_B is the Brillouin gain cross section and takes the value of about 4×10^{-9} cm/Watt.

Experimental studies using Dispersion Shifted Fibre Aand a very narrow spectral line width source at a wavelength in the region of 1500 nm showed the following results (Fig. 20)

We see that at low input powers, the transmission is linear and the output power rises linearly with input power. However, as the threshold for SBS is reached, the output power stabilises and further increases in input power are simply reflected back to the source. This threshold occurs in the region of 4 mW.

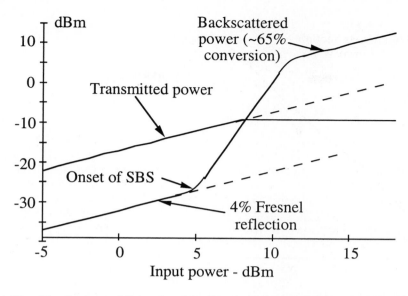

Fig. 20. Impact of Stimulated Brillouin Scattering on the transmitted power in a fibre as a function of launch power

Returning to the formula for the threshold power, we see that the value depends upon the ratio of the source line width to the Brillouin line width (20MHz) so that using a source with a natural line width of 200 MHz could be expected to increase the threshold to 40mW. In practice, most sources are broader than 20 MHz And this allows launch powers of perhaps 10-20 mW to be used. Special measures can also be taken to artificially broaden the source spectrum or otherwise suppress SBS such as to phase modulate the source carrier.

11 Stimulated Raman Scattering

This is a similar form of scattering but from transverse phonon vibrations in the glass and at very much higher frequencies. The gain cross section as a function of frequency shift is shown approximately below in Fig. 21.

It is immediately obvious that scattering from any carrier within the EDFA will scatter power throughout the window and/or produce gain at every other longer wavelength within the window. The result is that carriers become coupled and the potential exists for another form of

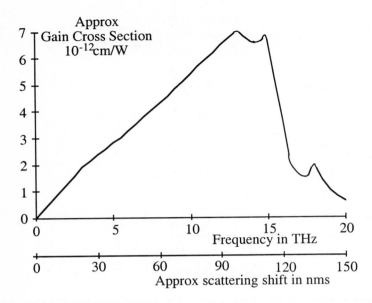

Fig.21. Stimulated Raman Gain Cross section versus frequency shift from the exciting carrier.

cross talk between data streams on different carriers in a WDM spectrum.

The equations describing the power coupling between two waves are given below :

$$\frac{dP_1}{dz} = -\left(\frac{\lambda_2}{\lambda_1}\right)\frac{\gamma}{2A_{eff}} P_1 P_2 - \alpha_1 P_1$$

$$\frac{dP_2}{dz} = \frac{\gamma}{2A_{eff}} P_1 P_2 - \alpha_2 P_2$$

where wave 1 is the higher frequency, shorter wavelength one. The gain coefficient γ is taken from the diagram above for the appropriate carrier separation while the alpha refer to the attenuation at each carrier frequency. The result of this is that pattern dependent cross talk occurs since the effect is only present when power is present in both waves (e.g. 1 pulses simultaneously on both carrier 1 & 2). This is shown schematically below in Fig. 22.

As with the CPM effect, a partial solution is to invoke high local dispersion so that pulses slide past each other. However, now we are

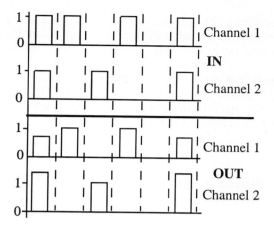

Fig. 22. Schematic representation of the effect of SRS crosstalk on a two carrier data stream.

concerned with a power transfer directly rather than a phase modulation that leads pulse distortion, so that this does nothing for the mean power transfer but it does tend to average out the pattern dependence of the transfer. Thus a power tilt occurs, with the waves at longer wavelength gaining power at the expense of those at shorter wavelengths and this can be corrected in principle by adjustment of the EDFA effective gain spectra. What cannot be corrected is the residual fluctuation form the randomness of the data patterns, which leads to a mean additional noise

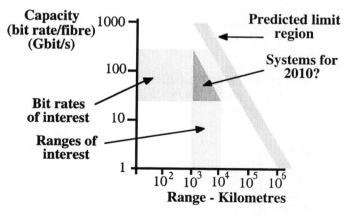

Fig. 23. Probable limits set by SRS crosstalk on long-haul amplified fibre systems.

fluctuation on all the carriers. There appears to be no way to avoid this at present and it is thought to set an ultimate limit to the transport capacity of a non-soliton system. This limit is shown above in Fig. 23.

Note that for the Transoceanic system operating over 3000-10000 km range it points to a limiting data rate in the region of 30-300 Gbit/s per fibre range. However, we may assume that the jury is still out on whether this is or is not a fundamental limit.

LECTURE III - Photonic Switching and All-Optical Networks

12 Background to switching

The simplest interpretations of the term switching centre around one or other of two operations. The first is used to describe one or other of the set of Boolean logical processes used in implementing digital logic e.g. Fig. 24

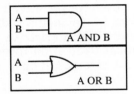

Fig.24. Switching in the Boolean logical sense.

Such basic operations can be cascaded in clocked digital processors to route data. For example, by combining a clocked routing signal at one port with the signal at the other of one AND and one NAND (NotAND) gates, a continuous series routing ONEs will send the signal in one

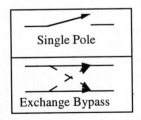

Fig.25. Analogue switching.

direction and continuous serious of ZEROs will send it the other. The same operation can be done in analogue switching using the devices shown below in Fig. 25.

Notice a crucial difference in the operation of two approaches. In the logical processor, it is fundamentally assumed that the data is in digital form and will be clocked at a precisely determined clock rate through the switch. The control (routing) data is indistinguishable from the message data, both being binary digital data streams and the route in the digital version can change on a bit by bit basis if required. In the case of the analogue switch, the control signal is completely in the correct position before any message data starts to flow.

This may seem pedantic, but it has a fundamental impact on how optics can be used in communications systems which have normally been designed to be switched using digital electronic logic and are thus of the former type. Whilst both technologies exist in optics, it is really only the latter that is competitive in any way yet it requires message data to contain time gaps suitably placed to allow switch resetting.

Note that real switches are not normally concerned with a single input port to be connected to one of two output ports as shown above but with many input ports to be connected to many outputs. Thus the switches above should be seen a building blocks to be assembled into much more complex arrays before they can perform a useful function.

13 Multiplexing

A further key issue to understand before one discuss switching is the nature of the multiplexing used. Fibres offer such large capacity that it is very rare for a single source-destination pair to be served exclusively by a single fibre with the result that many different tributary streams are normally multiplexed onto the single fibre in order to fill it. At a switching node, the tributary streams must be separated and regrouped for onward transmission towards their differing destinations. The method whereby they are grouped is of critical importance.

13.1 Time Division Multiplexing (TDM)

This is the normal format used in telecommunications for telephony traffic. The discrete telephone conversations each generate 8000 8-bit samples per second. The transmission system capacity is split into 8000 frames per second, each of 125 microseconds duration, and one 8-bit byte from each source conversation is placed in a preassigned position in each successive frame. Thus byte slot 23 is always the same conversation for the duration of the conversation. To switch such data at a node at the circuit (or conversation) level, it is necessary to break out the individual bytes, re-route them, time shift them and re-assemble them into the outgoing frame from the node. In a 10 Gbit/s system, one byte lasts 0.8 nanoseconds and within a 125 microsecond frame there would be 156000 separate bytes, each to be routed differently. Electronically, this is done by breaking the data flow down to many parallel more modest data-rate flows that can be handled in digital electronics. In optics, no really comparable operation exists, besides which handling a concentrated control flow for 156000 instructions every 125 microseconds would be challenging!

13.2 Time Division Multiple Access (TDMA)

Here multiple sources and destinations time share a single fibre highway but they do it by taking turns to monopolise its full capacity for more extended periods. Thus a single source might dump a 1 Gbyte file in a single operation before relinquishing the fibre to another user.

13.3 Packet Data

Packet Data is the format commonly used by computers iover the Internet as well as the physical letter-post! A bundle of message data is parcelled into a serial stream, a destination address fixed to its front and a marker to signal its end fixed to its tail. The resulting "packet" is then "posted" into the electronic network. At each node, the desired destination must be read, the packet placed into a buffer memory, the routing onward worked out, a suitable empty slot found to accomodate it and at just the right moment, the packet removed from buffer memory and posted onwards again. This format is clearly designed to exploit digital electronic logic. Sadly, as a result, it hits at all the greatest weaknesses of optics which is not noted for the excellence of its buffer memory or complex logic processing.

13.4 Asynchronous Transfer Mode (ATM)

ATM is a much favoured development now being widely deployed in telecommunications networks. It offers packet like transmission for both voice and data and mainly differs from earlier packet formats in having a shorter and fixed packet length (48 bytes of message + 5 bytes of header) and using preassigned pathways to minimise the node processing problems. The system receives advance warning that a user will be wishing to transmit ATM cells to a specific destination and allocates a short code for that complete route for the duration of that user's requirement. The header then only needs to carry that short code to identify its destination and the node processor only needs a relatively short look-up table to identify what to do with a given cell.

13.5 Frequency Division Multiplexing (FDM)

Used extensively in microwave and radio communications, different messages are modulated onto different frequency sub-carriers which in turn are then modulated onto a main carrier which might be optical. This has been used in some optical systems but again seems to imply full electrical demultiplexing before any form of regrouping can be attempted. Thus it offers little attraction for optical implementation.

13.6 Wavelength Division Multiplexing (WDM)

This is the closest equivalent in the optical domain to the use of different carriers in the radio spectrum. Each optical carrier is separately modulated with a high data rate signal (1-10 Gbit/s). At the node, complete carriers with their data intact are separated spectrally and re-routed. This is therefore switching only very large intact blocks of multiplexed data with no attempt to perform a finer grained operation. At present, it seems likely that this is one of the few forms of optical switching that will be used in real networks.

13.7 Space switching

Here all the signal(s) on a single bearer (fibre) are reconnected by a space switch to another bearer which carries only the single group of signals. However, with an N×N space switch, N input fibres can be connected in N! different ways to N output fibres using this approach. It requires no knowledge of the multiplexing adopted in any fibre but it does require that no transmission takes place while the connections are being established. This typer of switching is used for protection switching in telecommunications networks to redirect data from a broken fibre to an intact one.

14 Switch control and timing.

Factors frequently overlooked in discussions of optical switching are those of control and timing, both of which immensely complicate the problems of building real switches. In a complex network, one must not only identify a block of data but associate with it the correct desired destination, the correct route to get there and establish that route at the right point in time to allow the data to reach it. The difficulty of doing this escalates dramatically as one attempts to switch small blocks of TDM or Packet/ATM data at optical line rates. This is a major reason why the routing of complete WDM carriers or simple space switching between fibres look increasingly likely to be the first widely deployed optical switching technologies since the routing for such large blocks of data typically changes only slowly. In fact, such switching is done primarily for network configuration or protection purposes, thus bringing appropriate capacity on stream prior to largely predictable increases in traffic demand or to reroute large blocks of data following some equipment failure. Such applications fully exploit the optical capability to handle very large traffic flows and whilst side-stepping its extreme difficulty in carrying out very fast complex logical operations.

15 Switching Matrices and Blocking

We commented above that complex switches are constructed using arrays of simpler devices. A good example is the cross-bar switch shown schematically below as a two dimensional array of elements connecting 4 inputs to 4 outputs and with one of the analogue type switches above at each I/O line intersection (see Fig. 26)

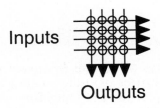

Fig.26. Schematic of a cross-bar switching array.

Such elements in optics are normally made using guided wave exchange-bypass four port switches and hence in reality look more like Fig. 27.

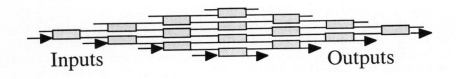

Fig. 27. Schematic layout of guided-wave crossbar switch.

Notice that in such a switch, only a single exchange-bypass needs to be set in order to make any desired input-output connection, assuming of course that neither line is already in use. In the off state, all need switches in the above matrix to be in the exchange state and making a single connection involves resetting one to bypass. However, an obvious problem with such switches is that they scale in complexity as the square of the number of ports and for large switches, this rapidly becomes a problem. Moreover, of the N^2 switches present, at best when all input ports are excited, only N are actually set, implying very inefficient use of resource (e.g. at 1/N level).

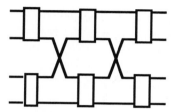

Fig.28. 4×4 Benes network.

As a result, many switch matrices are of different form. An example, the Benes network is shown below in 4×4 format (Fig. 28). It uses only 6 switches to perform the same function as 16 switches in the 4×4 cross bar. Note also that for such a switch, there are 4! = 24 possible connection patterns. With 6 switches, each having 2 possible states, there are 2^6 = 64 possible combinations of switch setting so this is still making less than 100% use of the hardware. However, the penalty of such a more efficient structure is that to establish a new set of pathways, many existing pathways will also have to be re-routed. Such a switch is therefore known as re-arrangeably non-blocking.

Some other switches are scaled down to a size where it is simply not possible to support all possible I/O connection patterns and these are known as fundamentally blocking or conditionally non-blocking, conditionally since subject to certain conditions (such as only half the inputs being in use) they are not blocking.

We that rather long pre-amble completed, let us now consider some optics again.

16. Digital Optical Logic

16.1 Free space "wiring".

During the 1980s and early 1990s, there was much discussion about Digital Optical Computing and. following from that Digital Optical

Switching often also known as Photonic Switching. The thesis generally adopted was that, by using simple imaging optics, a very large number of parallel data (ray) paths can be established from one plane to another so that if the planes were composed of two dimensional arrays of optically activated logic gates, powerful computing structures might become possible (see Fig. 29 for an example).

Computation of course requires the bringing together of data from different sources, making logical comparisons and then dispersing into yet different groupings. Thus a simple imaging interconnect as shown above does not suffice and much work has been done developing more complex interconnection patterns to that end. One very simple example is shown below in the form of a Perfect Shuffle Network (Fig. 30)

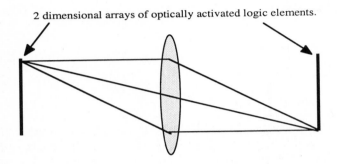

Fig. 29. A simple parallel "free space imaging" wiring harness.

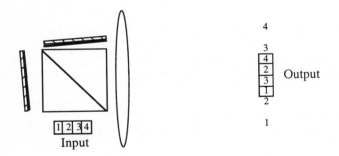

Fig.30. A free-space imaging implementation of a `Perfect Shuffle Network'.

One sees a linear array, magnified by a factor of two and, by means of the beam splitting prism and suitable aligned mirrors, with two such images overlaid on each other so that the upper half of one image exactly interlaces the lower half of the other. The interconnection pattern so obtained is known as a Perfect Shuffle since it corresponds to the reordering obtained when a pack of cards is plit in two and two halves are then shuffled or interleaved. (e.g. 1,2,3,4 becomes 1,3,2,4). It is one example of a connection pattern widely used in switching matrices and paralle processors that can be copied using imaging optics which are suitable for very regular array interconnections used in large array processors whilst being totally unsuited to the more typical connections found on the average printed circuit board.

16.2 SEEDs

A wide variety of different devices has been proposed for use in the optically triggered logic planes. One that has atttracted more interest than most is the Multiple Quantum Well (MQW) Self Electro-Optic Device (SEED).

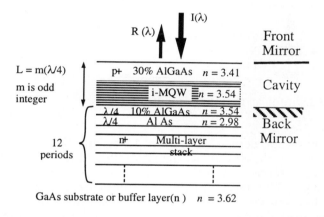

Fig.31. The structure of one particular member of the SEED family of devices.

It consists of a PIN diode, usually grown in the GaAs-AlGaAs system, in which the I region is formed from an MQW stack of layers. One such device structure is shown above in Fig. 31.

In such a structure under zero reverse bias, a strong exciton absorption is observed at the band edge of the material. As the reverse bias is increased, the exciton absorption moves to longer wavelengths, broadens and becomes less intense as shown schematically below in Fig. 32. This provides the basis for a bistable logic device. Consider such a device connected as shown below (Fig. 33).

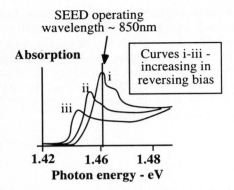

Fig.32. Absorption spectra for SEED devices versus bias voltage.

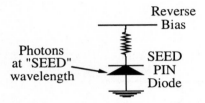

Fig. 33. The SEED device in an operating circuit.

When no light falls on the device, no current flows and the full bias voltage is dropped across the SEED and the absorption is low. As light is shone on the device, some current flows which in turn decreases the bias across the device. This increases the absorption which increases still further the photo-current. Under the right, conditions, positive feedback sets in and the device switches. A schematic curve showing the output power versus the input power is shown below in Fig. 34.

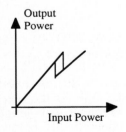

Fig. 34. Schematic input/output diagram for a SEED device in transmission showing a bistable "logic" region.

Turning this device into a workable digital logic element is more challenging when large arrays are required. The bistability threshold for the above device clearly depends upon the optical power level of the input beam and this is always difficult to control precisely.

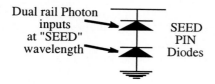

Fig. 35. The Symmetric SEED (S-SEED)

The contrast ratio of the ON and OFF states is poor and is susceptible to change with temperature etc. A clever solution to some of these problems was to introduce the Symmetric SEED (S-SEED) shown above (Fig. 35).

Two of the basic devices are connected in series between ground and bias voltage rail and they are now illuminated by two separate optical beams. The key factor here is that in such a device, switching is not initiated by the absolute level of the input power but by the ratio of the input powers. In an optical system, this is much more susceptible to control.

Using the S-SEED approach, large arrays (1000's of devices) have been constructed and complete prototype logical processors have been built around them using parallel imaging optical "wiring". However, it must be said that the technology appears to have no prospect of being used in anger for a number of rather fundamental reasons.

Self evidently, these devices are electronic in their operation albeit they are triggered optically. For light to be imaged successfully into such devices, their size must be large or comparable to the optical wavelength. This means that they are large by modern electronic standards, have large capacitance and need large amounts of charge to switch their state. As a result, they are power hungry compared modern digital electronic equivalent devices.

Another problem concerns the imaging optical wiring. Whilst it can be formed using special optical componentry, it is very cumbersome compared to the equivalent PCB approaches and hardly compatible one with the other. This is a problem since, as we noted earlier, in clocked digital switches it is axiomatic that data is accurately time-aligned (by bit-interval) and that the appropriate routing data enters in perfect synchonism also. In the parallel optical switch, it is assumed that data flows will be in byte parallel rather than serial form so that serial incoming data from a transmission system must be demultiplexed, converted from serial to parallel format, marshalled in time and launched into the optical parallel processor. The electronics necessary to do this is proibably as complex as the switch matrix itself, so the case for introducung an alaien technology fades still further. Finally, one must note that the selling point that the optics is massively-parallel and gains competitive power as a result never really stood up to critical examination since electronic systems were already adopting parallelism as way on increasing data flow. The overall conclusion seems to be that parallel digital optical logic systems have failed in their attempt to compete with complex electonic digital logic.

Remnants of this programme can still be found in studies of optical interconnects using free-space "wiring" for chip-to-chip and backplane application although whether anything will finally emerge that is really competive remains to be seen. In all these cases, the theoretical advantages of "simple" parallel imaging optical interconnections prove extremely difficult to realise in real electronic system environments since they are not usually assembled on massive, very stable air-bearing optical tables!

17 Optical Planar Space Switches

In 1969, Stewart Miller published a paper outlining ideas for a new Integrated Optics technology, so named by analogy to Integrated Electronics. The proposed the use of single mode waveguide structures formed in or on the surface of planar substrates by epitaxy or diffusion to construct optical analogues of many of the components already existing at microwave frequencies.

One of the components that has been studied extensisvely is the waveguide directional coupler shown schematically below (Fig. 36)

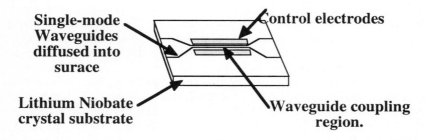

Fig.36. Schematic layout of a planar electro-optic directional coupler electrically operated switch.

Power entering one waveguide couples by means of the evanescent field overlap to the other guide once the region is entered where the two guides are close together. The degree of coupling is sensitively dependent upon the propagation constants for the modes in the two guides which in turn can be changed using the electro-optic effect (in the Lithium Niobate substrate) and a (DC) electric field applied using the surface electrodes. Thus by changing the voltage across the electrodes, the optical guided-wave power can be switched from one output port to another. Attaching such devices with fibre tails leads to simple fibre-compatible exchange-bypass switches. However, note that because the devices are planar they lack the perfect circular symmetry of the fibre and as a result, the two light polarisations are treated separately and tend to switch at different voltages. Making devices that are polarisation independent is difficult but possible.

It is the nature of these devices that they tend to be long (order mm.) and very narrow (order 10 µm). Building an array switch, say the cross bar shown earlier, thus leads to very long thin structures and the complexity that can be packaged onto a single chip tends to be limited by the number that fitted into the substrate length. As a result, only small switch arrays have been made, typically 8×8 port.

Similar size switching arrays can be constructed in semiconductor materials using similar principles or by using semiconductor laser amplifier arrays as ON-OFF switches with passive power splitters. The latter is particularly attractive since it allows power restoration as well as just ON-OFF switching but, as with all gain processes, carries a noise penalty. In this format, a simple 2×2 exchange-bypass switch now takes the form shown below(Fig. 37).

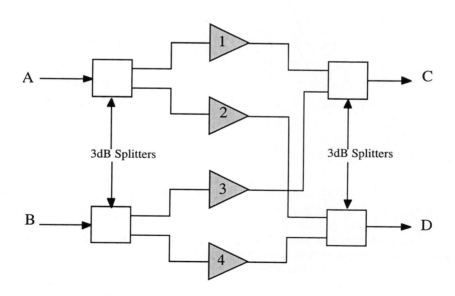

Fig.37. Directional coupler/splitter formed using semiconductor laser amplifiers.

Establishing the exchange connection pattern requires activation of Amplifiers 2 and 3 to connect Ports A to D and B to C. But notice that this structure also has an additional capability, namely that activating amplifiers 1 & 2 connects Port A to both outputs generating a "broadcast" operation which is useful in many system situations, such as CATV.

Notice also that with two 3dB splitters in every pathway, the minimum insertion loss for this structure is 6 dB which must be compensated for by a gain of 6 dB in each amplifier. Clearly, larger networks can be built using this element as a building block so that the 4×4 Benes Network shown earlier using 24 amplifiers.

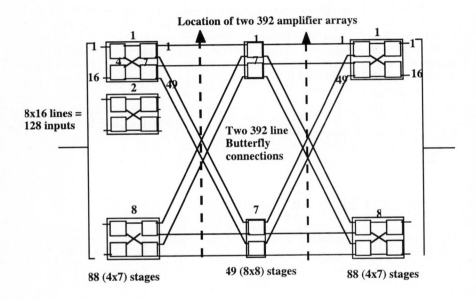

Fig. 38. Schematic layout of 128×128 guided wave optical switch.

Large space switches have been built using a variety of different approaches. For example, a group at NEC described a 128×128 space switch with the layout shown above (Fig. 38). The diagram both illustrates what has been proven possible but also the incredible complexity that emerges as such structures grow. The component count for the above switch is reavealing (Table 6)

Active component count
176 of the 4x7 switch matrix modules
49 of the 8x8 switch matrix modules
Each 4x7 stages contains 28 directional couplers
Each 8x8 stage contains 64 directional couplers

Total directional coupler count = 8064
Number of laser amplifiers = 784

Passive (single mode) connection count
Total number of guided wave connects = 4288

Optical insertion loss without amplifiers = 48dB

Table 6. Component count for the 128×128 switch of Fig. 38.

The problems of electrically wiring and controlling such a switching array give one pause for thought. 8064 directional couplers, plus 784 implying a minimum of nearly 9000 wires assuming that common ground connections can be used.

18 Optical Time Switches

Since most telecommunication switching today is done in the time domain, there has been a strong interest in implementing time switching in the optical domain. However, the digital time multiplex and associated time switching technology was developed around the characteristics of clocked digital electronic logic circuits and buffer memories and optics has no competitive equivalent technology at present.

However, much interest has centred on the use of fibre delay lines as analogue memories and one finds a variety of different approaches in demonstrators. For example, a delay line can be used to allow time for electronics to read and process a header signal on a packet prior to setting a switch position (Fig. 39)

The time delay is fixed and it is assumed in such a system that packets arrive with sufficiently large gaps between them to allow the switch state to be changed.

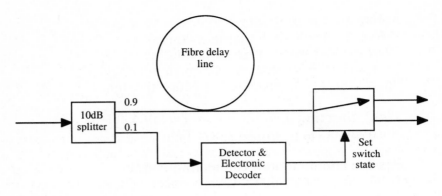

Fig. 39. Use of fibre delay line to deliver "thinking time" for electronic control.

The more complex situation of time slot interchange switching in which the order of bytes in a time sequenced signal has to be changed involves a more complex approach, such as that shown below (Fig. 40). In this sub-system, a number of new elements have been added. Access to the fibre delay lines is via switched exchange-bypass couplers making it possible for a packet to be injected into the ring-fibre and then to recirculate for several times. This is aided by the addition of a EDFA to maintain the power level in the ring. The Semiconductor Optical Amplifiers (SOAs) act as ON-OFF switches. The overall switch operation is then as follows.

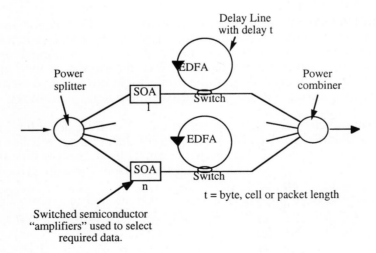

Fig. 40. A schematic optical time switch using recirculating loop delay lines.

The incoming stream of fixed length packets or bytes with guard bands between them is assumed to have been pre-identifed as to destination so the control is already in receipt of the requisite route information.

The data is broadcast to all the SOAs, one of which is activated to pass the data forward to the ring switch which is set to the exchange state. The data enters the ring after which the ring-switch reverts to bypass state, allowing the data to recirculate. After the desired delay, the ring switch is again switched to exchange, dropping the data out of the ring and passing to the output power splitter and hence to the output port. Using this approach a data stream containing N sub-unit messages can be resequenced using N delay lines. However, as N becomes large, the splitter losses also grow as do the timing and control problems and the system becomes increasingly inefficient. Note that in the worst case, to reorder an N message long sequence a delay of up 2N units can be required. Thus if there is an error in the analogue time delay of dt per revolution, this will be magnified to N.dt , implying sharply tightening tolerances as the switch size grows.

Thus far the analogue switching we have described largely sets out to re-engineer in the optical domain what was originally designed for use in the electronic domain. The problems of this approach are simply that it pits optics in head-on confrontation with digital electronics and generally it emerges looking severally bruised! Where the analogue approach offers a distinct advantage over the electronic equivalent is that, once a pathway is established, it offers virtually infinite bandwidth but still have taken a relatively long time to become established. Optics thus lends itself to applications where the whole broadband data-stream needs to be shifted intact and struggles where the data stream needs to be demultiplexed into a vast number of tributary channels. Switching of channels within WDM streams seems to be emerging as a prime application for this.

19 Wavelength Switching And Networks

In wavelength switching, there are two new operations that are required to perform wavelength routing in the optical domain. The first is that of wavelength multiplexing and demultiplexing to allow the individual wavelength channels within a fibre to be separated to individual processes and then to be recombined together again. During the reordering process, space switching and amplification will also be required using building blocks we have already discussed. Finally, the efficiency of wavelength channel reuse in a network can often be enhanced if a capability exists to wavelength shift or to change the carrier wavelength whilst leaving the data on it intact.

For calibration purposes, note that the emerging standard for wavelength carrier spacing is 100 GHz with some interest in interleaving a second set of carriers spaced by 50 GHz for local distribution purposes only. These correspond to line spacings of 0.85 and 0.375 nm respectively. The filter response should ideally be flat topped across the signal spectrum with a sharp cut-off at the band edge to prevent adjacent carrier breakthrough. Since data rates of up to 10 Gbit/s per carrier are considered, this implies very tight control of both filter shape, even with 100GHz carrier spacings (Fig. 41).

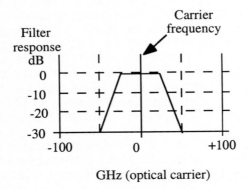

Fig. 41. Optical filter specification!

Note also that the centre frequency for such a filter probably needs to be controlled to better than 10 GHz or 1:20,000. When multiple filters are cascaded in an extended network, these design problems become especially acute since the channel is an entirely analogue one and the distortions of each filter add to those of earlier ones. For a filter to be sensitive to a change of 1:20,000 in wavelength or frequency, it must normally extend for a distance of order 20,000 wavelengths, which at 1500 nm wavelength corresponds to some 30 mm but it must also holds it dimensions stable to a fraction of wavelength over that distance. Herein lies the design problems of working with such devices, stability

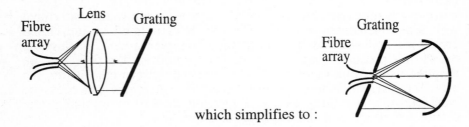

Fig. 42. Grating filter "spectrometer" constructed using bulk optical components.

and precision. Hence at present we find relatively simple extended WDM networks being built. However, some new classes of closed networks of the LAN type are beginning to emerge with interesting properties as a result of major component advances.

Filters fall into several categories. Many designs use bulk grating coupled with free-space optics to assemble what is in reality a small grating spectrometer with fibres at its entrance and outlet slits. Sometimes, these designs are integrated into a planar one dimensional guided wave structure as shown above (Fig. 42), which in turn integrates into a planar geometry as (Fig. 43)

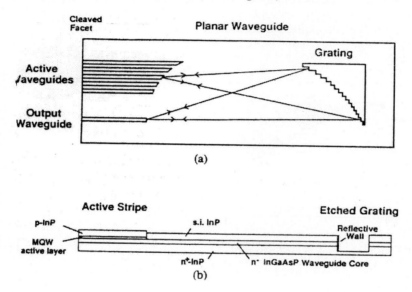

Fig. 43. The planar guided wave equivalent component.

Such devices offer limited scope opportunities for filter shaping but have been shown to provide impressive resolution and an ability to handle large numbers of wavelength channels.

A quite different approach relies upon writing refractive-index gratings by holography into single mode fibres using UV light. These gratings then need to extend over distance of order centimetres and to be written with great precision to generate the required filter shape. The underlying grating spacing corresponds to a half wavelength in the fibre so that the small amount of power back scattered from each undulation add in phase with that from the previous undulation. Such a pair of fibres is shown below in a coupler structure designed to provide ADD-DROP multiplexing functionality for a single wavelength channel (Fig. 44).

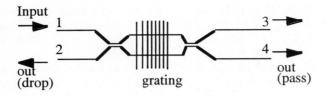

Fig. 44. Planar or fibre based impressed grating filter.

The properties of the structure are that the channel that is reflected by the grating wavelength emerges from Port 2 while the remainder of the channels emerge unscathed from Port 4. An attractive feature of such devices is that they can be cascaded to form a complete WDM (Fig. 45).

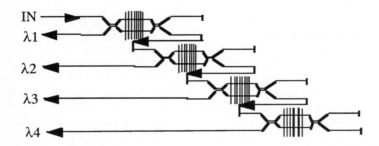

Fig. 45 Multichannel WDM assembled using sub-components of the Fig. 44 type.

However, it is immediately obvious that such an assembly will involve large lengths of fibre or planar waveguide all of which must be accurately temperature stabilised and vibration isolated if a stable precision device is to be obtained. To date devices with pass bands ranging from a few tenths of nanometres to a few nanometres have been reported, placing them well into the range of system interest.

In the quest for greater stability, it is natural to have explored the reproduction of such devices in planar geometry and the use of silica waveguides deposited on planar silicon substrates has proved to be a powerful combination for the purpose. Such an approach allows one to exploit all the planar lithographic technology developed for silicon circuits but incurs the additional complication that, having moved from circular to planar geometry, special designs of waveguide are now required to ensure that the two polarisation (TM & TE) behave in the same manner.

In planar implementations, a widely used design feature is that of the Array Waveguide Multiplexer shown schematically below (Fig. 46).

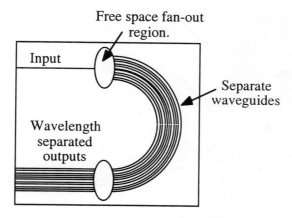

Fig. 46. Planar 1×N array-waveguide multiplexer schematic.

This elegant device relies upon two "free space" 1-D guide regions. Staring from the input, power enters the free space region from input guide and fans out across all the output guides. Each guide is exactly a length dL longer than it neighbour, starting from the smallest radius of curvature. When the separate wavelets emerge from the other end of the curved guides and enter the second free-space region, the relative phasing of the wavelets is such that they only add constructively at one of the output guides whilst an adjoining wavelength channels only adds at the adjoining output guide. In this manner, the separate input wavelength channels entering the device emerge from separate guides at the output. Evidently the device is reversible and can be used as a multiplexer as well as demultiplexer.

Many applications would benefit from the availability of a tuneable high resolution filter. Here the most widely used design is the bulk Fabry-Perot etalon with piezo-electrically driven mirror spacer although many designs make use of fibre components, some going as far as physically stretching the fibre to allow the whole structure to be fibre based.

Given tunable filters, it then become possible to consider assembling a more complex wavelength channel switch of the type shown schematically below (Fig. 47). The node accepts wavelength multiplexed inputs from each of three input fibres, with each carrying three wavelength channels. These are fanned out by power splitting to four 4×4 space switches after passing through tuneable channel selecting filters and power level controls (variable attenuators). Each filter can select just one wavelength channel. Since the output fibres carry the same four wavelength channels as each of the input fibres, the connection possibilities are complex to describe. If a wavelength channel A is selected from input fibre One and routed via switch One to output fibre Two, then this has three immediate effects.

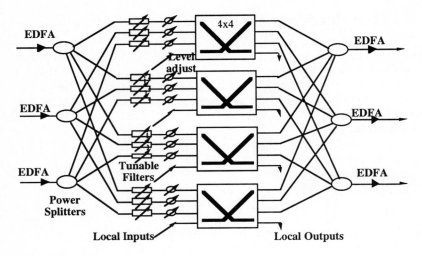

Fig. 47. A multi-fibre-port wavelength switch using an assembly of different building blocks.

- the connection from Input Fibre 1 to Switch 1 is occupied and cannot be used by any other channel from that input fibre.

- the output connection from Switch 1 to Output Fibre 2 is also used and cannot be reused by any other channel.

- the wavelength channel A in output fibre 2 is also used and cannot be used by any other signal.

So if it should transpire that Wavelength Channel A in input fibre 2 also wishes to be routed to output fibre 2, this is not possible unless the wavelength is shifted. In this switch design, this could be done by routing this signal to the local output port of Switch 2, detecting the signal and using a tuneable source tuned to a different carrier wavelength to re-transmit the data into the Local Input Port of the space switch. In this manner, a few wavelengths can be shifted albeit at the expense of blocking the scope for terminating data channels at the node. An obvious refinement on this design is then to interpose a wavelength shifter after every tuneable filter.

Wavelength shifting devices most commonly used follow the example above of linking a detection stage, perhaps with some electrical reshaping of the pulse stream, to an electronic driver and tuneable source. Monolithically integrated versions of this device offering similar functionality have also been made but combining the requirements of rapid tuning, tight spectral control, precision of central channel wavelength and data integrity remains a very tough problem.

20 Further developments in WDM Networks

The existence of the Array Waveguide Multiplexed (AWG) concept has led to a family of new and intriguing devices. For example, linking a AWG to an array of laser amplifiers with rear facet mirrors allows one to build an array laser source whose wavelengths are set by the properties of the AWG and hence corresponds those of the system it was designed for (Fig. 48).

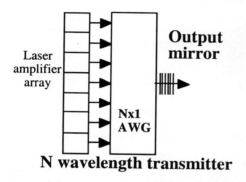

Fig. 48. Use of a 1×N AWG to generate an array WDM laser transmitter.

The output mirror provides the output reflection for all the N lasers whilst the AWG then splits their resonator paths to N separate gain chips shown on the left. At the receiver end of a system, a similar approach can be followed using an array of detector chips, each with its own front-end amplifier (Fig. 49).

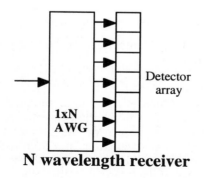

Fig. 49. Use of a 1×N AWG to generate an array wavelength receiver module.

Linking such components through a single fibre leads a simple N channel WDM transmission system. However, by introducing another components between them, a much more powerful network emerges. Building on the 1×N AWG concept, one can design and build N×N AWG as shown schematically below (Fig. 50).

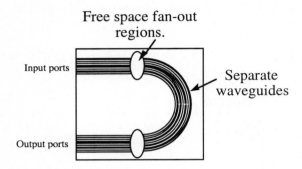

Fig. 50. A schematic diagram of a N×N port AWG multiplexer.

By careful design, the wavelength-channel-number connection matrix can take the following form (Table 7):

Input	1	2	3	4
Output				
1	1	2	3	4
2	4	1	2	3
3	3	4	1	2
4	2	3	4	1

Table 7. Connection matrix for 4×4 AWG. Numbers in the body signify wavelength channel making the connection.

The implication of this matrix is that Channel One is connected to Channel 2 by wavelength 2 but that Input 2 is connected to Channel 3 by the same wavelength. The "barrel-roll" nature of the connection matrix arises from making the AWG so that it operates in a high order (of order N). Using this device, it is now possible in principle to form a fully connected N×N network using only N wavelengths (Fig. 51).

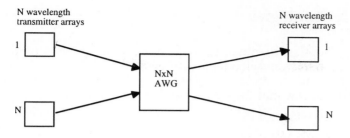

Fig. 51. Fully connected N×N network using AWG central multiplexer and array sources and receivers.

Since AWG Multiplexers (AWGM) of this type have been in sizes of 32×32 and even bigger, the potential for such networks is awesome. Assuming only 1Gbit/s per carrier were used, the above network using a 32×32 AWG would transport approximately 1 Tbit/s.

However, some problems do intervene in the use of AWGMs in this manner. One tough problem concerns the effect of interchannel interference. If two sources are operating simultaneously on the same wavelength channel, although they are using separate laser sources and are thus not phase coherent with each other, the fact that the optical spectrum has been so heavily filled with carriers and modulation implies that there virtually a certainty that power from one source will scatter and interfere with power from another similar wavelength channel source to generate modulation products within the signal bandwidth, showing up as cross channel interference. Since this is optically coherent interference, one must add optical field amplitudes, not powers. The result is that to hold the unwanted interference level below −20 dB, the scattered power must be held to below −40 dB implying very tough constraints on the allowable scatter levels within the device. And if all N sources are required to operate simultaneously, this will be even worse. As a result, networks of this type are usually operated in time-shared mode, with perhaps one transmitter array addressing all other terminals via its directed channels to be followed by the next transmitter array, cycling in turn.

There are then numerous other network configurations that exploit WDM technology in different ways, rings, stars with passive central splitters, using tuneable transmitters, receivers or both etc. Each has different traffic carrying and control characteristics and each is subject to endless variations through choice of componentry, cost-benefit trade-off, application etc.

21 Other technologies

This overview of optical communications has necessarily been partial and incomplete. Apart from selecting, often on a largely arbitrary basis, one device over another to illustrate a point, there are whole sectors that have been missed. A few that should be noted are given below.

21.1 The III-V lasers and detectors

These make all the systems possible and draw upon an immense amount of materials and device design & fabrication technology involved. However, we have chosen to use the available time and space to concentrate on the limitations imposed by the transmission medium and the impact this has of device specification.

21.2 The ultra-fast pulse area.

Optics has the ability to generate femtosecond or few-picosecond pulses and thus offers the possibility of exploiting fibre bandwidth via very high bit rate rather than multiple carrier WDM and work is in progress at rates in excess of 100 Gbit/s per carrier. Such data stream are usually formed from a series of tributary data stream that are bit interleaved so that simple bit multiplexing and demultiplexing can be done at the optical line rate leaving tributary streams at data rates that can just be handled electronically. However, such very high bit rates place very tough requirements on the linear dispersion properties of the fibre and may thus only be appropriate over short distances.

21.3 Free space optical systems

These are already widely used as low-technology communication links from lap-tops to printers etc. There may be much greater opportunity to use such technology to generate a truly wireless office that is broadband-connected etc.

21.4 Other modulation formats.

The use of binary ON-OFF modulation almost universally in optical communications systems has meant that other modulation formats have been relatively little explored. As the device sophistication increases and WDM systems approach ever closer to coherent systems, perhaps other options will emerge with, say, superior non-linear interference suppression characteristics.

21.5 Coherence

The coherence properties of the optical carriers are not being exploiting at present. Thus one might question the extent to which squeezed states are compatible with all other constraints under which systems operate.

22 Overall Conclusions

22.1. Single mode optical fibres are firmly established for all high performance applications. But being circularly symmetric, they actually support two degenerate polarisation modes and hence do NOT maintain polarisation state. This generates problems when interfacing to planar integrated-optics.

22.2. The wavelength of 1st preference over the last decade has been 1300 nm since it offered low attenuation and dispersion in an easily fabricated fibre. However, for extreme performance, the emphasis has now shifted firmly to 1550 nm operation to take advantage of the even lower attenuation and the EDFA. Dispersion problems can be fixed by using special fibre designs optimised specifically for this wavelength (e.g. DSF, DCF, Truewave etc).

22.3. Graded-index multimode fibres are still used where cost is a prime consideration rather than performance, probably at 1300 nm wavelength and typically at bit rates of 100 Mbit/s or less over distances of less than 10km. An excellent example is FDDI.

22.4. The choice of silica as the primary fibre material seems to be firmly established with negligible prospect of any major change in the foreseeable future. Lower attenuations at even longer wavelength are possible in principle in non-oxide glass materials but the other properties of these materials (toxicity, ruggedness, strength, durability, resistance to corrosion, purity etc) all militate heavily against them. Thus for high performance optical communications, the combination of the EDFA and 1550 nm silica transmission fibres seems established beyond question.

22.5. The push for ever greater bit rate over longer distance, up to 10,000 km transoceanic, has brought non-linear effects to the fore in advanced system design. The Kerr non-linearity gives rise to Intensity Dependent Refractive Index, solitons, self & cross phase modulation and four wave mixing. These effects are particularly important in WDM systems. Brillouin scattering can power-limit that power transmitted forward in the fibre whilst Raman Scattering gives rise to cross talk effects that may ultimately set the physical limit for WDM systems.

22.6. The fact that the fibre spectral bandwidth is hugely greater than that of electronic systems has added great impetus to the development of multi-carrier transmission WDM systems. Data rates of order 10Gbit/s per fibre are being deployed and although higher rates are possible, linear dispersion effects soon become a limiting factor on really long haul systems.

22.7. The development of the EDFA has effectively removed the distance limit for unrepeatered fibre systems, so that Trans-oceanic over 10,000 km is now possible. It also offer the attractive property of an almost transparent data pipe so that data formats can in principle be changed although (optical) noise filters in the system may prevent this.

22.8. The use of digital optical logic for time switching or processing data seems to be a completely dead subject since it has failed to demonstrate any ability to compete on level terms with electronic logic. The parallel-free-space imaging optical technique is still being considered for use as an interconnect technology but formidable engineering problems have to be solved in a context where price is a very critical factor if it is to find application.

22.9. Viable devices exist to assemble small space switches, perhaps up to 16×16 size, coupled with a large number of other devices which together allow the demonstration of different types of switch including packet and ATM, time slot interchange, as well as simple space switches. Despite offering huge bandwidth throughput, such devices are relatively slow to reconfigure and thus can only be operated in a circuit switched mode, resetting when no data flow is present, and this normally require data to be presented in non-standard formats. If this requires electronic processing between the transmission system terminal and the switch matrix, as it often will, the task of selling the optical switch is greatly increased.

22.10. A variety of switches are being use for protection or configuration switching where the task is to switch all the data on a given (fibre) bearer to another fibre following a break in the fibre or to allow the network operator to bring additional capacity on-stream on a give route.

22.11. A growing family of wavelength routing devices, multi-plexers, demultiplexers, wavelength shifters etc is becoming available and simple but very high capacity wavelength routed transport networks are now starting to be taken very seriously.

23 References for General & Further Reading

Note - a reference in the text to [2.4] implies Chapter 4 of Volume II. These volumes contain very extensive references to the research literature for the reader who requires further detail.

1. "Optical Fibre Telecommunications I", S E Miller & A G Chynoweth, Academic Press 1979

2. "Optical Fibre Telecommunications II", S E Miller & I P Kaminow Academic Press 1988

3. "Optical Fibre Telecommunications IIIA", I P Kaminow and T Koch, Academic Press 1997

4. "Optical Fibre Telecommunications IIIB", I P Kaminow and T Koch, Academic Press 1997

Thoughts on Quantum Computation

David P. DiVincenzo

IBM Research Division, T. J. Watson Research Center, PO Box 218, Yorktown Heights, NY 10598 USA

Abstract. If the improvement of computation continues at anything like the pace that it has set over the last fifty years, within a few decades we will have devices of atomic dimensions. Quantum computing, in which bits, in their atomic embodiment, can exist and be manipulated in a coherent superposition of computational states, is one possible method for fundamentally improving computing once the atomic scale is reached. I illustrated some of the design principles of quantum gate constructions, using the two-bit adder as a simple example. Some details of the Shor prime factoring algorithm are discussed.

1 Outlook

A large number of technologies—from steam locomotives to light bulbs—have experienced a period of exponential improvement, followed by a leveling off of performance and price, followed by an eclipsing by a better technology. Computers are unusual in that they have remained in this "initial" period of exponential improvement for a remarkably long time, now exceeding fifty years by some estimates. Through a remarkably creative succession of many sub-technologies which have arisen and been in turn eclipsed, the overall numbers—processor speed, processor memory, disk capacity, modem rate—have persisted in their steady and prodigious growth. These sub-technologies in computer switches began as mechanical relays, and continued through vacuum tubes, discrete transistors, integrated circuits with bipolar junction transistors, and now finally ICs with MOSFET transistors ("CMOS" is the current acronym). For mass storage they began with paper tapes and became magnetic drums, magnetic cores, bubbles, optical disks, and now spinning magnetic disks. The sub-sub-technologies, like the techniques of reading out magnetic data on a disk file, constitute a large part of the *raison d'etre* for much of solid-state physics research today.

But the development of computers is approaching a crisis, or, you might say, this development is preparing to go the way of all the steam locomotives and light bulbs of the past. That is, the end of the period of exponential improvement can now be seen. This end is inevitable because of basic physics: with a continuation of the present rate of exponential improvement, the logic gates in a computer logic or memory chip, and the bits on the surface of a magnetic disk, will reach atomic size in the next 15 years or so.[10]

So, what will eclipse computers? Nothing, think many of the seers at IBM and elsewhere. They will become ubiquitous and they will, for the large part, become invisible, hiding in everything from your toaster to your scuba mask. I will, however, address myself to a different question in these lectures: can the improvement of computers be put on a new track, can some new principle of improvement besides that of miniaturization be discovered?

2 Quantum Computers

The answer to this question may be yes, through the agency of quantum computers. At least, we have found, as I discuss in these lectures, that quantum mechanics sufficiently changes the rules of computing (which have been, up until now, based on what is possible in *classical* physics) that some computations which appear impossible on any computer based on ordinary digital computer become possible on a quantum computer. (For this part of the lecture, the reader should consult [3] for a more detailed review.)

The main change of rules that is produced by quantum mechanics for computer science is one involving data representation. In the world of boolean logic, a computer state is completely specified by a bit string 01100010101... of length n, for a computer containing n bits. If these bits are quantum mechanical, something different happens. This kind of bit is embodied in a quantum two level system, like the ground and excited states of an atom, vertical and horizontal polarization of a single photon, or the spin-up and spin-down states of a single electron. The most general quantum state of one such bit with energy levels $|0\rangle$ and $|1\rangle$ can be written as the Schrödinger wavefunction $|\Psi\rangle = \alpha|0\rangle + \beta|1\rangle$, where α and β are two complex amplitudes whose squares sum to one. The most general state of n quantum bits (qubits) is

$$\alpha_1|000...000\rangle + \alpha_2|000...001\rangle + \alpha_3|000...010\rangle + ... + \alpha_{2^n}|111...111\rangle. \quad (1)$$

Please note the enormous amount of information required to specify the state of these n bits. An exponentially large number of complex coefficients α_i are needed. This indicates that the information carrying capacity of an n quantum-bit systems is in some sense exponentially greater than that of a string specifying the state of n classical bits.

Only, I must emphasize, in some sense. For we know that by the rules of quantum measurement, if a quantum bit (a quantum two-level system) is measured, the outcome is still just a 0 or a 1. Holevo has shown that no more information can be extracted from the quantum bit by any measurement scheme whatsoever. So, most of the exponentially great amount of information required to *specify* the above state is inaccessible on read-out. Indeed, most states of the fully general form above are rather useless for data storage. If the superposition contains a large fraction of the possible states, then each bit will, upon measurement, give a completely random answer.

Nevertheless, quantum states are quite useful for computation. The paradigm for a useful quantum computation is (and indeed must be) the following:

Start in a simple initial state (all 0). Perform computational operations (to be discussed in a moment) which take the system into a complex superposition like Eq. (1) in the intermediate steps of computation. In the latter stages of computation, exploit quantum interference to make most of the elements in this superposition vanish (by a destructive interference) so that only a few members of the superposition are left. Then do a measurement, which now has a chance of giving some valuable (non-random) information.

3 Quantum Gates

This specification sounds rather abstract, hopefully a few of the examples I will mention will illustrate why this prescription is a good thing to do. First, though, I want to specify what I mean by a "computational operation" in quantum computation. It turns out that a good starting point is a type of computational operation which is studied in ordinary digital logic known as a *reversible* boolean logic gate. A reversible gate is one in which the output Boolean state is a unique function of the Boolean input state. For example, the inverter is a one-bit reversible gate, since if the output is 1 the input must have been a 0 and vice versa. A crucial reversible gate for quantum logic is the two-bit gate known as the controlled-NOT or reversible XOR gate, which has the input-output mapping $00 \to 00$, $01 \to 01$, $10 \to 11$, $11 \to 10$. The action of this gate may be explained by saying that the second bit (the "target") is inverted if the first bit (the "control") is a 1, and otherwise not.

There is a simple prescription for turning any reversible gate into a quantum logic gate, that is, a physically realizable operation on a quantum state like Eq. (1). It is that, if the quantum state is a pure bit state, one containing a single term in the expansion of Eq. (1), then the output is also a single term with a bit state specified by the boolean gate. So, for the controlled-NOT quantum gate, if the input quantum state is $|11\rangle$, the output state is $|10\rangle$. (If there are more bits in the state, their boolean value is untouched by the gate.) Then the mapping of the general input state is completely specified by the superposition principle; thus, if the input of the quantum controlled-NOT is $|00\rangle + |11\rangle$ (unnormalized), then the output is $|00\rangle + |10\rangle$.

We can make this sound more like traditional quantum mechanics by restating this specification as a time-development operator, from which we can ultimately work our way back to a Hamiltonian which any physicist would be comfortable with. The time-development operator is unitary (meaning that orthogonal quantum states are mapped to orthogonal quantum states); for the controlled NOT it can be written in a couple of alternative ways. In operator notation it is $U = |00\rangle\langle 00| + |01\rangle\langle 01| + |10\rangle\langle 10| + |11\rangle\langle 11|$; it should be easy for the reader to confirm that this maps general two-bit quantum states exactly as I have specified above. If the wavefunction coefficients of Eq. (1) are specified as a column vector, then the action of U can also be specified as a matrix:

$$U = \begin{pmatrix} 1 & 0 & 0 & 0 \\ 0 & 1 & 0 & 0 \\ 0 & 0 & 0 & 1 \\ 0 & 0 & 1 & 0 \end{pmatrix}. \tag{2}$$

Finally, we know that the time development operator is given by integrating the Schrodinger equation forward for a specified length of time (say from time 0 to time t), and this provides a connection to descriptions of quantum mechanics which are most familiar to most physicists. The formal solution of this equation, relating U to the Hamiltonian $H(t)$, is

$$U = T \int_0^t e^{iHt'} dt'. \tag{3}$$

I will not remind the readers about the technicalities of "time ordering" (T), which can be found in any good quantum textbook. The important point is that given some desired U acting on two quantum bits, it turns out to be relatively easy to invert this equation and find a simple $H(t)$ which will accomplish the desired logic operation. I should point out that I say $H(t)$ because it is quite important that the Hamiltonian be explicitly time dependent, for the simple reason that each quantum logic gate will be preceeded and followed by different ones involving different bits; so, a succession of turnings on and off of different contributions to the Hamiltonian should be envisioned to take the system through the desired discrete series of operations. Fortunately, such explicit time dependence is easily accomplished by, for example, turning on and off appropriate electric and magnetic fields, which can be done with lasers, voltage gates, current coils, etc. In fact, the controlled-NOT corresponds to rather well known spectroscopic manipulations of pairs of spins in double resonance experiments.

An apparently trivial observation is that all the same reasoning can be applied to operations acting on one bit at a time. This may appear trivial since there is only one single-bit boolean gate, the inverter, which is not by itself very interesting. The quantum operation of an inverter is represented by the two-dimensional unitary operator

$$U = \begin{pmatrix} 0 & 1 \\ 1 & 0 \end{pmatrix}. \tag{4}$$

It is at the level of such one-bit operations that it is most convenient to introduce the next, crucial generalization of quantum logic gate. This follows from the simple observation that the operator Eq. (4) is not the most general unitary operation on one qubit. Spectroscopy suggests several generalizations of Eq. (4); almost as easy to get as the inverter is

$$U = \begin{pmatrix} \cos\theta & \sin\theta \\ -\sin\theta & \cos\theta \end{pmatrix}. \tag{5}$$

or

$$U = \begin{pmatrix} \cos\theta & i\sin\theta \\ i\sin\theta & \cos\theta \end{pmatrix}. \tag{6}$$

Which involve different types of "tipping pulses" with tipping angle θ. While these operations have no meaning in boolean logic (although you could perhaps assign them an interpretation in fuzzy logic), they are very meaningful and very useful in quantum computation. They are important because they, unlike the quantum XOR gate, can (and do) change the number of terms in the quantum state. In fact they can increase or decrease the number of terms in the superposition by up to a factor of two. Thus, these gate are needed if the computational state is to progress from a one-term state (all 0s) to a complex multi-term superposition and back to one or a few final states. Note that an exponentially large number of terms in the superposition by only linearly many applications of such one bit gates in conjunction with XOR gates. So, these generalized one bit gates are very powerful, we have in fact proved that such one-bit gates plus just the one fixed two-bit gate (the XOR) are enough to implement any quantum time evolution, that is, any quantum computation.

4 Two-bit adder

Fig. 1 shows an example of a simple gate construction for a useful boolean arithmetic logic operation. It is a circuit for adding 1 modulo 4 to a two-bit number. Thus, the (reversible) transformation to be accomplished is $00 \rightarrow 01$, $01 \rightarrow 10$, $10 \rightarrow 11$, and $11 \rightarrow 00$. This circuit is not the most efficient one for accomplishing the desired task (exercise for the reader: implement the specified function with just XOR gate and one one-bit gate), but it is the simplest example of a scalable construction introduced by Chuang[2] for implementing "add n modulo 2^m" to an m-bit quantum register. His construction is very space efficient (no work bits) and reasonably gate efficient asymptotically for large m (the number of gates grows linearly with m, although with a fairly high constant prefactor). As the figure indicates, it first performs a discrete quantum Fourier transform (FT in the figure, to be described a bit more below), then a set of one-bit phase-change gates (s_1 and s_2), and finally the inverse of the Fourier transform (IFT).

To describe this gate array in more detail: The four two-bit gates are the XOR gates, with the control bit being the lower and the target bit being the upper. The h gates are the so-called Hadamard gates, they implement the unitary transformation

$$h = \frac{1}{\sqrt{2}} \begin{pmatrix} 1 & 1 \\ 1 & -1 \end{pmatrix}. \tag{7}$$

h_1 and h_2 indicate that the h gates act on the two different bits; as 4×4 matrices, these gates assume different kinds of block-diagonal structures:

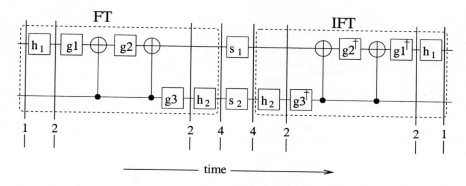

Fig. 1. Quantum gate array for the boolean logic function, "add 1 modulo 4." The numbers along the bottom indicate the increase and subsequent decrease of the number of terms in the superposition state during the course of the computation.

$$h_1 = \frac{1}{\sqrt{2}} \begin{pmatrix} 1 & 0 & 1 & 0 \\ 0 & 1 & 0 & 1 \\ 1 & 0 & -1 & 0 \\ 0 & 1 & 0 & -1 \end{pmatrix}, \quad h_2 = \frac{1}{\sqrt{2}} \begin{pmatrix} 1 & 1 & 0 & 0 \\ 1 & -1 & 0 & 0 \\ 0 & 0 & 1 & 1 \\ 0 & 0 & 1 & -1 \end{pmatrix}. \quad (8)$$

Gates g_1, g_2, and g_3, needed to implement the FT, are (in 2×2 form):

$$g_3 = \begin{pmatrix} 1 & 0 \\ 0 & e^{i\pi/4} \end{pmatrix}, \quad g_2 = \begin{pmatrix} e^{i\pi/8} & 0 \\ 0 & e^{-i\pi/8} \end{pmatrix}, \quad g_1 = g_2^\dagger. \quad (9)$$

The two phase-shift gates are

$$s_1 = \begin{pmatrix} 1 & 0 \\ 0 & -1 \end{pmatrix}, \quad s_2 = \begin{pmatrix} 1 & 0 \\ 0 & i \end{pmatrix}. \quad (10)$$

When computing the net unitary transformation produced by this gate array, it is helpful to recall that what reads from left to right in time order must be written in the reverse order in the rules of ordinary matrix arithmetic:

$$U_{tot} = h_1 \times g_1^\dagger \times U_{XOR} \times ...g_1 \times h_1. \quad (11)$$

The figure notes the feature that superposition states are created and then destroyed again in this process. Starting with any one-term classical state (e.g., $|00\rangle$), the first h gate takes the state to a two-term superposition, as indicated by the numbers over the bold vertical lines in the figure. The g gates are diagonal in the computational basis, so they do not change the number of terms in the superposition. Note that the $g1$ and $g2$ gates could not increase the number of terms on the superposition in any case, as they all are acting on the top bit along with the first h gate, and neither these not the XOR gate can change the state of the lower qubit. Finally, however, the h_2 gate doubles the number of terms in the superposition again, to four. Of course, four is the maximum number of terms possible, and this number is not changed by the s gates. the next h_2 gate, however, immediately begins the undoing of the superposition, reducing the number of terms from four to two (the other two are annihilated by destructive interference). Likewise the final h gate drops the number of terms to one, as required by the overall requirement that this gate array perform a classical boolean function.

5 Shor's prime factorization quantum algorithm

This growth and reshrinkage of the number of terms in the superposition of computational states is one thing that is going on, on a much larger scale, in the quantum algorithm of Peter Shor for prime factorization. I shall not give a complete review of Shor's ingenious algorithm, which is the subject of a complete review article in Reviews of Modern Physics[9]. I will highlight a few features which should further illuminate some of the points made above.

The Shor algorithm for factoring the number N relies on some properties of the (classical boolean) function

$$f(x) = a^x (\mathrm{mod}\ N). \qquad (12)$$

Here a and x are positive integers, a is chosen to be relatively prime to N, that is, to have no prime factors in common with N (such a number can be selected by performing a rapid classical calculation). The desired property is the period of this function, that is, the integer r for which $f(x+r) = f(x)$. r can also be identified as the first positive integer for which $f(r) = 1 (\mathrm{mod}\ N)$. This relatively simple equation for the period results because the function satisfies a one-term recurrence, i.e., $f(x+1) = af(x) (\mathrm{mod}\ N)$, so that once the function again becomes 1 it begins to repeat itself (note that $f(0) = 1$).

Here is how this period is related to prime factors. We can rewrite the equation

$$a^r = 1 (\mathrm{mod}\ N) \qquad (13)$$

as

$$a^r - 1 = kN, \qquad (14)$$

where k is some other integer (this last equation is not modulo N). If r is even, the left-hand side of this equation can be factored into two integers:

$$(a^{r/2} + 1)(a^{r/2} - 1) = kN. \qquad (15)$$

Unless the two factors on the left hand side are k and N (an unlikely occurrence), then at least one of these two numbers has a greatest common factor with N which is smaller than N; this common factor can be found by a rapid classical computation (Euclid's algorithm). While this factor is not necessarily prime, if it is not the procedure can be repeated until a prime factor is obtained.

So, factoring becomes easy if the period r is given. But computing r classically is very difficult; there is no much faster algorithm than examining each $f(x)$ one at a time until the period is found. But quantum computation does provide an efficient method of computing this classical constant r by the intermediate use of quantum superpositions.

As in the above toy example, the state of the computer begins in a classical state (one-term superposition), in fact the specific one for which all the bits are 0:

$$\Psi_0 = |000...\rangle. \qquad (16)$$

I will use the shorthand for this state of just indicating the integer value of the bit string, so that $|0\rangle = |0000\rangle$, $|13\rangle = |1101\rangle$, etc. To be more explicit, this means that, if we are using the state of four spins to encode numbers, then the number 13 is encoded by spin 1 being up, spin 2 begin up, spin 3 being down, and spin 4 being up.

The first step of Shor's algorithm is very simple: perform the h gate above on the first m_1 bits (we refer to these bits as the "first register"), leaving the remaining m_2 bits untouched (the "second register"). I will not be too specific about the values of m_1 and m_2 that are needed, see [9] for details; suffice it to say that m_1 and m_2 are simple polynomial functions of the size of the number N to be factored. Using the shorthand above and specifying the state of the first m_1 bits by one integer and the state of the final m_2 bits by a second integer, the state after performing the m_1 h gates is (leaving out normalizations)

$$\Psi_1 = \sum_{x=0}^{2^{m_1}-1} |x, 0\rangle = |0, 0\rangle + |1, 0\rangle + |2, 0\rangle + |3, 0\rangle + |4, 0\rangle + |5, 0\rangle + ... \qquad (17)$$

Note that the h gates have done their job very well. Each of them has achieved its best by the criterion of generating terms in the superposition; each has doubled the number of terms, so that the wavefunction has exponentially many terms as a function of the number m_1 of h gates.

The next step of the Shor algorithm involves the function $f(x)$. The high-level instruction is, "evaluate the function on the x given by the first register; place the result in the second register." This is a classical operation, so it does not further change the number of terms in the superposition. Constructing a gate array of XORs and one-bit gates to execute this function is mainly a job for conventional computer science; it can be done by recognizing that a compiler can code a high-level function like $f(x)$ into a set of primitive boolean logic operations like NOT and AND; then one simply needs to know that there is a simple, well known[1, 5] procedure for implementing each AND with a small number of XORs and quantum one-bit gates.

This function evaluation sets the wavefunction up so that meaningful destructive interference can take place. To illustrate how this next step goes, we pick some particular parameters for $f(x)$: $N = 15$, $a = 7$ ($a = 2$ is equally good and is left as an exercise for the reader). The new state is then:

$$\Psi_2 = \sum_{x=0}^{2^{m_1}-1} |x, f(x)\rangle = |0,1\rangle + |1,7\rangle + |2,4\rangle + |3,13\rangle + |4,1\rangle + |5,7\rangle + \ldots \quad (18)$$

It is clear from inspection of this state that the period $r = 4$, and plugging this into 15 reveals the factors of 15 immediately. In actual quantum computation, of course, we are not allowed to know this period "by inspection"; we must be able to do a measurement to determine it. Here is one way (but just a possible way) that the Shor algorithm can be completed. First, measure the second register. In the example above, the only possible outcomes of this measurement are 1, 7, 4, and 13 (that is, the binary expansion of these numbers as obtained by spin-up/spin-down measurements). Suppose the outcome is 7. Then the residual quantum state after this measurement is

$$\Psi_3 = |1,7\rangle + |5,7\rangle + |9,7\rangle + |13,7\rangle + \ldots \quad (19)$$

It is evident that we must do a further measurement to determine the periodicity in the first register. One way we can describe this is to say that we do a measurement in a kind of "plane wave" basis

$$\Psi_\phi = |0\rangle + e^{i\phi}|1\rangle + e^{2i\phi}|2\rangle + e^{3i\phi}|3\rangle + \ldots \quad (20)$$

Measuring in this basis would give ϕ, which would give the period. This is a bit awkward, as this "plane wave" basis is actually a very awkward multi-spin basis set. Fortunately, there is a further quantum computation which will change basis from these plane waves to the standard spin-up/spin-down basis in which measurements can be naturally made. This change of basis is accomplished by the quantum Fast Fourier Transform referred to above. I will not go into this here, as it is quite adequately described in the standard references[9, 3]; suffice it to say that a very simple array of XORs and one-bit gates will do the job.

6 Physical Implementation of Quantum Computation

All of what I have had to say at my Cargese lectures on the physical implementation of quantum computation I have written about elsewhere, so I will just close with a brief bibliography. I have presented a succinct list of five requirements for experimentalists to attempt to satisfy quantum computation, in whatever area of quantum mechanical research that they might be; the reader will find them described extensively in [4, 6]. A particular discussion of the problems associated with quantum decoherence may be found in [8]; methods for error correction to combat decoherence have been discovered, see [7] for my work and [12] for a review of the already vast amount of theoretical work that has been done in this area. We have proposed an architecture for quantum computation using the spin dynamics of single-electron quantum dots; see [11, 6] for details of this.

Acknowledgments

I thank that many scientists and future scientists with whom I had fruitful discussions during my stay in Cargese, and I thank Henri Benisty and Claude Weisbuch for inviting me to come and encouraging me to write these lecture notes.

References

1. Barenco, A., Bennett, C. H., DiVincenzo, D. P., Margolus, N., Shor, P., Sleator, T., Smolin, J. A., and Weinfurter, H. 1995: Phys. Rev. A **52**, 3457
2. Chuang, I. L., Gershenfeld, N., Kubinec, M. G., and Leung, D. W. (1998): Bulk quantum computation with nuclear magnetic resonance: theory and experiment. Proc. R. Soc. Lond. **454**, 257
3. DiVincenzo, D. P. (1995): Quantum Computation. Science **270**, 255
4. DiVincenzo, D. P. (1997): *Mesoscopic Electron Transport* (Vol. 345 of NATO Advanced Study Institute, Series E: Applied Sciences, eds. L. Sohn, L. Kouenhoven and G. Schoen, Kluwer, Dordrect), p. 657; http://xxx.lanl.gov/cond-mat/9612126
5. DiVincenzo, D. P. (1998): Quantum gates and circuits. Proc. R. Soc. Lond. **454**, 261
6. DiVincenzo, D. P., and Loss, D. (1998): Quantum information is physical. Superlattices and Microstructures **23** 419 Proc. R. Soc. Lond. **454**, 261
7. DiVincenzo, D. P., and Shor, P. W. (1996): Phys. Rev. Lett. **77**, 3260
8. DiVincenzo, D. P., and Terhal, B. M. (1998): Decoherence, the obstacle to quantum computation, Physics World **11**, (3), 53
9. Ekert, A. and Jozsa, R. (1996): Rev. Mod. Phys. **68**, 733
10. Keyes, R. W. (1988): IBM J. Res. Dev. **32**, 24
11. Loss, D. and DiVincenzo, D. P. (1998) Phys. Rev. A **57**, 120
12. Preskill, J. (1998): Proc. R. Soc. Lond. **454**, 385, and http://xxx.lanl.gov/quant-ph

Index

Atom-photon interaction
    bad cavity limit, 33, 314
    dressed atom model, 26
        atom field system, 28
        Rabi oscillation, 29
        time evolution, 28
    electric dipole approximation, 248
    Fermi's golden rule, 250
    Heisenberg representation, 9, 249
    Jayne-Cummings model, 26, 298,299
    oscillator strength, 4
    quasi-resonant approximation, 2
    Rabi oscillation, 4
    relaxation, influence, 29
    semi-classical approach, 4
    semi-classical model, 7
    strong coupling, 31
    weak coupling, 31

Bell inequalities, 15
Bloch Maxwell equations, 4

Cavity QED, 26, 298
    1D atom regime, 301
    anti-bunching, 35, 301
    atom transit 307
    bad cavity limit, 33, 314
    cold atoms, 304
    critical atom number, 303
    critical photon number, 303
    non-classical state generation, 34
    Purcell effect, 301
    shot noise reduction, 307
    squeezing, 35
    strong coupling, 301
        quantum box, 347
        reflection spectrum, 201, 204
    transmission spectrum, 306
    vacuum Rabi splitting, 301
    weak excitation, 300

Carrier dynamics
    carrier-carrier scattering, 125
    dephasing, 126
    detailed balance, 120
    LO phonons scattering, 121
    quantum Boltzmann equation, 120

Coulomb blockade
    electron injection, 316
    macroscopic junction, 314, 319
    mesoscopic junction, 320

Dipole emission
    DBR planar cavity, 68
    free space, 254
    planar cavity, 50
    radiation pattern, 51
    single mirror, 48
    (see also spontaneous emission)

Distributed Bragg Reflector (DBR), 59
    angular width, 67
    Brewster effect, 61
    microcavity, 130, 188
    penetration depth, 66, 189
    reflectivity, 65, 189, 190, 191
        phase, 189
    spectral width, 66
    total internal reflection, 61
    transfer-matrix method, 63

Electromagnetic field
    plane wave expansion, 42

Electron states
    anticommutation relations, 94
    band structure, 88
    Bloch theorem, 84
    Bloch wavefunction, 84
    Coulomb interaction, 91
    effective mass, 88
    envelope wavefunction, 92
    exciton, 91
    Fermi-Dirac distribution, 104, 119, 120, 123
    Fermions, 94
    Hartree-Fock approximation, 110
    lattice potential, 83
    Pauli exclusion, 91
    phase-space filling, 111
    quantum confinement, 92
    quantum dots, 92
    quantum wells, 92, 93
    quantum wires, 92
    renormalized transi tion energy, 111

second quantization, 94
tight-binding approximation, 85
(see also exciton)

Exciton, 91, 112
   absorption coefficient, 113
   bleaching, 140
   Bohr radius, 91
   Bose-Einstein condensation, 325
   boser, 324
   boson nature, 325
   composite boson scattering, 326
   continuum absorption, 114
   Elliott formula, 114
   in semiconductor microcavities,
      semi-classical theory, 129
   nonlinear saturation, 140
   oscillator strength, 229
   phase-space filling, 91
   quantum well
      broadening mechanism, 203, 213, 227
      inhomogeneous broadening, 229
      reflection, transmission 198
      spectral diffusion, 227
   Rydberg energy, 91
   saturation, 138
   second quantization, 217
   Sommerfeld enhancement factor, 114
   Wannier, 113
   (see also microcavity)

Fibers, see optical fibers

Laser
   dielectric photonic wire lasers, 291
   linewidth, 18
   phase noise, 18
   photonic wire, 291
   pump noise, 18
   semiconductor diode, 315

Light
   dielectric field, 9
   field momentum, 9
   photodetection, 16
   photon number, 9
   quantum description, 8
   quantum field operators, 8
   quantum measurements, 16

Light-Emitting Diodes, 38, 314
   extraction efficiency, 39, 393
      impact of source linewidth, 402
   microcavity (MC-LED), 40, 393

Lorentz model, 1

Mesoscopic p-i-n junction, 313

Microcavity
   cavity polariton (exciton-polariton), 199
      acoustic phonon relaxation, 226
      bottleneck effect, 226
      dispersion relation, 203, 207
      linewidth, 208
      photoluminescence, 225, 233
      radiative linewidth, 209
      radiative rate, 200, 229
      reflectance, 204
      transfer-matrix method for, 206
   DBR, 129, 188
      cavity linewidth, 192
      Fabry-Perot resonator, 182, 183
      leaky modes, 193
      mode dispersion, 195
      reflection, transmission coefficient, 218
   effective thickness, 224
   excitons in, 135
      linear susceptibility, 197
      luminescence, 144, 147
      optical response, 196
      Rabi splitting, 212
      strong coupling, 210
   Fabry-Perot
      dispersion curve, 187
   ideal planar
      extraction, 399
      modes, 395
   metallic mirrors, 184
   modes second quantization, 217
   normal mode coupling (NMC) or "vacuum field Rabi splitting", (VRS), 137, 173, 200, 201, 210
      saturation, 143
   photon escape rate, 224
   photonic microdisk, 335, 338, 420
   photonic wire, 335
   pillar microresonator, 335, 336
   quantum theory, 217
      Bose operators, 220
      exciton-photon interaction, 218
      Fano theory, 221
      Heisenberg equation, 221
      Kramers-Kronig relations, 223
      master equation, 224

quasi-mode approximation, 220
silica microsphere, 335
vs. macrocavity, 398

Noise
 Johnson-Nyquist, 315

Nonlinear optics
 coherence length, 368
 form birefringence, 370
 nonlinear coefficient $\chi^{(2)}$, 368
 phase matching, 368
 quasi-phase matching, 380
 second harmonic generation, 368

Optical coupler (beam splitter)
 linear, quantum theory, 19

Optical communications, 426
 all-optical networks, 454
 asynchronous transfer mode (ATM), 457
 dispersion budget, 433
  erbium doped fiber amplifiers (EDFA), 440
 frequency-division multiplexing, 457
 photonic switching, 444
  SEED (self-electro-optic device), 461
  switching matrices, 458
  optical time, 467
  planar switches, 464
 power budget, 433
 receiver sensitivity, 435
 time-division multiplexing (TDM), 455
 wavelength division multiplexing (WDM), 441, 457
  WDM networks, 475

Optical fibers
 attenuation, 430
 cross-phase modulation, 445
 dispersion-shifted, 436
 graded index, 417
 material dispersion, 429
 multi-mode dispersion, 428
 nonlinear effects, 442
 single mode, 427
 soliton propagation, 447
 stimulated Brillouin scattering, 450
 stimulated Raman scattering, 451

Optical transitions, 89
 direct transitions, 89

Parametric oscillator, 13

Photons
 antibunching, 24, 313, 321
 correlated, 14, 231
 noise spectral density, 17
 shot noise, 17
  suppression, 315
 twin, 24

Photon states
 coherent states, 12
 coincidence detection, 22
 density-of-states, 255
 g_{12}, 23
 $g^{(2)}$, 23, 231, 311
 Hanbury-Brown-Twiss experiment, 310
 heralded single photon states, 313
 homodyne detection, 19
 mean values, 11
 nonclassical states, 14, 310
 number squeezed states, 314
 number states 14
 mode volume factor 256
 noise spectral density, 20
 Poissonian distribution, 12, 312
 pump noise, 314
 quantum field states, second quantization, 9,10, 246
 quantum fluctuations, 11
 quantum noise, 11
 quiet, 312
 squeezed states, 12
 squeezed vacuum states, 13, 14, 21
 standard quantum noise, 12, 21
 subPoissonian distribution, 13, 18, 3
 vacuum state, 10
 white noise, 21

Photonic crystals, 150
 dielectric defect, 161
 line defect, 167, 418
 master equation, 150, 154
 microcavity, 166
 nonlinear optics, 366
 of $\chi^{(2)}$, 379
 point defect, 163
  monopole mode, 164
 reciprocal lattice, 153

rotational symmetries, 155
scale invariance, 155
spontaneous emission control, 164
two-dimensional
  design, 407
  diffraction, 414
  losses, 415
  reflection, experiment, 413
  transmission, experiment, 413, 417
  transmission, theory, 156
  two-dimensional cavity, 420
  waveguide configuration, 409
three-dimensional, 160
waveguide, 167
waveguide bends, 167
  reflection, 170
  transmission, 169
Yablonovite, 161

Planar cavity
  guided modes, 73

Purcell effect 272, 334
  cavity QED, 301
  in microdisks, 343
  in micropillars, 341

Quantum computation, 482
  physical implementation, 490

Quantum computers, 483
  quantum bits, 483
  quantum gates, 484
  Shor's algorithm, 488
  two-bit adder, 486

Quantum cryptography, 355

Rabi oscillation
  vacuum field Rabi splitting, see Microcavity, normal mode coupling

Refractive index
  complex, 42

Single photon generation, 352
  in quantum box microcavity, 348
  parametric down-conversion source, 361
  quantum information, 353
    quantum cryptography, 355
    single photon interference, 353
  single molecule source, 357
  weak laser source, 356

Semiconductors
  low-dimensional, 92
    quantum wells, 92, 198
  optical nonlinearities, 116
    Coulomb correlation effects, 116
    see also electron states, excitons, carrier dynamics
  optical response, 98
    Bloch equations, 107, 327
    free carriers, 102, 104
    Heisenberg equations, 117
    interband polarisation, 110
    Lindhard formula, 128
    Lorentz model, 98
    many-body hierarchy, 118
    Markov approximation, 117
    polarisation operator, 102
    quantum Boltzmann equation, 119
    screened Hartree-Fock, 128

Spontaneous emission
  bulk dielectric medium, 274
  decay rate, 252
  dielectric cavity, 276
  dielectric photonic wire, 282
  dielectric photonic wire lasers, 291
  effect of collisions, 2
  Einstein model, 3
  free space rate, 254
  in microcavity, 332
  lossless metallic cavity, 259
  lossless metallic photonic wire, 264
  lossy metallic photonic wire, 270
  modification, 72, 243
  Purcell factor, metallic cavities, 272
  theory, 245

Strong coupling (see microcavity, cavity QED)

Transfer-matrix method, 63, 131, 174
  transfer matrices, 176

Printing: Weihert-Druck GmbH, Darmstadt
Binding: Buchbinderei Schäffer, Grünstadt

Lecture Notes in Physics

For information about Vols. 1–494
please contact your bookseller or Springer-Verlag

Vol. 495: Y. Kosmann-Schwarzbach, B. Grammaticos, K.M. Tamizhmani (Eds.), Integrability of Nonlinear Systems. VII, 380 pages. 1997.

Vol. 496: F. Lenz, H. Grießhammer, D. Stoll (Eds.), Lectures on QCD. VII, 483 pages. 1997.

Vol. 497: J. P. Greve, R. Blomme, H. Hensberge (Eds.), Stellar Atmospheres: Theory and Observations. VIII, 352 pages. 1997

Vol. 498: Z. Horváth, L. Palla (Eds.), Conformal Field Theories and Integrable Models. Proceedings, 1996. X, 251 pages. 1997.

Vol. 499: K. Jungmann, J. Kowalski, I. Reinhard, F. Träger (Eds.), Atomic Physics Methods in Modern Research, IX, 448 pages. 1997.

Vol. 500: D. Joubert (Ed.), Density Functionals: Theory and Applications, XVI, 194 pages. 1998.

Vol. 501: J. Kertész, I. Kondor (Eds.), Advances in Computer Simulation. VIII, 166 pages. 1998.

Vol. 502: H. Aratyn, T. D. Imbo, W.-Y. Keung, U. Sukhatme (Eds.), Supersymmetry and Integrable Models. Proceedings, 1997. XI, 379 pages. 1998.

Vol. 503: J. Parisi, S. C. Müller, W. Zimmermann (Eds.), A Perspective Look at Nonlinear Media. From Physics to Biology and Social Sciences. VIII, 372 pages. 1998.

Vol. 504: A. Bohm, H.-D. Doebner, P. Kielanowski (Eds.), Irreversibility and Causality. Semigroups and Rigged Hilbert Spaces. XIX, 385 pages. 1998.

Vol. 505: D. Benest, C. Froeschlé (Eds.), Impacts on Earth. XVII, 223 pages. 1998.

Vol. 506: D. Breitschwerdt, M. J. Freyberg, J. Trümper (Eds.), The Local Bubble and Beyond. Proceedings, 1997. XXVIII, 603 pages. 1998.

Vol. 507: J. C. Vial, K. Bocchialini, P. Boumier (Eds.), Space Solar Physics. Proceedings, 1997. XIII, 296 pages. 1998.

Vol. 508: H. Meyer-Ortmanns, A. Klümper (Eds.), Field Theoretical Tools for Polymer and Particle Physics. XVI, 258 pages. 1998.

Vol. 509: J. Wess, V. P. Akulov (Eds.), Supersymmetry and Quantum Field Theory. Proceedings, 1997. XV, 405 pages. 1998.

Vol. 510: J. Navarro, A. Polls (Eds.), Microscopic Quantum Many-Body Theories and Their Applications. Proceedings, 1997. XIII, 379 pages. 1998.

Vol. 511: S. Benkadda, G. M. Zaslavsky (Eds.), Chaos, Kinetics and Nonlinear Dynamics in Fluids and Plasmas. Proceedings, 1997. VIII, 438 pages. 1998.

Vol. 512: H. Gausterer, C. Lang (Eds.), Computing Particle Properties. Proceedings, 1997. VII, 335 pages. 1998.

Vol. 513: A. Bernstein, D. Drechsel, T. Walcher (Eds.), Chiral Dynamics: Theory and Experiment. Proceedings, 1997. IX, 394 pages. 1998.

Vol. 514: F. W. Hehl, C. Kiefer, R. J. K. Metzler, Black Holes: Theory and Observation. Proceedings, 1997. XV, 519 pages. 1998.

Vol. 515: C.-H. Bruneau (Ed.), Sixteenth International Conference on Numerical Methods in Fluid Dynamics. Proceedings. XV, 568 pages. 1998.

Vol. 516: J. Cleymans, H. B. Geyer, F. G. Scholtz (Eds.), Hadrons in Dense Matter and Hadrosynthesis. Proceedings, 1998. XII, 253 pages. 1999.

Vol. 517: Ph. Blanchard, A. Jadczyk (Eds.), Quantum Future. Proceedings, 1997. X, 244 pages. 1999.

Vol. 518: P. G. L. Leach, S. E. Bouquet, J.-L. Rouet, E. Fijalkow (Eds.), Dynamical Systems, Plasmas and Gravitation. Proceedings, 1997. XII, 397 pages. 1999.

Vol. 519: R. Kutner, A. Pękalski, K. Sznajd-Weron (Eds.), Anomalous Diffusion. From Basics to Applications. Proceedings, 1998. XVIII, 378 pages. 1999.

Vol. 520: J. A. van Paradijs, J. A. M. Bleeker (Eds.), X-Ray Spectroscopy in Astrophysics. EADN School X. Proceedings, 1997. XV, 530 pages. 1999.

Vol. 521: L. Mathelitsch, W. Plessas (Eds.), Broken Symmetries. Proceedings, 1998. VII, 299 pages. 1999.

Vol. 522: J. W. Clark, T. Lindenau, M. L. Ristig (Eds.), Scientific Applications of Neural Nets. Proceedings, 1998. XIII, 288 pages. 1999.

Vol. 523: B. Wolf, O. Stahl, A. W. Fullerton (Eds.), Variable and Non-spherical Stellar Winds in Luminous Hot Stars. Proceedings, 1998. XX, 424 pages. 1999.

Vol. 524: J. Wess, E. A. Ivanov (Eds.), Supersymmetries and Quantum Symmetries. Proceedings, 1997. XX, 442 pages. 1999.

Vol. 525: A. Ceresole, C. Kounnas, D. Lüst, S. Theisen (Eds.), Quantum Aspects of Gauge Theories, Supersymmetry and Unification. Proceedings, 1998. X, 511 pages. 1999.

Vol. 526: H.-P. Breuer, F. Petruccione (Eds.), Open Systems and Measurement in Relativistic Quantum Theory. Proceedings, 1998. VIII, 240 pages. 1999.

Vol. 527: D. Reguera, M. Rubí, J. Vilar (Eds.), Statistical Mechanics of Biocomplexity. Proceedings, 1998. XI, 318 pages. 1999.

Vol. 528: I. Peschel, X. Wang, M. Kaulke, K. Hallberg (Eds.), Density-Matrix Renormalization. Proceedings, 1998. XVI, 355 pages. 1999.

Vol. 529: S. Biringen, H. Örs, A. Tezel, J.H. Ferziger (Eds.), Industrial and Environmental Applications of Direct and Large-Eddy Simulation. Proceedings, 1998. XVI, 301 pages. 1999.

Vol. 530: H.-J. Röser, K. Meisenheimer (Eds.), The Radio Galaxy Messier 87. Proceedings, 1997. XIII, 342 pages. 1999.

Vol. 531: H. Benisty, J.-M. Gérard, R. Houdré, J. Rarity, C. Weisbuch (Eds.), Confined Photon Systems. Proceedings, 1998. X, 496 pages. 1999.

Vol. 533: K. Hutter, Y. Wang, H. Beer (Eds.), Advances in Cold-Region Thermal Engineering and Sciences. Proceedings, 1999. XIV, 608 pages. 1999.

Monographs

For information about Vols. 1–14 please contact your bookseller or Springer-Verlag

Vol. m 15: N. Peters, B. Rogg (Eds.), Reduced Kinetic Mechanisms for Applications in Combustion Systems. X, 360 pages. 1993.

Vol. m 16: P. Christe, M. Henkel, Introduction to Conformal Invariance and Its Applications to Critical Phenomena. XV, 260 pages. 1993.

Vol. m 17: M. Schoen, Computer Simulation of Condensed Phases in Complex Geometries. X, 136 pages. 1993.

Vol. m 18: H. Carmichael, An Open Systems Approach to Quantum Optics. X, 179 pages. 1993.

Vol. m 19: S. D. Bogan, M. K. Hinders, Interface Effects in Elastic Wave Scattering. XII, 182 pages. 1994.

Vol. m 20: E. Abdalla, M. C. B. Abdalla, D. Dalmazi, A. Zadra, 2D-Gravity in Non-Critical Strings. IX, 319 pages. 1994.

Vol. m 21: G. P. Berman, E. N. Bulgakov, D. D. Holm, Crossover-Time in Quantum Boson and Spin Systems. XI, 268 pages. 1994.

Vol. m 22: M.-O. Hongler, Chaotic and Stochastic Behaviour in Automatic Production Lines. V, 85 pages. 1994.

Vol. m 23: V. S. Viswanath, G. Müller, The Recursion Method. X, 259 pages. 1994.

Vol. m 24: A. Ern, V. Giovangigli, Multicomponent Transport Algorithms. XIV, 427 pages. 1994.

Vol. m 25: A. V. Bogdanov, G. V. Dubrovskiy, M. P. Krutikov, D. V. Kulginov, V. M. Strelchenya, Interaction of Gases with Surfaces. XIV, 132 pages. 1995.

Vol. m 26: M. Dineykhan, G. V. Efimov, G. Ganbold, S. N. Nedelko, Oscillator Representation in Quantum Physics. IX, 279 pages. 1995.

Vol. m 27: J. T. Ottesen, Infinite Dimensional Groups and Algebras in Quantum Physics. IX, 218 pages. 1995.

Vol. m 28: O. Piguet, S. P. Sorella, Algebraic Renormalization. IX, 134 pages. 1995.

Vol. m 29: C. Bendjaballah, Introduction to Photon Communication. VII, 193 pages. 1995.

Vol. m 30: A. J. Greer, W. J. Kossler, Low Magnetic Fields in Anisotropic Superconductors. VII, 161 pages. 1995.

Vol. m 31 (Corr. Second Printing): P. Busch, M. Grabowski, P.J. Lahti, Operational Quantum Physics. XII, 230 pages. 1997.

Vol. m 32: L. de Broglie, Diverses questions de mécanique et de thermodynamique classiques et relativistes. XII, 198 pages. 1995.

Vol. m 33: R. Alkofer, H. Reinhardt, Chiral Quark Dynamics. VIII, 115 pages. 1995.

Vol. m 34: R. Jost, Das Märchen vom Elfenbeinernen Turm. VIII, 286 pages. 1995.

Vol. m 35: E. Elizalde, Ten Physical Applications of Spectral Zeta Functions. XIV, 224 pages. 1995.

Vol. m 36: G. Dunne, Self-Dual Chern-Simons Theories. X, 217 pages. 1995.

Vol. m 37: S. Childress, A.D. Gilbert, Stretch, Twist, Fold: The Fast Dynamo. XI, 406 pages. 1995.

Vol. m 38: J. González, M. A. Martín-Delgado, G. Sierra, A. H. Vozmediano, Quantum Electron Liquids and High-Tc Superconductivity. X, 299 pages. 1995.

Vol. m 39: L. Pittner, Algebraic Foundations of Non-Com-mutative Differential Geometry and Quantum Groups. XII, 469 pages. 1996.

Vol. m 40: H.-J. Borchers, Translation Group and Particle Representations in Quantum Field Theory. VII, 131 pages. 1996.

Vol. m 41: B. K. Chakrabarti, A. Dutta, P. Sen, Quantum Ising Phases and Transitions in Transverse Ising Models. X, 204 pages. 1996.

Vol. m 42: P. Bouwknegt, J. McCarthy, K. Pilch, The W3 Algebra. Modules, Semi-infinite Cohomology and BV Algebras. XI, 204 pages. 1996.

Vol. m 43: M. Schottenloher, A Mathematical Introduction to Conformal Field Theory. VIII, 142 pages. 1997.

Vol. m 44: A. Bach, Indistinguishable Classical Particles. VIII, 157 pages. 1997.

Vol. m 45: M. Ferrari, V. T. Granik, A. Imam, J. C. Nadeau (Eds.), Advances in Doublet Mechanics. XVI, 214 pages. 1997.

Vol. m 46: M. Camenzind, Les noyaux actifs de galaxies. XVIII, 218 pages. 1997.

Vol. m 47: L. M. Zubov, Nonlinear Theory of Dislocations and Disclinations in Elastic Body. VI, 205 pages. 1997.

Vol. m 48: P. Kopietz, Bosonization of Interacting Fermions in Arbitrary Dimensions. XII, 259 pages. 1997.

Vol. m 49: M. Zak, J. B. Zbilut, R. E. Meyers, From Instability to Intelligence. Complexity and Predictability in Nonlinear Dynamics. XIV, 552 pages. 1997.

Vol. m 50: J. Ambjørn, M. Carfora, A. Marzuoli, The Geometry of Dynamical Triangulations. VI, 197 pages. 1997.

Vol. m 51: G. Landi, An Introduction to Noncommutative Spaces and Their Geometries. XI, 200 pages. 1997.

Vol. m 52: M. Hénon, Generating Families in the Restricted Three-Body Problem. XI, 278 pages. 1997.

Vol. m 53: M. Gad-el-Hak, A. Pollard, J.-P. Bonnet (Eds.), Flow Control. Fundamentals and Practices. XII, 527 pages. 1998.

Vol. m 54: Y. Suzuki, K. Varga, Stochastic Variational Approach to Quantum-Mechanical Few-Body Problems. XIV, 324 pages. 1998.

Vol. m 55: F. Busse, S. C. Müller, Evolution of Spontaneous Structures in Dissipative Continuous Systems. X, 559 pages. 1998.

Vol. m 56: R. Haussmann, Self-consistent Quantum Field Theory and Bosonization for Strongly Correlated Electron Systems. VIII, 173 pages. 1999.

Vol. m 57: G. Cigogna, G. Gaeta, Symmetry and Perturbation Theory in Nonlinear Dynamics. XI, 208 pages. 1999.

Vol. m 58: J. Daillant, A. Gibaud (Eds.), X-Ray and Neutron Reflectivity: Principles and Applications. XVIII, 331 pages. 1999.